Lecture Notes in Computer Science   4832

Commenced Publication in 1973
Founding and Former Series Editors:
Gerhard Goos, Juris Hartmanis, and Jan van Leeuwen

Mathias Weske   Mohand-Saïd Hacid
Claude Godart (Eds.)

Web Information Systems Engineering – WISE 2007 Workshops

WISE 2007 International Workshops
Nancy, France, December 3, 2007
Proceedings

 Springer

Volume Editors

Mathias Weske
University of Potsdam
Hasso Plattner Institute for IT Systems Engineering
Business Process Technology
14482 Potsdam, Germany
E-mail: mathias.weske@hpi.uni-potsdam.de

Mohand-Saïd Hacid
Université Claude Bernard Lyon 1
LIRIS - UFR d'Informatique
69622 Villeurbanne, France
E-mail: mshacid@liris.cnrs.fr

Claude Godart
LORIA-ECOO
Campus Scientifique, BP 239
54506 Vandœuvre-lès-Nancy, France
E-mail: claude.godart@loria.fr

Library of Congress Control Number: 2007939830

CR Subject Classification (1998): H.4, H.3, H.2, C.2.4, I.2, H.5.1, J.1

LNCS Sublibrary: SL 3 – Information Systems and Application, incl. Internet/Web
and HCI

ISSN   0302-9743
ISBN-10   3-540-77009-7 Springer Berlin Heidelberg New York
ISBN-13   978-3-540-77009-1 Springer Berlin Heidelberg New York

Springer is a part of Springer Science+Business Media

springer.com

© Springer-Verlag Berlin Heidelberg 2007
Printed in Germany

Typesetting: Camera-ready by author, data conversion by Scientific Publishing Services, Chennai, India
Printed on acid-free paper   SPIN: 12197453   06/3180   5 4 3 2 1 0

Preface

Workshops are an effective means to discuss relevant and new research issues and share innovative and exciting ideas. Therefore, they are an essential part of scientific conferences. In highly dynamic fields of research and development with strong interaction between academia and industry, workshops are instrumental in sharing ideas, discussing new concepts and technologies that could finally lead to industrial uptake of research results.

The International Conference on Web Information Systems Engineering addresses issues that require focused discussions. In this year's WISE conference, held in Nancy, France, during December 3–6, and hosted by Nancy University and INRIA Grand-Est, the Program Committee selected six workshops, focusing on specific research issues related to Web information systems engineering. The workshops were organized by international experts in the respective fields; each workshop set up an International Program Committee that carefully selected the workshop contributions.

The Approaches and Architectures for Web Data Integration and Mining in Life Sciences, workshop (chaired by Marie-Dominique Devignes and Malika Smaïl-Tabbone) focused on the effective and efficient management and transformation of scientific data in the life sciences. By appropriate concepts and Web information systems, the bottlenecks for research in the life sciences that have shifted from data production to data integration, pre-processing, analysis/mining, and interpretation can be overcome.

In the Collaborative Knowledge Management for Web Information Systems workshop (chaired by Sergej Sizov and Stefan Siersdorfer), issues related to information acquisition through collaborative Web crawling, classification, and clustering were discussed and the relationship of these techniques to knowledge sharing through sharing of personal ontologies and their alignment was investigated. By bringing the respective communities to the workshop, interesting interdisciplinary discussions were sought.

Methods and techniques to support governance and compliance in Web information systems were considered in the Governance, Risk and Compliance in Web Information Systems workshop (chaired by Shazia Sadiq, Claude Godart and Michael zur Muehlen). These issues are currently emerging as a critical and challenging area of research and innovation. It opens new questions regarding, for instance, the modeling of compliance requirements, but existing challenges also have to be solved, for instance, extension of process and service modeling and enactment frameworks for compliance management.

The Human-Friendly (Web) Service Description, Discovery and Matchmaking workshop (chaired by Dominique Kuropka and Ingo Melzer) focused on annotations of services that facilitates service requestors to easily find and use them. It is based on the observation that existing technologies are either too complex to use or are just at the syntactic level, focusing on interface definitions. Finding the right level of specification detail is a challenging task that was at the center of this workshop.

The huge amount of information provided by the Web forces the designers of Web information systems to prevent users from experiencing the all-too-prevalent cognitive and informational overload. Elaborate personalization techniques are required to provide users with information that they are actually interested in. Models and mechanisms for personalization as well as personalized access and context acquisition were discussed in the Personalized Access to Web Information workshop (chaired by Sylvie Calabretto and Jérôme Gensel).

Given the ubiquity of Web information systems and the immense commercial interest of Web applications, usability issues become increasingly relevant. To broaden the user group, accessibility issues have to be considered. In the Web Usability and Accessibility workshop (chaired by Silvia Abrahao, Cristina Cachero and Maristella Matera), concepts, models and languages to improve Web information systems with respect to their usability and accessibility were investigated.

We would like to take this opportunity to thank all the workshop organizers who contributed to make WISE 2007 a real success.

We would like also to acknowledge the local organization, in particular Anne-Lise Charbonnier and François Charoy. We also thank Qing Li, Marek Rusinkiewicz and Yanchun Zhang for the relationship with previous events and the WISE Society, and Ustun Yildiz for his work in editing these proceedings.

September 2007   Mathias Weske
                                                          Mohand-Said Hacid
                                                            Claude Godart

Organization

General Chairs	Claude Godart, France
	Qing Li, China
Workshop Chairs	Mohand-Said Hacid, France
	Mathias Weske, Germany
Publication Chair	Claude Godart, France
Wise Society Representatives	Yanchun Zhang, Australia
	Marek Rusinkiewicz, USA
Local Organization Chair	François Charoy, France
Local Organization Committee	Anne-Lise Charbonnier, INRIA, France
	François Charoy, Nancy University, France
	Laurence Félicité, Nancy University, France
	Nawal Guermouche, Nancy University, France
	Olivier Perrin, Nancy University, France
	Mohsen Rouached, Nancy University, France
	Hala Skaf, Nancy University, France
	Ustun Yildiz, Nancy University, France

Program Committee

International Workshop on Approaches and Architectures for Web Data Integration and Mining in Life Sciences (WebDIM4LS)

Bettina Berendt, Germany
Olivier Bodenreider, USA
Omar Boucelma, France
Nacer Boudjlida, France
Bert Coessens, Belgium
Werner Ceusters, USA
Marie-Dominique Devignes,
   France (Co-chair)
Christine Froidevaux, France
Martin Kuiper, Belgium

Phil Lord, UK
Fouzia Moussouni, France
Amedeo Napoli, France
Peter Rice, UK
Paolo Romano, Italy
Mohamed Rouane Hacene, France
Malika Smaïl-Tabbone,
   France (Co-chair)

International Workshop on Collaborative Knowledge Management for Web Information Systems (WeKnow)

Paulo Barthelmess, USA
AnHai Doan, USA
Maria Halkidi, Greece
Joemon Jose, UK
Andreas Nürnberger, Germany
Daniela Petrelli, UK

Stefan Siersdorfer, UK (Co-chair)
Sergej Sizov, Germany (Co-chair)
Michalis Vazirgiannis, France
Jun Wang, The Netherlands
Yi Zhang, USA

International Workshop on Governance, Risk and Compliance in Web Information Systems (GDR)

Sami Bhiri, Ireland
Fabio Casati, Italy
Wojciech Cellary, Poland
Claude Godart, France (Co-chair)
Guido Governatori, Australia
Daniela Grigori, France
Marta Indulska, Australia
Olivier Perrin, France
Zoran Milosevic, Australia

Michael zur Muehlen, USA (Co-chair)
Michael Rosemann, Australia
Shazia Sadiq, Australia (Co-chair)
Andreas Schaad, Germany
Samir Tata, France
Paolo Torroni, Italy
Yathi Udupi, USA
Jan Vanthienen, Belgium
Julien Vayssiere, Australia

International Workshop on Human-Friendly Service Description, Discovery and Matchmaking (Hf-SDDM)

Marek Kowalkiewicz, Australia
Gennady Agre, Bulgaria
Anne-Marie Sassen, Belgium
Joerg Hoffmann, Austria
Tomasz Kaczmarek, Poland

Dominik Kuropka, Germany
 (Co-chair)
Ingo Melzer, Germany (Co-chair)
Ingo Weber, Germany
Massimiliano Di Penta, Italy

International Workshop on Personalized Access to Web Information (PAWI)

Rocio Abascal, Mexico
Maristella Agosti, Italy
Michel Beigbeder, France
Catherine Berrut, France
Sylvie Calabretto, France (Co-chair)
Jérôme Gensel, France (Co-chair)
Lynda Lechani-Tamine, France

Philippe Lopisteguy, France
Jian-Yun Nie, Canada
Gabriella Pasi, Italy
Béatrice Rumpler, France
Michel Simonet, France
Alan Smeaton, Ireland
Marlène Villanova-Oliver, France

International Workshop on Web Usability and Accessibility (IWWUA)

Silvia Abrahão, Spain (Co-chair)
Shadi Abou-Zahra, World Wide
  Web Consortium (W3C)
Nigel Bevan, UK
Cristina Cachero, Spain (Co-chair)
Coral Calero, Spain
Jair Cavalcanti Leite, Brazil
Maria-Francesca Costabile, Italy
Geert-Jan Houben, Belgium
Ebba Thora Hvannberg, Iceland
Emilio Insfran, Spain

Effie Lai-Chong Law, Switzerland
Maria Dolores Lozano, Spain
Maristella Matera, Italy (Co-chair)
Emilia Mendes, New Zealand
Luis Olsina, Argentina
Geert Poels, Belgium
Simos Retalis, Greece
Carmen Santoro, Italy
Corina Sas, UK
Marco Winckler, France

Sponsoring Institutions

Nancy-Université

emisa.org

Table of Contents

International Workshop on Governance, Risk and Compliance in Web Information Systems (GDR)

International Workshop on Human-Friendly Service Description, Discovery and Matchmaking (Hf-SDDM)

International Workshop on Personalized Access to Web Information (PAWI)

International Workshop on Web Usability and Accessibility (IWWUA)

International Workshop on Approaches and Architectures for Web Data Integration and Mining in Life Sciences (WebDIM4LS)

Workshop PC Chairs' Message

Marie-Dominique Devignes[1] and Malika Smaïl-Tabbone[2]

[1] LORIA, Orpailleur project, CNRS - National Scientific Research Center, France
[2] LORIA, Orpailleur project, Nancy University, France

Today, nobody can contest that the amount and complexity of scientific data are increasing exponentially. In life sciences, the unique conjunction of complexity, size and importance of available data deserves special attention in both Web Information Systems (WIS) and data integration and mining areas. In the last few years, WIS have become the favourite mean for offering open access to biological data. Hundreds of data sources (databases with either user or program interface), numerous web sites and peer-reviewed literature are currently covering multiple facets of biology (genomic sequences, protein structures, pathways, transcriptomics data, etc.). However, the scientists have enormous difficulties in keeping up with this data deluge. The effective and efficient management and use of available data (including omics data), and in particular the transformation of these data into information and knowledge, is a key requirement for success in scientific discovery process. In fact, the bottlenecks for life sciences have shifted from data production to data access/integration, pre-processing, analysis/mining, and interpretation.

The present volume contains five papers that were independently peer-reviewed (rate of acceptance : 60).

The accepted papers illustrate well a few distinct topics related to the workshop. Fouzia Moussouni, Laure Berti-Equille, G. Rozé and Emilie Guérin discuss in their paper the crucial issue of data quality when integrating multi-source biological data in a data warehouse. Related to the quality issues are the provenance of data and reproducibility of the biological workflows which are addressed in the paper written by Michel Kinsy, Zoé Lacroix, Christophe Legendre, Piotr Wlodarczyk and Nadia Yacoubi. They present their ProtocolDB system for managing scientific protocols with a domain ontology.

Gaelle Hignette, Patrice Buche, Juliette Dibie-Berthélemy and Ollivier Haemmerlé propose a method for automatic extraction of semantic information from data tables found on the web. They use a domain ontology just as Zhouyang Sun, Anthony Finkelstein and Jonathan Ashmore do in their paper where they combine the use of a domain ontology with a semantic web services infrastructure to provide a knowledge representation supporting systems biology modelling.

As an opening to other disciplines, André Schaaff presents in his paper how the astronomical community manages the data integration problem converging towards the virtual observatory on the web. He gives some common investigation fields that are relevant for life sciences and could lead to more interactions between the two communities.

M. Weske, M.-S. Hacid, C. Godart (Eds.): WISE 2007 Workshops, LNCS 4832, pp. 3–4, 2007.
© Springer-Verlag Berlin Heidelberg 2007

Finally, we would like to warmly thank the authors who submitted and presented their papers. Last but not least, we are grateful for the program committee members for their essential contribution to this workshop.

QDex:
A Database Profiler for Generic Bio-data Exploration and Quality Aware Integration

F. Moussouni[1], L. Berti-Équille[2], G. Rozé[1], O. Loréal, and E. Guérin[1]

[1] INSERM U522 CHU Pontchaillou, 35033 Rennes, France
[2] IRISA, Campus Universitaire de Beaulieu, 35042 Rennes, France
`fouzia.moussouni@univ-rennes.fr`

Abstract. In human health and life sciences, researchers extensively collaborate with each other, sharing genomic, biomedical and experimental results. This necessitates dynamically integrating different databases into a single repository or a warehouse. The data integrated in these warehouses are extracted from various heterogeneous sources, having different degrees of quality and trust. Most of the time, they are neither rigorously chosen nor carefully controlled for data quality. Data preparation and data quality metadata are recommended but still insufficiently exploited for ensuring quality and validating the results of information retrieval or data mining techniques.

In a previous work, we built a data warehouse called GEDAW (Gene Expression Data Warehouse) that stores various information: data on genes expressed in the liver during iron overload and liver diseases, relevant information from public databanks (mostly in XML), DNA-chips home experiments and also medical records. Based on our past experience, this paper reports briefly on the lessons learned from biomedical data integration and data quality issues, and the solutions we propose to the numerous problems of schema evolution of both data sources and warehousing system. In this context, we present QDex, a Quality driven bio-Data Exploration tool, which provides a functional and modular architecture for database profiling and exploration, enabling users to set up query workflows and take advantage of data quality profiling metadata before the complex processes of data integration in the warehouse. An illustration with QDex Tool is shown afterwards.

Keywords: warehousing, metadata, bio-data integration, database profiling, bioinformatics, data quality.

1 Introduction

In the context of modern life science, integrating resources is very challenging, mainly because biological objects are complex and spread in highly autonomous and evolving web resources. Biomedical web resources are extremely heterogeneous as they contain different kinds of data, have different structure and use different vocabularies to name same biological entities. Their information and knowledge contents are also partial and erroneous, morphing and in perpetual progress.

M. Weske, M.-S. Hacid, C. Godart (Eds.): WISE 2007 Workshops, LNCS 4832, pp. 5–16, 2007.

In spite of these barriers, we assist in bioinformatics to an explosion of data integration approaches to help biomedical researchers to interpret their results, test and generate new hypothesis. In high throughput biotechnologies data warehouse solutions encountered a great success in the last decades, due to constant needs to store locally, confront and enrich in-house data with web information for multiple possibilities of analyses.

A tremendous amount of data warehouse projects devoted to bioinformatics studies exists now in literature. These warehouses integrate data from various heterogeneous sources, having different degrees of quality and trust. Most of the time, the data are neither rigorously chosen nor carefully controlled for data quality. Data preparation and data quality metadata are recommended but still insufficiently exploited for ensuring quality and validating the results of information retrieval or data mining techniques [17]. Moreover, data are physically imported, transformed to match the warehouse schema which tends to change rapidly with user requirements, typically in Bioinformatics. In the case of materialised integration, data model modifications for adding new concepts in response to rapid evolving needs of biologists, lead to considerable updates of the warehouse schemas and their applications, complicating the warehouse maintainability.

Lessons learned from the problems of biomedical data sources integration and warehouse schema evolution are presented in this paper. The main data quality issues in this context with current solutions for warehousing and exploring biomedical data are shown [1,2]. An illustration is given using *QDex*, a Quality driven bio-Data Exploration tool that: *i)* provides a generic functional and modular architecture for database quality profiling and exploration, *ii)* takes advantage of data quality profiling metadata during the process of biomedical data integration in the warehouse and, *iii)* enables users to set up query workflows, store intermediate results or quality profiles, and refine their queries.

This paper is structured as follows: in Section 2, requirement analyses in bioinformatics and the limits of current data warehousing techniques with regards to data quality profiling are presented in the perspective of related work. In Section 3, an illustration with our experience in building a gene expression data warehousing system: system design, data curation, cleansing, analyses, and new insight on schema evolution, In Section 4, QDex architecture and functionalities to remediate to some of these limits are presented to provide database quality profiling and extraction of quality metadata, and *Section 6* concludes the paper.

2   Related Work

2.1   Data Integration Issues at the Structural Level

High throughput biotechnologies, like transcriptome, generate thousands of expression levels on genes, measured in different physiopathological situations. Beyond the process of management, normalization and clustering, biologists need to give a biological, molecular and medical sense to these raw data. Expression levels need to be enriched with the multitude of data available publicly on expressed genes: nucleic sequences, chromosomal and cellular locations, biological processes, molecular function, associated pathologies, and associated pathways. Relevant information on genes must be integrated from public databanks and warehoused locally for multiple possibilities of analyses and data mining solutions.

In the context of biological data warehouses, a survey of representative data integration systems is given in [8]. Current solutions are mostly based on data warehouse architecture (e.g., GIMS[1], DataFoundry[2]) or a federation approach with physical or virtual integration of data sources (e.g., TAMBIS [3], P/FDM [4], DiscoveryLink[5]) that are based on the union of the local schemas which have to be transformed to a uniform schema. In [3], Do and Rahm proposed a system called GenMapper for integrating biological and molecular annotations based on the semantic knowledge represented in cross-references. Finally, BioMart [18], which is a query-oriented data integration system that can be applied to a single or multiple databases, is a heavily used data warehouse system in bioinformatics since it supports large scale querying of individual databases as well as query-chaining between them.

Major problems in the context of biomedical data integration come from heterogeneity, strong autonomy and rapid evolution of the data sources on the Web. A data warehouse is relevant as long as it adapts its structure, schemas and applications to the constantly growing knowledge on the bio-Web.

2.2 Bio-data Quality Issues at the Instance Level

Recent advancement in biotechnology has produced massive amount of raw biological data which are accumulating at an exponential rate. Errors, redundancy and discrepancies are prevalent in the raw data, and there is a serious need for systematic approaches towards biological data cleaning. Biological databanks providers will not directly support data quality evaluations to the same degree since there is no equal motivation for them to and there are currently no standards for evaluating and comparing biomedical data quality. Little work has been done on biological data cleaning and it is usually carried out in proprietary or ad-hoc manner, sometimes even manual. Systematic processes are lacking. From among the few examples, Thanaraj uses in [14] stringent selection criteria to select 310 complete and unique records of Homo sapiens splice sites from the 4300 raw records in EMBL database.

Moreover, bio-entity identification is a complex problem in the biomedical domain, since the meaning of "entity" cannot be defined properly. In most applications, identical sequences of two genes in different organisms or even in different organs of the same organism are not treated as a single object since they can have different behaviours. In GENBANK data source for example, each sequence is treated as an entity in its own, since it was derived using a particular technique, has particular annotation, and could have individual errors.

Müller et al. [11] examined the production process of genome data and identified common types of data errors. Mining for patterns in contradictory biomedical data has been proposed in [10], but data quality evaluation techniques are needed for

---

[1] *GIMS, http://www.cs.man.ac.uk/img/gims/*
[2] *DataFoundry, http://www.llnl.gov/CASC/datafoundry/*
[3] *TAMBIS, http://imgproj.cs.man.ac.uk/tambis/*
[4] *P/FDM, http://www.csd.abdn.ac.uk/~gjlk/mediator/*
[5] *DiscoveryLink, http://www.research.ibm.com/journal/sj/402/haas.html*

structured, semi-structured or textual data before any biomedical mining applications. Although rigorous elimination of data is effective in removing redundancy, it may result in loss of critical information. In another example, a sequence structure parser is used to find missing or inconsistent features in records using the constraints of gene structure [12]. The method is only limited to detecting violations of the gene structure.

More specific to data quality scoring in the biomedical context, [9] propose to extend the semi-structured model with useful quality measures that are *biologically-relevant*, *objective* (i.e., with no ambiguous interpretation when assessing the value of the quality measure), and *easy to compute*. Six criteria such as stability (i.e., magnitude of changes applied to a record), density (i.e., number of attributes and values describing a data item), time since last update, redundancy (i.e., fraction of redundant information contained in a data item and its sub-items), correctness (i.e., degree of confidence that the data represents true information), and usefulness (i.e., utility of a data item defined as a function combining density, correctness, and redundancy) are defined and stored as quality metadata for each record (XML file) of the genomic databank of RefSeq . The authors also propose algorithms for updating the scores of quality measures when navigating, inserting or updating/deleting a node in the semi-structured record.

3   Lessons Learned from Building GEDAW

3.1   Database Design, Data Integration, and Application-Driven Workflow

The Gene Expression Data warehouse GEDAW [5] has been developed by the National Institute of Health Care and Medical Research (INSERM U522) to warehouse data on genes expressed in the liver during iron overload and liver pathologies. For interpreting gene expression measurements in different physiopathological situations in the liver, relevant information from public databanks (mostly in XML format), micro-array data, DNA chips home experiments and medical records are integrated, stored and managed into GEDAW. GEDAW aims at studying in-silico liver pathologies by enriching expression levels of genes with data extracted from the variety of scientific data sources, ontologies and standards in life science and medicine including GO ontology [6] and UMLS [7].

Designing a single global data warehouse schema (Fig 1) that integrates syntactically and semantically the whole heterogeneous life science data sources is still challenging. In GEDAW context, we integrate structured and semi-structured data sources and use a Global As View (GAV) schema mapping approach and a rule-based transformation process from a source schema to the global schema of the data warehouse (see [4] for details).

With the overall integrated knowledge, the warehouse has provided an excellent analysis framework where enriched experimental data can be mined through various workflows combining successive analysis steps.

GEDAW supports several functions that consist of analyses on demand made on a group of genes of interest upon a database selection query with one or more criteria. These analyses correspond to APIs that use OQL (Object Query Language) and java to retrieve multiple information items about the genes. Some external analyses

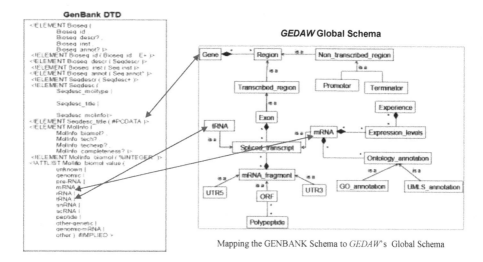

Mapping the GENBANK Schema to *GEDAW*'s Global Schema

Fig. 1. UML Class diagram representing the conceptual schema of GEDAW and some correspondences with the GENBANK DTD (e.g., Seqdes_title and Molinfo values will be extracted and migrated to the name and other description attributes of the class Gene in GEDAW schema)

that correspond to external bioinformatics tools have been applied on subsets of integrated data on genes, as clustering for example. These two kinds of analyses have been combined to connect successive steps, thus forming a workflow.

One of them (Fig 2) has been designed according to the hypothesis that genes sharing an expression pattern should be associated. The strategy consists in selecting a group of genes that are associated with a same disease and a typical expression pattern (steps 1 and 2 in Fig 2), and then extrapolate this group to more genes involved in the disease (step 5) by searching for expression pattern similarity (step 4). The genes are then characterized by studying the biological processes and the cellular components using integrated GO annotations (step 6).

This example, which is expert guided, has been used in order to extract new knowledge consisting of new gene associations to hepatic disorders [5]. The found genes are now biologically investigated by the expert for a better understanding of their involvement in the disease.

Requirement analysis from biologists and their associated workflows have been since rapidly evolving with a non-stop emergence on the Web of new complex data types like protein structures, gene interactions or metabolic pathways, urging to continuous evolution of the warehouse schema, contents and applications.

3.2 Bio-entity Identification

By using GAV mapping approach for integrating one data source at a time in GEDAW (e.g. Fig 1 with GENBANK), we have minimized as much as possible the problem of identification of equivalent attributes. The problem of equivalent instances identification is still complex to address. This is due to general redundancy of

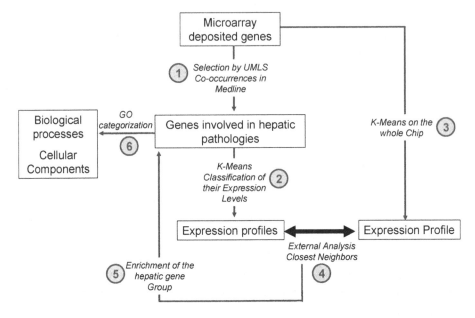

Fig. 2. Combining Biomedical Information within an Expert Guided Workflow

bio-entities in life science even within a single source. Biological databanks may also have inconsistent values in equivalent attributes of records referring to the same real-world object. For example, there are more than 10 ID's records for the same DNA segment associated to human HFE gene in GENBANK! Obviously the same segment could be a clone, a marker or a genomic sequence.

Anyone is indeed able to submit biological information to public databanks with more or less formalized submission protocols that usually do not include names standardization or data quality controls. Erroneous data may be easily entered and cross-referenced. Even if some tools propose clusters of records (like EntryGene for GENBANK) which identify the same biological concept across different biological databanks for being semantically related, biologists still must validate the correctness of these clusters and resolve the differences of interpretation among the records.

This is a typical problem of entity resolution and record linkage that is augmented and made more complex due to the high-level of expertise and knowledge it requires (i.e., difficult to formalize and related to many different sub-disciplines of biology, chemistry, pharmacology, and medical sciences). After the step of bio-entity resolution, data are scrubbed and transformed to fit the global DW schema with the appropriate standardized format for values, so that the data meets all the validation rules that have been decided upon by the warehouse designer.

Problems that can arise during this step include null or missing data; violations of data type; non-uniform value formats; invalid data. The process of data cleansing and scrubbing is rule-based. Then, data are migrated, physically integrated and imported into the data warehouse. During and after data cleansing and migration, quality metadata are computed or updated in the data warehouse metadata repository by pre- and post- data validation programs.

4 Database Profiling for Generic Data Exploration and Quality Aware Integration

4.1 Database Profiling

Database Profiling is the process of analyzing a database to determine its structure and internal relationships. It consists mainly of identifying: *i)* the objects used and their attributes (contents and number), *ii)* relationships between objects with their different kinds of associations including aggregation and inheritance, and *iii)* the objects behaviour and their relative functions. Database profiling is then useful when managing data conversion and data cleanup projects.

The XMI *(XML Metadata Interchange)* document (see Fig 3) that collects metadata information on the objects of the database is generated on-demand for profiling GEDAW. It has been quite useful to face the syntactic heterogeneity of the evolving schemas of data sources and the warehousing system during its life cycle. A generic data exploration has been made possible by the development of QDex tools (Quality based Database Exploration) that parse the XMI document detailing the database structure (in terms of class, attributes, relationships, etc.) and generate a model-based interface to explore the multiple attributes on genes description stored in the warehouse.

Fig. 3. The XMI document generated from GEDAW schema

4.2 Data Quality Profiling

Data quality profiling is the process of analyzing a database to identify and prioritize data quality problems. The results include simple summaries (counts, averages, percentages, etc.) describing for instance: completeness of datasets and the number of missing data records, the data freshness, and various data problems in existing records (e.g., outliers, duplicates, redundancies). During the process of data profiling, available data in the existing database are examined and statistics are being computed and gathered to track different summaries describing aspects of data quality. As a result, by providing QDex data profiling tools, one also provides data quality profiling tools.

A considerable amount of data quality research involves investigating and describing various categories of desirable attributes (or dimensions) of data quality. These lists commonly include accuracy, consistency, completeness, unicity (i.e., no duplicates), and freshness. Nearly 200 such terms have been identified in [15,16], regarding nature, definitions and measures of attributes.

Contradictory or ambiguous data is also a crucial problem as well, especially in bioinformatics where data are continuously speculative. Centralizing data in a warehouse is one of the initiatives one can take to ensure data validity.

Taking advantage of the stored XMI metadata information obtained by database profiling using the XMI document, QDex provides generic tools for bio-database exploration and data quality profiling. In developing QDex, we believe that profiling databases (both considering the structure of data sources and data warehouse) could be very useful for the integration process. Moreover, our work examines the extent of biological database profiling and proposes a way for flexibly building query workflows that follow the reasoning of biologists and assist them in the elaboration of their pioneer queries, including queries for data quality track.

5 QDEX Use Case: Application to GEDAW

5.1 Generic Bio-data Exploration

A global overview of QDex interface is given in Fig 4. Parts of the workflow that has been used to combine biological and medical knowledge to extract new knowledge on liver genes, has been flexibly reformulated using QDex GUI. The screen-shot below shows the central database of GEDAW as profiled using the XMI metadata document which gives an insight on the current warehouse schema. An overview of the extracted database profiling (classes and attributes) is browsed on the *Database Schema Viewer* frame. This includes the Gene, mRNA, ExpressionLevels, GOAnnotation and UMLSAnnotation classes. Based on these classes, the user built by himself, scenarios of queries on the objects, using his criteria, in the *Query Maker* frame.

By having an immediate glance on his intermediate or final results browsed on the *Result Viewer*, the user may modify and re-execute his queries when needed. He is also able to save a workflow for ulterior reuse on different data, or export effective

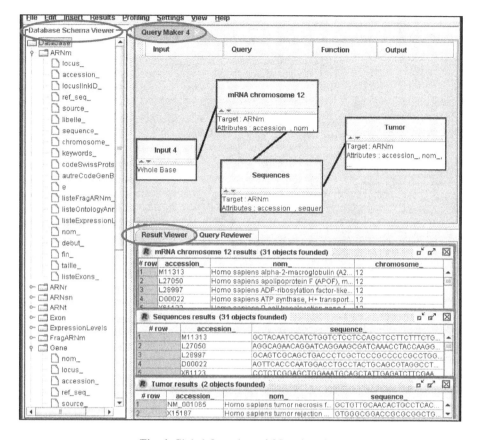

Fig. 4. Global Overview of QDex interface

resulting data for a future use on external tools (clustering, spreadsheet, etc,). This interface makes QDex quite flexible and attractive for the biologist.

To construct the Liver Disease Associated Genes Group, the Genes of the array that are annotated by "liver disease" concept and its descendants in UMLS are selected using UMLSAnnotation Class (See Fig igure4). Corresponding mRNA or Gene names are browsed on the Result Viewer sub-frame. Using the query maker, the selected objects are refined using successive queries on the group by adding boxes on demand, to look for information on their sequences, their expression levels, and their annotations in Gene Ontology making a more exhaustive workflow.

5.2 Preliminary Tools for Bio-data Quality Track

The completeness dimension of the result of a query workflow is computed by counting the number of missing values of queried objects (see Fig 5). Actually, by using QDex, much more possibilities are offered to the user to compose various workflows on integrated objects in GEDAW.

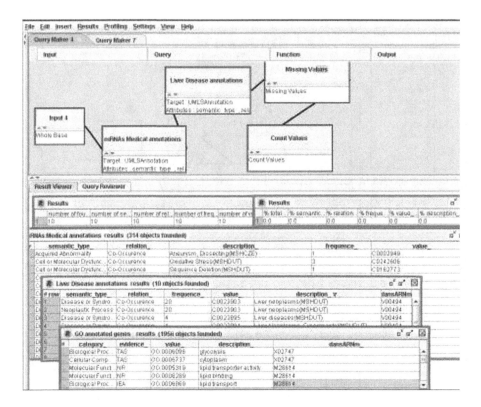

Fig. 5. Preliminary tools for tracking completeness of biomedical data

The user can have indicators associated to the datasets or query results by specifying various useful metrics to describe the aspects of database or query result quality. QDex project being still in progress, more tools will be provided to the user for evaluating the quality of the data that are being explored including redundancy, freshness, and inconsistency (by checking user-defined or statistical constraints).

6 Conclusion

In this paper, we have presented a database profiling approach for designing a generic biomedical database exploration tool devoted to quality aware data integration and exploration. QDex has been applied to GEDAW: an object oriented data warehouse devoted to the study of high throughput gene expression levels in the domain of hepatology. Metadata extracted from the XMI document of GEDAW have been used to provide a generic interface that supports tools for convivial building of query workflows using multiple profiled attributes on the genes and preliminary tools for data quality track. By developing QDex, data are supposed already being integrated. Using QDex, the user has the ability to make a clearer view of the database content and quality. As we have mentioned, QDex is under ongoing development and our perspectives are to keep on taking advantage of the extracted metadata information, and to provide more tools (such as a quality metric library) to be gradually integrated to the interface in order to evaluate the quality of the data that are being explored. Our main objective is to cover the main data quality dimensions by providing predefined analytical functions whose results (as computed indicators) will describe various aspects of consistency, accuracy, unicity, and freshness of data. Another important aspect of our future work is linked to data quality problems detection and concerns the design of pragmatic tools to help the expert to cleanse erroneous (or low quality) data within the QDex interface.

Finally, the original advantage of QDex resides in the fact that it can be generalized to any database schema outside bioinformatics. More specifically, we intend to apply QDex to the expected version of GEDAW which is being upgraded. This is for storing more actual bioinformatics data, like graph structures for gene pathways and system biology studies of genes expression profiles on the scale of a pangenomic DNA-Chip.

References

1. Anathakrishna, R., Chaudhuri, S., Ganti, V.: Eliminating Fuzzy Duplicates in Data warehouses. In: Proc. of Intl. Conf. VLDB (2002)
2. Batini, C., Catarci, T., Scannapiceco, M.: A Survey of Data Quality Issues in Cooperative Information Systems. In: Atzeni, P., Chu, W., Lu, H., Zhou, S., Ling, T.-W. (eds.) ER 2004. LNCS, vol. 3288, Springer, Heidelberg (2004)
3. Do, H.-H., Rahm, E.: Flexible Integration of Molecular-biological Annotation Data: The GenMapper Approach. In: Lindner, W., Mesiti, M., Türker, C., Tzitzikas, Y., Vakali, A.I. (eds.) EDBT 2004. LNCS, vol. 3268, Springer, Heidelberg (2004)
4. Guérin, E., Marquet, G., Burgun, A., Loréal, O., Berti-Equille, L., Leser, U., Moussouni, F.: Integrating and Warehousing Liver Gene Expression Data and Related Biomedical Resources in GEDAW. In: Ludäscher, B., Raschid, L. (eds.) DILS 2005. LNCS (LNBI), vol. 3615, Springer, Heidelberg (2005)
5. Guérin, E., Marquet, G., Chabalier, J., Troadec, M.B., Guguen-Guillouzo, C., Loréal, O., Burgun, A., Moussouni, F.: Combining biomedical knowledge and transcriptomic data to extract new knowledge on genes. Journal of Integrative Bioinformatics 3(2) (2006)
6. Harris, M.A., et al.: Gene Ontology Consortium. The Gene Ontology (GO) database and informatics resource. Nucleic Acids Res. (Database issue) 32, D258–D261 (2004)
7. Bodenreider, O.: The Unified Medical Language System (UMLS): integrating biomedical terminology. Nucleic Acids Res. (Database issue) 32, D267–D270 (2004)
8. Lacroix, Z., Critchlow, T. (eds.): Bioinformatics: Managing Scientific Data. Morgan Kaufmann, San Francisco (2003)
9. Martinez, A., Hammer, J.: Making Quality Count in Biological Data Sources. In: Proc. of the 2nd Intl. ACM Workshop on Information Quality in Information Systems (IQIS 2005), USA (June 2004)
10. Müller, H., Leser, U., Freytag, J.-C.: Mining for Patterns in Contradictory Data. In: Proc. of the 1rst Intl. ACM Workshop on Information Quality in Information Systems (IQIS 2004), France, pp. 51–58 (June 2004)
11. Müller, H., Naumann, F., Freytag, J.-C.: Data Quality in Genome Databases. In: Proc. of Conference on Information Quality (ICIQ 2003), USA, MIT, Cambridge (2003)
12. Overton, C.G., Haas, J.: Case-Based Reasoning Driven Gene Annotation. In: Computational Methods in MolecularBiology, Elsevier Science, Amsterdam (1998)

13. Rahm, E., Do, H.: Data Cleaning: Problems and Current Approaches. IEEE Data Eng. Bull 23(4), 3–13 (2000)
14. Thanaraj, T.A.: A clean data set of EST-confirmed splice sites from Homo sapiens and standards for clean-up procedures. Nucleic Acids Res. 27(13), 2627–2637 (1999)
15. Wang, R.Y.: Journey to Data Quality. In: Advances in Database Systems, Boston. 23, Kluwer Academic Press, Dordrecht (2002)
16. Wang, R., Kon, H., Madnick, S.: Data Quality Requirements Analysis and Modelling. In: Ninth International Conference of Data Engineering, Vienna, Austria Article (1993)
17. Pyle, D.: Data Preparation for Data Mining. Morgan Kaufmann, San Francisco (1999)
18. Durinck, S., Moreau, Y., Kasprzyk, A., Davis, S., De Moor, B., Brazma, A., Huber, W.: BioMart and BioConductor: a powerful link between biological databases and microarray data analysis. Bioinformatics 21(16), 3439–3440 (2005)

ProtocolDB: Storing Scientific Protocols with a Domain Ontology

Michel Kinsy, Zoé Lacroix, Christophe Legendre,
Piotr Wlodarczyk, and Nadia Yacoubi

Scientific Data Management Laboratory
Arizona State University
Tempe AZ 85287-5706, USA

Abstract. This paper addresses a systemic problem in science: although datasets collected through scientific protocols may be properly stored, the protocol itself is often only recorded on paper or stored electronically as the script developed to implement the protocol. Once the scientist who has implemented the protocol leaves the laboratory, this record may be lost. Collected datasets without a description of the process used to produce them become meaningless; furthermore, the experiment designed to produce the data is not reproducible. In this paper we present the ProtocolDB system that aims at assisting scientists in the process of (1) designing and implementing scientific protocols, (2) storing, querying, and transforming scientific protocols, and (3) reasoning about collected experimental data (data provenance).

1 Introduction

Public biological resources form a complex maze of heterogeneous data sources, interconnected by navigational capabilities and applications. Although this wide and valuable network offers scientists multiple options to execute their scientific protocols, selecting the resources suitable to obtain and exploit their data of interest is a tedious task. When designing a scientific protocol, they struggle to consolidate the best information about the scientific objects being studied and to implement it in terms of queries against biological resources. These protocols, although expressed at a conceptual level, are typically implemented using the resources the scientist is most familiar with, instead of the resources that may best meet the protocol's needs. This implementation-driven approach to express scientific protocols may significantly affect the outcome of a scientific experiment.

The biological semantic Web is diverse and offers multiple orthogonal viewpoints on scientific data. Each viewpoint is expressed by the way the data are organized (e.g., GenBank is sequence-centric when GeneCards is gene-centric), the access capabilities offered to scientists to retrieve data (e.g., to access gene descriptions in GeneCards, one can use a full-text search engine or provide a HUGO symbol), the applications, annotations, and links and indices to other relevant resources. In addition to this structural diversity, biological resources offer a rich semantic diversity characterized by data coverage (entries present

M. Weske, M.-S. Hacid, C. Godart (Eds.): WISE 2007 Workshops, LNCS 4832, pp. 17–28, 2007.

in the data source), identity, characterization and annotations (set of attributes pertaining to each entry), links and indices between entries, the domain, image, and cardinality of those links, quality, consistency, reliability, etc. All these semantic characteristics are metrics that may be used to predict the outcome of the execution of a scientific protocol on selected resources. Syntactic, semantic and cost characteristics all participate in the outcome of the execution of a scientific protocol. Indeed, the selection of a resource may dramatically affect the dataset collected at execution time [7].

To assist adequately scientists in the process of expressing scientific protocols, it is necessary to understand what scientific protocols are, how they are structured, how scientists express them, and how they are implemented for execution. While they are critical components of the scientific process that leads to discovery, scientific protocols have been poorly studied. A scientific protocol is the process that describes the experimental component of scientific reasoning. Scientific reasoning follows a hypothetico-deductive pattern, i.e., the successive expression of a causal question, a hypothesis, the predicted results, the design of an experiment, the actual results of the experiment, the comparison of the predicted results and the actual results, and the conclusion, which may or may not be supportive of the hypothesis [8]. Scientific protocols (or equivalently pipelines, workflows, or dataflows) are complex procedural processes composed of a succession of scientific tasks that express the way the experiment is conducted. Although there is no commonly used definition of what a scientific protocol really is, in January 2003 a brainstorming session devoted to scientific protocols[1] identified the following characteristic: a succession of steps (recipe) that describes a process that can be reproduced. A scientific protocol thus describes how the experiment is conducted and records all information necessary to reproduce the same experiment. In the context of digital scientific protocols each step of the protocol is a bioinformatics task [15,1] that records how biological data are produced from measurements, extracted from a data source or resulting from an application, etc.

In this paper we present the ProtocolDB system that aims at assisting scientists in the process of (1) designing and implementing scientific protocols, (2) storing, querying, and transforming scientific protocols, and (3) reasoning about collected experimental data (data provenance). A motivation example is presented in Section 2. Section 3 is devoted to conceptual protocols whereas the selection of resources and their integration are discussed in Section 4. We discuss related work in Section 5 and conclude in Section 6.

2   Motivating Example

Alternative Splicing (AS) is the splicing process of a pre-mRNA sequence transcribed from one gene that leads to different mature mRNA molecules thus to

---

[1] The session took place during the Dagstuhl Seminar 03051 devoted to Information and Process Integration: A Life Science Perspective. The material presented at the seminar is available at http://www.dagstuhl.de/03051/.

different functional proteins. Alternative splicing events are produced by different arrangements of the exons of a given gene. The Alternative Splicing Protocol (ASP) described as follows is currently supporting the BioInformatics Pipeline Alternative Splicing Services BIPASS [6].

> *The Alternative Splicing Protocol (ASP) takes a set of transcripts as input and returns clusters of transcripts aligned to a gene. The process of alignment consists of an alignment of each transcript sequence against each genomic sequence of a whole genome of one or more organisms. This step is executed with all known transcripts extracted from different public databases. A clustering step immediately follows the alignment step. That step allows delimiting the transcript region of a gene excluding its regulation region. A cluster normally represents or may be representative of all intermediate transcripts (from the Pre-messenger-RNA(s) to the mature messenger-RNA(s)) required to obtain one or several functional translated proteins from the same gene.*

Such a protocol description expresses the *design protocol*, i.e., its scientific aim. It specifies two scientific tasks with a conceptual description of their inputs and outputs illustrated in Figure 1.

1. Task 1 performs an *alignment* of transcripts against genomic sequences. The results (output) of this task is an alignment of the transcripts with respect to the genomic sequence.
2. Task 2 performs a *clustering* of the aligned transcripts. The result (output) of this task is a list of clusters of transcripts.

In general the description of a scientific protocol is a textual document that combines the scientific aim with the resources used to implement it. For example, ASP could be described as follows.

> *ASP performs a first alignment with BLAT, selects the 10% first ranked alignments, extracts the aligned transcripts and the aligned genomic sequences. Then it completes the extracted genomic sequences with 50,000 bases in upstream and downstream. It re-aligns the transcripts against the resulting genomic sequences with SIM4 and clusters the results.*

Such a protocol description is a poor record of the process.

– A textual document is difficult to parse to extract the exact tasks involved, their ordering, and all needed semantic and structural information exploited when retrieving, querying, and reasoning about scientific protocols.

Fig. 1. Alternative splicing design protocol

– It mixes the scientific aim and the resource selection (i.e., BLAT and SIM4). This makes it difficult to revise the protocol with new resources and compare their results. It also affects the ability to integrate data collected from different protocol implementations with similar scientific aim.
– It does not record the reasons why a simple scientific task such as a sequence alignment needs to be split into a sub-protocol involving five tasks: alignment, filter, extraction, collection, and alignment.
– It does not specify how the resources were integrated (schema mapping, variable binding).

In ProtocolDB, a scientific protocol is composed of a *design protocol* that captures its scientific aim expressed with respect to a domain ontology, and one or more *implementations* that specify the resources selected to implement each task and the dataflow expressed as ontology-driven schema mappings. The *design protocol* is mapped to *implementation protocols*, themselves mapped to experimental data collected after their execution as illustrated in Figure 2.

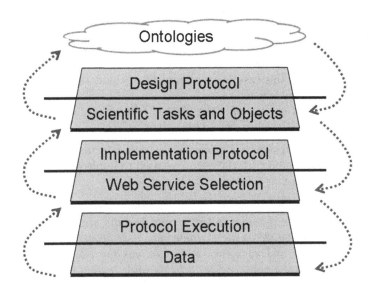

Fig. 2. Life Cycle of a scientific protocol

3   Design Protocol

A *design protocol* (or conceptual protocol) is a graph composed of connected scientific tasks whose inputs and outputs are collections of conceptual variables. Each scientific design task is a design protocol. Complex design protocols are composed of scientific design tasks connected with two binary operators • and ⊗, respectively denoting the *successor operator* and the *parallel composition* [4,3]. I (resp. O) denotes the input (resp. output) of the design task or protocol.

Definition 1. *The set of design protocols D is the closure of the set of design tasks T under two binary operators $\bullet$ and $\otimes$.*

- $T \subset D$
- $\forall P_1, P_2 \in D$ *if* $I_{P_2} = O_{P_1}$ *then* $P_1 \bullet P_2 \in D$, $I_{P_1 \bullet P_2} = I_{P_1}$ *and* $O_{P_1 \bullet P_2} = O_{P_2}$.
- $\forall P_1, P_2 \in D$ *then* $P_1 \otimes P_2 \in D$, $I_{P_1 \otimes P_2} = [I_{P_1}, I_{P_2}]$ *and* $O_{P_1 \otimes P_2} = [O_{P_1}, O_{P_2}]$.

A design protocol expresses the scientific aim of the protocol in terms of a domain ontology such as described in Figure 3. The design protocol is a conceptual protocol where each task expresses a conceptual scientific task or relationship. The dataflow captured by a design protocol is expressed in terms of collections of variables typed with respect to classes in an ontology. For example, the ASP design task *alignment* takes as input a record $[s_1, \{s_2\}]$, where s_1 and s_2 are two variables of type Sequence. Two design tasks (or protocols) may be composed sequentially if the input of the latter is the same as the output of the former.

3.1 Protocol Entry in ProtocolDB

The ProtocolDB entry interface allows scientists to edit and store scientific protocols. To enter a new protocol, a scientist registers in the system and records the context of the protocol (institution, author, etc.). Documents may be uploaded to the system. Once the overall scientific protocol is described, the user starts constructing the design protocol (conceptual workflow) from the interface shown in Figure 4. The initial step consists of a protocol blackbox (root) that represents the overall protocol.[2]

The user may rename the protocol (1), specify inputs (2) and outputs (3), and enter a description (4). Then the user may select one of two operators: *split sequential* or *split parallel* shown in the lower right square. By selecting the *split sequential* operator, the system splits the selected task box (root) into two successive tasks. The operator *split sequential* inserts a new task right after the selected task in the protocol branch. By default, the input (resp. output) of the new inserted task is a tuple composed of the input and output (resp. output) of the selected task. In contrast *split parallel* splits the protocol branch into two branches. One of the branches corresponds to the selected task whereas the second branch inserts a new task box with the same input and output than the selected task. The default specifications may be changed when documenting the new task box (input variables may be removed and new output may be produced).

Editing functions (currently under development) allow the user to remove tasks and upload existing protocols to instantiate a task box. Once the design protocol is entered in ProtocolDB, the user may map it to one or several implementations.

---

[2] Our approach follows a top-down approach splitting step-by-step the protocol into connected tasks.

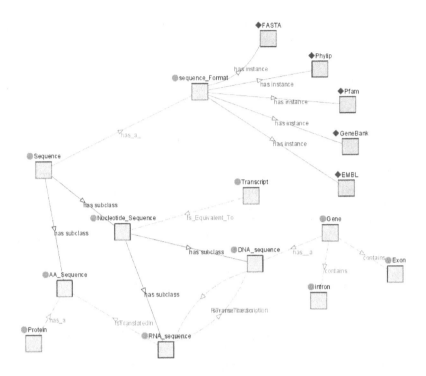

Fig. 3. Portion of a domain ontology

4   Implementation Protocol

An *implementation protocol* is a graph composed of connected scientific resources (database queries or tools) whose inputs and outputs are data types. A single bioinformatics service is an implementation protocol. Complex implementation protocols are composed of scientific resources connected with the same two binary operators • and ⊗ used to express design protocols. Each design task box of the design protocol is mapped to an implementation protocol. A single design task may be mapped to a complex implementation protocol invoking several bioinformatics resources, queries, and/or connectors to translate data formats (see Section 4.2). We first describe how services are selected to implement a design protocol in Section 4.1. We address the problem of service composition in Section 4.2.

4.1   Selecting Services

To enter an *implementation protocol* for a given design protocol, the user clicks on the New Implementation button that displays the tab implementation. Clicking on the implementation tab activates a new window where the user documents the implementation version to be entered. Then the implementation protocol

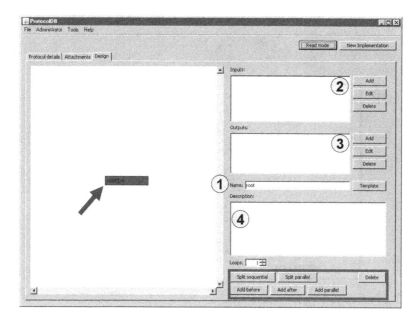

Fig. 4. ProtocolDB entry interface

may be entered with the window shown in Figure 5 opened by clicking on the `Implementation Diagram` tab.

Each selected design task (A) is first mapped to an implementation task box (B) that can be expanded with the two operators to create the corresponding implementation sub-protocol (red dark rectangle). Each implementation task may be instantiated with a bioinformatics resource by choosing a tool within a of resources (C).

4.2 Mapping Services

The selection of resources to implement a scientific protocol may lead to multiple successive attempts often failing because of syntactic, semantic, and efficiency reasons [5]. After discovering services that are relevant to the implementation of a protocol, the next step is to identify whether these services are *compatible*. Two services are semantically compatible when their respective input and output types refer to the same ontological class. However, semantic compatibility does not always correspond to syntactic interoperability. ProtocolDB relies on the use of domain ontologies to reconciliate conflicts occurring when a given scientific object is represented with different syntactical structures by bioinformatics resources. For instance, a scientist may select BLAT and SIM4 to implement the alignment design task identified in Section 2. The two alignment tools are semantically similar: their inputs and outputs are constructs of the same conceptual classes (Figure 3).

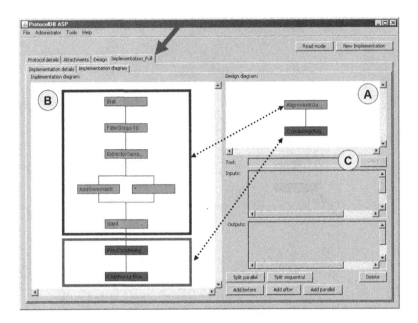

Fig. 5. ProtocolDB implementation protocol entry interface

The outputs and inputs of BLAT and SIM4 are closely related but cannot be composed (Figure 6(a)). An executable implementation would require a *connector* to compose the two services (Figure 6(b)). ProtocolDB relies on BioOnMap to generate mappings for data transformation and conversion [17,18] by exploiting domain knowledge expressed in an ontology. Because the description of the implementation of scientific protocols in ProtocolDB is mapped to a design protocol characterized by an ontology, the task of mapping bioinformatics services may exploit the domain knowledge of the ontology.

To generate a mapping, BioOnMap considers the *semantic type* of each input and output of services to identify whether the output of the former is compatible with the input of the latter. For instance, BLAT produces an output of semantic type aligned `Transcript` while SIM4 accepts as inputs a set of `DNA_sequence` and a set of `Sequence`. Exploiting the domain ontology depicted in Figure 3, BioOnMap may infer that a `Transcript` is semantically equivalent to a `Sequence` and a `DNA_sequence` is a sub-class of the same class, i.e., `Sequence`, which means that the two concepts `Transcript` and `DNA_sequence` are semantically equivalent. Consequently, BioOnMap infers that the two services are semantically compatible. On the syntactic side, the format expected by SIM4 is `FASTA` which is an instance of the `sequence_Format` class, the output format of BLAT is `PSL` which is also a `sequence_Format`. The mapping between the design and the implementation levels is given by the ontological relationship "a `sequence` has-a `sequence_Format`". `FASTA` and `PSL` are instances of the same conceptual class, they refer to different representations of the same scientific object, i.e., `Sequence`. In that case, the BioOnMap generates a connector service

(a) Implementation of alignment task in ASP

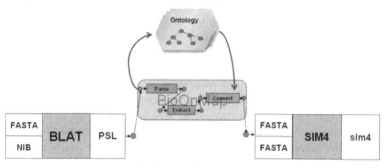

(b) Semantic mapping

Fig. 6. BLAT takes a set of transcripts in FASTA format and genomic sequences in NIB format and returns an alignment file in PSL format. SIM4 takes both inputs in FASTA format and produces an alignment in a format specific to SIM4.

that syntactically translates a PSL file into a FASTA file. Finally, each connector is invoked automatically in the implementation protocol. Once the implementation is completed, the protocol can be executed with a workflow system such as Taverna [14], Kepler [11], or Mobyle [12]. After execution, the data collected at each step of the protocol execution are stored in ProtocolDB and provide the fourth layer of our approach (bottom of Figure 2). The mapping of collected datasets to their corresponding implementation and design tasks provides the framework needed to reason about data provenance [2].

5   Discussion

The distinctive features of ProtocolDB presented in this paper include: 1) a two-layer model for the representation of protocols and 2) a light-weight semantic support by the use of domain ontologies that enhances significantly the composition and enactment of Web services.

ProtocolDB aims at providing support for designing, storing, and reasoning on scientific protocols. The two-layer representation of protocols with a *design protocol* mapped to one or several *implementation protocols* offers valuable functionalities to the user. The design protocol expresses the scientific aim in terms of classes and relationships of a domain ontology. An implementation protocol describes the way the scientific aim will be achieved. Because technology changes over time, it is a way to record within a laboratory the various ways a particular experiment is conducted identifying the machines, robots, and other technology involved in the process. Another benefit of the approach is to let the scientist explore and compare the performance of different implementations. ProtocolDB is not developed to execute scientific protocols but it is a system that offers the ability to reason about scientific protocols. BioOnMap allows support for composing services and generates protocols that can be executed. Future functionalities of the system include support to selection of resources suitable with the user's needs, prediction of the outcome of an execution (performance and quality of results), protocol re-use (query protocols, find similar protocols).

In contrast, workflow systems such as Taverna [13], Kepler [11], or Mobyle [12] enable the construction and execution of workflows over distributed Web services. These platforms are implementation-driven and they do not capture the scientific aim of the protocol. They do not provide a query language to query, compare, re-use scientific protocols stored in a repository. ProtocolDB aims at generating implementation protocols in a format compatible with these platforms so that once they have been entered, they can be easily uploaded and executed by these systems.

Expressing complex executable workflows remains a difficult, time-consuming, and expensive task. One of the reasons for this difficulty is the large number of bioinformatics resources. The paradigm of semantic Web services offers the possibility of highly flexible Web services architectures where new services can be quickly discovered, orchestrated, and composed into workflows [10]. Taverna relies on the Feta system to search semantically candidate services [9]. In the future we will integrate the Semantic Map [16] approach to ProtocolDB to guide scientists who wish to explore the maze of available resources to implement design tasks.

In the BioOnMap system [18], we present a formal framework of semantic descriptions for stateless services. From a syntactic point of view, our framework enhances resource description provided by OWL-S Service Profile (i.e., input and output description). Web services are being independently created by many parties worldwide, using different terminologies (ontologies) and datatypes, hindering their integration and reusability [19]. Taverna proposes a list of "shims", i.e., services that resolve basic syntactical mismatches in order to reconciliate closely related inputs and outputs. However, a new shim needs to be manually created for each pair of services that need to interoperate which make this manual approach not scalable. [11] describes a scalable framework that uses mappings to one or more ontologies for reconciling two services. Instead, the BioOnMap approach provides a uniform approach to workflow and data integration [17].

6   Conclusion

Recording scientific protocols together with experimental data is critical to scientific discovery. ProtocolDB presented in this paper is a system that assists scientists in the expression of protocols and provides a framework for protocol reuse and analysis, sharing, archival, and reasoning on provenance of experimental data. Our approach exploits a domain ontology to index semantically each protocol task and collected dataset. Scientific protocols are expressed as a pair of a design protocol that captures the scientific aim and one or more implementations where services are selected and composed into an executable workflow using BioOnMap. ProtocolDB provides the framework needed to record scientific protocols so that they can be reproduced, thus validating experimental results and to query, reuse, compare scientific protocols and their corresponding collected datasets. In the future we will include Semantic Map [16], a system designed to assist scientists in the selection of the bioinformatics resources to implement their protocols. ProtocolDB is available at http://bioinformatics.eas.asu.edu/siteProtocolDB/indexProtocolDB.htm.

Acknowledgments. This research was partially supported by the National Science Foundation[3] (grants IIS 0223042, IIS 0431174, IIS 0551444, IIS 0612273, and IIS 0738906). The authors would like to thank Louiqua Raschid, Maria Esther Vidal, Natalia Kwasnikowska, Jan Van den Bussche, Susan Davidson, and Sarah Cohen-Boulakia for their valuable input and exciting discussions.

References

1. Bartlett, J.C., Toms, E.G.: Developing a Protocol for Bioinformatics Analysis: An Integrated Information Behaviour and Task Analysis Approach. Journal of the American Society for Information Science and Technology 56(5), 469–482 (2005)
2. Cohen-Boulakia, S., Biton, O., Davidson, S.: Querying Biologically Relevant Provenance information in Scientific Workflow Systems with Zoom*UserViews. In: Proc. 33rd International Conference on Very Large Data Bases (demonstration paper), Vienna, Austria, pp. 1366–1369 (2007)
3. Kinsy, M., Lacroix, Z.: Storing Efficiently Bioinformatics Workflows. In: IEEE 7th International Symposium on Bionformatics and Bioengineering, Boston, MA (2007)
4. Kwasnikowska, N., Lacroix, Z., Chen, Y.: Modeling and Storing Scientific Protocols. In: Meersman, R., Tari, Z., Herrero, P. (eds.) On the Move to Meaningful Internet Systems 2006: OTM 2006 Workshops. LNCS, vol. 4277, pp. 730–739. Springer, Heidelberg (2006)
5. Lacroix, Z., Legendre, C.: Analysis of a Scientific Protocol: Selecting Suitable Resources. In: Proc. First IEEE International Workshop on Service Oriented Technologies for Biological Databases and Tools, Salt Lake City, UT, July 9-13, pp. 130–137 (2007)
6. Lacroix, Z., Legendre, C., Raschid, L., Snyder, B.: BIPASS: BioInformatics Pipelines Alternative Splicing Services. Nucleic Acids Research, Volume Web Services Issue 35, 292–296 (2007)
7. Lacroix, Z., Raschid, L., Eckman, B.: Techniques for Optimization of Queries on Integrated Biological Resources. Journal of Bioinformatics and Computational Biology 2(2), 375–411 (2004)

---

[3] Any opinion, finding, and conclusion or recommendation expressed in this material are those of the authors and do not necessarily reflect the views of the National Science Foundation.

8. Lawson, A.E.: The nature and development of hypothetico-predictive argumentation with implications for science teaching. International Journal of Science Education 25(11), 1387–1408 (2003)
9. Lord, P.W., Alper, P., Wroe, C., Goble, C.A.: Feta: A light-weight architecture for user oriented semantic service discovery. In: Gómez-Pérez, A., Euzenat, J. (eds.) ESWC 2005. LNCS, vol. 3532, pp. 17–31. Springer, Heidelberg (2005)
10. McIlraith, S.A., Son, T.C., Zeng, H.: Semantic web services. IEEE intelligent systems 16(2), 46–53 (2001)
11. McPhillips, T.M., Bowers, S., Ludäscher, B.: Collection-oriented scientific workflows for integrating and analyzing biological data. In: Leser, U., Naumann, F., Eckman, B. (eds.) DILS 2006. LNCS (LNBI), vol. 4075, pp. 248–263. Springer, Heidelberg (2006)
12. Néron, B., Tufféry, P., Letondal, C.: Mobyle: a Web portal framework for bioinformatics analyses. In: Network Tools and Applications in Biology (poster), Naples, Italy (2005)
13. Oinn, T., Greenwood, M., Addis, M., Alpdemir, M.N., Ferris, J., Glover, K., Goble, C., Goderis, A., Hull, D., Marvin, D., Li, P., Lord, P., Pocock, M.R., Senger, M., Stevens, R., Wipat, A., Wroe, C.: Taverna: lessons in creating a workflow environment for the life sciences. Concurrency and Computation: Practice and Experience 18(10), 1067–1100 (2006)
14. Oinn, T.M., Addis, M., Ferris, J., Marvin, D., Senger, M., Greenwood, R.M., Carver, T., Glover, K., Pocock, M.R., Wipat, A., Li, P.: Taverna: a tool for the composition and enactment of bioinformatics workflows. Bioinformatics 20(17), 3045–3054 (2004)
15. Stevens, R., Goble, C., Baker, P., Brass, A.: A Classification of Tasks in Bioinformatics. Bioinformatics 17(2), 180–188 (2001)
16. Tufféry, P., Lacroix, Z., Ménager, H.: Semantic Map of Services for Structural Bioinformatics. In: Tufféry, P., Lacroix, Z. (eds.) Proc. of the 18th International Conference on Scientific and Statistical Database Management, Washington DC, pp. 217–224 (2006)
17. Yacoubi, N., Lacroix, Z.: Resolving Scientific Service Interoperability With Schema Mapping. In: Proc. IEEE 7th International Symposium on Bionformatics and Bioengineering, Boston, MA, pp. 14–17 (2007)
18. Yacoubi, N., Lacroix, Z., Vidal, M.-E., Ruckhaus, E.: Deductive Web Services: an Ontology-driven Approach to Service Composition and Data Integration. In: Proc. International on Semantic Web and Web Semantics, Vilamoura, Portugal, pp. 29–30 (2007)
19. Zamboulis, L., Martin, N., Poulovassilis, A.: Bioinformatics Service Reconciliation By Heterogeneous Schema Transformation. In: Data Integration in the Life Sciences. LNCS, vol. 4544, pp. 89–104. Springer, Heidelberg (2007)

An Ontology-Driven Annotation of Data Tables

Gaëlle Hignette[1], Patrice Buche[1], Juliette Dibie-Barthélemy[1],
and Olivier Haemmerlé[2]

[1] UMR AgroParisTech/INRA MIA - INRA Unité Mét@risk
AgroParisTech, 16 rue Claude Bernard, F-75231 Paris Cedex 5, France
{hignette,buche,dibie}@agroparistech.fr
[2] IRIT - Université Toulouse le Mirail, Dpt. Mathématiques-Informatique
5 allées Antonio Machado, F-31058 Toulouse Cedex
ollivier.haemmerle@univ-tlse2.fr

Abstract. This paper deals with the integration of data extracted from
the web into an existing data warehouse indexed by a domain ontology.
We are specially interested in data tables extracted from scientific publications found on the web. We propose a way to annotate data tables from
the web according to a given domain ontology. In this paper we present
the different steps of our annotation process. The columns of a web data
table are first segregated according to whether they represent numeric or
symbolic data. Then, we annotate the numeric (resp.symbolic) columns
with their corresponding numeric (resp. symbolic) type found in the ontology. Our approach combines different evidences from the column contents and from the column title to find the best corresponding type in
the ontology. The relations represented by the web data table are recognized using both the table title and the types of the columns that were
previously annotated. We give experimental results of our annotation
process, our application domain being food microbiology.

Keywords: ontology-driven data integration, semantic annotation.

1 Introduction

In the scientific world, many experimental data are produced and continually
published on the web. It is often hard to keep track of all the experiments that
are conducted in a specific domain, and even harder to compile all the results
from different experiments at the time when one needs them. Our work is applied
to the domain of food microbiology. In a first system called MIEL [1], data on
food microbiology coming from scientific publications or industrial sources is
manually entered into a relational database: this database is indexed with a
domain ontology, so that data coming from different sources is represented with
a certain uniformity. Users can query the database using an interface that allows
them to select, within the domain ontology, the food product, microorganism
and relations they are interested in. The database is then queried to find data
corresponding to or close to the users selection criteria, and the answers are
ordered according to how close they are to the users criteria. However, manually

M. Weske, M.-S. Hacid, C. Godart (Eds.): WISE 2007 Workshops, LNCS 4832, pp. 29–40, 2007.
© Springer-Verlag Berlin Heidelberg 2007

feeding this database is very time-consuming, resulting in a largely incomplete database. Our work aims at building an XML data warehouse that completes the database with data gathered on the web that is automatically annotated with the same ontology, so as to provide a uniform way of querying. We focus on data that are presented in tables, since it is a usual way of presenting synthetic information in many scientific or economic domains, and in particular in food microbiology. Our goal is to annotate the table contents and recognize the relations represented in the tables. The annotated data tables can then be queried using the ontology. Our whole system relies on the domain ontology, so that changing the ontology is enough to change the application domain, provided that we are looking for data presented in tables.

There has been a lot of research work on table recognition, both for the recognition of tables within text and for the detection of the table orientation [2]. Our work is not focused on table recognition or orientation, but on annotating a table that has already been recognized and put in a standard orientation where relations are represented in lines, the first line of the table being the column titles and the next lines being the actual data we want to extract. In [3], frames are constructed from tables using semantic tools such as the WordNet ontology or GoogleSets. However, in their approach they create new relations with names according to the relation signature and a generic ontology, whereas we want to recognize predefined relations in an ontology specific to the targetted domain. In a more recent work [4], relations from an ontology are instanciated using various HTML structures including tables. However, they only identify binary concept-role relations between instances that are assumed to be already annotated (manually or using another information extraction system). Our work differs as we focus on the recognition of n-ary relations and we propose a step-by-step algorithm including the recognition of element types. From this point of view, the work presented in [5] is nearer to ours, as they transform tables of different structures into a common relational database schema with n-ary relations. However, their approach is dependent on the manual definition of an "extraction ontology" that defines extraction rules for each set of objects, giving all synonyms for an attribute name or defining the context of apparition of a string to extract. Our work differs as the extraction rules we build are independant from the ontology: the domain experts who build the ontology only define which data are of interest for their purpose, they don't have to concentrate on the way to find them. Other annotation techniques applicable to tables, based on learning wrappers using textual context and / or structure, such as BWI [6] or Lixto [7], are not interesting in our application for several reasons: there is no textual context for the apparition of terms in the tables, as we consider the entire content of a table cell for annotation; the table structures are not homogeneous; data is rare in our domain, so it is very difficult for us to rely on learning techniques, as obtaining data to train learning algorithms is a real challenge.

We first present the structure of the ontology that we use during our anno-tation process (Sect. 2). In the tables that we consider for annotation, semantic relations are represented as rows. Our aim is to recognize the signature of a known relation by looking at the different column types of the table. Our way of finding the type of a column differs depending on whether the column is numeric or symbolic. Thus, the first step of our annotation process consists in classifying columns as numeric or symbolic (Sect. 3). For the annotation of both numeric and symbolic columns, we use a similarity measure between the terms from the web and the terms from the ontology (Sect. 4). Then we present the way of finding the type of a numeric column (Sect. 5) and the type of a symbolic column (Sect. 6). We then show how the knowledge of the different column types is used to find the relations represented in the table (Sect. 7).

2   Structure of the Ontology

Figure 1 shows the structure of a simplified version of our domain ontology. The ontology is built in collaboration with domain experts, who ideally are the future users of the querying system that exploits the annotated tables. They define the numeric types, symbolic types and relations that are interesting for their domain.

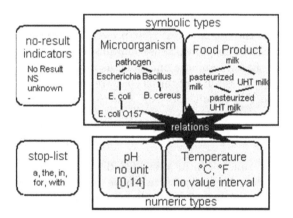

Fig. 1. Structure of the ontology used for our annotation process

Numeric types are used to define the numeric data that we want to extract from the tables. A numeric type in the ontology is described by the name of the type, the units in which data of this type is usually expressed, and the interval of possible values for this type. For example, the type named "Temperature" can be expressed in the units $\{°C, °F\}$ and has a range of $]-\infty, +\infty[$ while the type "pH" has no unit and has a range of $[0, 14]$.

Symbolic types are used when the data of interest is represented as a string. A symbolic type in the ontology is described by the name of the type and the type hierarchy (which is the set of possible values for the type, partially ordered by the subsumption relation). Such a hierarchy can be either a tree or a lattice, depending on whether it has sense for the domain that a concept has several parents: for example the type hierarchy attached to "Microorganism" is a biological taxonomy of selected microorganisms, additionally divided between the categories "pathogen", "spoiler" and "useful for food process".

Relations are used to represent semantic links between different types. A relation in the ontology is described by the name of the relation and its signature. Relations originate from a relational database schema: they were n-ary with a similar role for all types in the signature. However, we were able to redesign the relation definition to correspond to an application having for domain the set of controlled factors (numeric or symbolic types) in the experimental plan and for range the resulting experimental measure (always a numeric type). For example, the relation "Growth kinetics" of our ontology, which measures the growth of a microorganism over time, has for domain the set of types "Microorganism", "Food product", "Temperature", "Time", and for range the type "Colony count" (the microorganism concentration). As several measures of non-controlled factors can be taken in one experiment, tables often represent several relations. Note that the relations are not functions: due to biological variations and to other factors not taken into account in the model of a relation, the same set of values in the domain types can lead to distinct values in the range type.

In addition to the numeric or symbolic types and the relations defined in collaboration with domain experts, the ontology also presents two lists that are not really specific to the domain, but that present lexical knowledge useful for our annotation process. The stopword list contains a list of words that have more a grammatical role than a semantic one, such as articles and conjunctions. The "no-result indicator" list is a list of terms that are used to represent the absence of data, for example "No Result" or "NS" (for Not Specified).

3   Distinction Between Numeric and Symbolic Columns

The first step of our annotation process is to distinguish between numeric and symbolic columns. For that purpose, we have built a set of rules, using the ontology. Let *col* be a column of the table we want to annotate. We search *col* for all occurrences of numbers (in decimal or scientific format) and of units of numeric types described in the ontology. We also search *col* for all words, which are defined as alphabetic character sequences that are neither units nor "no result indicators".

Let c be a cell of the column *col*. We apply the following classification rules:

- if c contains a number immediately followed by a unit, or a number in scientific format, then c is numeric;
- else, if c contains more numbers and units than words, then c is numeric;
- else, if c contains more words than numbers and units, then c is symbolic;
- else (equal amount of words and numbers+units) the status of c is considered as unknown.

Once all cells of the column *col* have been classified using the above rules, *col* is classified as symbolic if there are more cells classified as symbolic than numeric.

Table 1. Classification results for numeric and symbolic column detection over 349 columns (as there is no unclassified column, precision is equal to recall)

computed / manual	classification using ontology		naive classification	
	numeric	symbolic	numeric	symbolic
numeric	261	2	229	34
symbolic	5	81	13	73
precision/recall	98%		87%	

Else, the column is classified as numeric (we have experimentally shown that when numbers of symbolic and numeric cells are equal, it usually corresponds to a high rate of absent data, which is more frequent in numeric columns).

We have experimented our method on 60 tables taken from publications in food microbiology. We have compared our classifier with a naive classifier, for which a cell is numeric if and only if it contains a number. Results are shown in Tab. 1. Our method gives much better results than the naive classifier because it is able to consider as non-numeric a cell that contains numbers (for example a microorganism with a strain number) as well as it is able to deal with unknown data: the "no result indicators" are not considered as words, thus leading to an unrecognized cell, whereas the naive method considers them as symbolic.

4   Similarity Measure Between a Term from the Web and a Term from the Ontology

Throughout our whole annotation process, we will be using a similarity measure that allows to compare a term from the web with a term from the ontology, a term being a set of consecutive words. Several similarity measures to compare two terms are presented in [9]. Semantic similarity measures imply the use of an ontology that contains both terms to compare. In our case, there is not such an ontology: we have tested both Wordnet[1] and AgroVoc[2], but the terms used on the web to represent food products are too specific and do not appear in those thesauri. Instead of using a semantic similarity, we thus use a word-by-word similarity measure.

Words in terms from the ontology are manually weighted, according to their importance in the meaning of the term. The non-informative words listed in the stopword list are given a weight of 0, and other words are weighted with only two possible weights: 1 for the most informative word(s) of the term, 0.2 for secondary words (the weight of 0.2 has been chosen after preliminary experiments that showed the results were better than when using 0.5). Determining the importance of a word in a term meaning is done by a domain expert (for example, "cooked"

---

[1] http://wordnet.princeton.edu/
[2] Thesaurus for agriculture and food industry used by the FAO, http://www.fao.org/aims/ag_intro.htm

Table 2. Vector representation of the term from the web "ground meat" and two terms from the ontology

coordinates \\ terms	ground	meat	fresh	beef	pork
ground meat	1	1	0	0	0
fresh meat	0	1	0.2	0	0
ground beef	0.2	0	0	1	0

in "cooked meat" is important if we are interested in microorganisms, while not if we are looking at chemical contaminants). We have shown in [10] that manual weight definition an improvement in annotation quality compared to giving a weight of 1 to each word, even if this improvement is quite small. Weighted terms are represented as vectors, in which the coordinates represent the different possible lemmatised words and their weight in the term (0 if the word does not belong to the term). Table 2 shows the vector representation of a term from the web and two terms from the ontology. The similarity between a term from the web and a term from the ontology is then a similarity between two weighted vectors: we have tested several similarity measures, such as the Dice coefficient [9] or the cosine similarity measure [11]. The choice of the similarity measure had no big impact on our results in the following steps of the annotation. We decided to keep the cosine similarity measure as it is the most popular one.

Definition 1. *Let w be a term from the web, represented as the weighted vector $\boldsymbol{w} = (w_1, \ldots, w_n)$ and o a term from the ontology, represented as the weighted vector $\boldsymbol{o} = (o_1, \ldots, o_n)$. The similarity between w and o is computed as:*

$$sim(w, o) = \frac{\sum_{k=1}^{n} w_k \times o_k}{\sqrt{\sum_{k=1}^{n} w_k^2 \times \sum_{k=1}^{n} o_k^2}} \tag{1}$$

Example 1. The similarity measures between the term "ground meat" from the web and the two terms from the ontology presented in table 2 are respectively:

- $sim(ground\ meat, fresh\ meat) = \frac{1 \times 0 + 1 \times 1 + 0 \times 0.2 + 0 \times 0 + 0 \times 0}{\sqrt{(1^2 + 1^2) + (1^2 + 0.2^2)}} \approx 0.57$
- $sim(ground\ meat, ground\ beef) = \frac{1 \times 0.2 + 1 \times 0 + 0 \times 0 + 0 \times 1 + 0 \times 0}{\sqrt{(1^2 + 1^2) + (0.2^2 + 1^2)}} \approx 0.11$

5   Numeric Column Annotation

When a column has been recognised as numeric (Sect. 3), we look for the numeric type of the ontology that corresponds to the column. For that purpose, we compute the score of each numeric type for the column. The score of a numeric type for the column is a combination of:

- the score of the numeric type for the column according to the column title. This score is the similarity measure (def. 1) between the column title and the numeric type name.

- the score of the numeric type for the column according to the units present in the column.

To compute the numeric type for the column according to the units in the column, we first compute the score of the numeric type for each unit that is present in the column: a numeric type has a score for each unit, depending on the number of numeric types that can be expressed in this unit.

Definition 2. *Let u be a unit and T_u the set of numeric types that can be expressed in this unit, the score of a type t for the unit u is $score(t, u) = 0$ if $t \notin T_u$, and $score(t, u) = \frac{1}{|T_u|}$ if $t \in T_u$.*

The score of a numeric type for the column is computed as the maximum of the scores of the type for each unit present in the column. If no unit from the ontology was identified in the column, then the column is considered as presenting the unit "no unit", which is treated as a normal unit in the ontology.

The final score of a numeric type t for the column col is a combination of the score of t for col according to the title of the column ($score_{title}(t, col)$), and the score of t for col according to the units in the column ($score_{unit}(t, col)$). However, types are filtered according to the numeric values presented in the column:

- if the column contains a numeric value outside the range of values for the type t, then $score_{final}(t, col) = 0$.
- if all values in the column are compatible with the range of type t, then $score_{final}(t, col) = 1 - (1 - score_{title}(t, col))(1 - score_{unit}(t, col))$. This method of combination of several scores is inspired by [12]: the score of the type t for the column col is much greater if there are several evidences (title of the column **and** units) that the type corresponds to the column.

Once the final score of each numeric type has been computed for the column, we choose the correct type for the column. Numeric types are ordered according to their final score for the column: let *best* be the type with the best score and *secondBest* be the type with the second best score. We compute the proportional advantage of *best* over *secondBest* on the column col:

$$advantage(best, col) = \frac{score_{final}(best, col) - score_{final}(secondBest, col)}{score_{final}(best, col)} \quad (2)$$

If $advantage(best, col)$ is greater than a threshold θ_{num} fixed by the user, then the column col is annotated with the type *best*. Else the type of col is considered as unknown.

We have experimented our method of numeric column annotation on the 261 columns that were correctly recognized as numeric (Sect. 3). The columns were manually annotated with the 18 numeric types of our domain ontology, and we compared the manual annotation with the types computed by our method, using the threshold $\theta_{num} = 0, 1$. On the 261 columns, 243 were correctly annotated, 9 were considered as unknown and 9 were annotated with a wrong type (i.e. 96% precision, 93% recall). By comparison, when considering the score from title as the final score with no use of the units defined in the ontology (as was done in [8]), precision was 96% but recall was only 83%. We consider the results of our annotation of numeric columns as good enough, the use of the units defined in the ontology allowing better recall with no more errors.

6   Symbolic Column Annotation

As it has just been presented for numeric columns, we have to choose the correct symbolic type for a column that has been recognised as symbolic during the first step of our annotation process (Sect. 3). We first compute the score of each symbolic type for the column according to the title of the column, as the similarity measure (def. 1) between the column name and the names of the symbolic types. Then we compute the score of each symbolic type for the column according to the contents of the column. For that purpose, we compute the similarity measure (def. 1) between each term in the cells of the column and each term in the hierarchies of the symbolic types. Let c be the term in a cell of a symbolic column col, let t be a symbolic type and $hier(t)$ the set of all terms in the hierarchy of the type t. The score of the type t for the cell c is:

$$score(t, c) = \sum_{x \in hier(t)} sim(c, x) \qquad (3)$$

We choose the type for the cell using the proportional advantage (Sect. 5). If the proportional advantage of the type having the best score for the cell is higher than a threshold $\theta_{symbCell}$, then the cell is considered as having this type, else the cell is considered as having an unknown type. When a type has been chosen for every cell in the column, the score of a type t for the column is the proportion of its cells that have been assigned the type t: let n be the number of cells in the column col and n_t the number of cells having the type t, then the score of t for the column col according to the column contents is:

$$score_{contents}(t, col) = \frac{n_t}{n} \qquad (4)$$

As for the numeric columns, the final score of a symbolic type t for the symbolic column col is a combination of the score according to the column title and the score according to the column contents:

$$score_{final}(t, col) = 1 - (1 - score_{title}(t, col))(1 - score_{contents}(t, col)) \qquad (5)$$

Once the final score of each symbolic type has been computed for the column, we choose the correct type for the column using the proportional advantage method (Sect. 5): if the proportional advantage of the best type for the column is greater than a threshold $\theta_{symbCol}$, then the column is annotated with this best type, else the column is considered as of unknown type.

We have experimented our symbolic column annotation using the 81 columns that were correctly recognized as symbolic (Sect. 3). In order to assess the quality of our results, we wanted to compare our method with a "standard" classifier. To our knowledge, there is no classifier that is dedicated to the classification of symbolic data using a domain ontology. We have thus decided to use the SMO classifier [13] implemented in Weka[3]: as it is an optimised version of the well-known SVM, it allows comparing our results with a "standard" method. It is thus a comparison between two alternatives: SMO uses no domain knowledge *but* uses learning, while our method is based on domain knowledge *but* has no learning phase. The 81 columns were manually annotated with the three symbolic types of the ontology ("Microorganism", "Food product", "Response") and a type "Other" for additional information too specific to be entered in the ontology. Our annotation method was run using the following thresholds: $\theta_{symbCell} = \theta_{symbCol} = 0.1$, columns of unknown type being classified under the type "Other". For the SMO classifier, we used the following pre-treatment: each distinct lemmatised word present in a column results in an attribute; the value of this attribute for a given column is the frequency of the word in the column. The SMO classifier was evaluated using a leave-one-out cross-validation, with default parameters of the Weka implementation. Results of this experiment are given in Tab. 3.

Table 3. Classification results on 81 symbolic columns

computed / manual	our method using the ontology				SMO			
	Food	Micro.	Resp.	Other	Food	Micro.	Resp.	Other
Food	19	0	0	27	45	0	0	1
Micro.	0	6	0	10	4	12	0	0
Response	0	0	0	1	0	0	0	1
Other	2	0	0	16	7	0	0	11

Over the three symbolic types of the ontology (we are not interested in the "Other" column type), we obtain a precision of 93% and a recall of 66% with our annotation method, while the SMO classifier gives a 84% precision and a 90% recall. Our method, which uses domain knowledge but no learning phase, gives a better precision but a lower recall than the learning classifier. This is mainly because we have biased our annotation technique towards precision: whenever we do not know for sure the type of a column, it is considered as unknown.

7   Finding the Semantic Relations Represented by the Table

Once the types of all columns of a table have been recognized, we look for the relation(s) of the ontology that are represented in the table. As for the column types recognition, the final score of a relation for the table is the combination of two scores: the score of the relation for the table according to the table title, and the score of the relation for the table according to the table signature (the set of its recognized columns).

---

[3] http://www.cs.waikato.ac.nz/ml/weka

The score of a relation for the table according to the table title is computed as the similarity measure (def. 1) between the table title and the relation name.

The score of a relation rel for the table tab according to the table signature is computed as follows:

- if the numeric type that constitutes the range of the relation rel was **not** recognized as a type of a column of the table, then $score_{signature}(rel, tab) = 0$
- else, the score of the relation for the table is the proportion of types in its signature that were recognized in the table columns. Let $Sign_{rel}$ be the set of types in the signature of relation rel (i.e. the types that constitute the domain and the type that cpnstitutes the range), and $Sign_{tab}$ the set of types that were recognized for the table columns, then $score_{signature}(rel, tab) = \frac{|Sign_{rel} \cap Sign_{tab}|}{|Sign_{rel}|}$

Then the final score of a relation rel for the table tab is computed as:

$$score_{final}(rel, tab) = 1 - (1 - score_{title}(rel, tab))(1 - score_{signature}(rel, tab)) \quad (6)$$

When the scores of all relations of the ontology have been computed for the table, we choose the relation(s) with which to annotate the table. A table can represent several relations at a time: for example, if a table gives the pH and the water activity of a food product, we will consider it as two separate relations: food pH and food water activity. If a relation has a non-zero score for the table and has no *concurrent relation*, this relation is used to annotate the table. Two relations are called *concurrent* if they have the same range. If there are several *concurrent relations* with non-zero scores for the table, then we only keep the one with the highest score for the annotation of the table (if several concurrent relations have the same highest score, we keep them all for the table annotation).

We have experimented this annotation method on the 60 tables that were used for all preceding experiments. Those tables were manually annotated with the 16 relations of the ontology: one table was typically annotated with 1 to 5 relations, which gives a total of 123 relations. We ran the different steps of our annotation system without validating the intermediate steps, i.e. even columns that were wrongly recognized in the symbolic versus numeric classification were further annotated and used for the relation annotation. Over the 123 relations in the manual annotation, 117 were correctly annotated with our annotation system, 6 were not recognized and there were 52 relations in the annotations that should not have been recognized. This gives a precision of 69% for a recall of 95%. Unfortunately, we cannot build a comparison of this method with another method to annotate tables, as we did not find other works on annotation of n-ary relations in tables using a domain ontology: the nearest work to ours is the one of [3] and it uses WordNet which is too general for our purpose.

In order to improve precision, we have tried to apply a score threshold on the relations: only relations with score greater than a threshold θ_{rel} are used for the annotation. The variation of precision and recall with the value of θ_{rel}, and the according F1-value is shown in Fig. 2. For $\theta_{rel} = 0.5$, there is a jump in precision to reach 95%, but also a fall in recall to 34%: increasing precision for our method means a too important drop in recall.

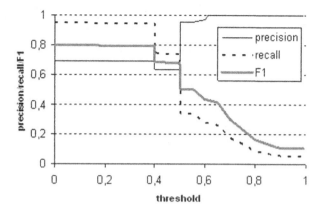

Fig. 2. Precision, recall and F1-value of relation annotation given the value of the threshold θ_{rel}

8   Conclusion and Perspectives

In this paper, we have shown the different steps of an annotation process that allows one to annotate data tables with the relations of a domain ontology. This annotation is entirely based on the domain knowledge given in the ontology, with no learning phase. The columns of a table are first segregated according to whether they represent numeric or symbolic data. Then, on top of the column titles, we use the units and interval of possible values defined in the ontology for numeric types to recognize the type of numeric columns, and we use the terms from the taxonomies of the symbolic types to recognize the type of symbolic columns. The relations represented by a table are recognized using both the table title and the types of the columns. When all steps are run one after the other, we obtain a high recall on relation recognition, with an acceptable precision level.

Our future works will aim at instanciating the relations according to the content of the cells in the table, not only annotating the symbolic contents with terms from the hierarchies of the corresponding types, but also analysing numeric values, dealing with intervals, standard deviations etc. Missing types that are in the relation signature but not in the table signature will be searched for: first we will try to identify them in the columns that have been annotated of unknown type, then we will also try and find if they are expressed as constants in the table title or in the sentences which reference the table. For example, a table named "Growth parameters for E. coli" is not likely to present a column of type "Microorganism" because the studied microorganism is a constant already given in the table title.

Fuzzy sets [14] will be used to represent similarities between symbolic cells and terms from the ontology, and to represent imprecise data for numeric cells. Then we will focus on the querying of the annotated data tables: the querying system must be integrated to the one already used in the MIEL system, which means that we will have to deal with user preferences, also expressed as fuzzy sets. Our querying system will have to take into account the different scores that we have computed during the annotation, that give hints about how sure we are of these annotations.

Acknowledgements

Financial support from the French National Research Agency (ANR) for the project WebContent in the framework of the National Network for Software Technology (RNTL) is gratefully acknowledged.

References

1. Buche, P., Dervin, C., Haemmerlé, O., Thomopoulos, R.: Fuzzy querying of incomplete, imprecise, and heterogeneously structured data in the relational model using ontologies and rules. IEEE T. Fuzzy Systems 13(3), 373–383 (2005)
2. Zanibbi, R., Blostein, D., Cordy, J.R.: A survey of table recognition: Models, observations, transformations, and inferences. International Journal on Document Analysis and Recognition 7, 1–16 (2004)
3. Pivk, A., Cimiano, P., Sure, Y.: From tables to frames. In: Third International Semantic Web Conference, pp. 116–181 (2004)
4. Tenier, S., Toussaint, Y., Napoli, A., Polanco, X.: Instantiation of relations for semantic annotation. In: International Conference on Web Intelligence, pp. 463–472 (2006)
5. Embley, D.W., Tao, C., Liddle, S.W.: Automatically extracting ontologically specified data from html tables of unknown structure. In: Spaccapietra, S., March, S.T., Kambayashi, Y. (eds.) ER 2002. LNCS, vol. 2503, pp. 322–337. Springer, Heidelberg (2002)
6. Freitag, D., Kushmerick, N.: Boosted wrapper induction. In: 17th National Conference on Artificial Intelligence, pp. 577–583 (2000)
7. Baumgartner, R., Flesca, S., Gottlob, G.: Visual web information extraction with Lixto. In: International Conference on Very Large Data Bases, pp. 119–128 (2001)
8. Gagliardi, H., Haemmerlé, O., Pernelle, N., Saïs, F.: An automatic ontology-based approach to enrich tables semantically. In: AAAI Context and Ontologies Workshop (2005)
9. Lin, D.: An information-theoretic definition of similarity. In: International Conference on Machine Learning, pp. 296–304 (1998)
10. Hignette, G., Buche, P., Dervin, C., Dibie-Barthélemy, J., Haemmerlé, O., Soler, L.: Fuzzy semantic approach for data integration applied to risk in food: an example about the cold chain. In: Proceedings of the 13th World Congress of Food Science and Technology, Food is Life (2006)
11. Van Rijsbergen, C.J.: Information Retrieval, 2nd edition. Dept. of Computer Science, University of Glasgow (1979)
12. Yangarber, R., Lin, W., Grishman, R.: Unsupervised learning of generalized names. In: International Conference on Computational Linguistics, pp. 1–7 (2002)
13. Platt, J.C.: In: Fast training of support vector machines using sequential minimal optimization, pp. 185–208. MIT Press, Cambridge (1999)
14. Zadeh, L.: Fuzzy sets. Information and control 8, 338–353 (1965)

Using Ontology with Semantic Web Services to Support Modeling in Systems Biology

Zhouyang Sun, Anthony Finkelstein, and Jonathan Ashmore

CoMPLEX, University College London, WC1E 6BT, London
{j.sun,a.finkelstein}@cs.ucl.ac.uk, j.ashmore@ucl.ac.uk

Abstract. Modeling in systems biology is concerned with using experimental information and mathematical methods to build quantitative models at different biological scales. This requires interoperation among various knowledge sources and services, such as biological databases, mathematical equations, data analysis tools, and so on. Semantic Web Services provide an infrastructure that allows a consistent representation of these knowledge sources as web-based information units, and enables discovery, composition, and execution of these units by associating machine-processable semantics description with them. In this paper, we show a method of using ontology alongside a semantic web services infrastructure to provide a knowledge standardisation framework in order to support modeling in systems biology. We demonstrate how ontologies are used to control the transformation of biological databases and data analysis methods into Web Services, and how ontology-based web services descriptions (OWL-S), are used to enable the composition between these services.

Keywords: Ontology, OWL, Semantic Web Services, OWL-S, Java EE.

1 Motivation

Systems biology is an emergent discipline that involves integrating biological knowledge across scales and domains, in order to understand the dynamics of diverse and interacting biological processes as integral systems [1]. In the study of systems biology, one of the essential tasks is to couple experimental biologists' observations with scientific models. This task encompasses several collaboration processes involving experimenters and modelers: using experimental observations as the ground for constructing models of biological entities and the relations among them; qualitatively and quantitatively analysing the resulting models and then comparing the analyses against experimental data for model validation; providing instructive feedback in order to refine both the models and experimental protocols. The progress of systems biology relies on the success of these experimenter-modeler collaboration processes.

Experimenter-modeler collaboration processes present a number of challenges. Firstly, both the content and the format of knowledge that need to be shared among participants are diverse. For instance, this knowledge can be experimental data with descriptions of laboratory settings, or mathematical models that quantitatively represent relations among biological entities. The languages used to represent models

M. Weske, M.-S. Hacid, C. Godart (Eds.): WISE 2007 Workshops, LNCS 4832, pp. 41–51, 2007.
© Springer-Verlag Berlin Heidelberg 2007

may not be directly compatible. The computing environments used to store experimental data are usually heterogeneous.

Secondly, knowledge creation in systems biology relies on combining knowledge from many various and distributed sources, in order to achieve a system-level understanding. For instance, when studying complex biological systems such as human liver, researchers may have to integrate knowledge from gene regulation level up to intercellular communication level for investigating certain physiological phenomena. Models on different levels may be developed independently by several groups using various modeling paradigms and computational environments.

Thirdly, models and information from experiments are reused in many different settings. When specifying parameters for an equation in a biological model, modelers need to interact with various biological data sources such as literature, biological databases, or data embedded in existing models. Then they need to give justification on the selection of data sources according to the context of the model, make judgment on which analysis methods and parameterisation approaches need to be applied. All the above information can be crucial for model reuse. Newcomers may build a model based on the same rationale of an existing model, but want to use alternative data sources and parameterisation methods. In this case, the reasoning involved in the model construction could rely on the previous cases.

2 Approach

In order to tackle these challenges, a common means of formally representing diverse knowledge in computer-based format is required. Also, in the context of combining computer-based resources, it is required for the individual pieces of knowledge to have a 'descriptive interface' to support computer-based communication, so that knowledge can be easily integrated. Moreover, the collaboration processes that mediate between distributed knowledge representation are part of the knowledge of modeling and should be formally represented for future model reuse.

We use ontology and semantic web services [15] as the primary means to meet these requirements. Ontology is the theory of conceptualisation. In computing, ontology provides a means of formally representing the structure of objects and relations in an information system and associating meaning with them [2]. Ontology can provide formal knowledge representation for distributed and heterogeneous computer-based biological information. Moreover, since ontologies explicitly define the content in information sources by formal semantics, they also enable the basis of interoperability between these sources. Further, as ontologies are able to separate domain knowledge from application-based knowledge, they can be used to define the collaboration processes among information-providing applications. Ontologies provide the benefits of reuse, sharing and portability of knowledge across platform [3].

Semantic Web Services are the conjunction of Semantic Web and service-oriented computing. The Semantic Web is a framework for creating a universal medium for information exchange by associating semantics with documents on the World Wide Web [4]. Service-oriented computing is a software architecture that allows information resources to be presented as platform-independent, self-describing, modular software units. The combination of both provides a web-based infrastructure for general knowledge sharing and reuse. Semantic Web Services allow the bringing together of knowledge sources as Web Services and describing interaction and workflows among them by means of a semantic markup language, and therefore is able to meet the need of modeling knowledge collaboration among systems biology practitioners. Constructing biological models in a portable exchange format is challenging because it requires assembling knowledge from heterogeneous sources. Semantic web techniques can help in building such model by supporting meaningful descriptions of the elements drawn from these sources and thus enhancing the interoperability and consistency among them.

In this paper, we will show how to help model construction by using ontology and ontology-based mapping of model components to Semantic Web Service architecture. We will describe a simple case of how the model is built in practice. Then we will describe an ontology for modeling in systems biology and how it is used to automate the transformation of biological databases, data analysis methods into web services, as well as the transformation of mathematical equation to OWL-S, the semantic markup for web services, for interactive parameterisation processes. We will describe our approach to ontology-based modeling in systems biology and the procedure for using an ontology model to control the transformation of experimental data and mathematical methods into Semantic Web Services, as well as the composition of these services for parameterising equations in the biological models.

3 A Simple Case of Modeling in Systems Biology

We describe a case of model construction for explaining the details of our approach. The model used is developed by Hudspeth, A.J. and Lewis, R.S. [5], and has been built to aid understanding electrical resonance in bull-frog hair cells. We chose this case because it encompasses all the major processes involved in the study of systems biology including data acquisition from experiments, abstract modeling, raw data analysis for parameterisation, and creation of simulations. Also, to construct this kind of models often requires integrating fairly complex sub-models of different types, which can demonstrate the complexity of the modeling tasks in systems biology.

To create the model, firstly an experiment is carried out using electrophysiological recording techniques. Data of cell responses to a series of electric stimuli are acquired and stored in a certain computer-based format. Meanwhile, abstract models are proposed which describe the system of interest as the entities and the relationships between them. In this case, the abstract models include a kinetic model for voltage-dependent calcium current, a diffusion model for the regulation of intracellular calcium ions, and a kinetic model for Ca2+-activated K+ current based on the kinetics of Ca2+-binding K+ channel. The relationships in these models are described by mathematical equations. For example, voltage-dependent calcium current is described by a third-order kinetic scheme [6],

$$I_{Ca} = \bar{g}_{Ca} m^3 (V_m - E_{Ca})$$

In which $\bar{g}_{Ca}$, m, and E_{Ca} are parameters; I_{Ca} and V_m are membrane current and potential variables.

In order to specify these models, experimental data are assessed for parameterisation. For example, to evaluate the conductance of voltage-dependent calcium channel, the data of tail currents are retrieved from datasheets and are fitted with a Boltzmann relation, $I = I_{max}/\left\{1 + e^{-(V_m - V_{1/2})/k}\right\}^3$, by a least-square-error criterion. Besides the parameters that are acquired by data analysis methods, some of the parameters are given from existing literature and some are given arbitrarily. No record will be kept on the data analysis and parameterisation processes. . Every time therefore, when the model needs to be modified or used for integration with other models, researchers have to go back to the data and analyse them again. This can be problematic, since models are shared among researchers but not everyone has access to the original data used for parameterisation, or the rationale for data source and mathematical methods selection. After the parameterisation, the models are fully instantiated. At this point, simulation can be created in certain computing languages such as Java etc.

4 A Simple Case of Modeling in Systems Biology

4.1 Implementation Details

For ontology model construction, we use Protégé 3.2.1 [7], an ontology editor by Stanford Medical Informatics. For generating web services, we use NetBeans 5.5 [8] by Sun Microsystems, a Java EE-enabled [9] integrated development environment (IDE) to generate Web Services. NetBeans 5.5 is an open-source IDE written entirely in Java featuring APIs including Java Persistence and JAX-WS [10], and bundles with Sun Java System Application Server Platform (SJSAP) 9.1 [11].

4.2 Create Abstract Biological Models by Using Ontology

In our approach, we construct an abstract model of the biological systems of interest as the starting point. We use OWL DL [12] as the format of our ontology models,

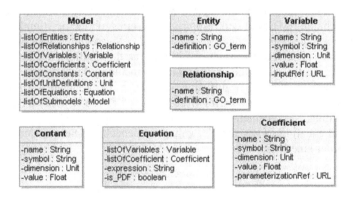

Fig. 1. UML Class diagram to represent the metadata of the OWL-based biological model

since OWL DL is a species of OWL [12] that provides the adequate expressiveness and has desirable computational properties for reasoning. By using ontology as the medium, the biological models we construct will be portable, integratable, and can be reasoned about using description logic.

We create an OWL abstract model as the meta-model for constructing biological models. In order to make it easy to understand, instead of using XML code we present it in a UML [13] diagram as bellow:

This model declares the information that needs to be assembled for constructing a valid biological model. The information includes the definitions of biological entities and biological processes, mathematical equations underlying the biological processes. This model is then used to embed or refer to external sources of parameters, data sources and analysis methods used for parameterisation.

The following is a sample fragment of an OWL model instantiated by the case of Lewis & Hudspeth hair cell model:

```
<Model rdf:ID=''regulation_of_intracellular_calcium_ion''>
   <listOfCoefficients
   rdf:resource=''#rate_constant_of_calcium_ion_migration''/>
   <listOfCoefficients
   rdf:resource=''#fraction_of_free_intracellular_calcium_ion''/>
   ...... <!--List of Coefficients -->
   <listOfVariables rdf:resource=''#concentration_of_calcium_ion''/>
   <listOfVariables rdf:resource=''#calcium_current''/>
   ...... <!--List of Variables -->
   <listofEntities>
      <GO_0005623 rdf:ID=''saccular_hair_cell_of_bull_frog''/>
   </listofEntities>
   ...... <!--List of Entities -->
   <listOfEquations
   rdf:resource=''#concentration_of_intracelluar_calcium_ion''/>
   ...... <!--List of Equations -->
   <Equation rdf:ID=''concentration_of_intracelluar_calcium_ion''>
      <Variable rdf:ID=''concentration_of_calcium_ion''>
         <symbol rdf:datatype=''http://www.w3.org/2001/XMLSchema#string''
         >c</symbol>
         <value rdf:datatype=''http://www.w3.org/2001/XMLSchema#float''
         >0.1</value>
         <unit rdf:resource=''#micromolar''/>
         <initialValue
 rdf:datatype=''http://www.w3.org/2001/XMLSchema#float''
         >0.0</initialValue>
      </Variable>
      ......
      <coefficient rdf:resource=''#fraction_of_calcium_accumulation''/>
      <coefficient rdf:resource=''#total_cell_volume''/>
      ......
      <is_PDE rdf:datatype=''http://www.w3.org/2001/XMLSchema#boolean''
      >true</is_PDE>
      <expression  rdf:datatype=''http://www.w3.org/2001/XMLSchema#string'
      >c = c + dtime * (- 1.0 *U*i_Ca/(z*F*C_vol*xi)-K_S*c)
      </expression>
   </Equation>
</Model>
```

4.3 From Experimental Data to Database Web Services

With the development of Java EE and its supporting APIs, we are able to transform experimental data stored in different formats into standard database web services. The general procedure of transformation is the following:

- Transform data source into relational database with a generic schema
- Generate Java Entity Classes from relational database
- Define generic operations to create Web services
- Deploy Web Services to generate WSDL
- Create semantic descriptions (OWL-S) for the generated web services

Experimental data exist in different formats such as data-containing documents or relational databases. Different kinds of data-containing documents can always be converted into text-based spreadsheets, which can be imported into relational database by using SQL statements [14]. These database web services are deployed in Java EE enabled application server and provide full flexibility of data retrieval, so that any analysis methods can be performed on them.

Since most of the procedure in this case has been simplified by Java EE platform, the remaining tasks are mainly about defining the meta-models of the database structure, the description of the web service competence, and the semantics for advertising the generated web services. These all can be modeled by using ontology. Moreover, as the operations of the database web services are dependent on the database structure, and the profiles of the web services are based on the description of the entities in the database, a single ontology model can be used to control the generation of the above information all at the same time.

In the case of Lewis & Hudspeth model, data are saved in a spreadsheet which contains both the specifications of the experiment and the recorded electrophysiological data. The specifications of the experiment may include the profile of the electric stimulus performed on the cells, the definitions of signals, the meaning of the columns in the tables of data, etc. The results of experiment are stored in a table whose columns are specified with names of entities and rows are sequences of recorded data.

In order to transform this kind of data sources into database web services, we defined an OWL model that describes the properties included in the settings, the entities that specify the columns of the data table, and the operations supported by the web services to be generated. By using the OWL model, database schema, semantic description of web services are automatically generated.

4.4 From Analysing Methods to Web Services

We use JAXWS API [10] in Java EE to transform mathematical methods into Web Services. It is a simple process. Any programming implementation of mathematical methods, such as calculation or best fitting methods etc., is transformed by using annotations as specified in A Metadata Facility for the Java Programming Language (JSR 175) and Web Services Metadata for the Java Platform (JSR 181), as well as additional annotations defined by the JAX-WS 2.0 specification. For example, the following is the implementation of the Boltzmann relation in our example, as a web service ($I = I_{max}/\left\{1 + e^{-(V_m - V_{1/2})/k}\right\}^3$).

```
Import javax.jws.WebService;
import javax.jws.WebMethod;
import javax.jws.WebParam;

@WebService
public class BoltzmannRel {
        public BoltzmannRel () {}

        @WebMethod(operationName= ''Boltzmann Relation '')
        public float BoltzmannRel (@WebParam(name = ''Membrane
Potential'') float var1, @WebParam(name = ''Peak Current'') float var2,
@WebParam(name = ''Steady State Potential'') float var3, @WebParam(name
= ''slope factor'') float var4) {
                float result = var2/math.pow((1+math.exp(-1*(var1-
var3)/var4)),3) ;
                return result;
        }
    }
```

Similar to the generation of database web services, we also define data analysis web services by using ontology models. The meta-model consists of the description of the inputs, outputs, preconditions, and post-conditions of the mathematical methods. The instantiations of this meta-model are then translated into web services implementation automatically by using an XML transformation stylesheet.

4.5   Use OWL-S to Specify Parameterisation in Computational Models

After experimental data and mathematical methods are transformed into web services and annotated with semantics, we orchestrate them together by using the web service composer developed by MINDSWAP. Web service composer is a software interface developed by Maryland Information and Network Dynamics Lab Semantic Web Agents Project (MINDSWAP) [16]. It can be used to guide users in the dynamic composition of web services by supporting OWL-S [17] standard. Using the composer one can generate a workflow of web services. The composition is done in a semi-automatic fashion where composer discovers web services by reasoning on their semantic descriptions, then presents the available web services at each step to a human controller to make the selection. The generated composition can be directly executable through the WSDL grounding of the services. Compositions generated by the user can also be saved as a new service which can be further used in other compositions.

In this work, we use OWL-S to specify the parameterisation process in model construction: we translate mathematical equations embedded in the OWL-based biological models into OWL-S files as template web service composition profiles.

For example, when we parameterise equation $\alpha_m = \alpha_0 e^{-(V_m + V_0)/k_B} + K_A$ in which α_0, K_A, K_B are parameters need to be specified, we create an OWL-S file that defines web service composition for the parameterisation of this equation. This OWL-S file can be handled by the composer and an interactive interface is then. To parameterize α_0, this interface searches all the available databases web services and analyser web services and enables orchestrating a selection of these services together. A new composition can be specified on what data to retrieve from database web services and then pass to analyser web services in order to obtain the value of α_0. After all the parameters are obtained, the values and the equation will be passed to an equation parser service to generate an instantiated equation (looks like $\alpha_m = 22800e^{-(V_m + V_0)/33} + 510$). When all the equations in a model are parameterised, they can then be passed to simulation generation service.

After the parameterisation process, the information of databases and analysis methods used is stored in OWL-S files and published on the web as new web services. We embedded the links to these OWL-S back to the OWL-based biological models, so that when users want to reuse these models, they are able to retrieve the parameterisation processes and modify them themselves.

4.6 Result

By these means we constructed a fully specified ontology model for the biological system of interest, i.e. the model of hair cell electrophysiology. All parameters that are specified interactively are annotated with external links to OWL-S files that represent the composition of database web services and data analysis web services used for parameterisation. This OWL-based biological model consists of all the essential information for generating a computational simulation.

The meta-model we proposed for modeling in systems biology contains the information specified by the XML schemas of SBML and CellML. Therefore, a subset of our model can be transformed into either SBML or CellML. This gives us the advantage of using any SBML or CellML-enabled software. For example, we have successfully transformed our OWL-based Lewis&Hudspeth model into CellML format and use it in Cell Electrophysiological Simulation Environment (CESE) [18]. A simulation is then generated automatically by transforming CellML model to JavaBeans programs.

5 Related Work

There are a number of XML-based specifications for modeling in systems biology. Representatives of such efforts are the Systems Biology Markup Language (SBML) and the Cell Markup Language (CellML). Both SBML and CellML are XML-based exchange formats that provide formal representation of main modeling components including biological entities, parameter definitions and the equations of the underlying biological processes such as reaction mechanisms, etc [19, 20]. These efforts have however, principally been focused on improving the exchanging models between simulation environments [21]. Both SBML and CellML do not contain information on parameterisation or associated rationale. Our model provides a novel way to combine semantic web services infrastructure, so that data sources and mathematical methods can be standardised in web-based units and then integrated together to achieve interoperation.

There are also attempts at creating biological Web services to enable e-Science in systems biology. An increasing number of tools and databases in molecular biology and bioinformatics are now available as Web Services. For example, *Nucleic Acids Research* describes 968 databases and 166 web servers available in molecular biology [3, 22]. Web service composition frameworks have also been developed for biological studies, including BioMoby [23] and myGrid [24], which support workflow design and execution, data management, and provenance collection among distributed biological web services. Almost all the existing efforts are however focused on publishing genomic data, such as DNA sequence, protein sequence, nucleotide sequence, and so on, and web services are mainly for sequence alignment or looking up definitions of biological terms. As far as we are aware, there is no web service available for physiological level simulation. Furthermore, our approach brings a collection of the latest semantic web techniques in a way that no others have proposed before.

6 Discussion and Future Work

In this work, we described an ontology-centered framework that uses a formal ontology language (OWL) to construct biological models in a portable exchange format, controls the transformation from biological data sources and data analysis methods into semantic web services, and defines the composition of these services in an OWL-based semantic description of web services (OWL-S). The main contribution of our work is the design of this novel framework. Currently no similar effort exists. This framework applies the latest advancement of information technology and is organised in a unique fashion specifically for the purpose of tackling the challenges of biological modeling. This framework allows users to build biological models from an abstract view, so that the biological entities and relationships underlying biological processes are based on the same shared vocabulary. The semantics in these models also enhance the interoperability of biological models, so that they can be easily integrated. Moreover, this framework directly connects with semantic web services infrastructure through ontology models, so that biological data sources and data analysis tools are published as web services. The transformation processes from the knowledge sources to web services are directly controlled by the ontology models. With the advantages of ontology, the automation of transformation can be achieved and consistency between different components can be maintained. Furthermore, this ontology-centered annotation/transformation framework allows users to control the parameterisation processes by semantically defining the workflow between these services. The parameterisation processes are saved in OWL-S models and can be retrieved and dynamically modified. This mechanism allows model specifications and the rationale of modeling processes to be associated with the representation of models, and therefore provides solid ground for further model reuse. The biological models written in OWL can also be translated into simulation models directly. This allows users to build models in abstraction and valid these models in real-time. This framework provides a basis for combining future development of semantic web techniques to support more sophisticated modeling tasks. For example, by constructing biological models in OWL format, reasoning engines that support description logic can be used to infer the relations between entities across different models. This can help the automatic discovery and matching of distributed models and control the consistency of model integration. Moreover, the OWL-S models that describe parameterisation processes can be combined with case-based reasoning engines so that modeling tasks can be solved automatically by reusing existing cases. Further, the semi-automatic transformation from abstract model to simulation provided by the framework can allow users to construct model heuristically and validate them against experiments in real time. All of these can be very beneficial to modeling in systems biology. We require further experience to be confident of the broader applicability of our work though clearly this has been an important design goal of the framework.

So far our approach has only been tested on a relatively simple example; therefore, the scalability of the framework has not been verified. Also, we have focused the modeling tasks on constructing electrophysiological models. It still needs to be determined whether this framework is sufficiently generic to handle other kinds of biological models. In future, we will focus on examining the framework with more complicated cases and different model types.

References

1. Finkelstein, A., et al.: Computational Challenges of Systems Biology. IEEE Computer 37(5), 26–33 (2004)
2. Noy, N.F., McGuinness, D.L.: Ontology Development 101: A Guide to Creating Your First Ontology - what is an ontology and why we need it, http://protege.stanford.edu/publications/
3. Baker, C.J.O., Cheung, K.-H.: Semantic Web: Revolutionizing Knowledge Discovery in the Life Sciences. Springer Science and Business Media, LLC 1st Edition. Springer, Heidelberg (2007)
4. Berners-Lee, Semantic Web on XML - Keynote presentation for XML 2000, http://www.w3.org/2000/Talks/1206-xml2k-tbl/slide1-0.html
5. Hudspeth, A.J, Lewis, R.S.: Kinetic Analysis of Voltage and Ion-dependent Conductances in Saccular Hair Cells of the Bull-frog, Rana Catesbeiana. J Physiol. 400, 237–274 (1998)
6. Hodgkin, A.L., Huxley, A.F.: A Quantitative Description of Membrane Current and Its Application to Conduction and Excitation in Nerve. J Physiol. 117, 500–544 (1952)
7. Protégé, http://protege.stanford.edu/
8. Netbeans, http://www.netbeans.org/index.html
9. Java EE, http://java.sun.com/javaee/
10. JAX-WS, https://jax-ws.dev.java.net/
11. Application Server, http://java.sun.com/javaee/community/glassfish/
12. OWL, http://www.w3.org/TR/owl-features/
13. UML, http://www.uml.org/
14. SQL, http://en.wikipedia.org/wiki/SQL
15. Semantic Web Services, http://en.wikipedia.org/wiki/Semantic_Web_Services
16. Web Service composer, http://www.mindswap.org/2005/composer/
17. OWL-S, http://www.daml.org/services/owl-s/1.1/
18. CESE, http://cese.sourceforge.net/
19. SBML, http://www.sbw-sbml.org/

20. CellMl, http://www.cellml.org/
21. Uhrmacher, A.M., et al.: Towards Reusing Model Component in Systems Biology. In: Danos, V., Schachter, V. (eds.) CMSB 2004. LNCS (LNBI), vol. 3082, pp. 192–206. Springer, Heidelberg (2005)
22. Nucleic Acids Research, http://nar.oxfordjournals.org/
23. BioMoby, http://biomoby.org/
24. myGrid and Taverna services, http://mygrid.org.uk/

Data in Astronomy: From the Pipeline to the Virtual Observatory

André Schaaff

Observatoire Astronomique, 11 rue de l'Université 67000 Strasbourg France
schaaff@astro.u-strasbg.fr

Abstract. Up to a recent time, numbers of projects did not make available to the whole community the data resulting from the spatial or terrestrial missions. Since the advent of new technologies, the interoperability between the data and service providers has become a priority because it allows an easy access and sharing of various data (catalogues, images, spectrum ...) and it leads to the concept of Virtual Observatory. The astronomer will in a near future have new "virtual" instruments useful for his research through simple Web interfaces or user friendly tools. Collaborations are essential in order to lead to the consensuses and the bases necessary for its construction. National (OV France, U.S. NVO...) and transnational (ESA, ESO) projects have joined together within the International Virtual Observatory Alliance to take part in the development of Recommendations in various fields (Data Model, Data Access Layer, Semantics, Grid, etc.) through biannual dedicated conferences.

Keywords: astronomy, massive data, Virtual Observatory, interoperability, value added services, Life Sciences.

1 Introduction

Since a long time the man was fascinated by the observation of the sky. This observation evolved from the simple poetic vision to the study of the stars and constellations trajectories until the modern astrophysical studies of the universe and the theories of its evolution. This spectacular development is due mainly to the evolution of the astronomical sensors directed towards the sky, namely the optical telescopes, the radio telescopes and the space telescope. These instruments make it possible to have a multi-wavelengths vision of space offering a spectral characterization of each observed zone.

In this paper we describe the different kinds of astronomical data, the challenges and their possible solution through the Virtual Observatory, which is involving a significant part of the astronomical community. The Virtual Observatory is an important challenge to make available the increasing amount of data to the astronomers and to produce scientific discoveries through the use of intelligent tools. A discussion about the possible application of/to Virtual Observatory concepts or developments to/from Life Sciences for the resolution of common challenges could be very interesting. Needs in Life Sciences seem to be very close from astronomy at several levels: public access to (massive) data, interoperability, Registry, Workflows,

M. Weske, M.-S. Hacid, C. Godart (Eds.): WISE 2007 Workshops, LNCS 4832, pp. 52–62, 2007.

efficient Web services, semantics, Grids, etc. As Life Sciences are a very large domain of investigation we will focus on precise sub domains.

2 Data and Services in Astronomy

In astronomy, the large observatories (ground and space) which explore in a systematic way the sky or a significant fraction of it, produce more and more data: the largest catalogue available online (USNO B2) contains more than one billion objects, another (SDSS) contains 200 million objects corresponding to a survey of 8,000 square degrees of the sky in five band passes. The online files of the observatories, distributed on the whole planet, contain tens, even hundreds, terabytes of data (images, spectra, time series, and articles in the electronic newspapers ...). An observation instrument is able to generate several terabytes, thousand gigabytes, per annum. In a near future it will be petabytes. The concept of "data avalanche" was introduced, in astronomy, in order to reflect the growing amount of data placed at the disposal of the astronomical community. As an illustration, the human analysis of the multi spectral images is rather tiresome and requires the individual examination of the images in the different wavelengths. It is of this point desirable to have automatic data processing sequences providing an effective summary of the images studied and making it possible to the astronomers to concentrate on areas of interest which appear more clearly.

In the astronomical community, the collaboration on data is a longtime work with results like FITS (Flexible Image Transport System) which was originally developed in the late 1970's as an archive and interchange format for astronomical data files and which is forty years later the most used format.

On the service side, Simbad [30] (Set of Identifiers, Measurements and Bibliography for Astronomical Data) is a very significant value added service. Born in 1972, this online service (~ 40,000 queries / day) offers an access to 3,800,000 astronomical objects (+ 150,000 / year), 210,000 bibliographic references (+ 10,000 / year) and 5,450,000 citations of objects in papers (+ 300,000 / year). The needs for drawing all the scientific potential from the great projects and the very great masses of data which they produce led these last years to the very fast development of the concept of "astronomical Virtual Observatory". We detail this concept in the paragraph 3.

3 Toward the Virtual Observatory (VO)

3.1 International Virtual Observatory Alliance

The IVOA [22] can be defined as an entity intended to make possible and to coordinate the development of the tools, protocols and collaborations necessary to carry out all the scientific potential of the astronomical data in the coming years. Two international conferences founders were held in 2000, Virtual Observatories of the Future [33] at California Institute of Technology in Pasadena, and Mining the Sky [6] at European Southern Observatory [20] in Garching, which allowed to clarify the scientific needs and to define the necessary bases.

Many projects were proposed and accepted in 2001. The principal ones were Astrophysical Virtual Observatory, AstroGrid (in the frame of e-Science project), National Virtual Observatory [23] in the United States (NSF funded). Other national projects started in Canada, France [24], Germany, India, Japan... and the construction of the Virtual Observatory is become a major topic for astronomy. Seven years later standards and tools are emerging from the huge work provided by the involved people all over the world. Data centres [5] are encouraged to participate through the organisation of dedicated workshops ("How to publish Data in the VO" workshop in Europe in June 2007 at ESA [21], NVO Summer School in United States every year).

The Virtual Observatory [9] project aims at allowing an optimal scientific use of the information, but many technological challenges have been/must be taken up. All the developments are Science driven but depend on technical evolution [15].

3.2 Architecture of the VO

The objective of the Virtual Observatory is to improve and unify access to astronomical data and services for professional astronomers, but also for other communities (amateurs, education, and outreach). VO tools like Aladin and Topcat (cf. Fig.1 on the top) are just some examples and the IVOA does not specify or recommend any specific one. In the IVOA architecture, the available services are divided into three main groups (cf. Fig.1 in the middle): Data Services (services providing access to data) [7], Compute Services (computation and federation of data) and Registry Services.

The Registry Services role is to facilitate publication and discovery of services. If a data centre puts a new dataset online or creates a new service it should publish that fact in a VO-compliant registry. This can be done through the filling of forms expressing who the curator is, where it is located and how to use the service. Nobody decides what is good and what is bad.

Data services [7] range from simple to sophisticated and return tabular data [10], image [14], or other data. At the simplest level, the request is a cone (conesearch) on the sky (direction/angular radius), and the response is a list of "objects". Similar services can return images and spectra [4] associated with sky regions.

Grid [11] [16] middleware (cf. Fig.1 on the bottom right) is used for high-performance computing, data transfer, authentication, and service environments. One of the most important parts of the IVOA architecture is "My Space" (cf. Fig.1 on the bottom center) which enables users to store data within the VO. "My Space" stores files and database tables between operations on services and it avoids the need to copy results to the user desktop. Using "My Space" establishes rights on access and over intermediate results and it allows the users to manage their storage remotely.

The Open SkyQuery [27] protocol allows querying of a relational database or a federation of databases. In this case, the request is written in a specific XML (ADQL, Astronomical Data Query Language) abstraction of SQL. The IVOA architecture will also support queries at a more semantic level, including queries to the registries and through data services. Development of standards for more sophisticated services is also expected (for federating and mining catalogues, image processing and source detection, spectral analysis, and visualization of complex datasets). These services are being implemented in terms of industry-standard mechanisms (like W3C Recommendation process), working in collaboration with the Grid [11] [16] community and at various levels of sophistication, from a stateless, text-based request-response, to a structured response from a structured request using authenticated, self-describing service. In the VO, services can be used not just individually, but also concatenated in a distributed workflow [12] [13][14]. Members of the IVOA are collaborating with IT people who are developing workflow software.

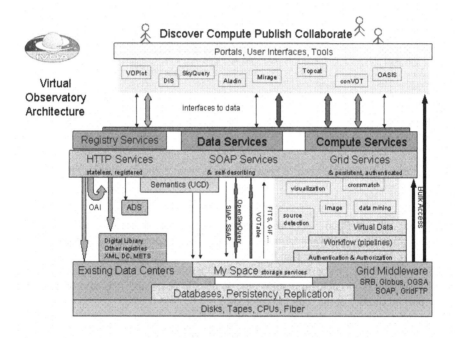

Fig. 1. VO Architecture

3.3   Exchange Protocols: The Case of VOTable [10][35]

In astronomy, many services are providing/exploiting tabular data (VizieR [34], the major catalogue service contains more than 6,000 tables and the largest table contains more than 10^9 rows). The first consensus was the definition of VOTable [10][35], an XML format which is since five years the mandatory format used in all the services and tools providing/exploiting tabular data. The format is simple, easy to implement and to use with parsers.

The following extract shows the definition of fields and the associate data part:

```
...

<FIELD name="Kmag" ucd="phot.mag;em.IR.K" datatype="float"
width="6" precision="3" unit="mag">

<DESCRIPTION>?Infrared K magnitude (from 2MASS)</DESCRIPTION>

<VALUES null=""/>

</FIELD>

<FIELD name="R" ucd="meta.note" datatype="char">

<DESCRIPTION>[R*] Recommended (R) or Problematic (*)
(4)</DESCRIPTION>

</FIELD>

<DATA>

<TABLEDATA>

<TR><TD>1.013</TD><TD>00 42 44.33</TD><TD>+41 16
08.4</TD><TD>13120013859</TD><TD>.M</TD><TD>010.6847117</TD><TD>+41.2
690039</TD><TD>C</TD><TD>6.9</TD><TD>4.6</TD><TD>-
0.4</TD><TD>4.1</TD><TD> </TD><TD> </TD><TD></TD><TD>
</TD><TD>6.970</TD><TD>C</TD><TD>9.453</TD><TD>8.668</TD><TD>8.475</T
D><TD> </TD></TR>

<TR><TD>2.959</TD><TD>00 42 44.56</TD><TD>+41 16
10.4</TD><TD>13120013873</TD><TD>.M</TD><TD>010.6856569</TD><TD>+41.2
695500</TD><TD>M</TD><TD>0.0</TD><TD>0.0</TD><TD>0.0</TD><TD>0.0</TD>
<TD></TD><TD> </TD><TD></TD><TD> </TD><TD></TD><TD>
</TD><TD>10.772</TD><TD>8.532</TD><TD>8.253</TD><TD> </TD></TR>

...
```

Remark: the attribute "ucd" has a semantic meaning, it expresses to which concept a column is linked (e.g. "phot.mag;em.IR.K" means that it corresponds to a photometric magnitude in infrared between 2 and 3 microns).

The success of this protocol is also due to the fact that just minor changes are done. The new version is at each time stable for at least 2 years and it is considered as a very important quality at the data provider side.

Other protocols dedicated to other kind of data are under development or amelioration: Simple Spectrum Access, Simple Image Access, etc.

3.4 The Registry

A *registry* function is a sort of yellow pages, or high-level directory, of astronomical catalogues, data archives and computational services. It is possible to search in the registry to find data of interest, review the data source descriptions, and it is then often possible to make direct position-based data requests. A *resource* is a general term referring to a Virtual Observatory element which describes who curates or maintains it and which can have a name and a unique identifier. Anything could be a resource like sky coverage or instrumental setup, or more concrete, a data collection. This definition is coherent with its use in the Web community as "anything that has an identity" (Berners-Lee 1998, IETF RFC2396). We extend this definition by saying

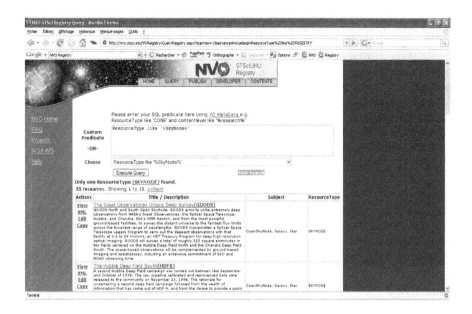

Fig. 2. NVO Registry [8]

that it is also describable. Several types of resources are published in the VO registries: standard VO data services (SIAP, Cone, SSAP, data collections, organizations ...). Other registries are also considered as resources.

Each registry has three kinds of interface to publish, query, and harvest. People can publish to a registry by filling in Web forms in a Web portal. In the future, the registry may also accept queries in one or more languages. As IVOA has not a centralized registry, all the registries (cf. Fig2 NVO example) harvest each other to know the new dataset and services added to other VO-registries. When another persons search a registry (by keyword, author, sky region, wavelength, etc.) they will also discover the published services in other registries.

A VO resource gets a universal identifier, a string starting with ivo://. Resources can contain links to related resources, as well as external links to the literature. The IVOA registry is compliant with digital library standards (Open Archive Initiative) for metadata harvesting and metadata schema. The aim is that IVOA-compliant resources can appear as part of every University library.

3.5 Semantics and Ontologies

The Semantic Web and ontologies are technologies that enable an advanced management and sharing of knowledge between scientists and also between Web services or more generally between software components. In the field of astronomy the first examples of ontologies [3] are being developed in the European VOTech [36] project. One of these ontologies describes knowledge about astronomical object types originally based on the standardization of object types used in the Simbad [30] database (with applications like the support of advanced resource queries in the VO Registries).

The UCDs [2] (Unified Content Descriptors) are a standardized vocabulary used to describe astronomical quantities and related concepts. But there is no formal representation structure, with syntax and semantics, describing the relationships and dependencies between the words, and it is not possible to perform automated reasoning on UCDs. This work was done at first in the frame of the VizieR [34] catalogue service and about 1500 UCDs were defined and linked to the 100,000 columns of the service corresponding to an average of 66 different column designations for the same concept.

The following UCDs are extracted from the set of 1500 UCDs:

...

src.ellipticity	|Source ellipticity
src.impactParam	|Impact parameter
src.morph	|Morphology structure
src.morph.param	|Morphological parameter
src.morph.scLength	|Scale length for a galactic component (disc or bulge)
src.morph.type	|Hubble morphological type (galaxies)
src.orbital	|Orbital parameters
src.orbital.eccentricity	|Orbit eccentricity
src.orbital.inclination	|Orbit inclination
src.orbital.meanAnomaly	|Orbit mean anomaly
src.orbital.meanMotion	|Mean motion
src.orbital.node	|Ascending node

...

4 Discussion About Possible Common Investigation Fields with the Life Sciences

4.1 Foreword

The consequences of Internet advent on scientific research are enormous, as well in Life Sciences as in astronomy. Life Sciences are a very large domain of investigation and it seems useful to try to choose a first set of sub domains. Molecular biology and genetics are very active domains for the online access to the data but this access is mostly dedicated to specialists of the domain and constitutes often a niche. Another domain like biodiversity is very interesting because like in astronomy the access to the data concerns not only professionals of the domain. The data can be exploited through very simple tools in Education for young people as well as for students. The possible outreach is very important like in astronomy. In the two following paragraphs we will try to give more details about these sub domains but we will not be too exhaustive.

4.2 Molecular Biology and Genetics

The Human Genome Project was an example for many biological databases available via the Internet which became essential resources to conduct researches. Multiple and heterogeneous data sources like sequence databases (UniProt Knowledgebase), gene expression data repositories (CIBEX, ArrayExpress), protein resources (PIR, PDB), literature databases (PubMed) are now available.

A difference compared to astronomy is that the information is not always accessible for free and this constitutes a fact to take into account. Different initiatives are trying to integrate multiple data sources (SRS [29], Entrez Gene, Ensembl) or to make complex bioinformatics queries (Biozon [19] is a unified biological resource on DNA sequences, proteins, complexes and cellular pathways) and analyses (MOBY [18], a system for interoperability between biological data hosts and analytical services) easier. Semantic Web technologies, like ontologies (cf. Ontology Lookup Service [26] at EBI) [1], will enable fast, context-sensitive retrieval of biological data. Web services will allow an automatization of bioinformatics processes through the build of complex workflows. Grid computing (Biogrid [17] in Japan, Life Sciences Grid Research Group [25] at the Open Grid Forum) will provide the power needed to solve the mystery of life as well as the mystery of the universe. EBI has participated to the development of Taverna [31] which is a tool for the execution of workflows. Taverna libraries have been used at ESO to develop Reflex [28], the ESO Recipe Flexible Execution Workbench. This work should be integrated in the frame of the Virtual Observatory. It is an example of work done in the frame of bioinformatics and applied to astronomy.

4.3   The GBIF [32] Initiative in Biodiversity

The Global Biodiversity Information Facility project started in 2001 and is a program which tries to gather all the data about biodiversity (species ...) and to place them at the disposal of the researchers and general public. Thus the biodiversity will be better known, better studied and also better used. The collections will also be more visible and the development will be better. Lastly, it will be easier to the researchers studying the biodiversity to prepare their research, to compare them and to put them in relation to former work. It should be also very useful for researchers in poor countries, giving them a possible access to a large set of data and tools.

GBIF's activities are divided in six thematic areas: Data Access and Database Interoperability, Digitisation of Natural History Collections Data, Electronic Catalogue of the Names of Known Organisms, Outreach and Capacity Building, Species bank, Digital Biodiversity Literature Resources.

The project is based on a main portal and on nodes in participating countries (similarity with the Virtual Observatory based on nodes like VO France, Japanese VO ...). Data providers can join the project; they are responsible of the quality of their data. A registration is needed and the Registry is based on UDDI. IVOA people have decided to develop their own Registries with a mechanism of harvesting because it is not easy to use just one Registry and to design one responsible of it!

It would be interesting to study the application of the Open SkyQuery [27] protocol to the GBIF's databases. The GBIF data providers have heterogeneous data and the UCDs[2] concept could be adapted to resolve interoperability problems concerning the tabular data.

5 Conclusion

After 7 years of work, the Virtual Observatory produced standards accepted by the community in many fields (access protocols to the images and the spectra, interchange format for the tabular data, various work within the framework of the use of grids, etc). The adhesion of the data centres and the services providers proves their growing interest and the need for them to become VO compliant. But that also raises problems of another kind: how to preserve the visibility of the data sources and the services in tools which precisely are designed to bring to the user an increasing transparency. In other words, how to make visible a contribution (e.g. Simbad [30]) to the total effort?

This type of problem will undoubtedly be much more important in the field of the Life Sciences because of non public data and literature in sub domains like molecular biology (existence of patents, etc.).

Technology is already present and its evolution should follow the constant rate of the growth of the data volumes produced by the various research programs. But like any instrument, the Virtual Observatory will require important material and manpower and consequently budgetary efforts. The multiplication of the scientific discoveries related to its use should justify and perpetuate its existence and the longtime efforts.

Concept and developments of Virtual Observatory developments could be transposed to / improved by other disciplines like the Life Sciences, at least in sub domains like biodiversity.

References

1. Anjum, A., et al.: The Requirements for Ontologies in Medical Data Integration: A Case StudyComputer Science - Databases. In: 11th International Database Engineering & Applications Symposium (Ideas 2007), Banff, Canada (September 2007)
2. Derriere, S., Preite-Martinez, A., Richard, A., Napoli, A., Nauer, E.: UCDs and Ontologies. In: Astronomical Data Analysis Software and Systems XV ASP Conference Series, vol. 351, Proceedings of the Conference Held 2-5 October 2005 in San Lorenzo de El Escorial, Spain, ASPC..351..449D (2006)
3. Derriere, S., Richard, A., Preite-Martinez, A.: An Ontology of Astronomical Object Types for the Virtual Observatory. In: The Virtual Observatory in Action: New Science, New Technology, and Next Generation Facilities, 26th meeting of the IAU, Special Session 3, 17-18, 21-22 August, 2006 in Prague, Czech Republic, IAUSS...3E..35D (2006)
4. Dobos, L., Budavari, T., Csabai, I., Szalay, A.: The Virtual Observatory Spectrum Services. In: The Virtual Observatory in Action: New Science, New Technology, and Next Generation Facilities, 26th meeting of the IAU, Special Session 3, 17-18, 21-22 August, in Prague, Czech Republic, IAUSS.3E.76D (2006)
5. Gaudet, S., Dowler, P., Goliath, S., Hill, N.: "Retooling" Data Centre Infrastructure To Support The Virtual Observatory. In: The Virtual Observatory in Action: New Science, New Technology, and Next Generation Facilities, 26th meeting of the IAU, Special Session 3, 17-18, 21-22 August, in Prague, Czech Republic, IAUSS.3E.22G (2006)
6. Genova, F., et al.: Mining the sky. In: Proceedings of the MPA/ESO/MPE Workshop Held at Garching, Germany, July 31 - August 4, 200 Mining the sky: Proceedings of the MPA/ESO/MPE Workshop Held at Garching, Germany, July 31 - August 4, ESO ASTROPHYSICS, misk.conf.674G (2000) (ISBN 3-540-42468-7)

7. Genova, F.: Data Centre. In: The Virtual Observatory. In: Astronomical Data Management. 26th meeting of the IAU, Special Session 6, 22 August, in Prague, Czech Republic, IAUSS.6E.4G (2006)

8. Graham, M.J., Williams, R.D., CARNIVORE.: an open source VO registry, in Astronomical Data Analysis Software and Systems XV ASP Conference Series.In: Proceedings of the Conference Held October 2-5, in San Lorenzo de El Escorial, vol. 351,Spain. p.421, ASPC.351.421G (2005)

9. Lawrence, A.: The Virtual Observatory: what it is and where it came from. In: The Virtual Observatory in Action: New Science, New Technology, and Next Generation Facilities, 26th meeting of the IAU, Special Session 3, 17-18, August 21-22,2006, in Prague, Czech Republic, IAUSS.3E.1L (2006)

10. Ochsenbein, F., et al.: VOTable: Tabular Data for the Virtual Observatory, in Toward an International Virtual Observatory. In: Proceedings of the ESO/ESA/NASA/NSF Conference Held at Garching, Germany, 10-14 June, ESO ASTROPHYSICS SYMPOSIA. ISBN 3-540-21001-6. p. 118, 2004tivo.conf.118O (2002)

11. Pasian, F., Taffoni, G., Vuerli, C.: Interconnecting the Virtual Observatory with Computational Grid infrastructures. In: The Virtual Observatory in Action: New Science, New Technology, and Next Generation Facilities, 26th meeting of the IAU, Special Session 3, 17-18, 21-22 August 2006, in Prague, Czech Republic, IAUSS.3E.33P (2006)

12. Schaaff, A., et al.: Work Around Distributed Image Processing and Workflow Management. In: Astronomical Data Analysis Software and Systems XV ASP Conference Series. In: Proceedings of the Conference Held 2-5 October, in San Lorenzo de El Escorial, vol. 351, Spain, p.323, ASPC.351.323S (2006)

13. Schaaff, A., Le Petit, F., Prugniel, P., Slezak, E., Surace, C.: Workflow Working Group in the frame of ASOV. In: SF2A-2006: Proceedings of the Annual meeting of the French Society of Astronomy and Astrophysics p.97, sf2a.conf.97S (2006)

14. Slezak, E., Schaaff, A.: Image Processing and Scientific Workflows in the Virtual Observatory Context. In: The Virtual Observatory in Action: New Science, New Technology, and Next Generation Facilities, 26th meeting of the IAU, Special Session 3, 17-18, 21-22 August 2006, in Prague, Czech Republic, IAUSS.3E.48S (2006)

15. Szalay, A.: The Science and Technology of the National Virtual Observatory, in 2007 AAS/AAPT Joint Meeting, American Astronomical Society Meeting 209, 187.01; Bulletin

16. of the American Astronomical Society 38, 1163, AAS...20918701S

17. Taffoni, G., et al.: Grid Data Source Engine: Grid gateway to the Virtual Observatory, in Astronomical Data Analysis Software and Systems XV ASP Conference Series. In: Proceedings of the Conference Held 2-5 October 2006, in San Lorenzo de El Escorial, Spain. Edited by Carlos Gabriel, Christophe Arviset, Daniel Ponz, and Enrique Solano. San Francisco: Astronomical Society of the Pacific, p.508, vol. 351, ASPC.351.508T (2005)

18. bioGrid, http://www.biogrid.jp/

19. BioMoby, http://biomoby.open-bio.org/index.php/what-is-moby/

20. Biozon unified biological resources, http://biozon.org/

21. European Southern Observatory, http://www.eso.org/public/

22. European Space Agency, http://esavo.esa.int/

23. International Virtual Observatory Alliance, http://ivoa.net/

24. National Virtual Observatory, http://www.us-vo.org/

25. Observatoires Virtuels France, http://www.france-ov.org/

26. Ontology Lookup Service, http://www.ebi.ac.uk/ontology-lookup/ontologyList.do

27. Open SkyQuery, http://openskyquery.net/Sky/SkySite/

28. Reflex, http://www.eso.org/sampo/reflex/
29. Sequence Retrieval System, http://srs.ebi.ac.uk/
30. Simbad, http://simbad.u-strasbg.fr/simbad/
31. Taverna, http://taverna.sourceforge.net/
32. The Global Biodiversity Information Facility project, http://www.gbif.org
33. Virtual Observatories of the Future, http://www.astro.caltech.edu/nvoconf/
34. VizieR Catalogue Service, http://vizier.u-strasbg.fr/viz-bin/VizieR
35. VOTable references, http://vizier.u-strasbg.fr/doc/VOTable/
36. VOTech project, http://eurovotech.org/

International Workshop on Collaborative Knowledge Management for Web Information Systems (WEKnow)

Workshop PC Chairs' Message

Stefan Siersdorfer[1] and Sergej Sizov[2]

[1] University of Sheffield, UK
[2] University of Koblenz-Landau, Germany

Introduction

Collaborative data/knowledge management methods aim to achieve improved result quality through combination or merging of results and models obtained from multiple users and sites. Typical application scenarios in the domain of Web information systems include collaborative methods and meta methods for information acquisition (e.g. collaborative Web crawling and search, tagging), organization of document collections (e.g. collaborative classification and clustering), and knowledge sharing (e.g. alignment and sharing of personal ontologies). These areas have received increasing attention in the Web Information Systems community; however, connections and relationships between them have largely been neglected.

The workshop on Collaborative Knowledge Management for Web Information Systems (WE.Know) aims at closing this gap by bringing together researchers and practitioners dealing with collaborative methods in distinct contexts, and discovering synergies between their research fields.

Overview

The workshop covers major areas of research associated with collaborative data and knowledge management including ontology alignment, XML schema matching, collaborative search, and sharing and management of personal multimedia collections. Below we provide very brief descriptions of the papers included, to give a sense of the range of themes and topics covered.

Mapping Metadata for SWHi: Aligning Schemas with Library Metadata for a Historical Ontology by Zhang, Fahmi, Ellermann, and Bouma describes research carried out in the context of the *Semantic Web for History* (SWHi) project. The authors create a Semantic Web compatible ontology for a large historical document collection by combining and aligning schemas from different sources such as existing topic hierarchies on the World Wide Web and various off-shelf ontologies.

Sequence Disunification and its Application in Collaborative Schema Construction by Coelho, Florido, and Kutsia addresses the problem of collaborative XML schema management: Multiple users aim at producing a common schema for XML data with different users imposing constraints on the schema structure. The authors represent the schemas in a properly chosen term algebra and solve the occurring disequation problems using sequence unification methods.

M. Weske, M.-S. Hacid, C. Godart (Eds.): WISE 2007 Workshops, LNCS 4832, pp. 65–66, 2007.
© Springer-Verlag Berlin Heidelberg 2007

How Do Users Express Goals on the Web? - An Exploration of Intentional Structures in Web Search by Strohmaier, Lux, Granitzer, Scheir, Liaskos, and Yu describes novel concepts towards exploring the role and structure of users' goals in web search engines. To this end, user goals such as "plan a trip" are represented by semi-formal graphs, which might be, e.g., constructed by analyzing the behavior of multiple users, and search is carried out using graph traversal.

Publishing and Sharing Ontology-based Information in a Collaborative Document Management System by Mitschick and Fritzsche presents a system architecture for sharing and collaborative management of annotated multimedia data. The contribution is primarily focused on access control and concurrency control methods. The publishing of resources is realized as a transfer between virtual "containers" that are associated with fine-grained access control policies, including access to the policies themselves.

Acknowledgements

We thank the organizers of the conference WISE 2007 for their support of the workshop. We also thank the members of the Program Committee (Paulo Barthelmess, AnHai Doan, Maria Halkidi, Joemon Jose, Andreas Nuernberger, Daniela Petrelli, Michalis Vazirgiannis, Jun Wang, Yi Zhang) for their efforts on behalf of the workshop. This workshop has been supported by the European project "Semiotic Dynamics in Online Social Communities" (Tagora, FP6-2005-34721) and the EU-funded project Memoir.

How Do Users Express Goals on the Web? -
An Exploration of Intentional Structures in Web Search

M. Strohmaier[1,3], M. Lux[2], M. Granitzer[3], P. Scheir[1,3], S. Liaskos[4], and E. Yu[5]

[1] Graz University of Technology, 8010 Graz, Austria
[2] Klagenfurt University, 9020 Klagenfurt, Austria
[3] Know-Center Graz, 8010 Graz, Austria
[4] York University, Toronto, Canada
[5] University of Toronto, Toronto, Canada
markus.strohmaier@tugraz.at, mlux@itec.uni-klu.ac.at,
mgrani@know-center.at, peter.scheir@tugraz.at, liaskos@yorku.ca,
yu@fis.utoronto.ca

Abstract. Many activities on the web are driven by high-level goals of users, such as "plan a trip" or "buy some product". In this paper, we are interested in exploring the role and structure of users' goals in web search. We want to gain insights into how users express goals, and how their goals can be represented in a semi-formal way. This paper presents results from an exploratory study that focused on analyzing selected search sessions from a search engine log. In a detailed example, we demonstrate how goal-oriented search can be represented and understood as a traversal of goal graphs. Finally, we provide some ideas on how to construct large-scale goal graphs in a semi-algorithmic, collaborative way. We conclude with a description of a series of challenges that we consider to be important for future research.

Keywords: information search, search process, goals, intentional structures.

1 Motivation

In a highly influential article regarding the future of the web [1], Tim Berners-Lee sketches a scenario that describes a set of agents collaborating on the web to address different needs of users – such as "get medication", "find medical providers" or "coordinate appointments".

In fact, many activities on the web are already implicitly driven by goals today. Users utilize the web for buying products, planning trips, conducting business, doing research or seeking health advice. Many of these activities involve rather high-level goals of users, which are typically knowledge intensive and often benefit from social relations and collaboration. Yet, the web in its current form is largely non-intentional. That means the web lacks explicit intentional structures and representations, which would allow systems to, for example, associate users' goals with resources available on the web. As a consequence, every time users turn to the web for a specific purpose they are required to cognitively translate their high-level goals into the non-intentional structure of the web. They need to break down their goals into specific search queries,

M. Weske, M.-S. Hacid, C. Godart (Eds.): WISE 2007 Workshops, LNCS 4832, pp. 67–78, 2007.
© Springer-Verlag Berlin Heidelberg 2007

tag concepts, classification terms or ontological vocabulary. This prevents users from, for example, effectively assessing the relevance and context of resources with respect to their goals, benefiting from the experiences of others who pursued similar goals and also prevents them from assessing conflicts or systematically exploring alternative means.

In a recent interview, Peter Norvig, Director of Google Research, acknowledged that understanding users' needs to a greater extent represents an *"outstanding"* research problem. He explains that Google is currently looking at *"finding ways to get the user more involved, to have them tell us more of what they want."* [2]. Having explicit intentional representations and structures available on the web would allow users to express and share their goals and would enable technologies and other users to explore, comprehend, reason about and act upon them.

It is only recently that researchers have developed a broad interest in the goals and motivations of web users. For example, several researchers studied intentionality and motivations in web search logs during the last years [3,4,5]. Because web search today represents a primary instrument through which users exercise their intent, search engines have a tremendous corpus of intentional artifacts at their disposal. We define intentional artifacts broadly to be electronic artifacts produced by users or user behaviour that contain *recognizable "traces of intent"*, i.e. implicit traces of users' goals and intentions.

This paper represents our initial attempt towards exploring the role and structure of users' goals in web search queries. We want to learn in detail *how* users express their goals on the web - as opposed to *what* goals they have, which is in the focus of other studies [3,4,5]. We also want to explore how search goals can be represented in an explicit, semi-formal way and we are interested in learning about the different ways in which explicit goal representations could be useful, and to what extent. From our preliminary findings of an exploratory study, we want to give a qualitative account of identified potentials and obstacles in the context of goal-oriented search.

2 State of the Art

We will discuss two main streams of research that are relevant in the context of this paper: The first stream of research focuses on identifying and understanding *what* goals users pursue in web search. The second stream focuses on developing goal-oriented technical solutions, i.e. solutions that depend on the explicit articulation of user goals or automatic inference thereof.

In the first stream, researchers have proposed categories and taxonomies of user goals [4,5] and automatic classification techniques to classify search queries into goal categories [3]. Goal taxonomies include, for example, navigational, informational and transactional categories [3]. Different categories are assumed to have different implications on users' search behaviour and search algorithms. To give some examples: Navigational search queries (such as the query "citeseer") characterize situations where a user has a particular web site in mind and where he is primarily interested in visiting this page. Informational search queries (such as the query "increase wine crop") are queries where this is not the case, and users intend to visit multiple pages to, for example, learn about a topic [3]. Further research aims to empirically assess the distribution of different goal categories in search query logs via manual classification and subsequent statistical generalization [4] and/or Web Query Mining techniques [3,6]. There is some evidence that certain categories of goals can be identified algorithmically based on different features of user behaviour, such as "past user-click behaviour" and an analysis of "click distributions" [3]. Recently, a community of researchers with an interest in Query Log Analysis has formed at the World Wide Web 2007 conference as a separate workshop.

A second stream of research attempts to demonstrate the principle feasibility of implementing goal-orientation on an operational level. *GOOSE*, for example, is a prototypical goal-oriented search engine that aims to assist users in finding adequate search terms for their goals [7]. *Miro*, another example, is an application that facilitates goal-oriented web browsing [8]. The Lumiere Project focused on inferring goals of software users based on Bayesian user modeling [10]. Work on goal-oriented acquisition of requirements for hypermedia applications [11] shows that it is possible to translate high-level goals of stakeholders into (among other things) low level content requirements for web applications. Another example [12] facilitates purposeful navigation of geospatial data through goal-driven service invocation based on WSMO. WSMO is a web service description approach that decouples user desires from service descriptions by modeling low-level goals (such as "havingATripConfirmation") and non-functional property constructs [13]. In addition to these approaches, there have been several studies in the domain of information science that focus on different search strategies (such as top-down, bottom-up, mixed strategies) of users [14].

Apart from these isolated, yet encouraging, attempts, current research lacks a deep understanding about *how* users express their goals, and what explicit representations could be suitable to describe them.

3   How Do Users Express Goals in Web Search?

We initiated an explorative study in response to the observation that there is a lack of research on *how* users express their goals in web search. In the following we will present preliminary findings from this study.

Data sources: We have used the AOL search database [15] as our main data source[1]. In addition to the AOL search database, several other web search logs are available [16]. We have used the AOL search database because it provides information about anonymous User IDs, time stamps, search queries, and clicked links. To our knowledge, the AOL search database is also the most recent corpus of search queries available (2006). We are aware of the ethic controversies arising from using the AOL search database. For example, although the User IDs are anonymous, a New York Times reporter was able to track back the identity of one of the users in the dataset [17]. As a consequence, we masked the search queries that are presented in this paper by maintaining their semantic frame structure, but exchanging certain frame element

---

[1] Because the AOL search database was retracted from AOL shortly after releasing it, we obtained a copy from a secondary source: http://www.gregsadetsky.com/aol-data/ last accessed on July 15th, 2007.

instantiations [19]. We will elaborate on this later on. In following such an approach, we aim to protect the real identity of the users being studied while retaining necessary temporal and intentional relations of search queries.

Methodology: In this study we were interested in how users express, refine, alter and reformulate their goals while searching. We have searched the AOL search database for different verbs that are considered to indicate the presence of goals, including verbs such as *achieve, make, improve, speedup, increase, satisfied, completed, allocated, maintain, keep, ensure* and others [18]. We subsequently annotated random results (different search queries) with semantic frame elements obtained from Berkeley's Framenet [19]. Framenet is a lexical database that aims to document the different semantic and syntactic combinatory possibilities of English words in each of its senses. It aims to achieve that by annotating large corpora of text. It currently provides information on more than 10.000 lexical units in more than 825 semantic frames [19]. A **lexical unit** is a pairing of a word with a meaning. For example, the verb "look" has several lexical units dealing with different meanings of this verb, such as "direct one's gaze in a specified direction" or "attempt to find". Each different meaning of the word belongs to a **semantic frame**, which is "a script-like conceptual structure that describes a particular type of situation, object or event along with its participants and props" [19]. Each of these elements of a semantic frame is called **frame elements**. Semantic frames are evoked by lexical units. To give an example, the semantic frame "Cause_change_of_position_on_a_scale" is evoked by a set of lexical units, such as decline, decrease, gain, plummet, rise, increase, etc, and has the core frame elements Agent [], Attribute [Variable], Cause [Cause] and Item [Item]. Agent refers to the person who causes a change of position on a scale, attribute refers to the scale that changes its value, cause refers to non-human causes to the change, and item refers to the entity that is being changed.

   Example: The search query "Increase Computer Speed" can be annotated with Frame Elements from Framenet's lexicon. The lexical unit "increase" evokes the frame "Cause_change_of_position_on_a_scale", which we can use to annotate "Increase Computer Speed" in the following way: "**Increase** [item Computer] [attribute Speed]". The frame elements Agent and Cause do not apply here.

Selected Results: One verb we were using to explore the dataset was "increase". The query history depicted in Table 1 below presents an excerpt of the search history of a single user that performed search queries containing the verb "increase". We picked this particular search log because it demonstrates several interesting aspects of the role of goals in web search. We do not claim that this user's search behaviour is typically or representative for a larger set of users or queries. In fact, the majority of search queries in the AOL search database is of a non-intentional nature. We discuss the implications of this observation in the Section 5.

   We obtained the complete search record of the selected user, frame-annotated his intentional queries based on the FrameNet lexicon and classified the queries from an intentional perspective (e.g. refinement, generalization, etc). The particular frame used during annotation was "Cause_change_of_position_on_a_scale", which is evoked by the verb "increase". For privacy reasons, we modified the search queries in the following way: We retained the verbs and attributes which were part of the original query, but modified the contents of the semantic frame element item (e.g. wine crop) and cause (e.g. fertilizer) as well as time stamps (maintaining relative time differences with an accuracy of +/- 60 seconds). We'd like to remark that the users' search history below was interrupted by other, non-intentional queries (queries such as "flickr.com") and also other more complex intentional queries. For reasons of illustration and simplicity, we leave these out in Table 1.

Table 1. Frame-based Annotation of Selected Queries from a Single Search Session

Nr.	Query	Frame Annotation	Time Stamp	Goal
#1	How to get more wine crop	How to **get more** [$_{item}$wine crop]	2006-03-30 19:29:59	Formulation
#2	Fertilizer or insecticide to increase wine crop	[$_{cause}$ Fertilizer] or [$_{cause}$ insecticide] to **increase** [$_{item}$wine crop]	2006-03-30 19:45:28	Refinement
#3	Fertilizer to increase wine crop	[$_{cause}$ Fertilizer] to **increase** [$_{item}$ wine crop]	2006-03-30 19:46:11	Refinement
[further non-intentional queries, not related to wine crop]				
#4	Increase wine crop	**increase** [$_{item}$ wine crop]	2006-03-30 19:48:25	Generali-zation
#5	How to get rich wine crop	How to **get rich** [$_{item}$ wine crop]	2006-04-07 06:29:19	Different Goal Formulation
[non-intentional query "wine crop"]				
#6	How to have good wine crop	How to **have good** [$_{item}$ wine crop]	2006-04-07 06:40:45	Re-formulation
[further non-intentional queries and further more complex intentional queries related to "wine crops"]				

From a semantic frame perspective, it is interesting to see that it is not possible to annotate all of the above queries consistently. While the verb **increase** evokes the corresponding frame "Cause_change_of_position_on_a_scale" in queries #2, #3 and #4, the other queries #1, #5, and #6 do not contain **increase** and therefore do not evoke the same frame. Although FrameNet contains lexical entries for the verbs **get** and **have** and the adjectives **good, rich** and **more**, the word senses *get more*, *get rich* and *have good* are not yet captured as lexical units in the FrameNet lexicon. However, it is easily conceivable that an expanded or customized version of FrameNet (possibly in combination with WordNet) would contain these units and that they could be associated with the same semantic frame.

From a goal-oriented perspective, we will use our findings to develop a set of hypothesis that we believe are relevant and helpful to further study the role and structure of users' goals on the web.

Several things are noteworthy in the search history of the above user: First, the user started off with a goal <u>formulation</u> (#1 how to get more wine crop) and then proceeded with a <u>refinement</u> of this goal in a second query (#2 Fertilizer or insecticide to increase wine crop). The provided time stamps reveal that in this case, the time difference between the two queries was more than 15 minutes! Although it is hard to assess the real cause for this time lag, the AOL search database provides a possible explanation by listing the websites that the user visited in response to query #1, which includes a discussion board website hosting discussions on different strategies to get more "wine crop" (including "insecticides" and "fertilizer"). This allows us to hypothesize that **H1: Goal refinement is a time-intensive process during search**.

In query #3, the user performed a further <u>refinement</u> of his goal to "fertilizer to increase wine crop" and in #4, he performs a <u>generalization</u> to "Increase wine crop". This is interesting again from a goal-oriented perspective: Instead of refining his goals in a strict top-down approach, the user alternates between top-down (refining) and bottom up (generalizing) goal formulations. We consider this observation in a hypothesis 2 that claims that, from a goal-oriented perspective, user search is neither a strict top-down, nor a purely bottom-up approach, but a combination of both. While we focus on informational queries only, previous studies have found that the type of approach does not only depend on the type of task, but also different types of users [14]. This leads us to hypothesize **H2: Users search by iteratively refining, generalizing and reformulating goals, in no particular order**.

In query #5 the user performs a <u>different goal formulation</u>: "How to get rich wine crop". Instead of focusing on quantity ("get more" / "increase"), the search now can be interpreted to focus on the quality of wine crop ("get rich"). In query #6, a <u>goal re-formulation</u> is performed. This can be regarded to represent the same goal, but articulated in a slightly different way ("get good" instead of "get rich" wine crop). Another very interesting observation is that there is a time span of more than 7 days between queries #1-#4 and queries #5-#6! Although we have no information about what the user might have done in between these search activities, we use this evidence to tentatively hypothesize that identifying different, but related, goals is difficult for users, and it involves significant time and potentially cognitive efforts. In a more intuitive way, we can say that it seems that, especially with high-level, knowledge intensive goals, users learn about their goals as they go. We formulate this observation in hypothesis **H3: Exploring related goals is more time-intensive than goal refinement**.

And finally, we can observe that a smaller amount of time is passing between search queries #5 and #6. The question that is interesting to ask based on this observation is whether goal refinements require more time and cognitive investments from users than goal re-formulations. One might expect that users with search experience become skilled in tweaking their queries based on the search engines' responses without modifying their initial goal. We express this question in our hypothesis **H4: Goal re-formulation requires less time than goal exploration or goal refinement**. Next, we will explore some implications of these observations.

Analysis: If hypothesis H1 would be corroborated in future studies, offering users possible goal refinements would be very likely to be considered a useful concept. If hypothesis H2 would be supported in further studies, goal-oriented search would not only need to focus on goal refinement, but also on providing a range of different intentional navigation structures, allowing to flexibly alternate between refining, generalizing and exploring goals. If the exploration of goals represents a very time intensive process (H3), then users can be assumed to greatly benefit from having access to the goals of other users. And finally, if goal re-formulation does not require significant amounts of time (H4), there might be little motivation for researchers to invest in semantic similarity of web searchers, but more motivation to invest in intentional similarity.

Surprisingly, when analyzing current search technologies such as Google, we can see that there is almost no support for any of these different goal-related search tasks (refinement, generalization, etc) identified. Although Google helps in reformulating search queries ("Did you mean X?"), this – at most – can be regarded to provide some support for users in goal re-formulation on a *syntactic* level, but not on a truly *intentional* level (help in goal refinement, generalization, etc).

These observations immediately raise a set of interesting research questions: Do the formulated hypotheses hold for large sets of search sessions? How can the hypotheses be further refined to make them amenable to algorithmic analysis? And how can the identified goals be represented in more formal structures? While we are interested in all of these questions, in this paper we will only discuss the issue of more formal representations in some greater detail.

4 Representing Search Goals as Semi-formal Goal Graphs

We have modeled the goals of a user who is interested in "wine crop" with the agent- and goal-oriented modeling framework i* [20]. When applying i*, we focused on goal aspects and neglected agent-related concepts such as actors, roles and others. The i* framework provides elements such as softgoals, goals, tasks, resources and a set of semantic relations between them. The goal graph in Figure 1 was constructed by one of the authors of this paper based on the frame-annotated goals depicted in Table 1. In the diagram, the goals of the users are represented through oval-shaped elements. Means-ends links are used to indicate alternative ways (means) by which a goal (ends) can be fulfilled. Goals represent states of affairs to be reached, and tasks, which are represented through hexagonal elements, describe specific activities that can be performed for the fulfillment of goals. Soft-goals, which are represented through cloud-shaped elements, describe goals for which there is no clear-cut criterion to be used for deciding whether they are satisfied or not. Thus, soft-goals are fulfilled or denied to a certain degree, based on the presence or absence of relevant evidence. In i* diagrams, links such as "help" or "hurt" are used to represent how a belief about the fulfillment or denial of a soft-goal depends on the satisfaction of other goals. From the goal-graph in Figure 1 we can infer that the goal "increase wine crop" can be achieved through a variety of means: Fertilizer, Insecticides and Irrigation all represent means to achieve the end of increasing wine crop. The goal "Increase wine crop" and the related goal "Improve wine crop" both have "help" contribution links to the overarching soft-goal "Winery be successful".

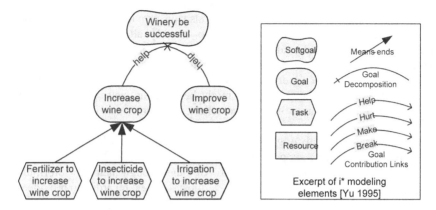

Fig. 1. Representing Users' Search Goals in a Semi-Formal Goal Graph

Assuming that such goal graphs can be constructed for a range of different domains (which is evident in a broad set of published examples from the domain of requirements engineering), it would be interesting to see how the different goal-related activities of users during search (such as goal formulation, goal refinement, goal generalization, etc) can be represented as a traversal of such a goal graph. We will explore this question next.

4.1 How Can Search Be Understood as a Traversal Through a Goal Graph?

Modifying search engines' algorithms to exploit knowledge about users' goals has a high priority for search engine vendors [5]. Being able to relate search queries to nodes in a goal graph could enable search engines to provide users goal-oriented support in search. This could mean that software could offer users to refine their search goals, generalize them or propose related goals from other users.

Figure 2, depicts the results of manually associating the search queries presented in Table 1 with the goal graph introduced in Figure 1. We can see that the user starts his search by formulating a version of the goal "increase wine crop" in query #1. This goal is refined in query #2 "Fertilizer or insecticides to increase wine crop" which can be mapped onto the two means "Fertilizer to increase wine crop" and "Insecticides to increase wine crop". Query #3 "fertilizer to increase wine crop" represents a further refinement. In query #4, the user generalizes his search goal to "increase wine crop" again. Query #5 and #6 relate to a different goal: "Improve wine crop". Query #5 and #6 can be considered to be re-formulations of the same goal.

Interestingly, the goal graph reveals that the user did not execute search queries related to the means "Irrigation to increase wine crop" or the soft-goal "Winery be successful", although one can reasonably expect that the user might have had a genuine interest in these goals too (although validation of this claim is certainly hard without user interaction).

As a consequence, a major benefit of having goal graphs available during search could be pointing users to refined goals or making sure that users do not miss related goals. But assuming that having such goal graphs would be beneficial, how can they be constructed?

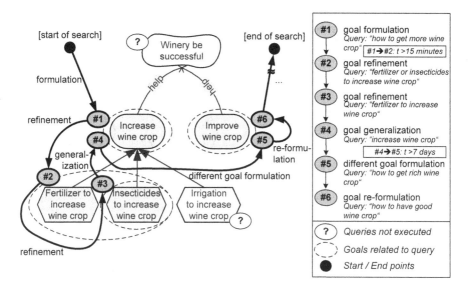

Fig. 2. Goal-Oriented Search as a Traversal of Goal Graphs

4.2 How Can Large-Scale Goal Graphs Be Constructed?

Mapping search queries onto goal graphs presumes the existence and availability of goal graphs. In our example, we have hand-crafted a goal graph for illustration purposes. However, manually constructing such goal graphs is costly, and anticipating the entirety, or even a large proportion, of users' goals on the internet would render such an approach unfeasible. So how can we construct large-scale goal graphs that do not rely on the involvement of expert modelers? Automatic user goal identification is an open research problem [6], and answering this question satisfyingly would go well beyond the scope of this paper, but we would like to discuss some pointers and ideas: The recent notion of folksonomies has powerfully demonstrated that meaningful relations can emerge out of collective behaviour and interactions [21]. We would like to briefly explore this idea and some of its implications for constructing large-scale goal graphs based on frame-analysis of intentional artifacts.

Let's assume that a system has the capability to come up with frame-based annotations of search queries. The search query "fertilizer or insecticide to increase wine crop" would then be annotated in a way that is depicted on the left side of Figure 3. Based on such annotations, a goal graph construction algorithm could use heuristics to construct a goal graph similar to the one depicted on the right side of Figure 3.

Heuristic rules could, for example, prescribe that the root goal is represented by the central verb ("increase") and its corresponding item ("wine crop"), and that the means to this end are represented by the frame elements cause ("fertilizer", "insecticide"). Each time a user formulates an intentional search query, the goal graph construction algorithm could construct such small, atomic goal graphs heuristically.

In a next step, these atomic goal graphs constructed from different users' search queries would need to be connected to larger whole. Considering hypothesis 2, this

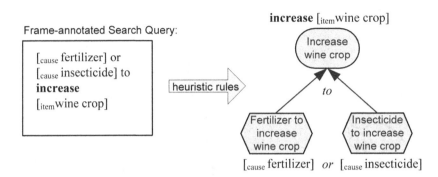

Fig. 3. Heuristic Construction of Atomic Goal Graphs via Frame-Annotation of Search Queries

appears to be a task that is hard to perform by algorithms alone. Nevertheless, usage data analysis, explicit user involvement or semi-automatic, collaborative model construction efforts (as e.g. pursued by the ConceptNet project [9]) might help to overcome this issue, which can be considered to represent a non-trivial research challenge.

5 Implications and Threats to Validity

We are aware that our particular research approach puts some constraints on the results of our work: Due to our focus, the search queries we analyzed were not required to be representative and, in fact, they are not. To obtain some quantitative evidence, two of the authors have categorized a pseudo-random sample (based on java.util.Random randomizer) of 2000 out of 21,011,340 queries into intentional and non-intentional categories, based on the criterion whether a query contains *at least* one verb (infinitive form, excluding gerund) and *at least* one noun. For each of these candidates, two authors of this paper judged whether it would be possible to envisage the goal a user might have had based on a specific query (such as "increase computer speed"). From our analysis, only 2.35% (47 out of 2000) of the searches from the AOL search database can be considered to be such "intentional queries". The probability of occurrence then results in a 95% confidence interval of [0.0169, 0.0301] for the probability of a query being intentional according to our criteria.

In contrast to these findings, related studies found somewhat higher numbers. A study reported in [4] suggests that 35% of search sessions have a general, high-level information research goal (such as questions, undirected requests for information, and advice seeking). The difference in numbers might be explained by different levels of analysis and a more relaxed understanding of goals in [4], which allows a broader set of queries (including queries that do not have verbs) to be labelled as goal-related.

There are several implications of this discrepancy: While users often have high-level goals when they are searching the web, they are currently not rewarded for formulating (strictly) intentional queries. In fact, one can assume that formulating non-intentional queries represents a (locally) successful strategy in today's search engine landscape. As a result, users might have adapted to the non-intentional mode in which Google, Yahoo and other search engines operate today. However, this situation makes it necessary for users to cognitively translate their high-level goals into search queries and perform reasoning about their goals in their mind. This potentially increases the cognitive burden of users and makes it hard for systems to connect them with other users who pursue similar goals or allowing them to benefit from the experiences made by other searchers.

We do not believe that these implications put constraints on our results: With a collaborative goal modeling approach, even a small percentage of strictly intentional queries could be used to construct large-scale goal graphs. Even if the percentage of intentional queries among the entirety of search queries would be as low as 1% or even lower, the sheer amount of queries executed on the World Wide Web would still provide algorithms with a rich corpus to construct large-scale goal graphs. On the web, such an approach is by far not unusual: For example, on wikipedia, a minority of users contributes content that is being used by a majority. However, the task of constructing large-scale goal graphs would obviously become much easier if users actually would be aware that search engines would interpret their queries as an expression of intent rather than an input that is being used for text string matching.

6 Conclusions

Based on our preliminary findings, we can formulate a set of interesting research challenges: First, how can large-scale goal graphs be represented and constructed? How can intentional artifacts (such as search queries) be associated with nodes in such goal graphs? How can goals and web resources be associated? And how can collaboration on the internet support the construction of such intentional structures?

Our work represents an initial attempt towards understanding the role and structure of goals in web search. We have demonstrated how search processes can be understood as a traversal through goal graphs and have provided some ideas on how to construct large scale goal graphs. In future work, we are interested in further investigating and shaping intentional structures on the web.

References

1. Berners-Lee, T., Hendler, J., Lassila, O.: The Semantic Web. Scientific American 284 (2001)
2. Greene, K., The Future of Search. MIT Technology Review (July 16, 2007) (last accessed on July 18th, 2007) , http://www.technologyreview.com/Biztech/19050/
3. Lee, U., Liu, Z., Cho, J.: Automatic Identification of User Goals in Web Search. In: WWW 2005. Proceedings of the 14th International World Wide Web Conference, pp. 391–400. ACM Press, New York (2005)
4. Rose, D., Levinson, D.: Understanding User Goals in Web Search. In: Feldman, S.I., Uretsky, M., Najork, M., Wills, C. (eds.) Proceedings of the 13th International World Wide Web Conference, pp. 13–19. ACM Press, New York (2004)
5. Broder, A.: A Taxonomy of Web Search. SIGIR Forum 36, 3–1 (2002)
6. Baeza-Yates, R., Calderon-Benavides, L., Gonzalez-Caro, C.: The Intention Behind Web Queries. In: Crestani, F., Ferragina, P., Sanderson, M. (eds.) SPIRE 2006. LNCS, vol. 4209, pp. 98–109. Springer, Heidelberg (2006)

7. Liu, H., Lieberman, H., Selker, T.: GOOSE: A Goal-Oriented Search Engine with Commonsense. In: De Bra, P., Brusilovsky, P., Conejo, R. (eds.) AH 2002. LNCS, vol. 2347, pp. 253–263. Springer, Heidelberg (2002)

8. Faaborg, A., Lieberman, H.: A Goal-Oriented Web Browser. In: CHI 2006. Proceedings of the SIGCHI Conference on Human Factors in Computing Systems, pp. 751–760. ACM Press, New York (2006)

9. Liu, H., Singh, P.: Conceptnet - A Practical Commonsense Reasoning Tool-Kit. BT Technology Journal 22, 211–226 (2004)

10. Horvitz, E., Breese, J., Heckerman, D., Hovel, D., Rommelse, K.: The Lumiere Project: Bayesian User Modeling for Inferring the Goals and Needs of Software Users. In: Proceedings of the Fourteenth Conference on Uncertainty in Artificial Intelligence, Madison, WI, pp. 256–265 (1998)

11. Bolchini, D., Paolini, P., Randazzo, G.: Adding Hypermedia Requirements to Goal-Driven Analysis. In: RE 2003. Proceedings of the 11th IEEE International Conference on Requirements Engineering, pp. 127–137. IEEE Computer Society, Los Alamitos (2003)

12. Tanasescu, V., Gugliotta, A., Domingue, J., Villarıas, L., Davies, R., Rowlatt, M., Richardson, M., Stincic, S.: In: Geospatial Data Integration with Semantic Web Services: the eMerges Approach (2007)

13. Roman, D., Keller, U., Lausen, H., de Bruijn, J., Lara, R., Stollberg, M., Polleres, A., Feier, C., Bussler, C., Fensel, D.: Web Service Modeling Ontology. Applied Ontology 1, 77–106 (2005)

14. Navarro-Prieto, R., Scaife, M., Rogers, Y.: Cognitive Strategies in Web Searching. In: Proceedings of the 5th Conference on Human Factors & the Web (1999)

15. Pass, G., Chowdhury, A., Torgeson, C.: A picture of search. In: Proceedings of the 1st International Conference on Scalable Information Systems, ACM Press, New York (2006)

16. Jansen, B., Spink, A.: How Are We Searching the World Wide Web? A Comparison of Nine Search Engine Transaction Logs. Information Processing and Management 42, 248–263 (2006)

17. Barbaro, M., Zeller Jr., T.: A Face Is Exposed for AOL Searcher No. 4417749, New York Times (August 9, 2006)

18. Regev, G., Wegmann, A.: Where Do Goals Come From: The Underlying Principles of Goal-Oriented Requirements Engineering. In: RE 2005. Proceedings of the 13th IEEE International Conference on Requirements Engineering, Washington, DC, USA, pp. 253–362. IEEE Computer Society, Los Alamitos (2005)

19. Ruppenhofer, J., Ellsworth, M., Petruck, M., Johnson, C., Scheffczyk, J.: FrameNet II: Extended Theory and Practice, International Computer Science Institute, University of California at Berkeley (2006)

20. Yu, E.: Modelling Strategic Relationships for Process Reengineering. PhD thesis, Department of Computer Science, University of Toronto, Toronto, Canada (1995)

21. Mika, P.: Ontologies Are Us: A Unified Model of Social Networks and Semantics. In: Gil, Y., Motta, E., Benjamins, V.R., Musen, M.A. (eds.) ISWC 2005. LNCS, vol. 3729, pp. 522–536. Springer, Heidelberg (2005)

Publishing and Sharing Ontology-Based Information in a Collaborative Multimedia Document Management System

Annett Mitschick and Ronny Fritzsche

Dresden University of Technology, Department of Computer Science
Chair of Multimedia Technology, 01062 Dresden, Germany
{annett.mitschick,ronny.fritzsche}@inf.tu-dresden.de

Abstract. Problems which users typically experience when dealing with search and management tasks within their personal document collections mainly result from lacking expressiveness and flexibility of the traditional file systems and data models to represent individual knowledge. Ontology-based approaches provide appropriate solutions, enabling people to semantically describe their multimedia items. Furthermore, suitable techniques for the support of information sharing and exchange in ontology-based systems are still missing or only weakly supported. In this paper we present a multi-user multimedia document management system based on Semantic Web technologies. We discuss requirements and preconditions with regard to user management and access control, and describe the technical implementation of our concept.

1 Introduction

Due to the development of powerful digital devices for the creation, processing, and storage of multimedia documents, the amount of available digital items and associated information has tremendously increased in a wide range of application fields. Via the Internet, multimedia documents such as images, music, etc. are spread out and exchanged between users rapidly and in large scale. The result is an increasing disorientation within heterogeneous document collections regarding origin, context, and interrelation of digital items. Thus, for the user, a mere syntactical description and storing is not sufficient. Today, common management systems and applications are typically limited to hierarchical navigation and storage of information. Problems and barriers which typically appear when users deal with search and management tasks within personal media collections mainly result from lacking expressiveness and flexibility of the traditional data models to represent individual knowledge.

Aim of the *K-IMM* project[1] is the development of a concept for semantic-based management of personal media collections, which allows the user to apply individual knowledge models and paths with preferably little effort. Therefore,

---

[1] http://www-mmt.inf.tu-dresden.de/K-IMM

M. Weske, M.-S. Hacid, C. Godart (Eds.): WISE 2007 Workshops, LNCS 4832, pp. 79–90, 2007.
© Springer-Verlag Berlin Heidelberg 2007

applying Semantic Web technologies to ensure machine-processability and interchangeability, a document collection is no longer an aggregation of separate items, but forms an individual, user-specific knowledge base. Furthermore, the user is now able to publish and share not only images and associated tags (e.g. on Flickr[2]), but interpretable knowledge about content in a community of people. By enabling other users to contribute to the annotation and contextualization of documents, the descriptions are semantically enriched with different individual views and conceptualizations. To facilitate the described scenario of community-driven knowledge engineering, one also has to deal with concepts of access control with respect to a multi-user document management system, granting rights to view, edit, or delete information resources to certain users or user groups. Related approaches of Semantic-Web-based collaborative systems are often Wiki-based (enabling everyone to edit everything) and thus are generally not focused in detail on these challenging issues.

In this paper we present an ontology-based multimedia document management system, developed within the *K-IMM* project, with regard to user management and access control. Section 2 illustrates background and state-of-the-art of collaborative knowledge modeling systems based on Semantic Web technologies or principles, referencing relevant related work. The focus of the second part (Section 2.2) lies on access control and concurrency issues. In Section 3 we introduce our ontology-based multimedia document management system, its basic architecture, and the extensions we made regarding multi-user support and knowledge publishing and sharing. Section 4 presents our showcase: a Web-based information sharing system on the basis of our multimedia document management architecture. Finally, a conclusion is given in Section 5.

2   Background and Related Work

Multimedia document management profits from principles of knowledge management theory. The tasks to be fulfilled by knowledge management systems comprise acquisition, organization, storage, disposal, and access (retrieval) of information. Reliable and expressive metadata can ensure that a multimedia archive is well-organized and retrieval jobs (distinct queries, browsing, question answering) are performed with convincing precision and recall rates. However, different users apply different conceptualizations, usually refinements or derivations of common concepts. The connection and integration of one or more user concepts within a community knowledge space for documents is so far not sufficiently supported. Solutions may arise from recent developments in the field of collaborative knowledge modeling and social software.

2.1   Collaborative Knowledge Modeling

Cooperative knowledge modeling has certain advantages over solitary work. It represents a process of knowledge construction resulting in what is often called

---

[2] http://www.flickr.com

"group memory" [1], comprising individual knowledge and experiences of the group members. This knowledge thus contains multiple views and can be disseminated and re-used. Some relevant techniques enabling informal or formal collaborative knowledge modeling are discussed in the following.

2.1.1   Social Tagging and Social Bookmarking

Social Web annotation tools (or resource sharing systems) have recently evolved into commonly used and very popular community portals, thanks to a very low entry barrier. In general, users are able to define "tags" (arbitrary words) for the description or categorization of resources, which might be, for instance, photos (e.g. Flickr), bookmarks (e.g. del.icio.us[3]), or bibliographic references (i.e. CiteULike[4]). These resources thus form a distributed and open knowledge base, interlinked by simple, informal keywords. A major drawback of this technique is of course the lack of formalism, allowing ambiguity or mistakes in writing, which complicates targeted search and retrieval. Therefore, recent work, like [2,3,4], focuses on the development of techniques to formalize "emergent semantics" from these so called "folksonomies", so that users can still profit from a lightweight knowledge modeling approach, but with a Semantic Web foundation.

2.1.2   Semantic Wikis

Also, Wikis as platforms to create, interlink, and share information within a community enjoy growing popularity. Several extensions to the classical Wiki approach have been presented recently to incorporate the advantages of formalization through Semantic Web technologies [5,6]. Using enhanced syntax, semantic statements can be integrated to describe and link information resources. Due to the typical characteristics of Wikis - any reader can edit any page [7] - existing projects are generally not focused on the definition of authoring roles and access rights. Thus, the delimitation of individual views or workspaces of users or groups is not supported.

2.1.3   Social Semantic Desktops

Based on the idea of Semantic Desktops, using Semantic Web technologies to improve personal information management on desktop computers, Social Semantic Desktops provide another technique for collaborative knowledge work. One representative is DBin [8], which uses Peer-to-Peer (P2P) technique to create interest groups and to allow exchange of RDF-based information (but also files) between P2P group members. The Nepomuk project[5] deals with the development of a standardized, conceptual framework for Semantic Desktops. A reference implementation of some parts is Gnowsis [9], though it does not provide collaboration or multi-user support.

---

[3] http://del.icio.us
[4] http://www.citeulike.org
[5] http://nepomuk.semanticdesktop.org

2.2   Multi-user Support on the Semantic Web

In the recent past, ontology engineering and knowledge work has often been accomplished by single users (experts). Now, with the increasing exertion of complex semantically-enriched applications, the support of teams working on large-scale knowledge bases has become an interesting research topic. Nevertheless, to our knowledge there has only been a small amount of research done on this field. In the following, we would like to focus on selected work on the two important topics Access Control and Concurrency for Semantic Web data stores.

2.2.1   Access Control

Much work has been done related to access control with regard to file systems and data bases. In general, access control can be realized using Access Control Lists (ACL), attaching lists of permissions (allowed actors and actions) to the resources, or Capability-based, assigning lists of resources and access rights to the agents. Another widely used mechanism is Role Based Access Control (RBAC) [10], where users or groups are assigned particular roles, which are associated with according permissions. All these techniques have certain advantages and disadvantages with regard to the specific application context.

Another important issue is the level of granularity. Access control to XML documents, for instance, can be specified at element level and propagated to lower levels according to their hierarchical structure [11]. Although, XML forms the foundation of the Semantic Web architecture, access control mechanisms for XML documents are not sufficient for RDF-based data. Qin and Atluri [12] propose an access control model for the Semantic Web, which facilitates granting of permissions on concept-level based on the propagation of semantic relationships among them. Thus, an agent is allowed to access instance data if he/she has permission for the according concept. Policy-based approaches as described in [13,14] make use of explicit rules to specify access rights, which gives administrators the opportunity to create simple or complex policies which are not necessarily restricted to concepts. A role-based approach (Role-based Collaborative Development Method RCDM) can be found in [15], which introduces five roles with different ranges of permission to mitigate conflicts: knowledge user, proposer, engineer, expert, and manager.

As far as we know, most research on access control for Semantic Web data starts from the assumption that permissions or prohibitions are managed by one or more administrators (role-based) and are primarily based on concepts. With regard to our application context it is necessary to have a decentralized, "flat hierarchy" approach, allowing all users to specify access rights on particular instances (e.g. a multimedia description, or a FOAF profile), and particularly not based on concepts. We introduce our approach in detail in Section 3.1.

2.2.2   Concurrency

Distributed authoring of semantic knowledge bases requires much more sophisticated solutions for concurrency problems than it is done in classical databases, using (conservative) locking or (optimistic) non-locking methods. Campbell et al. [16] argue that in case of traditional two-phase locking long-life transactions, which might take months, prevent other users from working on the locked data. Nevertheless, study of existing work [17,18] shows that, with regard to the higher complexity of semantic knowledge bases and higher probability and risk of conflicts, the preferred way to solve conflicts is preventing them before they occur. Most of the work actually focuses on collaborative ontology engineering, which necessitates appropriate solutions to select and propagate locks according to hierarchical relationships between concepts. If a user wants to edit concept A it is locked (*write lock*) for other users and furthermore all its superconcepts receive *read locks* (and their superconcepts recursively), as modifications to superconcepts of A must be prohibited as long as A is edited. In addition to that, [17] proposes locking of complete subtrees to allow ontology engineers to independently work on their parts and to reduce the number of lock checks with the server.

Concurrency problems within our application context have limited complexity as we focus on multi-user knowledge engineering on the instance level, as we will later show in Section 3.1. Thus the above-mentioned techniques are too extensive for our needs, but provide the appropriate solutions for future multi-user concept engineering facilities (Section 3.2).

3   Multi-user Support Within the *K-IMM* Ontology-Based Multimedia Document Management System

This work is based on the results and implementation within the *K-IMM* (Knowledge through Intelligent Media Management) project, which provides a system architecture for intelligent multimedia document management for private users (i.e. semi- or non- professionals) [19]. The intention of this project is to take advantage of ontology engineering and Semantic Web technologies in such a way that users without particular skills can manage and search a multimedia collection intuitively and without additional cost. Therefore, the system comprises components for automated import and indexing of media items (of different type) as background tasks. All of the components are realized as plug-ins (bundles). Thus, further plug-ins for specific media type analysis and processing or advanced components for visualization can easily be added to the system, and can be started and stopped dynamically at runtime. Hence, it is possible to run the system e.g. just for image management (starting only image analyzing and image semantics deducing components), or only with low-level indexing (without semantic modeling), if desired. The media analyzing components extract available properties and features and pass them to the knowledge modeling components. In our prototype implementation, RDF and OWL processing, storage, and reasoning is based on the Jena Framework[6] including the Jena Inference Support. Further components, which are also not subject of this paper, comprise knowledge instantiation and propagation [20], and context aggregation and modeling.

---

[6] http://jena.sourceforge.net

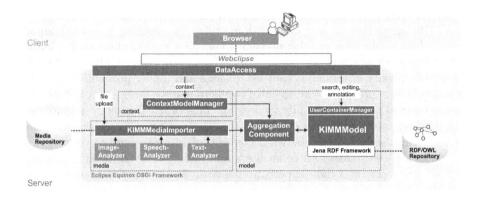

Fig. 1. The overall *K-IMM* System architecture

A conceptual overview of the overall architecture of the *K-IMM* System is depicted in Figure 1.

The foundational ontology (top-level) used for knowledge modeling is based on the ABC ontology described in [21], extended by certain base concepts and relations for the description of multimedia documents. Although replaceable, this foundational ontology is easy to use and understand, and provides a basis for individual extensions, as discussed in the following.

3.1 User Management and Access Control

As we described in Section 2.2, the study of existent work related to access control for ontology-based knowledge management systems showed that suitable techniques for the support of information sharing and exchange in a multi-user multimedia document management system are still missing. The following application scenario should illustrate our requirements and design principles:

User A stores a collection of personal photographs in the K-IMM system. The items are semantically annotated, and related real-world entities (people, locations, events, etc.) are formally described according to the K-IMM base ontology. This knowledge forms the user's individual "workspace" and is so far not visible to others or part of any community or group knowledge. Some of the photos have been assigned very private and confidential comments. As the user has the opportunity to extend the base ontology, he has established his own specializations of the base concept "Photo" for his needs as a semi-professional photographer: "Night-Shot", "CloseupView", "Landscape", etc., including specific attributes. Now he wants to publish a subset of his photo collection to User B, including semantic entities (creator, creation location, camera, etc.), but without the comments and the specialized instantiation. User B can add own comments to the photos (either visible to User A or only to herself), and is allowed to publish User A's disclosed photos and related entities to others.

Obviously, this application scenario necessitates a very fine-grained and sophisticated definition of access policies, including access to the policies themselves. As existing techniques (Section 2.2) turned out to be not sufficient, we developed a methodology which is based on the idea of shared workspaces or shared repositories. Such systems use a *Container* metaphor to store and retrieve collections of resources visible to a single user or a group. We adopted this approach as follows: Every user of the system has a *UserResourceContainer* which contains imported multimedia documents and their semantical descriptions. Every *Container* has a unique identifier used within the namespace for the RDF-based data. Within this namespace the user is allowed to create instances, relations, and specializations of base concepts. The publishing of resources is realized as a transfer between containers, i.e. from a *UserResourceContainer* to a *GroupResourceContainer*. The *UserResourceContainer* only retains a reference to the "relocated" resource. A *GroupResourceContainer* is a shared repository for more than one user and is initialized by a publishing process. As a matter of convenience and efficiency a *PublicResourceContainer*, as a specialized *GroupResourceContainer*, is used to publish resources to the group of all users. The following rules should be observed:

- Every container applies the base ontology. The base ontology itself stays stable. No user is allowed to edit or delete base concepts or properties.
- Every user can create subclasses (specializations) of base concepts and subproperties of base properties within his/her container.
- Every user can edit and delete specializations within his/her container.
- Every user can create instances within his/her container.
- Every user can access instances within the group containers he/she is member of.
- Every user can edit and delete instances, which he/she can access.
- Every user can publish instances, which he/she can access, to other users. He/she can select attributes and related resources which should be published supplementary.
- An instance of a user defined class is generalized to the next level base concept for publishing (as RDF allows instances to have multiple types, the instance is of the specialized type within the user's container and of general type outside).

The opportunity to pass published resources on to other groups (and namespaces) forms a decentralized and easy to use access mechanism, and is a lightweight approach. However, it is clear to see that a publishing process realized as a relocation of a resource can only be done once - thus, providing access to more than one group is not possible. Furthermore, the deletion of a published resource within a group container must not imply that references in several user containers get inconsistent. Therefore, we introduced a so called *HiddenResourceContainer*, which is a general, anonymous container, which holds resources published more than once or published to an "anonymous" group, which is not

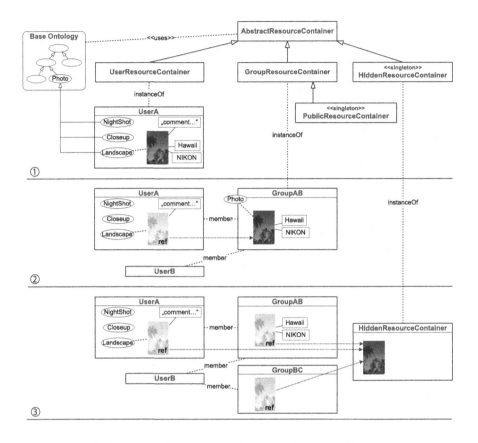

Fig. 2. An application scenario illustrating the publishing concept

permanent or fixed[7]. Users have no direct access to the *HiddenResourceContainer*, as it is the case with the other types of containers, only via references they possess. Please note that references to resources outside a user's container must be updated each time the original resource is moved or deleted from a group container (which is equivalent to moving the resource to the *HiddenResourceContainer* and deleting the reference within the group container).

Figure 2 illustrates the application of our approach using the example scenario given above. As user A publishes a photo (annotated with comments and information about location and camera), which he has specified within his workspace to be of type "Landscape", to user B, the resource is transfered to a group container to which both have access to share resources (2). User A's personal comments are left within his user container, connected to the resource's reference. As user B grants access to this photo to the members of another group she

---

[7] Sharing resources does not necessarily mean that users establish a formal collaboration group (explicitly labeled).

participates in, the resource is transfered to the *HiddenResourceContainer* and existing references are updated (3).

In our implementation the container paradigm is realized in the form of Java objects which encapsulate information about the container itself (as a resource, with a namespace), the owner resp. members, and the contained resources. Thus, they act as wrappers for the semantic models of these containers and provide the technical implementation of the above-mentioned policies and restrictions.

3.2 Concurrency and Locking

As described in Section 2.2.2, concurrency is a challenging issue in collaborative ontology engineering and primarily considered to be solved with conservative locking methods. In our case users do not collaborate on the concept level, but only on the instance level. User-defined ontology specializations are only allowed within *UserResourceContainers* and are based on a foundational ontology which itself is not editable. Therefore, it is necessary to use lock selection and propagation methods on the instance level. The following example should illustrate this issue:

User A has published a photo with information about the location and the camera (cf. Figure 2) to user B. Later on he wants to refine the location information ("Hawaii") within the group container. At the same time user B decides to publish the location information description to other users.

It is clear to see that user B should be prevented from modifying the location resource directly (publishing or deletion, i.e. transfer to other containers). Thus, the location resource obtains a *write-lock*, which means, that other users are not allowed to publish or delete it.

As an extension to our approach it would be possible to allow collaborative ontology engineering within the group containers. In this case the concurrency solutions described in Section 2.2 regarding lock propagation along concept hierarchies and relations should be applied.

4   The *K-IMM Wiki*

As an appropriate showcase we developed a prototype Web application as a user interface for the *K-IMM* service architecture which is operated on a server. Although the application is not a "real" Wiki in the strict sense (according to a page-based editing principle), we use this term for our Web application to emphasize the characteristic of quick and low-barrier access and publishing, and the absence of a differentiated role hierarchy.

The underlying *Webclipse* platform, which was developed within our working group, is a Rich Internet Application (RIA) framework based on *Eclipse* and the OSGi concept, and makes use of the Ajax framework *Echo2*[8] for the user

---

[8] http://www.nextapp.com/platform/echo2/echo/

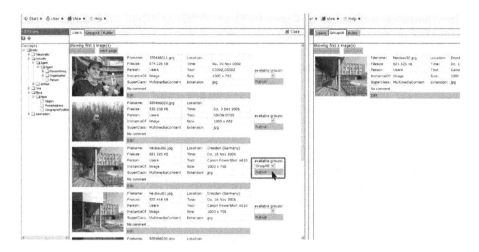

Fig. 3. User (left) and group aspect (right) within the K-IMM Wiki user interface

interface components. The K-IMM Wiki is a thin-client application, so that users only need a standard browser and are not required to install additional software.

Figure 3 presents two screenshots of the Web interface with exemplary data. The left one shows the user container view containing a number of photos with some semantic descriptions and menus for publishing them to available groups. In this example *UserA* is about to publish a photo to *GroupAB*. Thereafter, the photo appears in the workspace of this group (right clipping). The tabs allow for direct and quick switching between workspaces. The hierarchy on the left is a representation of the concepts of the base ontology and allows the user to create sub-concepts within his workspace. Please note that this prototypical interface does not yet provide sophisticated retrieval facilities as we focused on simple techniques to illustrate the publishing process.

So far we have tested and evaluated the concept using exemplary data sets (collection of 100 images and extracted descriptions) and virtual users (up to 100). Our results have shown that the approach is particularly suitable for small groups of up to ten concurrent users (fitting our target scenario of private document collections). Target of our future work will be a detailed evaluation of the user management concept within a real-world scenario, which includes usability and concurrency awareness. For this purpose, our designed Web application should help realizing and evaluating user interface concepts for annotation, retrieval, and sharing of multimedia documents, in particular regarding comprehension and benefits to the user.

5   Conclusion

In this paper we discussed multi-user support in the field of ontology-based knowledge and document management and examined a selection of existing solutions for access control and concurrency problems. We introduced our *K-IMM* system, a multimedia document management system based on Semantic Web technologies, and presented design principles and implementation of the made extensions to support user and group collaboration.

The realized *Container* metaphor gives users the opportunity to store and semantically annotate multimedia documents within a personal workspace and share resources with others by means of publishing to a group container. Thus, access control is not managed exclusively by an administrator and with the help of explicitly defined access rules, but defined and controlled by the users themselves and with an intuitive interaction technique.

Acknowledgment

We would like to thank Erik Peukert for the implementation work for the *Webclipse* platform as a foundation of our K-IMM Wiki.

References

1. Vasconcelos, J., Kimble, C., Gouveia, F.: A design for a group memory system using ontologies (2000)
2. Gendarmi, D., Lanubile, F.: Community-driven ontology evolution based on folksonomies. OTM Workshops 1, 181–188 (2006)
3. Angeletou, S., Sabou, M., Specia, L., Motta, E.: Bridging the gap between folksonomies and the semantic web: An experience report. In: Bridging the Gap between Semantic Web and Web 2.0 (SemNet 2007), pp. 30–43 (2007)
4. Abbasi, R., Staab, S., Cimiano, P.: Organizing resources on tagging systems using t-org. In: Bridging the Gap between Semantic Web and Web 2.0 (SemNet 2007), pp. 97–110 (2007)
5. Krötzsch, M., Vrandecic, D., Völkel, M.: Semantic mediawiki. In: Cruz, I., Decker, S., Allemang, D., Preist, C., Schwabe, D., Mika, P., Uschold, M., Aroyo, L. (eds.) ISWC 2006. LNCS, vol. 4273, pp. 935–942. Springer, Heidelberg (2006)
6. Schaffert, S.: Ikewiki: A semantic wiki for collaborative knowledge management. In: WETICE 2006: Proceedings of the 15th IEEE International Workshops on Enabling Technologies: Infrastructure for Collaborative Enterprises, pp. 388–396. IEEE Computer Society Press, Los Alamitos (2006)
7. Cunningham, W.: Wiki design principles, http://c2.com/cgi/wiki?WikiDesignPrinciples
8. Tummarello, G., Morbidoni, C., Nucci, M.: Enabling semantic web communities with dbin: An overview. In: Cruz, I., Decker, S., Allemang, D., Preist, C., Schwabe, D., Mika, P., Uschold, M., Aroyo, L. (eds.) ISWC 2006. LNCS, vol. 4273, pp. 943–950. Springer, Heidelberg (2006)
9. Sauermann, L., Grimnes, G.A., Kiesel, M., Fluit, C., Maus, H., Heim, D., Nadeem, D., Horak, B., Dengel, A.: Semantic desktop 2.0: The gnowsis experience. In: Cruz, I., Decker, S., Allemang, D., Preist, C., Schwabe, D., Mika, P., Uschold, M., Aroyo, L. (eds.) ISWC 2006. LNCS, vol. 4273, pp. 887–900. Springer, Heidelberg (2006)
10. Ferraiolo, D., Kuhn, R.: Role-based access controls. In: 15th NIST-NCSC National Computer Security Conference, pp. 554–563 (1992)

11. Damiani, E., di Vimercati, S.D.C., Paraboschi, S., Samarati, P.: A fine-grained access control system for xml documents. ACM Trans. Inf. Syst. Secur. 5, 169–202 (2002)
12. Qin, L., Atluri, V.: Concept-level access control for the semantic web. In: XMLSEC 2003: Proceedings of the 2003 ACM workshop on XML security, New York, NY, USA, ACM Press, pp. 94–103. ACM Press, New York (2003)
13. Dietzold, S., Auer, S.: Access control on rdf triple stores from a semantic wiki perspective. In: Sure, Y., Domingue, J. (eds.) ESWC 2006. LNCS, vol. 4011, Springer, Heidelberg (2006)
14. Reddivari, P., Finin, T., Joshi, A.: Policy-Based Access Control for an RDF Store. In: Proceedings of the IJCAI 2007 Workshop on Semantic Web for Collaborative Knowledge Acquisition (2007)
15. Li, M., Wang, D., Du, X., Wang, S.: Ontology construction for semantic web: A role-based collaborative development method. In: Zhang, Y., Tanaka, K., Yu, J.X., Wang, S., Li, M. (eds.) APWeb 2005. LNCS, vol. 3399, pp. 609–619. Springer, Heidelberg (2005)
16. Campbell, K., Cohn, S., Chute, C., Shortliffe, E., Rennels, G.: Scalable methodologies for distributed development of logic-based convergent medical terminology
17. Sure, Y., Erdmann, M., Angele, J., Staab, S., Studer, R., Wenke, D.: OntoEdit: Collaborative ontology development for the semantic web. In: Horrocks, I., Hendler, J. (eds.) ISWC 2002. LNCS, vol. 2342, pp. 9–12. Springer, Heidelberg (2002)
18. Seidenberg, A.R.J.: A methodology for asynchronous multi-user editing of semantic web ontologies. In: 5th International Conference on Knowledge Capture (K-CAP), Whistler, British Columbia, Canada (2007)
19. Mitschick, A.: Ontology-based management of private multimedia collections: Meeting the demands of home users. In: 6th International Conference on Knowledge Management (I-KNOW'06), Special Track on Advanced Semantic Technologies, Graz, Austria (2006)
20. Mitschick, A.: Meiß ner, K.: A stepwise modeling approach for individual media semantics. In: GI-Edition Lecture Notes in Informatics (LNI), Dresden, Germany (2006)
21. Lagoze, C., Hunter, J.: The ABC ontology and model. In: Dublin Core Conference, pp. 160–176 (2001)

Sequence Disunification and Its Application in Collaborative Schema Construction

Jorge Coelho[1], Mário Florido[2], and Temur Kutsia[3]

[1] Instituto Superior de Engenharia do Porto & LIACC Porto, Portugal
jcoelho@ncc.up.pt
[2] University of Porto, DCC-FC & LIACC Porto, Portugal
amf@ncc.up.pt
[3] RISC, Johannes Kepler University Linz, Austria
tkutsia@risc.uni-linz.ac.at

Abstract. We describe procedures for solving disequation problems in theories with sequence variables and flexible arity symbols, and show how to use them for Collaborative Schema Construction.

1 Introduction

In this paper we define and present algorithms for collaborative XML schema management. We call such a management process *Collaborative Schema Construction*, which is based on the following idea: Several people are interested in producing a common schema for XML data and each of those people may impose some constraints on the schema structure. For example in the Super-Journal Project [17], the goal was *"to produce a cluster of journal content to make it worth the author submitting multimedia content and the reader doing useful searching and browsing in the electronic field with sufficient content that is relevant."*. SuperJournal brings a consortium of publishers to develop models for network publishing by gathering the content from several different journals. The consortium members propose a desired schema for their own domain. Collaboratively they could unify all their schemas in one general schema which satisfies everyone's requisites. After this all the information could be integrated in one general journal and distributed.

The integration of different schemas into a common, unified one, usually called *Schema Integration*, has been a prominent area of research for the database community over the past years [9,16,18]. With the widespread adoption of XML has the standard syntax to share data, new attention was given to schema integration for XML schemas [19,11]. In all these previous works, schema integration means a semantic integration, i.e., in the different schemas, one may have different names for the same semantic concept, and this semantic knowledge is used to match schemas. Thus these works necessarily rely on domain specific information to perform matching. In our work, we assume that the syntax used in the different XML schemas is the same and that the several schemas are incomplete specifications where only some domain specific part is defined. Thus in this

M. Weske, M.-S. Hacid, C. Godart (Eds.): WISE 2007 Workshops, LNCS 4832, pp. 91–102, 2007.

paper, when we refer to collaborative schema construction we mean *syntactic* collaborative schema construction.

The approach we follow is to represent schemas by equations in a properly chosen term algebra and try to find the most general common schema by solving equations in that algebra. We use an algebra of terms with function symbols of flexible arity (suitable to represent XML) and variables that can be substituted by sequences of terms, called sequence variables. Unification in theories with sequence variables and flexible arity function symbols (sequence unification or SU in short) is a recent topic of research with applications in several areas in computer science. In [12,14] decidability of sequence unification was shown, and a minimal complete solving procedure was introduced. Recently, unification of terms with flexible arity functions symbols has been successfully applied to XML-processing and website content verification [15,13,4,3,6,5]. Sequence unification is infinitary, that means that for some problems the minimal complete set of solutions (*mcs*) may be infinite, which implies that any complete procedure is in general nonterminating. Several finitary fragments of sequence unification have been identified in [14].

With the motivation of Collaborative Schema Construction, here we define a terminating procedure that covers larger fragments than those from [14]. More-over, we incorporate negative information in the specification, for instance, spec-ifying that two sequences do not share a common element. It leads to sequence disunification that is a new development. It extends sequence unification with disjointness equations, that for terms without sequence variables coincides with standard disunification [7,8,2].

Thus, the main contributions of this work are:

- A minimal complete procedure for sequence disunification (dealing with dis-jointness equations).
- A minimal complete algorithm for the case where each sequence variable occurs maximum twice in the disunification problem.
- Applying these techniques to Collaborative Schema Construction in XML where disunification of schemas is used to create more general ones and a parser for the general schema is generated.

Due to space limitations, proofs are presented in the extended version of this paper available at `http://www.ncc.up.pt/xcentric/cfk.pdf`.

The rest of the paper is organized as follows: in Section 2 the preliminaries. In Section 3 a complete and terminating procedure for disunification is described. Section 4 presents the algorithm for quadratic sequence disunification problems. In Section 5 we discuss the application of the defined procedure to XML. Con-clusions are presented in Section 6.

2   Preliminaries

We assume that the reader is familiar with the basic notions of unification the-ory [1] and automata theory [10]. Here we introduce preliminary notions about unification along the lines presented in [14]. We consider an alphabet consisting of the following mutually disjoint sets of individual variables $\mathcal{V}_I$, sequence variables $\mathcal{V}_S$, and flexible arity function symbols $\mathcal{F}$. The set of variables $\mathcal{V}_I \cup \mathcal{V}_S$ is denoted by $\mathcal{V}$. *Terms* over $\mathcal{F}$ and $\mathcal{V}$ are defined by the grammar $t ::= x \mid X \mid f(t_1, \ldots, t_n)$, where $x \in \mathcal{V}_I$, $X \in \mathcal{V}_S$, $f \in \mathcal{F}$, and $n \geq 0$. If $n = 0$, we will write f instead of $f()$. The set of terms over $\mathcal{F}$ and $\mathcal{V}$ is denoted by $T(\mathcal{F}, \mathcal{V})$. We will use x, y, z for individual variables, X, Y, Z for sequence variables, and s, t, r for terms. The letters a, b, c will denote terms with the empty list of arguments. A pair of terms, written $s \doteq t$, where $s, t \notin \mathcal{V}_S$, is called an *equation*. For readability purposes, we often write sequences of terms in the parentheses, like, e.g. $(t_1, \ldots, t_n)$, or $()$ for the empty sequence. We also use an abbreviated notation $\tilde{t}$ for sequences. The set of variables of a syntactic object O (i.e. of a term, a term sequence, an equation, or a set of equations) is denoted by $vars(O)$.

A *substitution* is a mapping from individual variables to terms that are not sequence variables, and from sequence variables to finite sequences of terms so that all but finitely many individual variables are mapped to themselves, and all but finitely many sequence variables are mapped to themselves considered as singleton sequences. For example, $\{x \mapsto f(a, Y), X \mapsto (), Y \mapsto (a, f(b, Y), x)\}$ is a representation of a substitution. We will use lower case Greek letters for substitutions, with ε for the empty substitution. The notions of *instance* of a term or a sequence under a substitution, and substitution *composition* are standard.

Example 1. $f(f(X), X, a, f(Y, x), b)\{X \mapsto (), Y \mapsto (a, b, g(x)), x \mapsto f(X)\} = f(f, a, f(a, b, g(x), f(X)), b)$.

A substitution ϑ is *more general* than σ, written $\vartheta \leq \sigma$, iff there exists a substitution φ such that $\vartheta\varphi = \sigma$. Respectively, ϑ is *strongly more general* than σ with respect to a set of sequence variable variables $\mathcal{X}$, iff there exists a substitution φ such that $\vartheta\varphi = \sigma$ and $X\varphi \neq ()$ for all $X \in \mathcal{X}$.

Example 2. Let $\vartheta = \{X \mapsto Y\}$ and $\sigma = \{X \mapsto (), Y \mapsto ()\}$. Then $\vartheta \leq \sigma$, $\vartheta \trianglelefteq^{\{X\}} \sigma$, and $\vartheta \ntrianglelefteq^{\{Y\}} \sigma$.

A set of substitutions S is *minimal* (resp. *almost minimal* with respect to a set of sequence variables $\mathcal{X}$) iff for any substitutions $\vartheta, \sigma \in S$, if $\vartheta \leq \sigma$ (resp. if $\vartheta \trianglelefteq^{\mathcal{X}} \sigma$) then $\vartheta = \sigma$. A *disjointness equation*, d-equation in short, is a pair of term sequences, written $\tilde{s} \asymp \tilde{t}$. We say that a disjointness equation $\tilde{s} \asymp \tilde{t}$ is true if the sequences $\tilde{s}$ and $\tilde{t}$ do not have a common element. For instance, $(a, f(x)) \asymp (f(b), X, f(y))$, but $(a, f(x)) \not\asymp (f(x), X, f(y))$. D-equations generalize disequations $s \neq t$ that can be seen as d-equations between single terms.

A *Sequence Disunification problem* (SD problem in short) is a set of equations and d-equations $\{s_1 \doteq^? t_1, \ldots, s_n \doteq^? t_n, \tilde{s}_1 \asymp \tilde{t}_1, \ldots, \tilde{s}_m \asymp \tilde{t}_m\}$.[1] A substitution σ is called its *solution* iff $s_i\sigma = t_i\sigma$ for all $1 \leq i \leq n$ and $\tilde{s}_i\sigma \asymp \tilde{t}_i\sigma$ for all $1 \leq i \leq m$. Note that not all the instances of a solution of a disunification

---

[1] We write $\doteq^?$ to indicate that the equations has to be solved.

problem are again its solution (in contrast to unification problems). Rather, one has to pick only those instances that keep the disjointness equations true. We use upper case Greek letters to denote disunification problems. For a disunification problem Γ we denote by $\Gamma^{\doteq}$ its maximal subset that consists of equations only, and by $\Gamma^{\times}$ its maximal subset that consists of d-equations only. The set of function symbols in Γ is called its *signature* and is denoted by $sig(\Gamma)$.

A *complete set of solutions* of a SD problem Γ is a set of substitutions S such that each $\vartheta \in S$ is a solution of Γ, and for any solution σ of Γ there exists $\vartheta \in S$ such that $\vartheta \leq \sigma$. A *minimal complete set of solutions* of Γ is a set of solutions that is complete and minimal. An *almost minimal complete set of solutions* of Γ with respect to a set of sequence variables $\mathcal{X}$ is a complete set of solutions of Γ that is almost minimal with respect to $\mathcal{X}$. For each $\mathcal{X}$, a almost minimal set of solutions of Γ with respect to $\mathcal{X}$ is unique up to the equivalence associated with $\lhd^{\mathcal{X}}$. We denote such a unique almost minimal complete set of solutions of Γ with respect to $vars(\Gamma)$ by $amcs(\Gamma)$. In the similar way we have the notation $mcs(\Gamma)$ for the minimal complete set of solutions of Γ.

Example 3. Let Γ be a SD problem $\{f(X) \doteq^? f(Y)\}$. Then $mcs(\Gamma) = \{\{X \mapsto Y\}\}$ and $amcs(\Gamma) = \{\{X \mapsto Y\}, \{X \mapsto (), Y \mapsto ()\}\}$. If we add to Γ a d-equation $X \times ()$ then $mcs(\Gamma) = amcs(\Gamma) = \{\{X \mapsto Y\}\}$. For $\Delta = \{f(X, a) \doteq^? f(a, X)\}$ these sets are infinite: $mcs(\Delta) = amcs(\Delta) = \{\{X \mapsto ()\}, \{X \mapsto a\}, \{X \mapsto (a, a)\}, \ldots\}$. If we add $X \times a$ to Δ, then $mcs(\Delta) = amcs(\Delta) = \{\{X \mapsto ()\}\}$.

3   Sequence Disunification

Now we introduce rules for solving SD problems. They are very similar to the rules for sequence unification from [14]. The difference is an extra condition in some rules related to satisfying d-equations, and one extra rule.

For an SD problem Γ, the set of *projecting substitutions* $\Pi(\Gamma)$ is defined as $\Pi(\Gamma) = \{\sigma \mid \mathcal{D}om(\sigma) \subseteq \mathcal{V}_S(\Gamma^{\doteq}), X\sigma = ()$ for all $X \in \mathcal{D}om(\sigma)$, and $\Gamma^{\times}$ is true$\}$. For instance, the set of projecting substitutions for the problem $\{f(X, a) \doteq^? f(a, Y)\}$ is $\{\{X \mapsto ()\}, \{Y \mapsto ()\}, \{X \mapsto (), Y \mapsto ()\}\}$, and for $\{f(X, a) \doteq^? f(a, Y), Y \times ()\}$ it is $\{\{X \mapsto ()\}\}$.

The transformation rules are the following:

P: **Projection**
$$\Gamma \Longrightarrow_\sigma \Gamma\sigma, \qquad \text{where } \sigma \in \Pi(\Gamma).$$

T: **Trivial**
$$\{s \doteq^? s\} \cup \Gamma \Longrightarrow_\varepsilon \Gamma.$$

O1: **Orient 1**
$$\{s \doteq^? x\} \cup \Gamma \Longrightarrow_\varepsilon \{x \doteq^? s\} \cup \Gamma, \qquad \text{if } s \notin \mathcal{V}_I.$$

O2: **Orient 2**
$$\{f(s, \tilde{s}) \doteq^? f(X, \tilde{t})\} \cup \Gamma \Longrightarrow_\varepsilon \{f(X, \tilde{t}) \doteq^? f(s, \tilde{s})\} \cup \Gamma, \qquad \text{if } s \notin \mathcal{V}_S.$$

S: Solve

$$\{x \doteq^? t\} \cup \Gamma \Longrightarrow_\sigma \Gamma\sigma, \qquad \text{if } x \notin \mathcal{V}_I(t), \ \sigma = \{x \mapsto t\}, \text{ and } \Gamma^{\succeq}\sigma \text{ is true.}$$

FAS: First Argument Simplification

$$\{f(s, \tilde{s}) \doteq^? f(t, \tilde{t})\} \cup \Gamma \Longrightarrow_\varepsilon \{s \doteq^? t, f(\tilde{s}) \doteq^? f(\tilde{t})\} \cup \Gamma,$$
if $f(s, \tilde{s}) \neq f(t, \tilde{t})$ and $s, t \notin \mathcal{V}_S$.

SVE1: Sequence Variable Elimination 1

$$\{f(X, \tilde{s}) \doteq^? f(X, \tilde{t})\} \cup \Gamma \Longrightarrow_\varepsilon \{f(\tilde{s}) \doteq^? f(\tilde{t})\} \cup \Gamma, \qquad \text{if } f(X, \tilde{s}) \neq f(X, \tilde{t}).$$

SVE2: Sequence Variable Elimination 2

$$\{f(X, \tilde{s}) \doteq^? f(t, \tilde{t})\} \cup \Gamma \Longrightarrow_\sigma \{f(\tilde{s})\sigma \doteq^? f(\tilde{t})\sigma\} \cup \Gamma\sigma,$$
if $X \notin \mathcal{V}_S(t)$, $\sigma = \{X \mapsto t\}$, and $\Gamma^{\succeq}\sigma$ is true.

W1: Widening 1

$$\{f(X, \tilde{s}) \doteq^? f(t, \tilde{t})\} \cup \Gamma \Longrightarrow_\sigma \{f(X, \tilde{s}\sigma) \doteq^? f(\tilde{t}\sigma)\} \cup \Gamma\sigma,$$
if $X \notin \mathcal{V}_S(t)$, $\sigma = \{X \mapsto (t, X)\}$, and $\Gamma^{\succeq}\sigma$ is true.

W2: Widening 2

$$\{f(X, \tilde{s}) \doteq^? f(Y, \tilde{t})\} \cup \Gamma \Longrightarrow_\sigma \{f(\tilde{s}\sigma) \doteq^? f(Y, \tilde{t}\sigma)\} \cup \Gamma\sigma,$$
where $\sigma = \{Y \mapsto (X, Y)\}$ and $\Gamma^{\succeq}\sigma$ is true.

Suc: Success

$$\Gamma \Longrightarrow_\varepsilon \top, \qquad \text{if } \Gamma^{\doteq} = \emptyset \text{ and } \Gamma^{\succeq} \text{ is true.}$$

We denote the set of these rules by $\mathfrak{R}$, and the set $\mathfrak{R} \setminus \{\mathsf{P}\}$ by $\mathfrak{R}^-$. (We could have added couple of more rules to simplify the d-equation part of SD, but decided to keep the list of rules as short as possible.) The substitution σ used to make a step in the transformation rules is called a *local substitution*. Sometimes we may use the rule name abbreviation as subscripts and write, for instance, $\Gamma \Longrightarrow_{\sigma,\mathsf{W2}} \Delta$ to indicate that Γ was transformed to Δ by W2. A *derivation* is a sequence $\Gamma_1; \sigma_1 \Longrightarrow \Gamma_2; \sigma_2 \Longrightarrow \cdots$ of transformations. A *selection strategy* $\mathcal{S}$ is a function which given a derivation $\Gamma_1 \Longrightarrow_{\sigma_1} \cdots \Longrightarrow_{\sigma_{n-1}} \Gamma_n$ returns an equation, called a *selected equation*, from Γ_n. A derivation is via a selection strategy $\mathcal{S}$ if in the derivation all choices of selected equations, being transformed by the transformation rules, are performed according to $\mathcal{S}$.

Definition 1. *A syntactic sequence disunification procedure $\mathfrak{D}$ is any program that takes a SD problem Γ and a selection strategy $\mathcal{S}$ as an input and uses the transformation rules to generate a tree of derivations via $\mathcal{S}$, called the* disunification tree for Γ via $\mathcal{S}$, *denoted $\mathcal{DT}_{\mathfrak{D}}^{\mathcal{S}}(\Gamma)$, in the following way:*

1. *The root of the tree is labeled with Γ;*
2. *Each branch of the tree is a derivation via $\mathcal{S}$ of the form $\Gamma \Longrightarrow_{\sigma_0,\mathsf{P}} \Gamma_1 \Longrightarrow_{\sigma_1,\mathsf{R}_1}$ $\Gamma_2 \Longrightarrow_{\sigma_2,\mathsf{R}_2} \cdots$ or of the form $\Gamma \Longrightarrow_{\sigma_0,\mathsf{R}_0} \Gamma_1 \Longrightarrow_{\sigma_1,\mathsf{R}_1} \Gamma_2 \Longrightarrow_{\sigma_2,\mathsf{R}_2} \cdots$, where $R_i \in \mathfrak{R}^-$ for $i \geq 0$.*

3. *If different instances of the projection rule are applicable to the root node in the tree, or if several rules are applicable to the selected equation in a node in the tree, they are applied concurrently.*
4. *If a node in the tree contains an SD problem Δ such that $\Delta^{\doteq}$ is unsolvable (this can be decided by the decision algorithm for sequence unification), or if no rule can be applied to Δ, then the branch is extended in a unique way by $\Delta \Longrightarrow_{\varepsilon} \bot$.*

The leaves of $\mathcal{DT}_{\mathfrak{D}}^{\mathcal{S}}(\Gamma)$ are labeled either with $\top$ or with $\bot$. The branches that end with $\top$ are called *successful branches*, and those with the leaves $\bot$ are *failed branches*. With each successful branch we associate the substitution obtained by composing local substitutions along the branch starting from the root towards the leaf. The set of all such substitutions is the *solution set for Γ computed by $\mathfrak{D}$*. It can be shown that this set coincides with $amcs(\Gamma)$. Hence, the procedure enumerates $amcs(\Gamma)$ and terminates if this set is finite. The finitary fragments of sequence unification from [14] give finitary fragments of SD. Here we just mention two other ones that have not been discussed in [14] but whose finitary property can be easily established. One is linear SU (each variable occurs at most once), and the other one can be described as a problem that can be represented with a single equation $\{s \doteq^? t\}$ where t is linear and does not contain sequence variables. Representation as a single equation is not a restriction because any general SU problem $\{s_1 \doteq^? t_1, \ldots, s_n \doteq^? t_n\}$ is equivalent to $\{f(s_1, \ldots, s_n) \doteq^? f(t_1, \ldots, t_n)\}$. These SU problems induce finitary SD problems.

Because of d-equations, an SD problem can be finitary even if its equational part is not. We showed one such problem in Example 3.

4   Quadratic Disunification Problems

In this section we present an algorithm for a fragment of SD that has an infinite *amcs* (i.e. an infinite search space), but it can be represented by a finite means. The algorithm works on a *quadratic fragment* of SD, denoted by QSD and requiring that no variable occurs in its equational part more than twice, and returns a deterministic finite automaton that models the search space. The automaton accepts a language of substitutions that is exactly the *amcs* for a given QSD problem.

Before describing the algorithm we recall the definition of a (deterministic) *finite automaton*, DFA: It is a 5-tuple $(Q, \Sigma, \delta, q_0, F)$ where Q is a finite set of states, Σ is a finite input alphabet, $q_0 \in Q$ is the initial state, $F \subseteq Q$ is the set of final states, and δ is a transition function mapping $Q \times \Sigma$ to Q.

We consider QSD problems as sequences of equations followed by d-equations, with the leftmost selection. Hence, the derivations in the algorithm below are organized in the *last in, first out* way.

Algorithm 1 (Solving QSD Problems)
Input: A QSD problem Γ.
Output: A DFA $(Q, \Sigma, \delta, q_0, F)$.

```
 1   function SolveQSD(Γ)
 2       Σ := Π(Γ)
 3       Q := {Γσ | σ ∈ Σ} ∪ {Γ, ⊤, ⊥}
 4       δ := {δ(Γ, σ) → Γσ | σ ∈ Σ}
 5       q₀ := Γ
 6       F := {⊤}
 7       U := {(Γσ, {Γ}) | σ ∈ Σ \ {ε}} ∪ {(Γ, ∅)}
 8       while U ≠ ∅ do
 9           Select a pair (Δ, h) from U.
10           U := U \ {Δ, h}
11           for each Δ' obtained from Δ by R ∈ ℜ⁻ via leftmost selection:
12               Δ ⟹_{σ,R} Δ' do
13               Σ := Σ ∪ {σ}
14               Q := Q ∪ {Δ'}
15               δ := δ ∪ {δ(Δ, σ) → Δ'}
16               if R ∈ {W1, W2} then
17                   h' := h ∪ {Δ}
18                   if Δ'^{≐} ∉ h'^{≐} modulo symmetry of ≐ then
19                       U := U ∪ {(Δ', h')}
20                   else U := U∪{(Δ'\{e}, ∅)} where e is the leftmost equation in Δ'
21               else
22                   if R ∈ {O1, O2} then h' := h else h' := ∅
23                   if Δ' ≠ ⊤ then U := U ∪ {(Δ', h')}
24       return (Q, Σ, δ, q₀, F)
```

On line 18, $\Delta'^{\doteq} \notin h'^{\doteq}$ means that the equational part of Δ' is not the same as the equational part of any element of h'. Roughly, the idea behind the SolveQSD algorithm is to try to reduce a given QSD problem to $\top$ by applying the transformation rules in all possible ways. In the derivations, with each QSD problem we associate its history (the set of QSD problems that appear in the same derivation before the given problem). The purpose of the history is to perform a cycle check, whether the equational part of a new QSD problem has already appeared in the same derivation before. To keep the history small, only necessary QSD's are kept there. It gets cleaned if a rule is applied that strictly decreases a certain measure, because all the problems generated after this rule application are supposed to be strictly smaller than the ones currently in the history, and it does not make sense to keep junk in the history anymore.

Note that the construction behind SolveQSD can be used to develop analogous algorithms for quadratic fragments of certain infinitary unification problems, e.g. for word and context unification. The main idea is to identify unification rules that do not change the size of the problem, and those that strictly reduce it. The first ones should be treated like W1 and W2 in SolveQSD.

Example 4. Given the SD problem $\{f(X, a) \doteq f(a, X)\}$, then:

- $\Sigma = \{\epsilon, \{X \mapsto ()\}, \{X \mapsto (a, X)\}, \{X \mapsto a\}\}$
- $Q = \{\{f(a) \doteq f(a)\}, \{f(X, a) \doteq f(a, X)\}, ∅, \top, \bot\}$

– δ is defined by the following transitions: $\{\delta(\{f(X,a) \doteq f(a,X)\}, \varepsilon) \rightarrow$
$\{f(X,a) \doteq f(a,X)\}, \delta(\{f(X,a) \doteq f(a,X)\}, \{X \mapsto ()\}) \rightarrow \{f(a) \doteq f(a)\},$
$\delta(\{f(X,a) \doteq f(a,X)\}, \{X \mapsto (a,X)\}) \rightarrow \{f(X,a) \doteq f(a,X)\}, \delta(\{f(X,a) \doteq$
$f(a,X)\}, \{X \mapsto a\} \rightarrow \{f(a) \doteq f(a)\}, \delta(\{f(a) \doteq f(a)\}, \varepsilon) \rightarrow \emptyset, \delta(\emptyset, \varepsilon) \rightarrow \top\}\}$
– $q_0 = \{\{f(X,a) \doteq f(a,X)\}\}$
– $F = \top$

The generated automaton is shown in figure 1

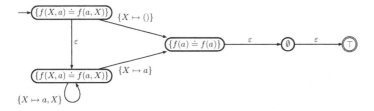

Fig. 1. Automaton generated for $\{f(X,a) \doteq f(a,X)\}$

Theorem 1 (Termination). SolveQSD *terminates on any input.*

It is not hard to see that the automaton returned by the SolveQSD algorithm
is a DFA. The fixed selection rule (the leftmost selection) guarantees that for
each pair of an QSD problem (i.e. state) and a substitution (i.e. input symbol)
there exists exactly one transition to a next QSD problem (state). The empty
substitution, like any other, is considered to be an input symbol.

Let $\mathcal{A}(\Gamma)$ be an automaton generated by SolveQSD for a QSD problem Γ.
The input accepted by the $\mathcal{A}(\Gamma)$ has usually the form $\sigma_1 \cdots \sigma_n$, where σ's are
substitutions. We can compose them from left to right, obtaining a substitution
σ. Slightly abusing the terminology, we say that σ is also accepted by $\mathcal{A}(\Gamma)$.

Given a QSD problem Γ with $\Gamma^{\asymp} = \emptyset$, there exists one-to-one correspondence between the runs of the automaton $\mathcal{A}(\Gamma)$ generated by SolveQSD and the
derivations generated by the SD procedure $\mathfrak{U}$ in [14] via the leftmost selection
strategy (considering Γ ordered), because every run step (transition) corresponds
to a derivation step (transformation) and vice versa. Since $\mathfrak{U}$ is sound, complete,
and almost minimal, we have that the set of substitutions accepted by $\mathcal{A}(\Gamma)$ is
$amcs(\Gamma)$. If $\Gamma^{\asymp} \neq \emptyset$, the DFA $\mathcal{A}(\Gamma)$ accepts a subset of $amcs(\Gamma^{\doteq})$ that satisfies
$\Gamma^{\asymp}$. Hence, we get the following result:

Theorem 2 (Main Theorem). *Let $\mathcal{A}(\Gamma)$ be a DFA generated by SolveQSD*
for a QSD problem Γ, and let S be the set of substitutions accepted by $\mathcal{A}(\Gamma)$.
Then $S = amcs(\Gamma)$.

The set $amcs(\Gamma)$ can be infinite as, for instance, it is for $\Gamma = \{f(X, z, Y, z) \doteq^?$
$f(g(X,a), g(a), Y), X \asymp Y\}$. For it, $amcs(\Gamma) = \{\{X \mapsto (), Y \mapsto g(a), z \mapsto g(a)\},$
$\{X \mapsto (), Y \mapsto (g(a), g(a)), z \mapsto g(a)\}, \ldots\}$. SolveQSD gives its finite representa-
tion in the form of a DFA. From it we can obtain another finite representation,

in terms of regular expressions on substitutions, which often might be more intuitive: $\{X \mapsto (), z \mapsto g(a), Y \mapsto (g(a), Y)\}\{Y \mapsto (g(a), Y)\}^*$.

Since the language accepted by a DFA is a regular language, we conclude that *amcs* for a QSD problem is a regular language of substitutions where the concatenation operation is the substitution composition.

Note that the SolveQSD algorithm can be used also on certain problems that are not quadratic. First of all, it is clear that it can be used on any SD problem that is finitary. Some more fragments of SD amenable to SolveQSD that we will use later are the following: (a) An SD that contains equations of the form $x_i \doteq^? t_i$ and $f(X_i) \doteq^? f(\tilde{t}_i)$, where x's and X's are not necessarily distinct, and all t's and $\tilde{t}$'s are variable disjoint and linear. Such problems can be easily reduced to QSD problems by introducing extra variables. (b) If t's and $\tilde{t}$'s are not necessarily variable disjoint but all occurrences of the same sequence variable in different t's and $f(\tilde{t})$'s have the same prefix,[2] then the problem is finitary. (c) Any Γ that can be represented as $\Gamma_1^{\doteq} \cup \Gamma_2^{\doteq} \cup \Gamma^{\asymp}$ where $\Gamma_1^{\doteq}$ is a finitary part of $\Gamma^{\doteq}$, $\Gamma_2^{\doteq}$ is a quadratic part of $\Gamma^{\doteq}$, and $\Gamma^{\asymp}$ requires all variables that occur in $\Gamma_2^{\doteq}$ twice, to be disjoint from each other. An example of such a problem is $\Gamma = \{f(a, Y) \doteq^? f(X, Z), \; f(X, a) \doteq^? f(a, X), \; f(b, Y) \doteq^? f(Y, b), X \asymp Y\}$. It can be represented as $\Gamma_1^{\doteq} \cup \Gamma_2^{\doteq} \cup \Gamma^{\asymp}$ with $\Gamma_1^{\doteq} = \{f(a, Y) \doteq^? f(X, Z)\}$, $\Gamma_2^{\doteq} = \{f(X, a) \doteq^? f(a, X), \; f(b, Y) \doteq^? f(Y, b)\}$, and $\Gamma^{\asymp} = \{X \asymp Y\}$. SolveQSD will return a DFA that accepts $\{X \mapsto (), Z \mapsto (a, Y)\}\{Y \mapsto (b, Y)\}^*\{Y \mapsto b\}$ and $\{X \mapsto a, Y \mapsto Z\}\{Z \mapsto (b, Z)\}^*\{Z \mapsto b\}$.

5   Application in Collaborative Schema Construction

One application of this work is what we call Collaborative Schema Construction. It is based on the following idea: Several people are interested in producing a common schema for XML data and each of those people may impose some constraints on the schema structure.

We start with defining a relation between DTD's and conjunctions and disjunctions of equations. Then we will translate the schema requirements into such formulae (adding disjointness equations, if necessary), bring them into conjunctive normal form, represent each conjunct as an SD problem and try to solve them. The (representations of the) solutions for each such problem then can be combined together, from which we can read off the desired schema.

To establish a relation between DTD's and equations, we need translation rules from sets of regular membership atoms to disjunction and conjunction of equations. A regular membership atom is an expression of the form $X \in [\![R]\!]$, where R is a regular expression built from ground terms and regular operators $*, +, ?, |, ,$. For instance, $f(a) \mid (b, g(a, b))$ and $f(a^* \mid b)+, g(a?)$ are such expressions. The translation rules are given below: (R is a regular expression with at least one regular operator, t is a ground term, and f is a dummy function.)

---

[2] The occurrences of a sequence variable X in the subterms $f_1(s_1, \ldots, s_n, X, \ldots)$ and $f_2(t_1, \ldots, t_m, X, \ldots)$ have the same prefix if $f_1 = f_2$, $n = m$, and $s_i = t_i$ for each $1 \le i \le n$.

$$\{X \in [\![R^*]\!]\} \cup R; E \Rightarrow \{Y \in [\![R]\!]\} \cup R; E \wedge f(X, Y) \doteq f(Y, X).$$
$$\{X \in [\![R+]\!]\} \cup R; E \Rightarrow \{Z \in [\![R]\!]\} \cup R; E \wedge f(X) \doteq f(Z, Y) \wedge f(Y, Z) \doteq f(Z, Y).$$
$$\{X \in [\![R?]\!]\} \cup R; E \Rightarrow \{Y \in [\![R]\!]\} \cup R; E \wedge (f(X) \doteq f() \vee f(X) \doteq f(Y)).$$
$$\{X \in [\![R_1 \mid R_2]\!]\} \cup R; E \Rightarrow \{Y \in [\![R_1]\!], Z \in [\![R_2]\!]\} \cup R;$$
$$E \wedge (f(X) \doteq f(Y) \vee f(X) \doteq f(Z)).$$
$$\{X \in [\![R_1, R_2]\!]\} \cup R; E \Rightarrow \{Y \in [\![R_1]\!], Z \in [\![R_2]\!]\} \cup R; E \wedge f(X) \doteq f(Y, Z).$$
$$\{X \in [\![f(R)]\!]\} \cup R; E \Rightarrow \{Y \in [\![R]\!]\} \cup R; E \wedge f(X) \doteq f(Y).$$
$$\{X \in [\![t]\!]\} \cup R; E \Rightarrow R; E \wedge f(X) \doteq f(t).$$

We now demonstrate our approach with an example:

Example 5. Three entities need to share a custom address book document. All of them agree that the document should have root element *addrbook* and one element *name*, and each of them imposes other constraints in the document content (note that places representing sequences which are not of interest for a given entity are represented by an anonymous variable '_'):

1. The document must have a sequence of zero or more elements *address* after *name* and does not matter what comes next. The proposed document has the simplified structure: *addrbook(name, address*, _)*.
2. The document must have a sequence of one or more elements *email* somewhere, with the simplified structure: *addrbook(name, _, email+, _)*
3. The document must have one or more *address* elements after *name* and an optional *telephone* element somewhere, with the following simplified structure: *addrbook(name, address+, _, telephone?, _)*

These requirement can be described by the following equations where f is a dummy function symbol and $X's$ are fresh distinct variables. (In fact, we could extend the rules in $\mathfrak{R}$ to work with the anonymous variables directly, taking into account all its subtleties, but because of space limitation we do not discuss it here.) The equations are obtained by the translation rules, with a simple further simplification to save space. (e.g., removing equations of the form $f(X) \doteq f(t_1, \ldots, t_n)$ and changing X everywhere by $t_1, \ldots, t_n$).

1. $x \doteq addrbook(name, A, X_1) \wedge f(A, address) \doteq f(address, A).$
2. $x \doteq addrbook(name, X_2, E, email, X_3) \wedge f(email, E) \doteq f(E, email).$
3. $(x \doteq addrbook(name, A, address, X_4, telephone) \vee$
   $D \doteq addrbook(name, A, address, X_4)) \wedge f(A, address) \doteq f(address, A).$

Taking the conjunction of these sequence equations and bringing it to the cnf we obtain two SD problems, one of which we denote by Γ:

$$\{x \doteq addrbook(name, A, X_1), x \doteq addrbook(name, X_2, E, email, X_3),$$
$$x \doteq addrbook(name, A, address, X_4, telephone),$$
$$f(A, address) \doteq f(address, A), f(email, E) \doteq f(E, email)\}.$$

(Due to space limitation we will not consider the other SD problem Δ here). We can try to solve the problem with $\mathfrak{U}$, but if we modify it a bit to be able to apply algorithms for quadratic problems.

The modification starts by noting that $\Gamma^{\doteq}$ can be represented as the union $\Gamma_1^{\doteq} \cup \Gamma_2^{\doteq}$, where $\Gamma_1^{\doteq} = \{x \doteq addrbook(name, A, X_1), x \doteq addrbook(name, X_2, E,$ $email, X_3), x \doteq addrbook(name, A, address, X_4, telephone)\}$ is finitary and $\Gamma_2^{\doteq} = \{f(A, address) \doteq f(address, A), f(email, E) \doteq f(E, email)\}$ is quadratic. Next, we impose the disjointness restriction on A and E (when at least one of them is not empty), which intuitively is well justified since addresses and emails should be different. We obtain the SD problems $\Gamma\{A \mapsto (), E \mapsto ()\}$ and $\Gamma \cup \{E \times A\}$. The first one is finitary. The second one falls in one of the fragments described at the end of Section 4 and, hence, can be solved by SolveQSD, obtaining the DFA that describes the solution set of $\Gamma \cup \{E \times A\}$. Taking the union of the DFA's for these two problems with the one that can be obtained in the same way from Δ and further minimizing it, we obtain a DFA from which one can read solutions:

$$\{\{x \mapsto addrbook(name, address, email, telephone), A \mapsto (), E \mapsto ()\},$$
$$\{x \mapsto addrbook(name, A, address, email, telephone), E \mapsto ()\}$$
$$\{A \mapsto (address, A)\}^* \{A \mapsto address\}\},$$
$$\{x \mapsto addrbook(name, address, E, email, telephone), A \mapsto ()\}$$
$$\{E \mapsto (email, E)^* \{E \mapsto email\}\},$$
$$\{x \mapsto addrbook(name, A, address, E, email, telephone)\}$$
$$\{A \mapsto (address, A)^* \{A \mapsto address\} \{E \mapsto (email, E)\}^* \{E \mapsto email\}\},$$

and the same without *telephone*. The DFA also describes the following DTD:

```
<!ELEMENT addrbook (name,address+,email+,telephone?)*>
<!ELEMENT name (#PCDATA)>
<!ELEMENT address (#PCDATA)>
<!ELEMENT telephone (#PCDATA)>
<!ELEMENT email (#PCDATA)>
```

To summarize, with the techniques described in this section we can

1. Define a common schema (or DTD) for several incomplete specifications for XML by reducing it to a SU problem;
2. The resulting DFA can be used for parsing of valid documents if the δ function uses the pattern matching in the same fashion as in the XCentric language [6] to match ground documents to the sequence variables.

6 Conclusion and Future Work

In this paper we formulated sequence disunification problem that besides sequence equations contains disjointness equations, and described a complete procedure to solve such problems. We demonstrated application of sequence disunification in collaborative schema construction, which consists of representing incomplete schemas by equations and using disunification to get the more general schema along with a parser.

Future work is two-fold. First, from the theoretical point of view, we plan (a) to study sequence disunification with regular constraints which, in our opinion, will be very useful in XML-related applications, and (b) investigate a possibility of bringing in certain higher-order features like e.g. context variables. Second, we will work on integration of these techniques in the XML-processing logic language XCentric [6].

References

1. Baader, F., Snyder, W.: Unification theory. In: Handbook of Automated Reasoning, pp. 445–532 (2001)
2. Buntine, W.L., Bürckert, H.-J.: On solving equations and disequations. J. ACM 41(4), 591–629 (1994)
3. Coelho, J., Florido, M.: Unification with flexible arity symbols: a typed approach. In (UNIF 2006) Proc. 20th Int. Workshop on Unification, Seattle, USA (2006)
4. Coelho, J., Florido, M.: VeriFLog: Constraint Logic Programming Applied to Verification of Website Content. In: Shen, H.T., Li, J., Li, M., Ni, J., Wang, W. (eds.) Int. Workshop XML Research and Applications (XRA 2006). LNCS, vol. 3842, Springer, Heidelberg (2006)
5. Coelho, J., Florido, M.: Type-based static and dynamic website verification. In: 2nd Int. Conf. on Internet and Web Applications and Services, IEEE Computer Society Press, Los Alamitos (2007)
6. Coelho, J., Florido, M.: XCentric: A logic-programming language for XML processing. In: PLAN-X (2007)
7. Comon, H.: Disunification: A survey. In: Computational Logic - Essays in Honor of Alan Robinson, pp. 322–359 (1991)
8. Comon, H., Lescanne, P.: Equational problems and disunification. J. Symb. Comput. 7(3/4), 371–425 (1989)
9. Doan, A., Halevy, A.Y.: Semantic integration research in the database community: A brief survey. AI Magazine 26(1), 83–94 (2005)
10. Hopcroft, J.E., Motwani, R., Ullman, J.D.: Introduction to Automata Theory, Languages, and Computation. Adison-Wesley (2007)
11. Kensche, D., Quix, C., Chatti, M.A., Jarke, M.: Gerome: A generic role based metamodel for model management. J. Data Semantics 8, 82–117 (2007)
12. Kutsia, T.: Unification with sequence variables and flexible arity symbols and its extension with pattern-terms. In: Calmet, J., Benhamou, B., Caprotti, O., Henocque, L., Sorge, V. (eds.) AISC 2002 and Calculemus 2002. LNCS (LNAI), vol. 2385, Springer, Heidelberg (2002)
13. Kutsia, T.: Context sequence matching for XML. ENTCS 157(2), 47–65 (2006)
14. Kutsia, T.: Solving equations with sequence variables and sequence functions. J. Symb. Comp. 42, 352–388 (2007)
15. Kutsia, T., Marin, M.: Can context sequence matching be used for querying XML? In: Proc (UNIF 2005), pp. 77–92 (2005)
16. Lee, Y., Sayyadian, M., Doan, A., Rosenthal, A.: eTuner: tuning schema matching software using synthetic scenarios. VLDB J. 16(1), 97–122 (2007)
17. Pullinger, D.J.: SuperJournal Project. IOP Publishing Ltd, Bristol, UK (1994)
18. Quix, C., Kensche, D., Li, X.: Generic schema merging. In: CAiSE. LNCS, vol. 4495, pp. 127–141. Springer, Heidelberg (2007)
19. Rahm, E., Hai-Do, H., Massmann, S.: Matching large XML schemas. SIGMOD Rec. 33(4), 26–31 (2004)

Mapping Metadata for SWHi: Aligning Schemas with Library Metadata for a Historical Ontology

Junte Zhang*, Ismail Fahmi, Henk Ellermann, and Gosse Bouma

Humanities Computing Department and University Library
University of Groningen
Oude Kijk in't Jatstraat 26, 9712 EK Groningen, The Netherlands
j.zhang@uva.nl

Abstract. What are the possibilities of Semantic Web technologies for organizations which traditionally have lots of structured data, such as metadata, available? A library is such a particular organization. We mapped a digital library's descriptive (bibliographic) metadata for a large historical document collection encoded in MARC21 to a historical ontology using an out-of-the-box ontology, existing topic hierarchies on the World Wide Web and other resources. We also created and explored useful relations for such an ontology. We show that mapping the metadata to an ontology adds information and makes the existing information more easily accessible for users. The paper discusses various issues that arose during the mapping process. The result of mapping metadata to RDF/OWL is a populated ontology, ready to be deployed.

1 Introduction

The Early American Imprints Series I[1] are a microfiche collection of all known existing books, pamphlets and periodical publications printed in the United States from 1639-1800, and gives insights in many aspects of life in 17th and 18th century America, and are based on Charles Evans' American Bibliography. This bibliography has been created in MARC21. Identifying and characterizing a resource and placing it in an intellectual context is expensive. The 'expensive' metadata are not fully used by the library's users, and hence the resources are not fully disclosed using the existing metadata as an extra supporting layer.

This paper will present a method to make these existing bibliographic (descriptive) metadata more easily accessible to (casual) users using Semantic Web technologies. The Semantic Web builds on information that is machine-readable and allows links to be created with relationship values [1]. Ontologies are the backbone of this idea, because these are used to organize and store information. Ontologies are built independently of a given application, and ensure that there is a common understanding of a domain (interoperability) [3].

There is related work in other domains, where thesauri such as the Arts and Architecture Thesaurus (AAT) are being used to create an ontology [8], and there

---

* Current affiliation is Archives and Information Studies, University of Amsterdam.

[1] http://library.truman.edu/microforms/early_american_imprints.htm

M. Weske, M.-S. Hacid, C. Godart (Eds.): WISE 2007 Workshops, LNCS 4832, pp. 103–114, 2007.

is more merging of datasets and vocabularies in the Cultural Heritage domain in [6]. Similar research has been conducted by [7] where a medical thesaurus called Medical Subject Headings (MeSH) and WordNet are converted to RDF/OWL.

However, these papers do not specifically deal with digital libraries' ample bibliographic metadata formatted in MARC21, which is often 'noisy', and what characteristic issues arose while aligning MARC21 metadata with schemas or vocabularies and its mapping process. Much metadata of libraries or archival institutions is encoded in this format, and other organizations with semi-structured data may face similar challenges. We also wondered whether the idea of mapping metadata, storing it in repositories using Semantic Web techniques, and advanced querying with inferencing, is feasible for our library or others, and could really be useful for people interested in the history of the United States.

This research was carried out as part of the *Semantic Web for History* (SWHi) project, of which a system description is presented in [2].[2] An objective of this project is to explore how a historical ontology can be built using library metadata, allowing libraries to push the Semantic Web forward, with real use for historians and other users. For example, the system should be able to answer questions such as:

1. When was George Washington born, and when did he die?
2. What events have occurred in the Early American History?
3. What did George Washington publish?

These questions retrieve factoid answers, and the three questions could be synthesized in this form:

4 Did George Washington publish about the events that have only occurred in his life?

On the one hand, we wonder whether the populated historical ontology can answer the first 3 questions, and use our historical ontology to infer the answer of the fourth question. On the other hand, an objective is to find out whether an ontology can offer more focused access to a specific nugget of information that captures the user's information need. The methodology of our approach is explained in Section 2. The results are presented in Section 3, where we query the ontology with a few examples. We conclude with our findings and point to future work in Section 4.

2   Methodology

2.1   Data Exploration

Our bibliographic MARC21 metadata have been created by librarians. To know what is in the data, and to know what information is needed, the MARC21[3] file

---

[2] http://semweb.ub.rug.nl/
[3] http://www.loc.gov/standards/marcxml/

was analyzed in detail. We started by encoding this file in XML. A MARC21 record encoded in XML consists of a set of datafield blocks, and each of these datafields has got numerous subfields as attributes, which makes MARC21 an expressive and rich metadata standard. The combination of a datafield and a subfield indicates the semantics of the metadata value. In this example, the datafield is *100* and the subfield code is *a*, which means the name of an author.

```
<datafield tag="100" ind1="1" ind2="">
  <subfield code="a">Gardiner, John Sylvester John,
  </subfield>
  <subfield code="d">1765-1830.              </subfield>
</datafield>
```

The metadata of the Evans bibliography (150 Mb, 36,305 records) are the single source of input for populating the ontology. There are 772,258 datafield items, classified in 35 types (using the TAG attribute). There are 1,647,280 subfield items (of which almost 40% is subfield *a*), classified in 27 different types (using the CODE attribute). If we also take the type of the datafield into account, there are 190 different combinations of datafield type/subfield type pairs. The most frequent combinations are given in Table 1.

Table 1. Snippet of the mapping table

MARC21	#	%	$\sum$	Description	Schema	Used
035 a	72610	4.41	4.41	SYS. CONTROL NO	N/A	N/A
510 c	69262	4.20	8.61	CITATION	dc:relation	Y
510 a	69262	4.20	12.82	CITATION	dc:relation	Y
500 a	68244	4.14	16.96	GENERAL NOTE	dc:description	Y
...	...	...	...	...	...	...
600 k	1	0.00	100.00	PERSONAL NAME		N

Note that the metadata consists of mappable *descriptive* (bibliographic), *structural*, and *administrative* metadata. The latter 2 are not mapped, because it does not have meaning in the historical domain of the ontology, and it is not useful for (non-librarian) users. We give a quantitative overview of the portion of metadata that was mapped, and discuss the contribution of each schema.

2.2   Reusing Existing Resources

We do not convert the MARC21 metadata directly one-to-one to RDF, which would be trivial, but try to use extra knowledge, links and descriptives provided by different resources and align them together comprehensively. A plethora of existing resources can be reused and merged to create a single historical ontology. We have decided to use an existing historical ontology as the base, and modify and enrich it with existing resources. By reusing existing knowledge structures, there is greater acceptance for the ontology. Prefixes and namespaces are essential to align schemas and map ontologies, because it indicates where an instance (individual), class (concept) or relation is derived from. It is also a matter of accountability and accuracy, which is important for a historical ontology.

The new ontology SWHi Ω can be seen as the union of different subsets or $\Omega = \{V \cup D \cup S \cup F \cup N\}$, where V stands for the VICODI core ontology[4], D for the Dublin Core predicates for describing *documents*[5], F for the FOAF predicates and classes for describing the *social networks*[6], N for the Newsbank Topic Hierarchy (taxonomy) classes for *topic classification*[7], and S for the SWHi predicates and classes. An instance i, where $i \in \Omega$, is described by combining the predicates from these different subsets.

VICODI. A history-specific ontology was built by [4], because there were no suitable existing ontologies available. The VICODI structure is intuitive, simple and allows for the uploading of instances and relations (representing historical facts) into the ontology in large numbers. Although VICODI is intended for European history, it can also be used for American history. Any shortcomings can be dealt with by modifying or enriching the ontology.[8]

Newsbank Topic Hierarchy. (NTH) We have substituted VICODI's Category hierarchy with new classes based on the taxonomical structure of the webpage of NewsBank's Readex archive of the Evans dataset. This taxonomical structure was ready made and is depicted in Fig. 1.

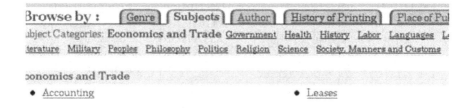

Fig. 1. Screen Caption of Taxonomical Structure of Newsbank Evans Portal

All the subclasses of the tab *Subjects* are manually extracted as the categories of the Imprints. 16 categories in total are extracted from the NTH. Each of the categories has numerous (1909) topics listed. These are saved as HTML files and semi-automatically fetched and fed as instances to each of the 16 corresponding categories. For example, the subject *Accounting* is an instance of class *Economics and Trade*. The subjects are mentioned in the documents (imprints) as topics, so each document can be related to the NTH.

---

[4] http://www.vicodi.org/about.htm

[5] http://dublincore.org

[6] http://www.foaf-project.org

[7] http://infoweb.newsbank.com

[8] The latest beta version of the SWHi ontology uses PROTON (http://proton.semanticweb.org/) as its core, but the same kind of relations as in VICODI are used as discussed in this paper.

Type of Imprints. In the metadata, each MARC21 record has the properties of an imprint. In many imprints, it is known what kind of imprint it is. This piece of code shows that an imprint can both be categorized as *Broadsides* and *Hymns* with datafield 655 and subfield a.

```
<datafield tag="655" ind1="" ind2="7">
   <subfield code="a">Broadsides.</subfield>
   <subfield code="2">rbgenr        </subfield>
</datafield>
<datafield tag="655" ind1="" ind2="7">
   <subfield code="a">Hymns.</subfield>
   <subfield code="2">rbgenr        </subfield>
</datafield>
```

This structure was not present in the VICODI ontology, so it will be automatically extended with these. 131 types of imprints are extracted from the metadata and used as classes for the ontology.

Dublin Core. (DC) It is an annotation vocabulary for metadata. There is a distinction between a qualified and unqualified (or simple) version, because the former has been intended for finer semantic distinctions and more extensibility, while the latter is simple and concise. Unqualified DC contains 15 elements. Qualified DC uses qualifiers to narrow the scope of an element, e.g. *dc:date.created* is more refined than *dc.date* alone. The 'Dumb-Down Principle' is applicable here, because values of qualified DC can always be mapped to unqualified DC. That is why we have decided to use qualified DC wherever possible.

Friend of a Friend. (FOAF) This vocabulary is used for describing social networks. It is a suitable schema, because predefined elements of persons or organizations exist in the metadata and can be mapped to RDF/OWL using the FOAF vocabulary, such as names of authors or publishers. FOAF-properties have FOAF-classes as their domain (or range), so FOAF-classes are added to the ontology in order to use FOAF properties.

2.3   Adding Class Hierarchies

- **Semi-automatic.** The classes, properties and their subclass relations are defined and stored on top of the OWL file, which is illustrated in this code.

```
<owl:Class rdf:about="http://semweb.ub.rug.nl/newsbank/Science">
   <rdfs:subClassOf rdf:resource="http://vicodi.org/ontology#Category"/>
   <rdfs:label rdf:datatype="http://www.w3.org/2001/XMLSchema#string">
     newsbank:Science</rdfs:label>
</owl:Class>
```

Science is made a subclass of *Category*. We automatically added all 131 subclasses of the *Type of Imprints* to the class *foaf:Document*.

– **Manual.** Protégé[9] arguably has become the most widely used ontology editor for the Semantic Web [5]. We have used it to define classes and their
hierarchies, relationships between classes, and properties of these relations.
The 16 NTH classes have been manually added, as previously discussed.
FOAF-classes and properties have also been added manually using Protégé.
The properties of FOAF have FOAF-classes as their domain. *foaf:Group*,
foaf:Organization, and *foaf:Person* are subclasses of *foaf:Agent*.

2.4 Properties and Relations

Schemas are used to add properties (predicates) to existing classes, and to create
relations between the classes. These properties have been manually added using
Protégé and are saved as RDF/OWL. We have created relations between classes,
besides the *subClassOf* relations in RDF. Inference is important: all classes inherit the properties (attributes or slots) of the superclass. RDF resources can
either be a literal or another resource (object). In the latter case it is of type
instance or type *class*. We have used the latter type of properties to create such
relations, and enrich the historical ontology. These are depicted in Fig. 2. The
depicted ISA-relations are equivalent to *subClassOf* relations.

1. **Time Properties** make any object in the ontology temporal:
   – *vicodi:exists*, which contains general descriptions about time,
   – *dc:coverage.temporal* is used for a document to describe the timeframe
     the imprint is covering (e.g. American Revolutionary War),
   – and *dc:date.publication* which describes when a document was published.
   The domain of *dc:coverage.temporal* is the *vicodi:Time-Dependent* class, and
   with inheritance, all its subclasses have this property as well, and it refers
   to *swhi:Ontology*, so it can take any instance in the ontology as value. It
   does not link directly to the instances of *vicodi:Time*, because this property
   should also be allowed to take as object instances of *vicodi:Event*, which is a
   different 'leaf' in the subclass hierarchy of the ontology. All other resources
   can be made temporal with *vicodi:Time-Dependent*.
2. **Agent Properties** are related to persons or institutions, and described with
   – *dc:creator*, which refers to an agent that created an object (document),
   – *dc:publisher*, which refers refers to an agent of type Publisher,
   – and *foaf:knows* which relates a person to another agent.
   It is assumed that an author expressed with *100 a* knows the names of the
   persons that he or she has covered in his publications, e.g. with the datafields
   600 and *700* in the same MARC21 record. This implicit knowledge in the
   MARC21 data is made explicit by mapping it using the *foaf:knows* property.
   Moreover, incoming and outgoing links can be detected for persons.
3. **Topic Properties** classify and cluster objects in the ontology:
   – *vicodi:hasCategory*, which makes clear that anything in the historical
     ontology can have a topic, and hence be classified with the *NTH*.

---

[9] http://protege.stanford.edu/

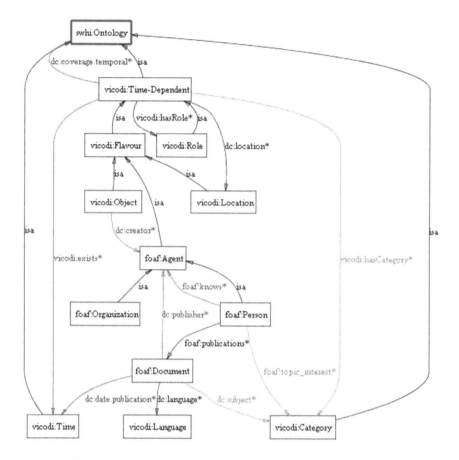

Fig. 2. Some classes and key relations in the SWHI Ontology

- *foaf:topic_interest*, which specifically refers to the topics that a person is interested in, because he or she has published about it,
- and *dc:subject* which point to the topics from the document collection.

For example, this makes it possible to retrieve all persons who are also interested in a certain topic of an imprint, which could give a user a list of possibly interesting authors as query expansion.

4. **Remaining Properties** are:

- *vicodi:hasRole*, which makes temporal objects (like persons) having roles,
- *dc:location* links an object besides time also with space,
- *foaf:publications* gives a list of publications for each author pointing to the instances of *foaf:Document* and its subclass hierarchy,
- and *dc:language* makes clear which language is used.

The *vicodi:hasRole* relation takes a class as its range, as it refers to subclasses (e.g. *vicodi:Author* and *vicodi:Publisher*) of the class *vicodi:Role*.

2.5 Automatically Populating the Ontology

This section presents what steps were needed to use the library metadata for automatically populating the ontology and the reasoning behind it.

Process and Cleanup Data. The metadata was already encoded in XML using MARC21 elements, and thus we could have used XSLT for the conversion. However, as the conversion required a substantial amount of *string processing*, *regular expressions* and *text normalization*, we opted for Perl. String processing is useful a.o. for dissecting temporal intervals and for cleaning up the noisy metadata. Instances are automatically extracted and printed as one line. Duplicates are removed by retrieving only unique lines, after removal of leading and trailing whitespace, squeezing of multiple spaces, and removal of some punctuation.

Mapping Metadata. The process of mapping the MARC21 to RDF/OWL is done by checking out Table 1 from top to bottom, which is sorted by frequency (descending). The purpose was to reuse multiple existing schemas for mapping knowledge in the metadata, in this case DC, FOAF and VICODI, as much as possible, because existing schemas are accepted by other people and widely used. Another reason is for instance that Dublin Core is not expressive enough to capture the semantics of MARC21. RDF allows us to create our project-specific SWHi schema as the fourth schema. We identified four ways to do the mapping.

1. **One-to-one mapping (1:1).** Some values in the metadata can be mapped directly as a value to a property of an instance. For example, the *topic* of an imprint is depicted as code *650 a* in the metadata. This topic is mapped 1:1 by making it an instance of category like *newsbank:Economics_and_Trade*, and depicted here in the automatically generated RDF/XML code.

   ```
   <newsbank:Economics_and_Trade rdf:ID="Accounting">
     <rdfs:comment rdf:datatype="http://www.w3.org/2001/XMLSchema#string">
       Accounting</rdfs:comment>
     <swhi:subject xml:lang="en">Accounting</swhi:subject>
   </newsbank:Economics_and_Trade>
   ```

2. **One-to-many mapping (1:m).** Sometimes, one value can be split up into multiple properties. This is the case for the instances of *Time* and the names of a person. For example, a name written in the form `<Lastname, Firstname>` with code *100 a* can be split up by mapping the `Lastname` to *foaf:surname* and the `Firstname` to *foaf:firstName*.
3. **Many-to-one mapping (m:1).** Sometimes, information contained in several subfields can be concatenated to one value of one property. Datafield *300*, for instance, describes the physical properties of an imprint. The values of these subfields can be concatenated and given as value for *dc:format*.
4. **Filtering redundant and non-descriptive knowledge.** This example shows the location *Boston* with *260 a*.

```
<datafield tag="260" ind1="" ind2="">
   <subfield code="a">[Boston] :</subfield>
   <subfield code="b">N. Coverly, Jr. printer,
      Boston.,</subfield>
   <subfield code="c">[between 1810 and 1814]
   </subfield>
</datafield>
```

However, that information has also been entered by a librarian in datafield *752* of the same record in a more informative way, i.e. it has a country, state, city combination. This means that code *260 a* is redundant in our case and does not need to be mapped. Besides redundant information, there is non-descriptive knowledge in the form of administrative metadata, which is not useful for historians or other non-librarian users.

Time and Events. Since we have a historical ontology, the method to link any object with *time* is crucial. Time can be presented in a 'discrete' and 'conceptual' view. The former is expressed in the metadata with the unit *year*, and the latter is expressed as an event. For example, *Queen Anne's War* is a conceptual expression of time, standing for the linear temporal interval of 1702–1713, which is identified as an *event* for the ontology, thus instance of *vicodi:Event*. Both temporal views are linked together. The 'discrete' temporal intervals are further disseminated into a starting and ending year as properties of the instances. Besides code *651 y* in this example, we identified other candidate instances of *vicodi:Event* with code *650 a*, where the same procedure was applied.

```
<datafield tag="651" ind1="" ind2="0">
   <subfield code="a">United States</subfield>
   <subfield code="x">History</subfield>
   <subfield code="y">Queen Anne's War, 1702-1713</subfield>
   <subfield code="v">Personal narratives.
      </subfield>
</datafield>
```

The algorithm to extract *Time* and *Events* for code *651 y* is:

Algorithm *processTime*(T)
($*$ Extracting *Time* for code *651 y* $*$)
1.   **if** T contains a question mark
2.   **then** T is instance of 'vicodi:FuzzyTemporalInterval'
3.   **else**
4.   **if** T contains the pattern 'dddd-dddd' or 'between dddd and dddd'
5.   **then** T is instance of 'vicodi:TemporalInterval'
6.   **else**
7.   **if** T has pattern 'dddd'
8.   **then** T is instance of 'swhi:Year'
9.   **else**
10.   **if** T begins with string, followed up with a number
11.   **then** T is instance of 'vicodi:Event'
12.   **else** T is instance of superclass 'swhi:Time'

Eventually, some values do not meet the condition of these rules. These are fetched with the last **else** statement as instances of superclass *swhi:Time*. This information can be further processed by making them instances of the more concrete subclasses of *swhi:Time*. Examples of the 'residue' using this algorithm:

– *in the year one thousand seven hundred and seventy-five*
– *MDCCLXXXIV.*

3   Results

Table 2 depicts the distribution of all the instances over the populated (used) classes of the SWHi ontology, including the populated subclasses, if applicable. It shows that 44,298 instances of *foaf:Document* were created, which describe the *Imprints*. There are actually 36,305 imprints in the metadata, but an imprint was also classified using multiple *types of imprints* (e.g. the topics 'Broadsides' AND 'Hymns') in the metadata.

Table 2. Number of instances for a class

Class	Instances #	%
`foaf:Document`	44298	48.60
`foaf:Organization`	26634	29.22
`foaf:Person`	10225	11.22
`vicodi:Time`	7093	7.78
`vicodi:Category`	1909	2.09
`vicodi:Location`	818	0.90
`vicodi:Event`	163	0.18
`vicodi:Language`	11	0.01
Total	**91151**	**100**

The table shows that 1909 subjects are covered by the *Imprints*, and these subjects are grouped together in 16 categories. The *document collection* is classified and clustered using this topic hierarchy. There are 163 events in our historical ontology, and many more 'discrete' *time* instances. Thousands of names have been extracted (*foaf:Person*), as well as hundreds of locations (*vicodi:Location*). The former type of instance can be regarded as a short *biography* (names, dates, topics of interests, publications, etc), whereas the latter type of instance can also be seen as a very simple *gazetteer*, since it lists combinations of a country, province (or state) and capital. Besides the 91,151 instances, there are 334 direct classes, 46 direct properties and in total 1,003,180 statements. About 50 MARC21 codes and up to 46 properties are used, and it is about 100 MB big.

The quality of the populated historical ontology is also evaluated by exploring its potential to answer user queries. We stored the ontology in Sesame[10], which

---

[10] http://www.openrdf.org/

Table 3. Results of query: *Did George Washington publish about the events that have only occurred in his life?*

#	Event
1×	"French and Indian War 1755-1763"
1×	"Washington's Expedition to the Ohio, 1st, 1753-1754"
8×	"Revolution 1775-1783"

Table 4. Query: *Who did also publish about the American Revolutionary War (1775-1783) besides George Washington?* The first and last results of 366 answers are depicted

Name of Author	Lifespan	Event
"Bancroft, Edward"	"1744-1821"	"Revolution 1775-1783"
"..."	"..."	"..."
"Mansfield, Isaac"	"1750-1826"	"Revolution 1775-1783"

is an open source framework for storage, inferencing and querying of RDF data. It was queried for a number of conceivable questions about the Early American Period and the Evans dataset in the RDF query language SPARQL using the subject-predicate-object principle[11]. The ontology allows us to answer the question posed in the beginning of this paper as case in point: *"Did George Washington publish about the events that have only occurred in his life?"*

The results of this query are shown in Table 3. It shows that in our dataset, George Washington mostly published about the American Revolutionary War (1775-1783) during his life, which is not surprising, because he was a key actor in that event. *And who did also publish about this event besides George Washington?* The ontology returns 366 results, which were retrieved in 139 ms. A subset of the relevant nuggets of information is depicted in Table 4. While we query for factual knowledge, we can link to the full texts of the imprints at any time. So we can continue providing context to the answers by traversing the RDF graphs and linking all relevant nuggets of information together with inference.

4   Summary and Future Work

Libraries increase their amount of metadata each day, but much of these metadata is not used as aid to disclose subjects. We have used these metadata to create a Semantic Web compatible "historical" ontology. An advantage for historians and other users is that these technologies will make implicit knowledge explicit, and new perspectives can be gained. By aligning schemas and vocabularies, historical objects are described with more semantics, additional (implicit) relationships can be explored.

For digital libraries, there are various potential benefits to adopting Semantic Web technology. Adding an ontological layer to metadata makes this information

---

[11] http://www.w3.org/TR/rdf-sparql-query/

much more valuable and accessible. For the development of the Semantic Web, metadata can be valuable. The metadata are carefully entered by librarians, and provides information which is hard to obtain otherwise. Existing library metadata repositories can be made useful in the (near) future in a Semantic Web context by mapping them as we have done. New ontologies can be populated using the vast amounts of already existing metadata.

The SWHi project is still in progress. We are developing semantic services based on our ontoloy, and continue to improve our ontology by populating it further with term extraction from full texts. We are also planning to evaluate our Semantic Web application with experimental empirical user studies.

References

1. Berners-Lee, T., Hendler, J., Lassila, O.: The Semantic Web. A new form of Web content that is meaningful to computers will unleash a revolution of new possibilities. Scientific American 284(5), 34–44 (2001)
2. Fahmi, I., Zhang, J., Ellermann, H., Bouma, G.: SWHi System Description: A Case Study in Information Retrieval, Inference, and Visualization in the Semantic Web. In: Franconi, E., Kifer, M., May, W. (eds.) ESWC. LNCS, vol. 4519, pp. 769–778. Springer, Heidelberg (2007)
3. McGuinness, D.L.: Ontologies Come of Age. In: Fensel, D., Hendler, J., Lieberman, H., Wahlster, W. (eds.) Spinning the Semantic Web: Bringing the World Wide Web to Its Full Potential, pp. 171–194. MIT Press, Cambridge (2005)
4. Nagypál, G., Deswarte, R., Oosthoek, J.: Applying the Semantic Web: The VICODI Experience in Creating Visual Contextualization for History. Literary and Linguistic Computing 20(3), 327–394 (2005)
5. Noy, N.: Order from chaos. Queue 3(8), 42–49 (2005)
6. Schreiber, G., Amin, A., van Assem, M., de Boer, V., Hardman, L., Hildebrand, M., Hollink, L., Huang, Z., van Kersen, J., de Niet, M., Omelayenko, B., van Ossenbruggen, J., Siebes, R., Taekema, J., Wielemaker, J., Wielinga, B.J.: Multimedian e-culture demonstrator. In: International Semantic Web Conference, pp. 951–958 (2006)
7. van Assem, M., Menken, M.R., Schreiber, G., Wielemaker, J., Wielinga, B.: A method for converting thesauri to RDF/OWL. In: McIlraith, S.A., Plexousakis, D., van Harmelen, F. (eds.) ISWC 2004. LNCS, vol. 3298, pp. 17–31. Springer, Heidelberg (2004)
8. Wielinga, B.J., Schreiber, A.T., Wielemaker, J., Sandberg, J.A.C.: From thesaurus to ontology. In: K-CAP 2001. Proceedings of the 1st international conference on Knowledge capture, pp. 194–201. ACM Press, New York (2001)

International Workshop on Governance, Risk and Compliance in Web Information Systems (GDR)

Workshop PC Chairs' Message

Shazia Sadiq[1], Claude Godart[2], and Michael zur Muehlen[3]

[1] School of Information Technology and Electrical Engineering
The University of Queensland, St Lucia QLD 4072, Brisbane, Australia
shazia@itee.uq.edu.au
[2] LORIA, Campus scientifique, BP239, 54506, Vandeuvre Cedex, France
claude.godart@loria.fr
[3] Director, SAP/IDS Scheer Center of Excellence in Business Process Innovation,
Stevens
Institute of Technology, Castle Point on Hudson, Hoboken, NJ 07030
Michael.zurMuehlen@stevens.edu

Importance of governance and associated issues of compliance and risk management are well recognized in enterprise systems. This importance has dramatically increased over the last few years as a result of recent events that led to some of the largest scandals in corporate history. Compliance related software and services is expected to reach a market value of over $ 27billion this year. At the same time, there is an increasing complexity on how to facilitate compliant business processes due to frequent and dynamic changes as well as shared processes and services executing in highly decentralized environments.

Due to the distributed nature of business operations, there is a need to develop tools, methods and techniques to support governance and compliance in web information systems (WIS). This is emerging as a critical and challenging area of research and innovation. It opens new questions e.g. modeling approaches for compliance requirements as well as challenge existing ones e.g. extension of process and service modeling and execution frameworks for compliance and risk management.

The papers presented in this workshop represent the diverse challenges of the area. They include A Conceptual model of risk: towards a risk modeling language, A Critical Analysis of Last Advances in Building Trusted P2P Networks Using Reputation Systems, Deriving XACML Policies from Business Process Models, Enforcing Policies and Guidelines in Web Portals: A Case Study, Workflows abstraction for privacy preservation, Using Control Patterns in Business Processes Compliance, A Framework for Evidence Lifecycle Management, and Collaboration for Human-Centric eGovernment Workflows.

The quality of the papers presented is indicative of the strong support provided by the program committee. We would like to thank all program committee members as well as secondary reviewers Jan Mendling, Stijn Goedertier, Marco Prandini, and Marco Montali.

M. Weske, M.-S. Hacid, C. Godart (Eds.): WISE 2007 Workshops, LNCS 4832, p. 117, 2007.
© Springer-Verlag Berlin Heidelberg 2007

Conceptual Model of Risk: Towards a Risk Modelling Language

Amadou Sienou[1], Elyes Lamine[1], Achim Karduck[2], and Hervé Pingaud[1]

[1] École des mines d'Albi-Carmaux, Centre de Génie Industriel
Campus Jarlard Route de Teillet, 81 013 Albi Cedex 09, France
{sienou,lamine,pingaud}@enstimac.fr
[2] Hochschule Furtwangen, Faculty of Computer Science
Robert-Gerwig-Platz 1, 78120 Furtwangen, Germany
karduck@hs-furtwangen.de

Abstract. Nowadays organisations are subjects to frequent changes requiring continuous strategic alignment of business processes subject to increasing compliance requirements. We suggest a holistic integration of process management and risk management supporting a robust management of business processes while improving organisation's resilience. The integration is based on a conceptual integration of risks and processes through meta-models. This paper is about a unified conceptual model of risk, which is a foundation for defining a semi-formal risk modelling language.

Keywords: risk modelling, meta-model, conceptual model, modelling language.

1 Introduction

The management of business processes is a paradigm that consists in designing, controlling and improving business processes in order to adapt to the changing business environment. It is an approach to cope with innovations, mutations of customers' expectations and increasing competition [1]. A process is a resource of value for organisations [1, 2]. It may be subject to variations due to unexpected events, which may even cause business interruption. A process may also affect other resources (internal as well as external). Consequently, practitioners of business continuity [3] and enterprise risk management [4] suggest a global approach to risk management in organisations.

There are various definitions of risk management. Consider the following: *« Risk management is a systematic approach to setting the best course of action under uncertainty by identifying, assessing, understanding, acting on and communicating risk issues. »* [5]. The authors define risk management as a recurring management practice (systematic approach) with a strong relation to decisions (setting the best course of action) which consists of a sequence of activities (identifying, assessing,…). Risk management follows therefore a lifecycle and helps decision makers to balance proactively the impact of unexpected events and the effort required to manage these events in order to have confidence in the future.

M. Weske, M.-S. Hacid, C. Godart (Eds.): WISE 2007 Workshops, LNCS 4832, pp. 118–129, 2007.
© Springer-Verlag Berlin Heidelberg 2007

Since a business process is a resource that "keeps in sync" [1] all other enterprise resources for value creation and value preservation, we believe that integrating risk management and business process management contributes to the global management of enterprise risks. This approach enables horizontal (along the value chain) as well as vertical (along decision time decomposition) risk management within organisations.

2 Approach

The integration of risk and process management is a synchronisation of the life cycles of both processes into a single coordinated process (Fig. 1).

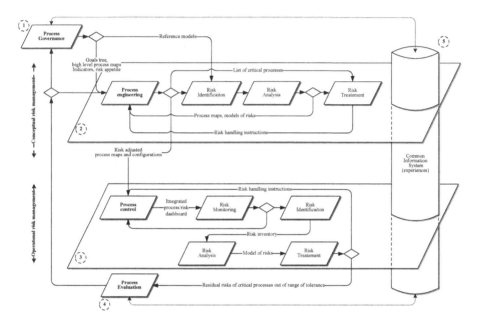

Fig. 1. Life cycle integration of risk and process management

- The first step is the fusion of the planning phases of risk and process management into a governance phase leading to the definition of a charter common to both processes.
- At the conceptual level, information is exchanged between the risk manager and the process owners using a common vocabulary.
- At the operational level, processes are monitored and adjusted by acting on control variables. Actually, a cybernetic control loop is applied [6]. Unpredicted events are reported for conceptual risk management.
- Based upon the history of the process and the history of enterprise risks, the process is evaluated in order to make continuous improvement or reengineering decisions.
- The lifecycle synchronization is based on information exchanges at different levels, which is supported by a common information system.

The even described lifecycle integration should be further detailed at different levels and phases. We suggest the integration at three levels:

1. Integration of the lifecycles of process management and risk management.
2. Integration of concepts of risks and processes into a common vocabulary supporting argumentations between stakeholders.
3. Integration of representation formalisms of risks and processes.

As supported in [22], there is a "close link" between process and risk management: process driven risk management and risk aware process models. We adopt an approach, which is exactly at the interface between both paradigms. On the one hand we integrate concepts common to both communities supporting therefore process aware risk models. On the other hand, we suggest lifecycle integration leading to a synchronized process, which is managed.

This paper handles the second and third integration levels from the perspective of risks. First, we shall define the concepts, and then propose a unified model of risk, which serves as a foundation for defining a graphical semi-formal modelling language. Finally, we propose sample applications before providing an overview of related work and concluding.

3 A Conceptual Model of Risk

Consider first the risk management process [7, 8] of Fig. 2. The process starts with a preparation step leading to the definition of the context of analysis. Next, risks are identified and reported. Then the risk analysis step follows. Here structural and quantitative models of risks are defined. Subsequently, risk handling activities are defined. The last step is a monitoring phase based on a risk dashboard. Thus, along

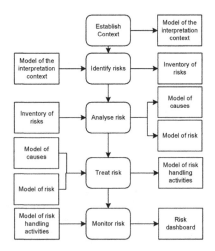

Fig. 2. Risk management - selected input/output

the risk management process, interdependent models are defined and enriched. While building models, constructs of the domain of interest and rules, which are needed, are set up in a conceptual model or meta-model. This section provides a conceptual model (Fig. 3) the instances of which are generated while managing risks.

A literature analysis and effort of modelling the main concepts of risk from various perspectives allowed the definition of the conceptual model of Fig. 3.

- Risk is the possibility of a situation (risk situation) affecting abstract or physical objects, which are of value for defined stakeholders (concern).
  - o RiskIdentity is a concept that allows the description of a risk [9].
  - o Risk is multi-stated (RiskState): possible states are latent, appeared, treated, disappeared [9] or achieved, increasing, decreasing, static [10].
  - o A RiskSituation [9, 11-16] is a cindynic situation [11] defined in three dimensions: a set of time and space, a set of actors and a set of hyperspaces of danger. A risk situation is evaluated by stakeholders.
  - o The evaluation of the effect of a risk situation on a stakeholder's concern is the impact (RiskImpact) [13, 14, 16]. Depending on the interpretation context, the impact is a positive (opportunity) or negative (threat).
  - o A CausalEvent is the event leading directly to a risk situation [9, 11, 12, 14-16]. An event is a happening with a probability. The probability of the CausalEvent is a characteristic of the risk to which it is associated.
- It is possible to identify three kinds of relations between risks: (1) abstraction relation (generalisation) to express that a risk is more/less generic than another risk. this relation supports the definition of domain specific taxonomies and model-reuse; (2) influence relation to express that the CausalEvent of a risk affects the probability of the CausalEvent of another risk; it is therefore a causality relation between causal events of risk situations; (3) As supported in [17] risks are decomposed into sub risks in order to improve management and estimation. This structure is expressed as a composition relation where risks are aggregated.
- Risk may be classified into categories (RiskCategory) according to various criteria (for example [18] people risk, process risk, relationships, technology, and external risks).
- Internal as well as external characteristics of the system under analysis may affect the probability of CausalEvent or the impact of a risk. They are RiskFactor [9, 11, 13, 14, 16].
- Once identified and evaluated risks shall be handled by defining and planning handling activities (HandlingActivity).
- Evaluating a situation requires an interpretation context. In fact, to understand a given risk, one needs a stakeholder whose concern is exposed to it and a system (entity of analysis [14]) which is analysed. Furthermore, risks are evaluated with regard to a given perspective (management, government, engineers, etc.), and a given unit of measure (time, money, health, etc.) called view. This interpretation

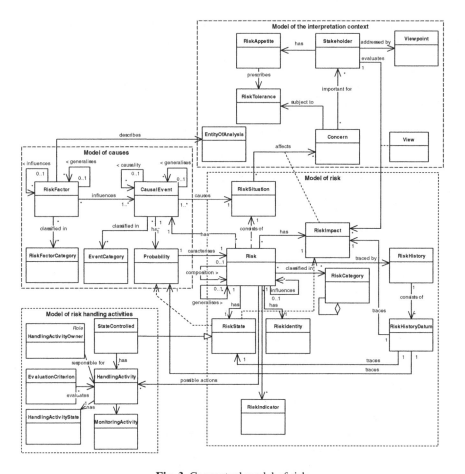

Fig. 3. Conceptual model of risk

context has three main advantages: (1) understanding the relations between risk factors and causal events; (2) the relations between risk situation and impacts, and (3) the relations between the impacts and the stakeholders.

In the practice, approaches to risk modelling range from analytics methods, simulation methods, statistical methods, structural methods. [19, 20] classifies these approaches into statistical modelling approaches and structural modelling approaches.

Statistical approaches are based on statistical observations of random variables and statistical correlations whereas structural approaches such as Bayesian networks consider explicitly cause-effect relationships between variables.

It is possible to instantiate the even proposed conceptual model by using a structural method of risk modelling such as fishbone diagram or event tree analysis; however some aspects like multi-stakeholder's views, interface between model instances along the risk management process, knowledge management, considerations to risk handling activities will cause challenges.

4 Towards a Risk Modelling Language

4.1 Motivation

As stated in the introduction, we are concerned with business process risks. These are risks related to the execution of business processes. Process risks are classified as operational risks [17]. Experiences [19, 20] show some limitations of statistical approaches with regard to these kind of risks: (1) since operational risks depend from a specific business context and are subject to high variety and dynamic, the quality of statistical data is affected by changes in business and operations affecting therefore the reliability of risk models. (2) Operational risks are driven through operational events and business decisions. Their nature is rather causality and decision based than data based. Structural approaches to risk modelling seem therefore to be adequate for operational risks.

Structural approaches such as Bayesian networks and the fishbone diagrams focus on causalities. In fact, an influence diagram (extension of Bayesian network) is a graphical notation providing information about probability dependence, and time precedence between uncertain variables, decisions and information flow [21]. Commonly used in decision analysis, this method can support the analysis of risk scenarios. In contrast to influence diagrams, fishbone diagrams support a systematic analysis of root causes of a problem. Others useful approaches to industrial events analysis such as Failure Modes and Effects Analysis (FMEA), Fault Tree Analysis (FTA), Event Tree Analysis (ETA) also focus primarily on causal analysis.

Risks are however complex structures: (1) a causal component, which is analysed by using approaches to event analysis, (2) an impact component, (3) an interpretation context, and (4) a decision (action) component. With regard to operational risks, considering the variety of process modelling languages and risk modelling techniques, we suggest a high level graphical modelling language. The later shall provide facilities for linking the components of risks and support the coupling between primitives of process modelling languages and elements of risk modelling. The following advantages are expected:

- Representing and understanding risk situations through simulations.
- Support of risk identification with knowledge management and automation.
- A single formalism and a set of modelling rules supporting communication between heterogeneous stakeholders (risk manager, process owner, and analysts). Providing a graphical representation, which is a more natural way to represent relations between risks and support the selection of risk handling activities.
- Linking models defined at different stages of risk and process management.
- Support for validation and verification of business processes with regard to risks and compliance.

4.2 Fundamental Graphical Elements

Table 1 illustrates the visual representation of the concepts defined in Fig. 3. Given our intention to couple processes and risks, an effort is made to re-used primitives of process modelling languages, mainly from Even-driven Process Chain (EPC).

Table 1. Visual representation of concepts

Concept	Representation	Description
Risk	Risk name / Proba / Impact / State	A risk is represented by an ellipse divided in two parts: (1) an upside part with the identity of the risk; (2) a downside part with the probability, the impact (value and view) and the state of the risk.
Risk situation	Risk situation	A risk situation is represented by a rounded rectangle with the identity of the situation.
Event	Event	An event is represented by a hexagon with the identity of the event. A causal event is any event that is the direct cause of a risk situation. This element is also known as event in the EPC formalism.
Risk factor	Risk factor	A risk factor is represented by a rectangle.
Stakeholder concern	Name/ Criticity/ Tolerance	A stakeholder concern or asset. This element is a variant of the objective-element in EPC.
Stakeholder	Stakeholder	This element is a variant of the stakeholder-element in EPC.
Handling activity	Handling Activity (state)	Risk handling activities are represented by a blocked arrow. Preventive risk handling activities (actions affecting the probability of risks) are represented in different colour than protective handling activities (actions affecting impacts of the risks). Risk handling activities are processes.
Category	Type / Category name	Categories are represented by a small rectangle on the top a big one. The same representation is used for risk categories, event categories and factor categories. The text in the small rectangle indicates the type of the category (risk, event, …)
Note	Document (note)	Note that may be associated to any element of a diagram.
Generic relation	————	A generic undirected relation between any elements of a diagram.
Generalisation (inheritance)	———▷	A generalization relation between risks, factors, and causal events. The direction indicates the more generic concept.

Table 1. (*continued*)

Composition	⟶◇	A composition relation between two risks. The direction indicates the composite risk, which is a composition of other risks.
Causality	⟶▸▸	Causality relation between a causal event and a risk situation. The direction indicates the risk situation.
Influence	⟶	Probabilistic influence relation: the direction indicates the affected object.
Classification	⋯⋯▷	A classification relation between an object and the categories. The direction indicates the category.
Or operator	(∨)	Or-operation between: – 2 events: simulates Event Tree Analysis. – 1 risk and 2 risk handling activities: alternative handling activities. – Risk handling activities: indicates alternative activities.
And operator	(∧)	And-operation between: – 2 events: simulates Event Tree Analysis. – 1 risk and 2 risk handling activities: AND combination of handling activities. – Risk handling activities: indicates AND combination of activities.
Exclusive Or Operator	(XOR)	XOR-operation between: – 2 events: simulates Event Tree Analysis. – 1 risk and 2 risk handling activities: XOR alternatives of handling activities. – Risk handling activities: indicates XOR combination of activities

5 Sample Application

For illustrative purpose, let us show how selected elements from Table 1 can be applied for modelling different views of risks in a simple order processing activity in a manufacturing company.

Example: *a customer sends an order, which is checked with regard to manufacturing capacity, inventory and schedule. In case these constraints are satisfied, the order is approved and the customer is informed. Otherwise a rejection notification is sent.*

(1) Establish context: to establish the interpretation context, one defines the system under analysis, identifies the stakeholders and describes the later in term of assets, which are important to them. In the following table, we illustrate the model of the system in an adapted SADT/IDEF formalism and propose a model of selected stakeholders and asset.

Table 2. Partial models of the interpretation context

Adapted SADT/IDEF model of the entity of analysis.	Model of stakeholders and interests.

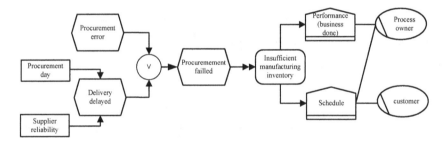

(2) Analyse risks: Suppose an inventory of potential risk has been defined. The possibility of a deficient manufacturing inventory may affect the performance of the business process and the delivery schedule of eventual order requests. The following scenario diagrams (internal view risk) illustrates how risk factors and events (internal and external) lead to risk situations and who is concerned.

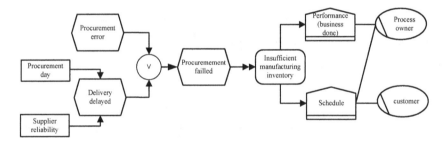

Fig. 4. Sample risk scenario diagram: risk due to lack in resources of the process

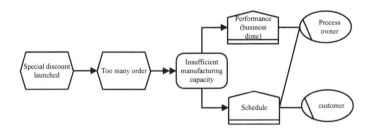

Fig. 5. Sample risk scenario diagram: risk due to lack in control factors of the process

Note that in Fig. 4 and Fig. 5 both risks affect the same assets. We illustrate this by adopting an external view of risks (represented as an ellipse) Fig. 6. In this case, it is possible to aggregate the risk into one risk as shown (aggregation relation) in Fig. 6.

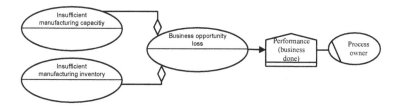

Fig. 6. Sample risk relationship diagram: risk composition

(3) Treat risks: Fig. 7 exemplifies relations between risks and handling activities. In this case, the insufficient manufacturing capacity can be handled by limiting the special discount in time and quantity. The insufficient manufacturing inventory is treated by automating the procurement process or by introducing a process to improve supplier reliability.

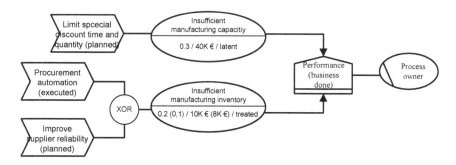

Fig. 7. Sample risk scenario handling diagram

6 Related Work

The COSO proposed a framework for Enterprise Risk Management (ERM) [4]. Approaches to Business Continuity Management [3] also address process level risks. In [22], the authors integrated risk models in process models while modelling risk as an error occurrence. An approach to the analysis of operational process risks is also handled by [23]. From the modelling point of view, in [24], the authors also proposed model based risk management approaches for security critical systems.

Commonly used in decision analysis influence diagrams and fishbone diagrams are related to our approach. In contrast to influence diagrams, which emphasise scenario analysis, fishbone diagrams and other approaches to industrial events analysis (FMEA, FTA, ETA) supports systematic analysis of root causes of problems.

We adopt a model based approach and consider risk as a complex structured concept, immerged in an environment, which may allow a positive or negative interpretation depending on the objects of interest. In addition to causalities, we consider perspectives of different stakeholders and provide a possibility to associate to each risk the corresponding risk handling activities which may be automated.

7 Conclusions and Future Work

This paper was about a conceptual model of risk and a visual modelling language for risk serving as foundation for modelling, documenting, interchanging risks, and automating risk identification, risk monitoring and the execution of handling activities.

Since relations of the language are defined in term of arc between well defined sets (input and output), the modelling language is qualified as a semi-formal language. In the future we plan a formalisation of the language, the development of a risk modelling tool and the experimentation of the method. Subsequently, we shall handle the conceptual integration of risk and business processes and consider the integration of representation formalisms.

References

1. Burlton, R.T.: Business Process Management: Profiting From Process. Sams publishing, Indianapolis (2001)
2. Hammer, M., Champy, J.: Reengineering the Corporation: A Manifesto for Business Revolution. Harper Business, New York (1993)
3. The Business Continuity Institute: Good Practice Guidelines - A Framework for Business Continuity Management. In: Smith, D.J. (ed.) The Business Continuity Institute (2005)
4. COSO: Enterprise Risk Management - Integrated Framework. Committee of Sponsoring Organizations of the Treadway Commission (2004)
5. Robillard, L.: Integrated Risk Management Framework. Treasury Board of Canada Secretariat (2001)
6. zur Muehlen, M.: Workflow-based Process Controlling. Foundation, Design, and Application of workflow-driven Process Information Systems. Logos Verlag, Berlin (2004)
7. AS/NZS: AS/NZS 4360:2004: Risk management. Australian / New Zealand Standard for Risk Management (2004)
8. Scheherazade, B.: Contribution á une démarche d'intégration des processus de gestion des risques et des projets: étude de la fonction planification. École doctorale Systémes. Ecole des Mines d'Albi-Carmaux, PhD thesis, Institut National Polytechnique de Toulouse (2003)
9. Gourc, D.: Vers un modéle général du risque pour le pilotage et la conduite des activités de biens et de services: Propositions pour une conduite des projets et une gestion des risques intégrées. Ecole des Mines d'Albi-Carmaux, HDR. Institut National Polytechnique de Toulouse (2006)
10. Office of Government Commerce: Management of Risk: Guidance for Practitioners. The Stationery Office Books (2002)
11. Kervern, G.-Y.: Éléments fondamentaux des cindyniques. Economica, Paris (1995)
12. RBDM: Risk-based Decision-making guidelines. U.S. Coast Guard, Homeland Security (1997)
13. Kontio, J.: Software Engineering Risk Management: A Method, Improvement Framework, and Empirical Evaluation. Department of Computer Science and Engineering, Laboratory of Information Processing Science, PhD Thesis, Helsinki University of Technology, Finland (2001)

14. Bernard, J.-G., Aubert, A.B., Bourdeau, S., Clément, E., Debuissy, C., Dumoulin, M.-J., Laberge, M., de Marcellis, N., Peigner, I.: Le risque: un model conceptuel d'integration. In: CIRANO: Centre interuniversitaire de recherche en analyse des organisations, Montréal (2002)
15. Stamatelatos, M.: Probabilistic Risk Assessment Procedures Guide for NASA Managers and Practitioners, Version 1.1. Office of Safety and Mission Assurance, NASA, Washington, DC, 323 (2002)
16. Alberts, C.J.: Common Elements of Risk. Acquisition Support Program. Carnegie Mellon University, Software Engineering Institute, Pittsburgh, Pennsylvania (2006)
17. Basel Committee on Banking Supervision: Operational Risk - Consultative Document. Bank for International Settlements (2001)
18. Álvarez, G.: Operational Risk Quantification: Mathematical Solutions for Analyzing Loss Data. Basel Committee on Banking Supervision, Bank for International Settlements (2001)
19. Miccolis, J., Shah, S.: Enterprise Risk Management: An Analytic Approach. Tillinghast-Towers Perrin (2000)
20. Miccolis, J., Shah, S.: RiskValueInsights: Creating Value Through Enterprise Risk Management-A Practical Approach for the Insurance Industry. Tillinghast-Towers Perrin
21. Shachter, R.D.: Evaluating Influence Diagrams. Operations Research 34, 871–882 (1996)
22. zur Muehlen, M., Rosemann, M.: Integrating Risks in Business Process Models. In: ACIS 2005. Proceedings of the 2005 Australasian Conference on Information Systems, Manly, Sydney, Australia (2005)
23. Alberts, C.J., Dorofee, A.J.: Mission Assurance Analysis Protocol (MAAP): Assessing Risk in Complex Environments. Carnegie Mellon University, Software Engineering Institute, Pittsburgh, Pennsylvania (2005)
24. Lund, M.S., Hogganvik, I., Seehusen, F., Stølen, K.: The CORAS framework, the CORAS UML profile for security assessment, and the CORAS library of reusable elements. Information Society Technologies, European Commission (2003)

A Critical Analysis of Latest Advances in Building Trusted P2P Networks Using Reputation Systems

Xavier Bonnaire and Erika Rosas

Departamento de Informática Universidad Técnica Federico Santa María - Avenida
España 1680 - Valparaíso - Chile
Xavier.Bonnaire@inf.utfsm.cl, erosas@inf.utfsm.cl

Abstract. Building trusted P2P networks is a very difficult task, due
to the size of the networks, the high volatility of the nodes, and potential byzantine behaviors. Interest in this area of research is ever increasing, because of the popularity of P2P networks, and their capability to
provide a high availability of resources and services. In the last years,
reputation systems have shown to be a good solution to build trust in
large scale networks. Many papers have been published on this subject,
and having a complete understanding on how they work is not easy. In
this paper, we make a critical analysis of the most relevant results about
reputation systems in the last three years, and propose a classification
to clearly discriminate their common advantages and specific problems.

1 Introduction

P2P networks[1] are more and more popular and will be widely used in a near
future. Building such networks uses a so called P2P overlay. The overlay provides
all the basic services needed to build the network, a name space, a message
routing function, and self-organization. The self organization is one of the most
interesting characteristic of a P2P network, as it allows the network to scale
without any central or human management. Therefore, the size of such systems
is an important aspect. It can go from thousands of nodes up to million nodes.
One of the best well known examples, is the Gnutella network that has grown
to million users.

P2P networks are divided into two main categories, structured and unstructured overlays. A structured overlay usually relies on techniques like DHTs (Distributed Hash Tables) to build a global structure into the overlay name space
(mostly a ring). Well-known structured overlays are PASTRY[2], CHORD[3],
and VICEROY[4]. This structure allows the overlay to maintain efficient routing tables and to provide powerful search algorithms (complexity in $O(\log n)$
where n is the number of nodes). On the other hand, unstructured overlays only
rely on a local structure, where a peer have a restricted view of the whole network
(a few neighbor peers). Well-known unstructured overlays are GNUTELLA[5],
or KAZAA[6]. Each category of overlay provides interesting properties like high

M. Weske, M.-S. Hacid, C. Godart (Eds.): WISE 2007 Workshops, LNCS 4832, pp. 130–141, 2007.
© Springer-Verlag Berlin Heidelberg 2007

availability of the data and services, a huge storage capacity, a very low implementation cost, and users anonymity.

Building trust in P2P networks is a very difficult task. This mainly comes from two typical characteristics of such networks:

– The number of nodes and the high dynamism of the network make very difficult to use a central set of servers to manage trustworthiness. A set of central servers will also bring some difficulties to manage accesses bottlenecks, and will potentially require costly consistency mechanisms among servers. Moreover, a set of servers is a solution that will not necessarily have a good resistance to common failures in P2P overlays like network partitions, and that will also not scale.
– The ability of a user to change its identification and the inherent need of anonymity make very difficult to use traditional authentication and trusting techniques.

The need of trust always comes from an application requirement. If we think about a music or movie sharing P2P application, a user will share a set of files, and will access to files from others users. There is no guarantee that a user will effectively provide some requested files, or that a user will provide the real files and not fake ones. Before downloading the file, we must trust the provider. Another example where we need trust is in a P2P ECommerce system where users can sell objects. The vendor can perfectly get the money without sending the merchandise, or he can perfectly send something that does not correspond to what has been bought. We also need to trust the vendor before buying from him. Thus, trust is needed when we have to rely on a user behavior that can potentially be prejudicial during a transaction.

Reputation systems have shown to be a very good way to implement trust in P2P networks [7,8,9,10,11,12,13,14]. The goal of a reputation system is to provide a way for a peer, to decide if another peer, can be trusted or not. The key idea is to maintain a reputation for each peer. A reputation can be seen as a probability for a peer to be or not to be a trusted peer. When a peer wants to make a transaction with another one, it queries the reputation system about the other peer's reputation. If it has a good reputation, the transaction will take place, and if it has not, the transaction won't take place.

In Sect. 2, we present a list of 7 criteria that will be used to evaluate existing reputation systems. Section 3 gives a critical analysis of representative systems about latest advances in the area. Section 4 is a discussion about the results of the evaluation, and Sect. 5 concludes with some hints for a next generation of reputation systems.

2   Criteria to Evaluate Existing Reputation Systems

The purpose of this paper is to define a set of qualitative criteria to evaluate reputation systems, and to propose a critical analysis of the latest advances in this research area. Each criterion is focused on traditional problems that reputation systems have to handle. This analysis will result in a classification of the reputation systems according to these criteria. The number of reputation systems presented in this survey has been reduced to 5 because of space limitation. However, they represent the latest advances in the area, including systems that have not been covered in some previous surveys[15].

Therefore, we propose to use the following seven criteria to evaluate existing reputation systems:

1. **Overlay Category:** A reputation system can be designed for a structured P2P network, or for an unstructured one. A structured P2P overlay usually makes easier the construction of a more robust reputation system because of the easy and fast data management offered by the Distributed Hash Tables (DHTs). An unstructured overlay will imply a higher cost in data management as in search algorithms.

2. **Rumors Propagation:** A rumor[16] in a reputation system consists in one or more peers that propagate false reputation information about others peers. A false negative reputation propagation about a peer P can help another peer to artificially appear as a better option than P. The benefit of false positive reputation propagation is to artificially increase the reputation of peers. A strong reputation system should be resistant to rumors propagation.

3. **Identity Changes and Usurpation:** Changing their identity, or usurping a peer's identity are common behaviors of malicious peers in a P2P network with a reputation system. Changing its identity is a good way of a peer to reappear with a clean new identity. Usurping another peer's identity is on the other hand a good and fast method to gain a good reputation and to appear to be a good choice for transactions in the network. A strong reputation system should prevent, detect or reject identity modifications and usurpation.

4. **Untrusted Recommendations:** We call an untrusted recommendation a false or partially false recommendation that a peer will provide about another peer. A peer that has just completed a successful transaction with another peer can lie, providing a negative recommendation for a peer that actually have behaved correctly. The difference with a rumor, is that a rumor can be propagated even if a peer never had any transaction with the concerned peer. A strong reputation system should detect or avoid untrusted recommendations from peers.

5. **Positive and Negative Discrimination:** A peer that has behaved correctly and has made a lot of transactions will have a very good reputation and therefore will be more interesting than other ones. A negative discrimination consists in artificially decrease the real reputation of a given peer to avoid accesses bottlenecks. Bottlenecks generated by a top rated peer will result in a poor availability, and potentially a capacity overflow of the peer for a given service. However, a peer that has made only a small number of transactions, but that has always behaved correctly, will logically have an average or low reputation. A positive discrimination consists in artificially increase

the peer's reputation to give it a chance to make more transactions, and therefore to increase its reputation. A strong reputation system should take into account some positive and negative discrimination to avoid bottleneck and peers penalization.

6. **Security Issues:** There are several security issues in a reputation system. One of the most important security problems is the integrity of the data used to manage the peer's reputations. The way of storing the reputation information is an important issue for a reputation system. Another issue can also be called cooperating attacks, where several nodes can make some joint attacks to prevent peers from being able to give recommendations, like Deny Of Service Attack (DoS), or putting down the whole reputation system with distributed rumors propagation. A peer that stores the reputation information of another one, can make a deal with this peer to give it a better reputation than its real one. A strong reputation system should try to provide mechanisms to prevent or to resist this kind of security issues.

7. **Free Riders:** Free riders are especially present in information sharing P2P networks. The behavior of a free rider is very simple, he highly download information from other peers, but never provide information to others. It has been demonstrated that in well known file sharing networks[17], an 80% of the users are free riders, and only a 2% of them provide more than 95% of the files. In networks where this is required, a strong reputation system should detect and give free riders bad reputations.

3   Critical Analysis

In this section, we make a critical analysis of several of the latest reputation systems for P2P networks. The following reputation systems are far from being an exhaustive list of all existing systems, but they are representative of the latest advances in the area. We give a short description of each system, and make a discussion according to the previous seven criteria.

3.1   PRIDE (R1)

PRIDE[8] is a reputation system which uses an elicitation storage schema to store peer's reputations. Each peer has a self digital certificate that is used to sign the peer recommendations. The certificate also contains an IP address range from where this identity can be used. An IP Based Safeguard mechanism (IBS) allows a peer to mitigate the vulnerability of peers to liar farms. A client peer obtains a list of provider peers that have the requested service or information. The client normalizes the received recommendations using the IBS mechanism and choose the provider peer that has the best reputation. The client asks the provider for recommendations it received, and recalculates the provider recommendation. After a transaction between a client and a provider has been completed, the client makes a recommendation about the provider. The recommendation can be a positive one or a negative one. The client peer digitally sign its recommendation and sends it to the provider. The provider peer updates its reputation according to the received recommendation.

1. **Overlay Category:** PRIDE is for unstructured P2P networks.
2. **Rumors Propagation:** This reputation system is not affected with rumors propagation as the peer itself stores and gives its own reputation directly to the requester. There is no indirection in the reputation request process.
3. **Identity Changes and Usurpation:** PRIDE has a pretty efficient way to control the identity changes and identity usurpation of a peer using the IBS mechanism. However this mechanism could be broken with a moderate effort, using some network addresses translation and source routing techniques.
4. **Untrusted Recommendations:** PRIDE is very concerned by the untrusted recommendations. When a provider accepts a signed recommendation from a client, the provider does not take into account the client's reputation. Therefore, the client may lie about the provider reputation sending a bad recommendation when the provider has behaved correctly. This allows a client to artificially and arbitrarily decrease the reputation of another peer. Moreover, a client can intentionally request a transaction that it does not need, to explicitly give a negative recommendation to the provider.
5. **Positive and Negative Discrimination:** PRIDE does not take into account peer discrimination. The reputation of a peer increases when it receives good recommendation from clients, and decreases when it receives bad ones. Therefore, a peer's reputation can rapidly increase to a very high value that could lead in bottleneck problems. Positive discrimination is not easily applicable in PRIDE, as the reputation value does not much relies on the number of transactions already completed by a given peer. A peer that has made only some few transactions appears to be a very good peer.
6. **Security Issues:** PRIDE does not provide any mechanism to avoid some kind of cooperative attacks. It is for example easy for some peers to cooperate to provide bad recommendations to another one. It is also easy for some peers to make a deal to provide some very good recommendations for each others. However, PRIDE uses a challenge/response mechanism to verify the peer's identity. PRIDE could also be vulnerable to DoS attacks.
7. **Free Riders:** The system does not take into account the presence of free riders. The reputation is only used to evaluate the provider and never to evaluate the client peer. Thus, this system is not very adapted to a file sharing P2P system where free riders is a major concern.

3.2 Developing Trust in Large-Scale Peer-to-Peer Systems (R2)

In this paper[14], Bin Yu, M. Singh and K. Sycara have proposed a very interesting way to manage reputation. The reputation does not use a binary evaluation of a peer, but a probabilistic one. The peer reputation is then a probability [0,1] for the peer to behave correctly. The system uses a certification and encryption mechanism (PKI) to have secured communications between peers. The key idea of this reputation system is to use the notion of witnesses and the notion of local ratings. To evaluate the reputation of a provider peer, a client peer asks to its local neighbors to search for some witnesses of the provider peer. Witnesses are peers that have already had some transactions with the provider. To compute the reputation value of the peers it uses an average metric or a exponential metric that allows to forget the past behavior of the peers, in order to have a more up to date reputation evaluation. The local rating is composed of the recommendations given by local peers about the provider. It is slightly difficult for a peer to gain reputation, but it is quite easy to lose it.

1. **Overlay Category:** This reputation system is made for unstructured P2P systems like GNUTELLA or KAZAA.
2. **Rumors Propagation:** The system is not concerned by rumors propagation. The system builds a Trust Graph for a target peer. The trust graph is compose of witnesses of the target peer, that is peers that already had a previous experience with the target peer. Thus the reputation of a provider is directly obtained from a witness that already had transactions with it.
3. **Identity Changes and Usurpation:** Even if the peers use a certification technique (PKI), it is not very clear if a peer can or cannot forge another identity to try to appear as a new clean peer in the network. This is very interesting for peers that have a low reputation, because it takes much more time to build a good one doing several correct transactions.
4. **Untrusted Recommendations:** The proposition is not concerned by untrusted recommendations, as the system assigns weight to each peers when collecting recommendations. This weight tries to minimize the effect of intentional false recommendations.
5. **Positive and Negative Discrimination:** This proposition does not takes into account the possibility to obtain a very high probability of being a trusted peer. The system is thus concerned by the accesses bottlenecks problem due to the reputation management.
6. **Security Issues:** As the peers uses PKI to have secure communications, it is quite impossible for a peer to modify some reputation data during the communications between other peers. The system is also concerned by DoS attacks or cooperative attacks where peers can make a deal to jointly increase or decrease their mutual reputations when queried.
7. **Free Riders:** The presence of free riders is not addressed by this proposition even if it targets unstructured networks usually adapted to file sharing. A provider never takes into account the reputation of the client peer before initiating a transaction.

3.3 A Reputation Management System in Structured Peer-to-Peer Networks (R3)

In this paper[11], S. Young Lee, O. Kwon, J. Kim and S. Je Hong have proposed a reputation system for the well-known CHORD structured P2P overlay. The key idea is to use the efficiency of DHT based overlays to store a peer reputation, or a file reputation, into a *manager peer*. The system computes the hash function

for the peer or for the file identifier to find which peer will be responsible of
this reputation. It uses a simple replication method to ensure the reputation
availability. The reputation of a file is stored using the hash of the file content
and not only the file identifier. The system only manage recommendations that
are issued from a direct knowledge for a peer, that is from a peer that already
had a direct transaction with a given peer.

1. **Overlay Category:** This system is clearly adapted to structured networks.
2. **Rumors Propagation:** This proposition is not concerned by the rumors
   propagation problem as there is no indirect reputation recommendation between peers. A client peer always obtain a recommendation from a peer that
   has previously had an experience with the targeted peer.
3. **Identity Changes and Usurpation:** For the file sharing point of view, the
   system is very resistant to the identity changing problem, as the computation
   of the reputation of a file also relies on the file content. Using a good hash
   coding function makes practically impossible to change the content of a
   given file, preserving the same reputation. The same occurs with a peer's
   identification where it is impossible for a peer to change it in the P2P overlay
   conserving the same result when applying the hash function on it.
4. **Untrusted Recommendations:** Even if the system has a very good resistance to the rumor propagation problem, it is concerned by untrusted
   recommendations. There is no guarantee that a peer will not provide false
   recommendation when queried by another peer. A peer can perfectly complete a correct transaction with another peer but will give it a bad recommendation.
5. **Positive and Negative Discrimination:** The system handles positive and
   negative discrimination to avoid high reputations bottlenecks, and to give
   an opportunity to peers that has only made very few transactions, but that
   have behaved correctly.
6. **Security Issues:** The system is concerned by both cooperative and DoS
   attacks. The fact to store a peer's reputation into another peer using the
   hash coding function of the overlay tends to minimize cooperatives attacks,
   but it is still possible for a peer to make deals to provide false reputations.
7. **Free Riders:** The system does not take into account the presence of free
   riders but it would be relatively easy to do so. As the reputation of a peer
   is not handle by the peer itself, it is relatively reliable for a provider to ask
   for the client reputation before initiating a transaction.

3.4 A Reputation-Based Trust Management System for P2P Networks (R4)

In this paper[10], A. Selcuk, E. Ezun and M. Pariente have proposed a reputation
system for unstructured P2P networks, based on a Trust Vector. A peer has a
fixed length binary vector for each peer it has already realized a transaction. If
the transaction has been successful, the bit has a value of 1, and 0 otherwise. For
each provider peer, a peer can compute a trusting value between 0 and 1. If a peer

does not have all the required information to compute a reputation, it can ask to another peers about their previous experience with the target provider. The querying algorithm to search for previous experiences uses a polling technique with a fixed TTL value. Moreover, for each peer that has provided information about a previous experience, the client peer keeps a credibility record. The credibility uses the same vector method as the reputation. If a recommendation from a peer appears to be correct, the client peer gives a value of 1 to the credibility of this peer, and 0 otherwise. All the communications between peers use encryption and signature techniques (PKI) to provide authentication and data integrity.

1. **Overlay Category:** The system is clearly made for unstructured P2P networks, and especially for file sharing P2P networks.
2. **Rumors Propagation:** The system is not concerned by rumors propagation as a client looks for peers that have a direct experience with a target peer.
3. **Identity Changes and Usurpation:** The system is not concerned by identity usurpation as every communications between peers uses digital certificates and encryption techniques. However, a peer can forge a new identity to appear as a clean new peer in the system.
4. **Untrusted Recommendations:** The system is using a direct reputation schema where a peer only collects reputation data from peers that already had direct transaction with the target peer. The use of credibility vectors prevents a node from doing a transaction with a peer that is lying.
5. **Positive and Negative Discrimination:** The proposition does not take into account the possibility to have a bottleneck due to a very high reputation of a given peer. Positive discrimination for peers that have few correct transactions does not exists.
6. **Security Issues:** The use of PKI for certification and encryption guarantee the integrity of the communication between peers. DoS attacks are solved using a puzzle challenge to slow down possible queries flooding. This technique can be rather efficient, but introduces a significant overhead when there is no attack. An increasing complexity of the puzzle can be used to moderate the effect during normal behaviors, but this makes the reputation system significantly more complex.
7. **Free Riders:** As the reputation control is only applied to the providers, the system accepts the presence of free riders.

3.5   PeerTrust (R5)

PeerTrust [13] is a reputation solution proposed by L. Xiong and L. Liu. made for the P-Grid environment. The key idea in PeerTrust is to evaluate the trust of a peer using 5 parameters: the other peers feedback, the number of transactions that have already been completed by a peer, the credibility of the peer that provides the reputation, the context of the actual transaction, and a community context factor to encourage peers to participate to the system.

The reputation of a peer is stored on a manager peer chosen according to its identification. The system uses a temporal floating window to identify sudden variations in a peer's behavior. The feedback allows to evaluate if a peer has effectively behaved according to its reputation. The number of transactions already made by a peer is very useful to make a distinction between peers that have already correctly completed a lot of transactions, and peers that have a very low number of past transactions. The credibility of a peer is an important aspect as it allows to know how we can trust a recommendation provided by a queried peer. Combined with the feedback, a peer can know if another one usually gives credible recommendations, or if it usually lies. The context of the actual transaction is also an important point as all applications on the overlay don't necessarily need the same level of trust and therefore may not have the same interpretation of the reputation of a peer.

1. **Overlay Category:** This proposition is made for the P-Grid system which is a structured network.
2. **Rumors Propagation:** The system is resistant to rumors propagation as the feedback used by the client peer allows to minimize the effect of intentional bad recommendations.
3. **Identity Changes and Usurpation:** As the system uses PKI for authentication of the peers and communications encryption, it is not vulnerable to identity usurpation. However, it should be possible for a peer to change its identity to appear as a clean new peer.
4. **Untrusted Recommendations:** The system is not concerned by untrusted recommendations as it uses an evaluation of the credibility of the peer which provides the recommendation. The credibility is maintained using feedback of the peer behavior after a transaction.
5. **Positive and Negative Discrimination:** The proposition does not solve the problem of peers that can produce bottlenecks because of their high reputation. A peer with a very high reputation should be negatively discriminated. On the other hand, the credibility of a peer could be used to positively discriminate new peers that have a good credibility.
6. **Security Issues:** The system provides data integrity during the communication between peers using PKI. Peers with high credibility could jointly provide some false recommendation for a given peer. The consequences for the own credibility of a peer won't be much important, if a peer does not provide more than one false recommendation.
7. **Free Riders:** The system has a good way to handle free riders. The community context factor, that forms a part of the reputation computation, encourages peers to participate in the networks. A peer with a low community factor will have a bad reputation, and peers that provide resources could reject all transactions coming from peers with low participation.

3.6   Global Evaluation

The following table shows the global evaluation. The last column, presents a global evaluation for each reputation system. The last row gives an idea of how the reputation systems globally fulfil each evaluation criterion.

R	Overlay	Rumors	Identity	U. Rec.	Discr.	Secu.	F.Riders	Global
R1	unstruct.	++	+	−	−	+	−	-
R2	unstruct.	++	+	++	−	+	−	++
R3	structured	++	++	−	++	−	−	O
R4	unstruct.	++	+	+	−	+	−	+
R5	structured	++	+	++	-	+	++	++++
Global	X	++	+	+	−	+	−	X

4   Discussion and Future Work in Reputation Systems

Even if reputation systems have shown to be a good solution to build trust in P2P networks, our previous analysis brings out that actual reputation systems still have some important deficiencies. The two criteria where existing reputation systems are strongest are the management of rumors propagation and the identity usurpation. These two criteria are very important for strong reputation systems. If we do have propagation of rumors, then it becomes very difficult to conclude about a given reputation. How can we trust a reputation if this reputation can be a false rumor? We should be able to trust the trusting system itself.

A very strange noting, is that all of the presented reputation systems that claim to work in an unstructured overlay networks do not address the problem of Free Riders. However, free riders is one of the best well known problem of unstructured P2P network like GNUTELLA. A reputation system should be able to prevent free riders from heavily use the network, assigning them a poor reputation, and using the peer's reputation not only on the client side, but also on the provider side.

Nevertheless, here are the four points where we think that a major effort must be done:

1. **Management of mixed P2P networks:** The presence of mixed P2P networks will increase in the near future. They involve fixed and Ad-Hoc mobile P2P overlays to provide a good information sharing layer for fixed computers as well as mobile ones. However, all of the previous reputation systems only focus on one of the P2P overlay category (structured or unstructured). None of them can manage mixed P2P networks where reputation maintenance is not the same when dealing with a fixed or a mobile node. Future P2P reputation systems should be able to manage mixed P2P networks to provide trust for the next generation of information sharing or ECommerce applications.
2. **Whitewashers:** The presence of whitewashers is one of the most difficult problems for a reputation system. Whitewasher are users or node that can leave the system and then enter with a new identity. This allows them to clear some potential bad reputation information in the system. Only one for the 5 reputation systems that we have analyzed tries to propose a solution to the whitewashers problem, introducing a weak IP based control for the

nodes. Nevertheless, this solution cannot handle the users identity changes. Future propositions will have to propose new solutions to this problem.

3. **Positive and Negative Discrimination:** Only a few number of reputation systems try to tackle this problem. However, this is a important issue to make efficient reputation systems. Otherwise, we can easily fall into a situation where the reputation system itself can be prejudicial to nodes that have always behaved correctly, or for new nodes that are genuine ones. Positive discrimination is particularly important to give a chance to a new node to participate to the system. It is also a good way to avoid bottlenecks due to nodes with a very high reputation, providing an alternative to them.

4. **Deny of Services:** The security issue is the most worrying aspect of actual reputation systems. Near all of them use encryption and certificates techniques for communications, but near none of them can properly support cooperative attacks where some peers make a deal to provide false information. The D.O.S. problem is without any doubt the most well known security problem for every computing system. Almost all existing reputation systems are vulnerable to DoS like attacks. In most cases, the problem comes from the P2P overlay for which it is relatively easy to build a DoS attack to prevent the overlay from being able to provide the requested information. It is particularly easy in a structured P2P network where the nodeIDs are assigned by the nodes themselves. DoS like attacks are a major concern for reputation systems, as building trust is only possible when we can trust the reputation system itself, that is, if we can assume that the reputation system is always able to give an acceptable answer to a reputation request (high availability of the reputation information). Proposing efficient solutions to the DoS like attacks will probably be one of the most difficult tasks for future reputation systems.

In addition to the previous intrinsic points, future reputation systems should become flexible. Most of the existing ones are designed for a given application, or for a given type of applications. A reputation system should be more flexible in order to be used in several types of applications. Information sharing applications and ECommerce applications do not have the same requirements. The flexibility of the system should be about the way to compute the reputation value of a node, the way to manage whitewashers or the way to handle very high reputation nodes, etc... This could allow to run several applications on the same P2P overlay, using a unique reputation system.

5   Conclusion

Reputation systems have shown to be good solutions to build trust in P2P systems. They are able to introduce an important level of trustworthiness in an environment where all actors are unknown. Nevertheless, existing reputation systems still have important deficiencies. Most of the work should be focused on the four points in Sect. 4. We especially think that an important effort has to be done on the security aspects. Using encryption and certificated to authenticate peers is a very good practice for a reputation system, but we must not forget that DoS attacks are the most common attacks on the Internet. Front of daily growing use of unstructured P2P network on the Internet, designing reputation systems with strong resistance to this kind of problem is a major need.

References

1. Aberer, K., Hauswirth, M.: An overview of peer-to-peer information systems. In: WDAS 2002, Paris, France, pp. 171–188. Carleton Scientific (2002)
2. Rowstron, A., Druschel, P.: Pastry: Scalable, decentralized object location and routing for large-scale peer-to-peer systems. In: Guerraoui, R. (ed.) Middleware 2001. LNCS, vol. 2218, pp. 329–350. Springer, Heidelberg (2001)
3. Stoica, I., Morris, R., Karger, D., Kaashoek, M., Balakrishnan, H.: Chord: A scalable peer-to-peer lookup service for internet applications. In: SIGCOMM 2001, pp. 149–160. ACM Press, New York (2001)
4. Malkhi, D., Naor, M., Ratajczak, D.: Viceroy: a scalable and dynamic emulation of the butterfly. In: PODC 2002, pp. 183–192. ACM Press, New York (2002)
5. Ripeanu, M.: Peer-to-peer architecture case study: Gnutella network. In: P2P 2001, Washington, DC, USA, p. 99. IEEE Computer Society Press, Los Alamitos (2001)
6. Liang, J., Kumar, R., Ross, K.: Understanding kazaA (2004)
7. Buchegger, S., Boudec, J.L.: A robust reputation system for p2p and mobile ad-hoc networks. In: P2PEcon-04, Cambridge MA, USA (2004)
8. Dewan, P., Dasgupta, P.: Pride: peer-to-peer reputation infrastructure for decentralized environments. In: WWW Alt. 2004, pp. 480–481. ACM Press, New York (2004)
9. Liang, Z., Shi, W.: PET: A PErsonalized trust model with reputation and risk evaluation for P2P resource sharing. In: HICSS 2005, Washington, DC, USA, p. 201. IEEE Computer Society Press, Los Alamitos (2005)
10. Selcuk, A., Uzun, E., Pariente, M.: A reputation-based trust management system for p2p networks. In: CCGRID 2004, Washington, DC, USA, pp. 251–258 (2004)
11. Lee, S., Kwon, O., Kim, J., Hong, S.: A reputation management system in structured peer-to-peer networks. In: WETICE 2005, Washington, DC, USA, pp. 362–367. IEEE Computer Society Press, Los Alamitos (2005)
12. Ravichandran, A., Yoon, J.: Trust management with delegation in grouped peer-to-peer communities. In: SACMAT 2006, pp. 71–80. ACM Press, New York (2006)
13. Xiong, L., Liu, L.: Peertrust: Supporting reputation-based trust in peer-to-peer communities. IEEE Transactions on Knowledge and Data Engineering 16(7), 843–857 (2004)
14. Yu, B., Singh, M., Sycara, K.: Developing trust in large-scale peer-to-peer systems. In: IEEE MAS&S2004, Philadelphia, Pennsylvania, USA, pp. 1–10 (2004)
15. Suryanarayana, G., Taylor, R.: A survey of trust management and resource discovery technologies in peer-to-peer applications. ISR Technical Report (2004)
16. Pittel, B.: On spreading a rumor. SIAM. J. Appl. Math. 47(1), 213–223 (1987)
17. Hughes, D., Coulson, G., Walkerdine, J.: Free riding on gnutella revisited: The bell tolls? IEEE Distributed Systems Online 6(6), 1 (2005)

Deriving XACML Policies
from Business Process Models

Christian Wolter[1], Andreas Schaad[1], and Christoph Meinel[2]

[1] SAP Research
Vincenz-Priessnitz-Str. 1, 76131 Karlsruhe, Germany
{christian.wolter,andreas.schaad}@sap.com
[2] Hasso-Plattner-Institute (HPI) for IT Systems Engineering
University of Potsdam, Germany
meinel@hpi.uni-potsdam.de

Abstract. The Business Process Modeling Notation (BPMN) has become a de-facto standard for describing processes in an accessible graphical notation. The eXtensible Access Control Markup Language (XACML) is an OASIS standard to specify and enforce platform independent access control policies.

In this paper we define a mapping between the BPMN and XACML meta-models to provide a model-driven extraction of security policies from a business process model. Specific types of organisational control and compliance policies that can be expressed in a graphical fashion at the business process modeling level can now be transformed into the corresponding task authorizations and access control policies for process-aware information systems.

As a proof of concept, we extract XACML access control policies from a security augmented banking domain business process. We present an XSLT converter that transforms modeled security constraints into XACML policies that can be deployed and enforced in a policy enforcement and decision environment. We discuss the benefits of our modeling approach and outline how XACML can support task-based compliance in business processes.

Keywords: Policy Definition, Integration, Enforcement, Separation of Duties Business Process Modeling, eXtensible Access Control Markup Language.

1   Introduction

The control and audit of activities is a fundamental principle in systems with a high degree of human interaction [1]. In the domain of information systems security and compliance, access control models are used to decide on the ways in which the availability of resources is managed and company assets are protected.

A multitude of access control models and related specification languages have emerged over the last decades. Requirements, such as "access control models must allow high level specification of access rights, thereby better managing the increased complexity [..]" [2], are representative of today's collaborative, service-integrated, and process-aware information systems. What is evident is the need for business experts to able to define their compliance requirements at a business process level, while the corresponding access control policies need to be specified and enforced at the backend- and

M. Weske, M.-S. Hacid, C. Godart (Eds.): WISE 2007 Workshops, LNCS 4832, pp. 142–153, 2007.

service-level of an enterprise information system [3]. In the domain of access control engineering Crampton *et al.* state "[..] that existing approaches to the specification of authorization constraints are unnecessarily complicated" [4]. For instance, this applies to the well defined eXtensible Access Control Markup Language [5]. This XML-based notation is very expressive, but policy definition is cumbersome and tool support is scarce. The manual definition of XACML policies is slow, error prone and may lead to policy inconsistencies and disruption of related business process execution.

We base our work on two key observations with respect to business process driven security administration and compliance management for service-based integrated enterprise environments. First, there is a set of well defined and accessible business process modeling notations used to describe enterprise service orchestration based on Web Services that are widely accepted and standardized. Second, only the security specialists specify security policies for enterprise services on a very technical level to create related security policies instead of the actual stakeholder which is the business process domain expert.

Based on these observations, as an initial step, we had enriched the semantics of the Business Process Modeling Notations (BPMN) in [6] to leverage the specification of access control security policies (e.g., separation of duty) for process-aware information systems onto the level of the process model definition itself. In this paper, we now address the issue of XACML policy specification usability and present a novel approach to derive concrete security policies, from a business process model, transform them into a dedicated policy specification language, such as RBAC-extended XACML policy sets, and deploy these policies in a policy enforcement enriched process execution environment based on Web Services. In essence, in this paper we provide:

- An evaluation of the BPMN and XACML meta-model and description of which entities must be mapped to each other in order to derive XACML security policies from BPMN models.
- A mapping to define a model transformation that extracts policy information from an XML-based business process model description to automate the generation of according XACML policies.
- A proof of concept prototype XSLT transformation script, applied on some example separation of duty constraints taken from a banking process [1].

The rest of this paper is organized as follows. The next section provides some background information about the Business Process Modeling Notation along with our security augmentation for task-based authorization. In Section 3 we discuss the underlying meta-models of BPMN and XACML and identify a mapping between entities of both models in order to define a model transformation to derive XACML policies. In Section 4 we demonstrate the feasibility of our mapping proposal by presenting an XSLT transformation applied to a given example banking process. The process is stored as an XML document and parsed by an XSLT transformer to generate XACML policies. Section 5 presents some related work in the area of policy modeling and model-based policy generation. The last section discusses our current approach and outlines future work, such as policy analysis that goes beyond syntactical validation of policy sets.

2   Background

In this section we provide some short background information about the Business Process Modeling Notation and the eXtensible Access Control Markup Language.

2.1   Business Process Modeling Notation

The Business Process Modeling Notation has emerged as a standard notation for capturing business processes, especially at the level of domain analysis and high-level systems design. The notation inherits and combines elements from a number of other proposed notations for business process modeling, including the XML Process Definition Language (XPDL) [7] and the Activity Diagrams component of the Unified Modeling Notation (UML). Figure 1 depicts an excerpt of the BPMN meta-model. For the sake of simplicity we omitted some elements that are not relevant in the course of this paper. BPMN process models are composed of flow objects such as routing gateways, events, and activity nodes. Activities, commonly referred to as tasks, represent items of work performed by software systems or humans, i.e. human activities. Activities are assigned to pools and lanes expressing organizational institutions, roles and role hierarchies. Routing gateways and events capture the flow of control between activities. Control flow elements connect activity nodes by means of a flow relation in almost arbitrary ways [8]. BPMN supports so called artifacts that enrich the process model by information entities that do not affect the underlying control flow and are a dedicated extension points to add additional information to the model. We provided an extension for security semantics in BPMN by adding an authorization constraint artifact.

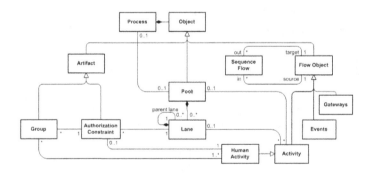

Fig. 1. BPMN Entitiy Diagram Excerpt

2.2   Constrained Business Process

The authorization constraint artifacts we proposed in [6] basically consist of two arguments n_u and m_{th} (cf. Figure 2). The first denotes the number of different users that must perform at least one activity of the set of activities indicated by a group or lane element the constraint is applied to. The latter argument defines the maximum number of activity instances of a given set of activities a single user may perform. This threshold value is necessary to restrict the number of possible task invocations of looped or

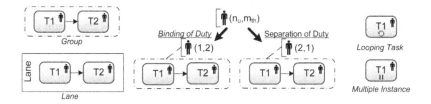

Fig. 2. Modeling of Authorization Constraints [6]

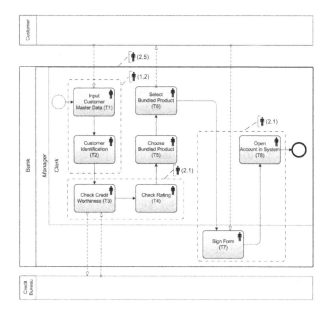

Fig. 3. Example Business Process Model

multiple-instance tasks. Crampton *et al.* defined such constraints as cardinality constraints [4]. The authorization constraint artifact can be assigned to groups, lanes, and repetitive activities.

We applied our notation to a banking process illustrated in Figure 3 and modeled security constraints for this process that are discussed in [1]. The process describes the necessary steps for opening an account for a customer. Therefore, the customer's personal data is acquired. The customer is identified and the customer's credit worthiness is checked by an external institution. Afterwards, one of several product bundles is chosen. A form is printed for the selected bundle that is signed by the customer and the bank.

A role-based authorization constraint for a role *Clerk* is expressed by assigning the set of tasks $\{T1, T2, T3, T4, T5, T6, T8\}$ to the lane labeled *Clerk*. A second role-based authorization constraint for the role *Manager* is expressed by combining the tasks of the role clerk with the task $T7$. The nesting of the lane *Clerk* within the lane *Manager* expresses a role hierarchy. We defined two separation of duty constraints for the pairs of conflicting tasks $\{T3, T4\}$ and $\{T7, T8\}$ by adding the related authorization constraints to both groups. In the same way we expressed a binding of duty constraint for tasks $\{T1, T2\}$. A last operative separation of duty constraint is directly assigned to the lane *Clerk* that restricts the executive power of a single clerk for a process instance to five activities.

2.3   eXtensible Access Control Markup Language

XACML is an OASIS standard that allows the specification of XML-based access control policies, primarily applied to the domain of Web Services. Referring to Figure 4, XACML specifies a request-response protocol and a data flow model between a service requester, a policy enforcement point (PEP), a context handler, and a policy decision point (PDP). Each access request is send to the PEP and forwarded to the context handler. The context handler creates a request context that is unique for each access request. The request context holds a snapshot of the overall system state. For instance audit log information, timestamps, or a subject's organisational context, such as user role. This information is collected from diverse backend systems, e.g. directory services, database systems, or workflow management systems. This request-dependent information is stored as context attributes and is used by the policy decision point to decide whether to grant or deny an access request. A detailed description of the data flow is given in [5].

The meta-model of XACML is shown in Figure 5. The root of all XACML policies is a *policy* or a *policyset* element. *Policysets* may hold one or more *policies* or other *policysets*. Each *policy* contains *rule* elements which are evaluated by the PDP. *Target* elements specify the context a rule applies to or not. A *target* is composed of *subject*, *resource*, and *action* elements. A *subject* element defines a human interacting with the system. A *resource* element defines a protected entity, such as a Web Service, file or process task in the context of a workflow management system. An *action* element defines the operation that is performed on the protected *resource* element. If more than one *rule* applies a rule combining algorithm defines the outcome of the overall decision request.

Condition elements further restrict the overall decision based on the contextual attributes of the access request. A *condition* contains a predicate which evaluates to either *true* or *false*, e.g. a subject's role attribute must be *clerk*. If the *condition* returns *true* the rule's *effect* is returned. The effect may result in a *permit*, *deny*, or *not applicable*

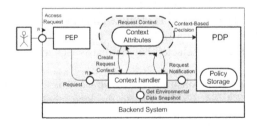

Fig. 4. XACML Overview

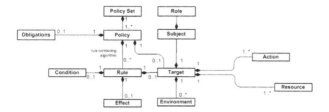

Fig. 5. XACML Entity Diagram Excerpt [5]

statement. An *obligation* is an additional activity (e.g. sending a notification email) that must be performed by the policy enforcement point in case a policy applies to a specific request and rule effect.

3   Mapping of BPMN to XACML Elements

In order to automate the extraction of security policies from process models we discuss in this section the relationship and respective mapping between selected meta-model entities of BPMN and XACML. The dotted lines in Figure 6, indicate the mappings we are discussing in the following:

XACML subject *role* attributes of the policy *target* can be derived from lanes and nested lanes respectively. A lane represents an organisational role, thus the semantic of embedded human activities is interpreted as a role-task assignment.

The BPMN elements *activity* and *human activity* are mapped to XACML *resource* elements as part of the policy *target*. In the context of a workflow management system we consider a process task as a resource. In the domain of process management an activity has several possible states that are related to human interaction, namely *ready*, *activated*, *completed*, or *canceled*. Therefore, for each task we derive three XACML *action* elements for the XACML policy *target*, each related to a state transition that can be performed on a task.

BPMN *group*, *lane*, and *authorization constraint* elements are mapped onto an XACML *condition* element of an XACML policy *rule*. A *condition* describes under which circumstances a rule applies for a matching *target* or not. This supports fine-grained access control and is essential to express separation of duty or binding of duty constraints. In case of separation of duty authorization constraints an XACML *condition* is generated, where the condition must not be met. For instance a specific task must not be performed by the same subject. Binding of duty constraints result in an XACML *condition* element that must be met. For instance, a subject must have executed a previous activity in order to perform the requested activity.

BPMN *authorization constraint* elements also map onto an XACML *obligation* element. *Obligation* elements represent meta-information for an XACML enabled policy enforcement point and contain activities that must be performed by the enforcement point depending on the outcome of the policy evaluation. For instance in case of operational separation and binding of duty constraints potential obligations for a policy enforcement point are extended audit information housekeeping in a backend system to store a subject's activity history information.

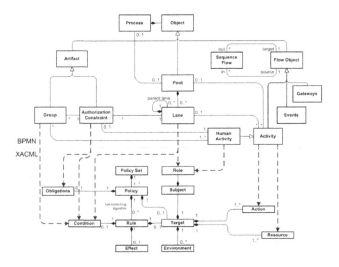

Fig. 6. Mapping of BPMN to XACML

4   Extraction of XACML Elements from BPMN

We developed an XSLT transformation script that automates the generation of XACML policies and stores them in an XML format file that can be read by an XACML based policy decision point. Basically, the script performs the following steps on a given BPMN process source stored in XML-format (cf. Figure 7):

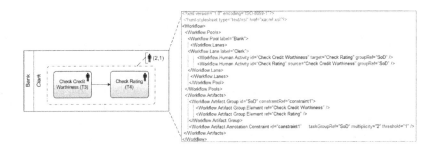

Fig. 7. BPMN Process Source

1. **Role Engineering**

   The process model is parsed for lane elements. For each lane the script generates a role-based policy set. Each human activity that is assigned to that lane results in a policy and three rules. These rules allow the actions *activate*, *complete*, and *cancel* for each human activity by a subject that holds a role attribute with the corresponding role represented by the lane. If the transformer detects a nested lane a new role-based policy set is created and its contained policies are referenced by the parent lane. According to [5] this allows to express role-hierarchies in terms of role seniority.

2. **Condition Definition**

   If the transformer detects an authorization artifact it adds a condition element to the affected policy rule. In the case of separation of duty constraints the logical function of the condition is *NOT*. For binding of duty constraints the logical function *AND* is used. This logical function will be applied to all arguments within the condition element. The authorization constraint references a set of activities either by pointing to a group artifact or a lane element.

3. **Argument Generation**

   For each human activity defined in a referenced task group or lane element an attribute function is applied to the condition element. This element takes a subject identifier, a process identifier, a set of task identifier (i.e. argument bag), and an optional threshold value as arguments. We defined an abstract *check:history* function for the XACML context handler able to query audit log information from workflow systems. Depending on the utilized system this function must be adapted. The *check:history* function returns *true* or *false* if the subject has performed any of the tasks referenced by the set of tasks in the argument bag.

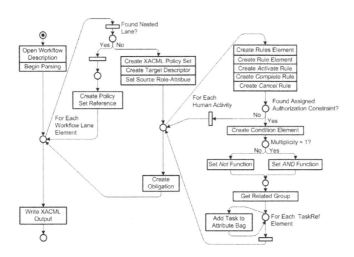

Fig. 8. Policy Generation as a Petri-Net

The described steps are implemented by an XSLT script. The overall algorithm is shown in Figure 8 as a Petri-Net. This script extracts information from a BPMN source process, such as shown in Figure 7, and stores them as XACML policies. Please note, due to the length of the overall transformation we can only provide a small script extract in this paper. Parts of the resulting XACML policy set that is generated based on our example (cf. Figure 7) is depicted in Figure 9. The extract shows a separation of duty condition between the two tasks *Check Credit Worthiness* and *Check Rating* as well as a simple task-role assignment of the task *Check Credit Worthiness* to the role *Clerk*. Regarding the syntactical complexity of a simple separation of duty policy between two tasks in XACML, it is comprehensible why XACML lacks a wide spread deployment in

```
<Condition FunctionId="urn:oasis:names:tc:xacml:1.0:function:not">
  <Apply FunctionId="urn:process:access:control:check:history">
    <SubjectAttributeDesignator AttributeId="urn:subject:id" DataType="http://www.w3.org/2001/XMLSchema#string"/>
    <EnvironmentAttributeDesignator AttributeId="urn:process:id" DataType="http://www.w3.org/2001/XMLSchema#string"/>
    <Apply FunctionId="urn:oasis:names:tc:xacml:1.0:function:string-bag">
      <AttributeValue DataType="http://www.w3.org/2001/XMLSchema#string">Check Credit Worthiness</AttributeValue>
      <AttributeValue DataType="http://www.w3.org/2001/XMLSchema#string">Check Rating</AttributeValue>
    </Apply>
  </Apply>
</Condition>
</Rule>
<Rule xmlns="urn:oasis:names:tc:xacml:2.0:polic, urn:oasis:names:tc:xacml:2.0:context" RuleId="permission:to:Check Credit
ss:complete" Effect="Permit">
  <Target>
    <Subjects>
      <Subject>
        <SubjectMatch MatchId="urn:oasis:names:tc:xacml:1.0:function:anyURI-equal">
          <AttributeValue DataType="http://www.w3.org/2001/XMLSchema#anyURI">rn:oasis:names:tc:xacml:2.0:subject:role:
Clerk</AttributeValue>
          <SubjectAttributeDesignator AttributeId="rn:oasis:names:tc:xacml:2.0:subject:role;" DataType="
w.w3.org/2001/XMLSchema#anyURI"/>
        </SubjectMatch>
      </Subject>
    </Subjects>
    <Resources>
      <Resource>
        <ResourceMatch MatchId="urn:oasis:names:tc:xacml:1.0:function:string-equal">
          <AttributeValue DataType="http://www.w3.org/2001/XMLSchema#string">Check Credit Worthiness</AttributeValue>
          <ResourceAttributeDesignator AttributeId="urn:oasis:names:tc:xacml:1.0:resource:resource-id" DataType="
w.w3.org/2001/XMLSchema#string"/>
        </ResourceMatch>
      </Resource>
    </Resources>
    <Actions>
      <Action>
```

Fig. 9. Resulting XACML Policy Set Extract

enterprise environments and how our model transformation approach to automatically derive security policies from business process models may mitigate this situation.

While this description is tailored for the BPMN meta-model, our approach is general in nature and can be applied to other process modeling notations, for instance XPDL [7] or jPDL [9].

5   Related Work

Our integration approach of BPMN and XACML on the meta-model level is related to some areas of model-driven security and general modeling of authorizations in workflow systems.

In the domain of UML-based security modeling several approaches exist, such as UMLSec [10], SecureUML [11], or a modeling methodology presented by Dobmeier and Pernul in [12]. These approaches allow to model fine-grained access control on the application level for accessing services in open systems, such as Web Services, by terms of generating code fragments from UML models, but they do not address access control in the context of workflow management systems. In SECTET [13] a novel approach for the specification of model-driven access rights for service-oriented architectures is presented. In SECTET security policies are expressed in a predictive language and are translated into platform independent XACML permissions interpreted by a security gateway. A similar approach is taken in [14] by deriving XML-based security policies from policies described in natural language. They created a set of grammars which execute on a shallow parser that are designed to identify rule elements in natural language policies. In the domain of graphical security policy specification Hoagland *et al.* developed LaSCO [15] a graph-based specification language for security policies. LaSCO policies are specified as expressions in logic and as directed graphs, giving a visual view of the policy. A LaSCO specification can be automatically translated into executable code that checks an invocation of a Java application with respect to a policy. In [16], Neumann *et al.* discussed a graphical role engineering and administration tool to define authorization constraints that can be enforced as part of RBAC context constraints.

In [17] and [18] various workflow meta models are analysed in order to evaluate the capabilities to model a workflow as a set of relating roles. Roles are defined in term of goals, qualifications, obligations, permissions, protocols, etc. Nevertheless, they did not address the problem of task-based authorization and their enforcement in a process-aware environment. In [19], Atluri *et al.* define role allocation constraints in the context of workflows and then assign one or more users to each role. They are able to express security policies, such as separation of duties, as constraints on users and roles and developed a constraint consistency analysis algorithms, but provide no translation to actually enforce their policies in an information system. In the area of model transformation for business processes Strembeck *et al.* [20] presented a transformation algorithm to extract RBAC policies from process models expressed in BPEL. They present an approach to integrate Role-Based Access Control (RBAC) and BPEL on the meta-model level and automate steps of the role engineering process. Transforming process models into a machine executable notation is also discussed in [21] by Aalst *et al.* They defined a mapping to convert the control-flow structure of a BPMN source model into BPEL, a language supported by several process execution platforms. The workflow meta-model described in [22] by Leymann and Roller also contained access control elements, but omitted a detailed discussion and extraction of enforceable security policies.

6 Conclusion

In this paper we presented an approach for the automated derivation of authorization constraints from BPMN model annotations to enforceable XACML policies. This was based on an analysis of the two corresponding meta-models and the definition of an appropriate transformation algorithm. Our work is motivated by two main observations.

First, the definition of XACML policies is overly complex, cumbersome, and time consuming. This may result in syntactical and semantical errors. The automated generation of XACML policies from a modeling notation that is more accessible by humans than directly editing XACML policies will speed up the whole policy engineering process. It allows for the direct definition of policies in the context of the underlying business process, avoiding potential inconsistencies between defined security policies and process models based on model changes.

Second, the usability and benefits of our compliance extensions for the Business Process Modeling Notation are strengthened by the mapping to XACML policies that can be directly enforced in an enterprise environment based on Web-Service orchestration and demonstrate how the complexity of XACML-based policy administration can be reduced by defining them on a more abstract level. A reduced complexity of XACML policy administration may lead to a wider roll-out of the XACML standard in enterprise information systems and overall increased general acceptance.

In future work we plan to integrate an XACML-based policy enforcement and decision point into a process execution engine that works in conjunction with an user management service. As mentioned in Section 4, we will implement the XACML context handler extension *check:history* capable of accessing audit logs and activity specific history information stored in workflow management systems along with a new XACML obligation method used for policy-based auditing. In conjunction, both will support task-based compliance for business critical processes that goes beyond traditional role-based security and access control lists. Another interesting aspect would be to apply a consistency checking algorithm, such as proposed in [4], to avoid the creation of contradicting policies either on the XACML model or the BPMN model itself. From a model transformation perspective, we also think about bi-directional model transformation approaches to enable the import of existing XACML policies into a given process model in order to support system migration scenarios.

References

1. Schaad, A., Lotz, V., Sohr, K.: A Model-checking Approach to Analysing Organisational Controls in a Loan Origination Process. In: SACMAT 2006. Proceedings of the eleventh ACM symposium on Access control models and technologies (2006)
2. Tolone, W., Ahn, G.-J., Pai, T., Hong, S.-P.: Access control in collaborative systems. ACM Comput. Surv. 37(1), 29–41 (2005)
3. Schreiter, T., Laures, G.: A Business Process-centered Approach for Modeling Enterprise Architectures. In: Proceedings of Methoden, Konzepte und Technologien für die Entwicklung von dienstebasierten Informationssystemen (EMISA) (2006)
4. Tan, K., Crampton, J., Gunter, C.: The consistency of task-based authorization constraints in workflow systems. In: CSFW 2004. Proceedings of the 17th IEEE workshop on Computer Security Foundations (2004)
5. Anderson, A.: Core and hierarchical role based access control (RBAC) profile of XACML v2.0. OASIS Standard (2005)
6. Wolter, C., Schaad, A.: Modeling of Authorization Constraints in BPMN. In: BPM 2007. Proceedings of the 5th International Conference on Business Process Management (2007)
7. The Workflow Management Coalition.: Process Definition Interface – XML Process Definition Language (2005), http://www.wfmc.org
8. Dijkman, R.M., Dumas, M., Ouyang, C.: Formal Semantics and Automated Analysis of BPMN Process Models. In: ePrints Archive (2006)
9. Red Hat Middleware.: JBoss jBPM 2.0 jPdl Reference Manual (2007), http://www.jboss.com/products/jbpm/docs/jpdl
10. Jürjens, J.: UMLsec: Extending UML for Secure Systems Development. In: UML 2002. Proceedings of the 5th International Conference on The Unified Modeling Language, pp. 412–425 (2002)
11. Basin, D., Doser, J., Lodderstedt, T.: Model Driven Security for Process-Oriented Systems. In: SACMAT 2003. Proceedings of the eighth ACM symposium on Access control models and technologies, pp. 100–109 (2003)
12. Dobmeier, W., Pernuk, G.: Modellierung von Zugiffsrichtlinien für offene Systeme. In: Tagungsband Fachgruppentreffen Entwicklungsmethoden für Informationssysteme und deren Anwendung (EMISA 2006) (2006)

13. Alam, M., Breu, R., Hafner, M.: Modeling permissions in a (u/x)ml world. In: ARES 2006. Proceedings of the First International Conference on Availability, Reliability and Security, Washington, DC, USA, pp. 685–692. IEEE Computer Society Press, Los Alamitos (2006)
14. Brodie, C.A., Karat, C.-M., Karat, J.: An empirical study of natural language parsing of privacy policy rules using the sparcle policy workbench. In: SOUPS 2006. Proceedings of the second symposium on Usable privacy and security, pp. 8–19. ACM Press, New York (2006)
15. Hoagland, J.A., Pandey, R., Levitt, K.N.: Security Policy Specification Using a Graphical Approach. In Technical report CSE-98-3. The University of California, Davis Department of Computer Science (1998)
16. Neumann, G., Strembeck, M.: An approach to engineer and enforce context constraints in an rbac environment. In: Proc. of the 8th ACM Symposium on Access Control Models and Technologies (SACMAT) (2003)
17. Muehlen, M.z.: Evaluation of workflow management systems using meta models. In: HICSS 1999. Proceedings of the Thirty-second Annual Hawaii International Conference on System Sciences, Washington, DC, USA, vol. 5, p. 5060. IEEE Computer Society Press, Los Alamitos (1999)
18. Yu, L., Schmid, B.: A conceptual framework for agent-oriented and role-based work ow modeling. In: Proc. of the 1st Int. Workshop on Agent-Oriented Information Systems (1999)
19. Bertino, E., Ferrari, E., Atluri, V.: The Specification and Enforcement of Authorization Constraints in Workflow Management Systems. ACM Transactions on Information and System Security 2, 65–104 (1999)
20. Mendling, J., Strembeck, M., Stermsek, G., Neumann, G.: An approach to extract rbac models from bpel4ws processes. In: Proceedings of the 13th IEEE International Workshops on Enabling Technologies: Infrastructures for Collaborative Enterprises (WETICE), Modena, Italy (June 2004)
21. Ouyang, C., van der Aalst, W.M.P., Marlon, D., ter Hofstede, Arthur, H.M.: Translating BPMN to BPEL. In: BPM Center Report BPM-06-02 (2006)
22. Leymann, F., Roller, D.: Production Workflow: Concepts and Techniques. Prentice Hall PTR, Upper Saddle River (2000)
23. Object Management Group.: Business Process Modeling Notation Specification (2006), www.bpmn.org

Enforcing Policies and Guidelines in Web Portals: A Case Study

Siim Karus[1] and Marlon Dumas[1,2]

[1] Institute of Computer Science, University of Tartu, Estonia
siim04@ut.ee
[2] Faculty of IT, Queensland University of Technology, Australia
m.dumas@qut.edu.au

Abstract. Customizability is generally considered a desirable feature of web portals. However, if left uncontrolled, customizability may come at the price of lack of uniformity or lack of maintainability. Indeed, as the portal content and services evolve, they can break assumptions made in the definition of customized views. Also, uncontrolled customization may lead to certain content considered important by the web portal owners (e.g. advertisements), to not be displayed to end users. Thus, web portal customization is hindered by the need to enforce customization policies and guidelines with minimal overhead. This paper presents a case study where a combination of techniques was employed to semi-automatically enforce policies and guidelines on community-built presentation components in a web portal. The study shows that a combination of automated verification and semantics extraction techniques can reduce the amount of manual checks required to enforce these policies and guidelines.

Keywords: web portal, customization, policy, guideline.

1 Introduction

Continuing advances in web technology combined with trends such as Web 2.0 are generating higher expectations for user participation and customized user experiences on the Web [4]. These heightened expectations entail additional maintenance costs for community-oriented web sites, such as web portals. A natural way for web portal owners to balance higher expectations with the imperative of keeping a manageable cost base, is to "open up the box" by allowing the community to contribute content, services and presentation components into the portal. This way, the portal owner can focus on developing and maintaining the core of the portal instead of doing so for every service and presentation component offered by the portal. Also, increased openness and community participation has the potential of promoting fidelity, by motivating end users and partner sites to continue relying on the portal once they have invested efforts into customizing it or contributing to it. On the other hand, this increased openness needs to be accompanied by a sound governance framework as well as tool support to apply this framework in a scalable manner. Indeed, a manual approach to reviewing and correcting user-contributed components would easily offset the benefits of accepting such contributions in the first place.

M. Weske, M.-S. Hacid, C. Godart (Eds.): WISE 2007 Workshops, LNCS 4832, pp. 154–165, 2007.

This paper considers the problem of allowing third parties to contribute presentation components to a web portal, while enabling portal administrators to enforce a set of policies and guidelines over these components in a scalable manner.[1] Specifically, the paper presents a case study where a team of web portal administrators needed to enforce a number of such policies and guidelines. Central to the approach adopted in this case study is a language, namely *xslt-req*, that allows portal administrators to capture the impact of policies and guidelines on the XML transformations that presentation components are allowed to perform. As a result, the portal administrators do not need to inspect and to fix every single submission in all its details; instead, most of the enforcement is done by a set of tools based on *xslt-req*. A key feature of *xslt-req* is that it builds on top of well-known web standards, specifically XML Schema and XSLT, thus lowering the barriers for its adoption.

The paper is structured as follows. Section 2 introduces the case study, including the policies and guidelines that needed to be enforced. Next, Section 3 discusses the techniques used to specify and to enforce these policies and guidelines. Related work is discussed in Section 4 while Section 5 draws conclusions.

2   Case Study: VabaVaraVeeb

VabaVaraVeeb (http://vabavara.net) is an Estonian portal for freeware.[2] A key feature of the portal is its high degree of customizability. Specifically, the portal allows third-parties to layer their own *presentation components* on top of the portal's services. Third parties may introduce custm-built presentation components for various features of the portal, such as the 'mailbox' feature, the 'user menu' feature or the 'statistics' feature. Once a third-party has registered a presentation component in the portal, they can re-direct users into the portal in such a way that users will consume VabaVaraVeeb's services through this presentation component. This allows third parties to loosely integrate services from VabaVaraVeeb with their own services at the presentation level, while enabling VabaVaraVeeb to retain some control over the delivery of its services. For example, a third-party web site that maintains a catalogue of security software, namely Securenet.ee, has added a presentation component on top of VabaVaraVeeb, to match its own presentation style. This way, users of Securenet are transferred to VabaVaraVeeb and then back to Securenet transparently, since the presentation style remains the same when moving across the two sites. The degree of customizability has been pushed to the level where the portal's services can be rendered not only through traditional HTML web pages, but also through alternative technologies such as XAML (eXtensible Application Markup Language) and XUL (XML User Interface Language). Presentation components are defined as XSL transformations [3] while the data delivered by the portal is represented in XML.

Figure 1 shows the default interface of VabaVaraVeeb while Figure 2 displays a modified interface. One can see that Figure 2 has a different main menu that appears at the top on the page (below the banner), and a repositioned right pane and search

---

[1] In this paper, the term policy refers to a rule that must be followed and for which violation can be objectively defined, while the term guideline refers to a rule that should generally be followed, but for violations can not always be objectively asserted.

[2] The first author of this paper is one of the co-founders and administrators of this portal.

Fig. 1. VabaVaraVeeb default interface

Fig. 2. Modified VabaVaraVeeb interface with new main menu and other layout changes

box. There are other less visible differences. For example, the presentation component corresponding to Figure 2 uses JavaScript to a larger extent than the default one.

Even though the portal gives significant freedom to third parties, all presentation components are required to conform to a set of *portal policies and guidelines* in order to achieve a certain level of uniformity and manageability. Originally, user-defined presentation components were manually verified for conformance against the portal policies and guidelines. However, as the portal grew, it became clear to the administrators that this manual approach would not scale up. Indeed, each contributed presentation component has to be checked against each policy and guideline, and each check is time-consuming. In addition, new versions of existing presentation components are submitted from times to times, and need to be verified again and again. This led to the need to automate the conformance verification of submitted presentation components against policies and guidelines.

The portal's policies and guidelines can be classified as follows:[3]

1. Certain content is mandatory and must always be presented to the users. Although the portal does not currently generate revenue through advertisement, it is foreseen this will happen sooner or later. Hence, it is important to ensure that paid advertisements and special announcements produced by the portal are always displayed, regardless of the presentation components in use. Some of this mandatory content must be presented to the users "as is", while other content may be presented in alternative ways. For example, promotional announcements may have different presentation requirements than other announcements.
2. Conversely, certain content must never be presented to the user. Presentation components are applied directly to the XML documents managed by the portal. Some information contained in these XML documents (e.g. user's access rights) may be sensitive or may only be needed by internal procedures. This information should therefore not be included in the generated pages.
3. Certain content must/may be delivered in certain output formats. For example, for banners we have both the URL of the banner and the URL of the advertised service. While rendering in graphical format, both URLs are marked compulsory, while in text mode the banner's URL is not marked as compulsory, however, the advertisement's alternate text is marked as compulsory.
4. Some content may need to be hidden or shown depending on the values of certain elements/attributes in the XML documents.
5. Styles should use existing or common controls when possible. This avoids duplicate code and contributes to forward compatibility [1], which is one of the design goals in VabaVaraVeeb.
6. Generic services must be preferred over internal components to expose similar aspects of objects or types of objects. Generic services are services used to perform common tasks like presenting simple dialogues or notifying about errors.
7. Complex services should be composed of individually addressable and "subscribable" services. Services built this way lower communication overhead and follow service-based approach [2] to enable forward compatibility. This service-based approach also allows portal owners to bill usage of every service

---

[3] For detailed specifications of these policies and guidelines, the reader is referred to [6].

separately (hence the requirement for "subscribability"). Users only need access to services and sub-services they are subscribed to.

8. The need for presentation components to access additional metadata to render a document should be minimized. On the other hand, the metadata already contained in the document should be used extensively. The names and values of XML elements often give valuable hints about their underlying semantics. For example, an element name with a suffix 's' in English, usually denotes multiple items, and this knowledge can be exploited to render the element's contents as a list. Also, in some cases, the XML document contains URLs and by inspecting such URLs, we can derive valuable metadata and use it for presentation purposes.

Rules 1 to 4 above correspond to policies while rules 5 to 8 represent guidelines. Guideline 7 does not relate to presentation components but to actual services provided by the portal. At present, third parties are not allowed to contribute such services, but it is foreseen that this will happen in the future, thus the guideline has been introduced and is being applied to all services internally developed by the VabaVaraVeeb team.

In principle, the conformance of a submitted presentation component against the first four policies can be automatically determined if the document structure is rigidly defined, i.e. not allowing unqualified nodes and nodes of type "any", and not allowing cyclic constructs in the document schema. Indeed, if the structure of the document is rigid, we can compute all possible source document classes that lead to different output document in terms of their structure and we can test the presentation component on sample documents representing each class. For each page generated by these tests, we can then automatically check if the policy is violated or not.

However, constantly evolving web portals can not rely on strict definitions of document structures as these definitions change too often and styles would need to be updated with every change. Due to this continuous evolution, document schemas must be designed in a forward-compatible manner by making them as loose as possible. This in turn makes the automatic enforcement of policies difficult. As explained below, we have found techniques to enforce these policies to some extent, but endless possible rulesets and document structures make it impossible to enforce in all cases.

Guideline 5 can be enforced by removing the users' ability to create custom basic controls. This may, however, result in lower performance of the solution as some simple tasks might have to be addressed using complex components. In some cases common controls are not present and have to be created beforehand.

Guideline 6 is difficult to enforce automatically as it requires detection of semantically similar code portions.

Guideline 7 is subjective as there is no metrics to decide whether or not a service should be divided into sub-services. It is still possible to use some metrics for approximation and compile-time warnings can be displayed at chosen value ranges.

By removing the ability for presentation components to access information in the portal – other than the presented document – we can easily enforce Guideline 8. However, if we enforced this guideline too strictly, we would lose forward compatibility. In Section 3.2, we discuss a technique to enforce this guideline while achieving forward compatibility, by following conventions in the naming of XML elements and exploiting these conventions to derive semantic information.

3   Defining and Enforcing Policies and Guidelines

To facilitate the enforcement of the policies and guidelines introduced above, several techniques are currently employed by VabaVaraVeeb's administration team. Central to these techniques is a language for capturing requirements over stylesheets, namely *xslt-req* [5]. Some of the guidelines are not crisply defined, so their enforcement can not be fully automated. Hence, other strategies are used to complement *xslt-req*. This section provides an overview of *xslt-req* and the techniques and strategies used for policy and guideline enforcement in VabaVaraVeeb.

3.1   The XSLT Requirements Definition Language (xslt-req)

In the context of VabaVaraVeeb, restrictions over the transformations that presentations are allowed to perform are treated as a natural extension of restrictions over the structure of documents over which these transformations are applied. Accordingly, *xslt-req* was defined as an extension of XML Schema and the syntax of these extensions is similar to XSLT. In addition to providing an integrated framework for expressing document structure and allowed transformations, this design choice has the benefit that the portal developers and third-party contributors are familiar with XML Schema and XSLT, and it is thus straightforward for them to learn *xslt-req*.

The aim of *xslt-req* is to capture allowed and required data transformations. For each element, attribute or group in a schema, *xslt-req* provides extensions to specify:

1. whether the value of the element or attribute may be used in the output directly;
2. whether the value of the element or attribute may be used in the output indirectly;
3. whether the value of the element or attribute may be ignored; and
4. whether the values of the element's children may be ignored.

xslt-req also supports the specification of conditions that determine when should these rules be applied. These conditions are captured as XPath expressions over the source document. They may also depend on the requested output format. Additionally *xslt-req* can be used to limit the set of allowed output formats. Finally, *xslt-req* supports versioning and allows developers to explicitly designate the root element of the source document and to attach default policies to the document's nodes.

As mentioned earlier, *xslt-req* was designed to be easy to learn for developers familiar with XML Schema and XSLT. All *xslt-req* directives are in XML Schema *appinfo* sections. The similarity with XSLT can be seen in the following listing.

```
1  <?xml version="1.0" encoding="UTF-8"?>
2  <xs:schema  xmlns:xs="http://www.w3.org/2001/XMLSchema"
3  xmlns:xr="xslt-req" xr:schemaLocation="xslt-req.xsd">
4    <xs:annotation>
5     <xs:appinfo>
6      <!-- Declaration of a new ruleset -->
7      <xr:xslt-requirements id="näide" version="1"
8          rootName="root" ignoreChildsDefault="false" />
9      <!-- Output formats used in transformation rules-->
10     <xr:output-format name="xhtml" method="xml"
11         namespace="http://www.w3.org/1999/xhtml" />
12     <xr:output-format name="rss" method="xml"
13       namespace="http://backend.userland.com/rss2" />
```

```
14        <xr:output-format name="text" method="text" />
15      </xs:appinfo>
16    </xs:annotation>
17    <xs:element name="root">
18      <xs:complexType>
19        <xs:sequence>
20          <xs:element name="child">
21            <xs:annotation>
22              <xs:appinfo>
23                <xr:choose>
24                  <xr:when matchOutputFormat="text|rss">
25                    <!— Transformation rules -->
26                    <xr:transformation-rules
27                  ignoreChilds="true" outputIgnore="true"
28                  outputValue="false" outputCondition="false" />
29                  </xr:when>
30                  <xr:otherwise>
31                    <xr:transformation-rules
32                  ignoreChilds="true" outputIgnore="false"
33                  outputValue="true" outputCondition="false" />
34                  </...>
```

The *xslt-requirements* element in lines 7-8 of this listing specifies the version of the *xslt-req* ruleset, its identifier, and the name of the root element. XML Schema allows multiple root elements in a document. However, *xslt-req* requires that there is a single root element. This feature makes it easier to verify stylesheets. The *ignoreChildsDefault* boolean attribute specifies that unless elsewhere overridden, presentation component may not ignore child nodes of any element. Boolean attributes *outputValueDefault*, *outputConditionDefault*, *outputIgnoreDefault* specify other default values of *xslt-req* transformation rules, the meaning of which is explained below. By default, these attributes are false except for *outputValueDefault* which is true by default. Rules for unqualified elements and attributes can be specified in the *xslt-requirements* element as well. One can specify multiple *xslt-req* rulesets in one file by using different namespace prefixes.

Lines 10-14 specify the output formats affected by the ruleset. Output formats are given a name, which is used later to specify rules specific to that output format. The element for specifying output formats also includes an attribute called *method*, corresponding to the *method* attribute of XSLT's *output* element. Optionally, the URL to the document schema can be supplied with attribute *schemaLocation*.

Lines 23 to 34 illustrate the *choose-when-otherwise* construct. The syntax of this construct is defined in Figure 3. It is similar to XSLT's *choose-when-otherwise* construct. The main difference is that the element *when* in *xslt-req* contains an output format selector attribute, namely *matchOutputFormat*. Lines 24-29 formulate rules that apply only to "rss" or "test" output formats. Meanwhile, lines 30-34 formulate rules that must be followed in other cases. Element *when* also allows using attribute *test* as selector on the source document. This is similar to how XSLT elements *when* attribute *test* is used. It is possible to have multiple *when* elements (i.e. multiple selectors). It is also possible to use *choose* or *if* elements as children of *when* or *otherwise* elements (i.e. nested conditional statements are allowed).

Lines 26-28 and lines 31-33 show examples of xslt-req transformation rules. The syntax of such rules is given in Figure 4. Attribute ignoreChilds states whether or not an element's children may be ignored. Attributes outputIgnore, outputValue and

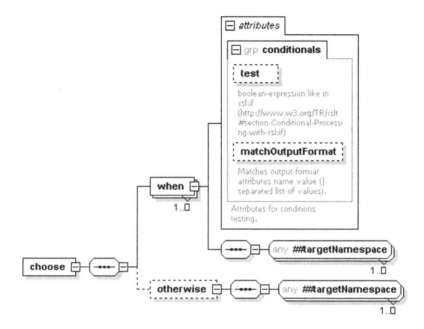

Fig. 3. Syntax of *choose-when-otherwise* construct in xslt-req

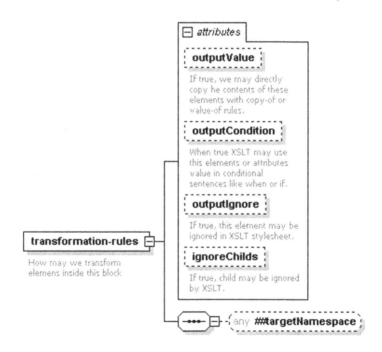

Fig. 4. Syntax of transformation-rules in of xslt-req

outputCondition state, respectively, whether or not the element or attribute may be ignored in the output, whether or not the nodes value may be included in the output (i.e. used in XSLT value-of or copy-of expressions) and whether or not the node's value may be used for formatting (i.e. used in XSLT when or if statements).

Even though the portal core is small and the structural descriptions and core services amount to less than 1% of the internal data structures, they have a significant impact on how presentations are built. This is the reason why about half the customizations made by users are to the presentation of the core services. Therefore, defining the internal data structures and using XML Schema with *xslt-req* to verify user stylesheets has significantly reduced the human workload of verifying stylesheets manually. This also allowed community-built presentation components to remain functional despite changes in the core services.

3.2 Exploiting Implicit Metadata

The workload for maintaining stylesheets was also lowered by using metadata implicitly contained in the source XML documents. In the early days of VabaVaraVeeb, the portal developers noted that the names of elements and attributes gave valuable hints for their presentation, and this was used to make the stylesheets simpler, forward-compatible and in line with the guidelines. In particular, the portal developers found they could use the suffixes of element names to detect lists of items. This enables the use of a single template for catalogue entries, search results, message lists and other list-like structures within the portal. When an element is identified to be a list, special attributes attached to this element are used to display information like the total number of items, the number of displayed items, the list name, page number (for multi-page lists) and buttons for navigating to previous and next page. The names of these special attributes tend to follow certain naming conventions. If some of the special attributes are missing, the features expressed by the missing attributes are not displayed. Using this simple detection made it possible to implement a presentation component for the search feature of one of the modules of the portal, and then reuse this presentation component for other modules. Another usage scenario for this technique is detecting modules themselves to display them in a special area of the page. Also, user input areas can be identified similarly for presentation purposes.

Initially, the portal administrators applied this technique to simplify their stylesheets and to increase reuse. After successful trials, third parties were also encouraged to use implicit metadata when contributing presentation components. This was achieved by emphasizing the transient nature of any internal structure and by supplying examples of detection rules.

This technique, in conjunction with the hard-coded requirement that all user-contributed stylesheets must have a fall-back to the default style, has made it possible to enforce policy number 8 and indirectly helped to enforce policies 5 and 6. This technique is, however, applicable only manually. It is interesting to note that users apply this technique naturally in about 80% of cases.

This technique is not error fail-proof. Rare oddities of general rules can be expressed by using templates with higher priority level than the priority level of the general templates thus ensuring that specialized templates are chosen over general ones. This kind of 'overriding' has been used in several cases to fine-tune the detection rules. For example the template for displaying lists of messages overrides the general template for rendering lists.

3.3 Additional Techniques

Additional techniques included combining similar code snippets and updating the portal's core to manage services in an addressable and subscribable way.

Similar code snippets were identified automatically using code profiling scripts that collect information on certain portions of code. The collected information included function parameters or code block input values, code block length in lines of code (with and without comments). This information was used to review code blocks that were similar according to the profiler script output. If these code blocks did turn out to have similar functionality, they were extracted and generalized functions were made to replace these code blocks. The profiler found only a small portion (less than 0.5% of lines of code) of the whole code-base to have similar code blocks. Of these, less than half were false positives. This technique is now used for all new services.

The new version of the portal's core features improved support for tracking service access. This allows one service to be responsible for tracking service usage and therefore manage subscriptions to services. As the portal core manages the addressing of portal services, services only need to handle their internal addressing and subscription to their sub-services.

3.4 Outcomes

The guidelines put into effect have reduced the complexity of the presentation layer significantly. Management of all the presentation components (i.e. styles) has become a task requiring minimal manual intervention. And because the guidelines are designed to ensure forward compatibility, styles written once can be and are often used to display new features and services without modification. This has led to shorter development effort for adding new features into the portal, since these new features tend not to break the existing styles, and when they do, the source of the problem is easier to find. Furthermore, guideline 8 has reduced the number of lines of code in the presentation layer by almost half, lowering the human effort for verifying styles.

The impact of automatic tools used to verify the styles, based on *xslt-req*, is as high as that of using code examples and suggestions on how to follow the guidelines. Also, deviations from policies which are not automatically verified are often accompanied by deviations from policies that are automatically verified. This might be caused by the different levels of experience in programming among the contributors.

It is estimated that the cost of maintaining the portal has been reduced by more than 50% thanks to the contributions from third parties, the enforcement of the policies and guidelines, and the lowered costs of enforcement.

4 Related Work

There is extensive literature dealing with the enforcement of access control policies on XML content [7]. Policy definition languages proposed in this area allow one to attach access control policies to an XML document node and its descendants. In this sense, these policy definition languages share commonalities with xslt-req. However, they differ in several respects: First, access control policy languages focus on capturing under which conditions can a given XML node be read or updated by a user. Thus, they cover cases such as the one in the second policy outlined in Section 2. In contrast, they do not allow one to capture obligations such as "a given element must be displayed" or "an element must be displayed only in certain formats", both of which are key features of xslt-req. Nevertheless, xslt-req may benefit from ideas in [7] and in similar work, to improve its ability to capture access control requirements.

There is also significant literature related to enforcing accessibility and usability guidelines on web sites. For example, Vanderdonckt & Beirekdar [8] propose the Guideline Definition Language (GDL) which supports the definition of rules composed of two parts. The *structural* part designates the HTML elements and attributes relevant to a guideline. This part is expressed in a language corresponding to a limited subset of XPath. The second part (the *evaluation logic*) is a boolean expression over the content extracted by the structural part. In contrast, xslt-req operates over XML documents representing the internal data managed by the portal, so that policies and guidelines are checked before the HTML code is generated.

The work presented in this paper is also related to the integration of services (possibly from multiple providers) at the presentation layer. This integration is supported by various portal frameworks based on standard specifications such as Java Portlets or WSRP [10]. More recently, Yu et al. [9] have proposed an event-based model for presentation components and a presentation integration "middleware" that enables the integration of services from multiple providers at the presentation layer without relying on specific platforms or APIs. However, these frameworks do not consider the enforcement of policies and guidelines as addressed in this paper.

5 Conclusion and Future Work

The paper discussed various techniques for enforcing guidelines in community-built web portals. Some of these techniques involve automatic verification of user-contributed components, others were merely suggestive, meaning that they give suggestions to the portal administrator regarding potential deadline violations, but still, the administrator has to manually verify the suggestion. The usefulness of automatic verification techniques was found to be at the same level as that of suggestive techniques. Therefore, these latter techniques should not be undervalued. Even though there are still verifications that require human intervention, much of the enforcement occurs before the components reach the portal administrators.

The automatic verification techniques can be costlier to the portals maintenance team as these techniques usually required human proofing or solving of the problems. Even if the output of automatic verification were presented directly to the authors of presentation components, (s)he might not be able to understand and solve the problems reported without help from the portals developers.

Less than half of policy and guideline deviations detected by automatic verification techniques were false positives, verification against *xslt-req* does not lead to false positives due to its design. Automatic verification was used to detect less than 30% of all identified types of policy/guideline deviations. However, this 30% of types of deviations contained the most common deviations experienced.

The techniques presented in this paper have room for improvement along several directions. For example *xslt-req* could be extended to support the specification of rules based on patterns or XPath expressions. This way, *xslt-req* could be applied to more than just the static core structure. This would make it possible to allow all business layer services to have mandatory content or hidden content. In addition to extending *xslt-req*, the verification methods that use xslt-req need to be reviewed, as they currently assume that the document base structure is rigid.

As discussed in Section 3, templates that automatically extract semantics from XML element names and values have been successfully used to achieve forward-compatible stylesheets and to enhance reuse, despite the fact that these techniques are not fail-proof. Making these techniques more robust and studying their applicability in a wider setting is an avenue for future work.

References

1. Armbruster, C.: Design for Evolution. White paper (1999), available at: http://chrisarmbruster.com/documents/D4E/witepapr.htm
2. Bennett, K., Layzell, P., Budgen, D., Brereton, P., Macaulay, L., Munro, M.: Service-Based Software: The Future for Flexible Software. In: APSEC. Proceedings of the 7th Asia-Pacific Software Engineering Conference, Singapore, December 2000, pp. 214–221. IEEE Computer Society, Los Alamitos (2000)
3. Clark, J.(ed.): XSL Transformations (XSLT), W3C Recommendation (1999) http://www.w3.org/TR/xslt
4. Jazayeri, M.: Some Trends in Web Application Development. In: FOSE 2007. Future of Software Engineering, May 2007, pp. 199–213. IEEE Computer Society, Los Alamitos (2007)
5. Karus, S.: Kasutajate poolt loodud XSL teisendustele esitavate nõuete spetsifitseerimine (Specifying Requirements for User-Created XSL Transformations). Bachelors Thesis, Faculty of Mathematics & Computer Science, University of Tartu, Estonia (in Estonian) (2005), http://math.ut.ee/ siim04/b2005/bak1.0_word2.doc
6. Karus, S.: Forward Compatible Design of Web Services Presentation Layer. Masters Thesis, Faculty of Mathematics & Computer Science, University of Tartu, Estonia (2007), http://www.cyber.ee/dokumendid/Karus.pdf/
7. Fundulaki, I., Marx, M.: Specifying Access Control Policies for XML Documents with XPath. In: Proceedings of the 9th ACM Symposium on Access Control Models and Technologies (SACMAT), Yorktown Heights, NY, USA, June 2004, pp. 61–69. ACM Press, New York (2004)
8. Vanderdonckt, J., Beirekdar, A.: Automated Web Evaluation by Guideline Review. Journal of Web Engineering 4(2), 102–117 (2005)
9. Yu, J., Benatallah, B., Saint-Paul, R., Casati, F., Daniel, F., Matera, M.: A Framework for Rapid Integration of Presentation Components. In: WWW. Proceedings of the 16th International World Wide Web Conference, Banff, Alberta, Canada, May 2007, ACM Press, New York (2007)
10. Wege, C.: Portal Server Technology. IEEE Internet Computing 6(3), 73–77 (2002)

Workflow Abstraction for Privacy Preservation

Issam Chebbi and Samir Tata

GET/INT CNRS UMR SAMOVAR
Institut National des Télécommunications,
9, Rue Charles Fourrier, 91011 Evry cedex, France
{Issam.Chebbi,Samir.Tata}@int-edu.eu
http://www-inf.int-edu.eu/~tata

Abstract. This work is in line with the *CoopFlow* approach dedicated for workflow advertisement, interconnection, and cooperation in the context of virtual enterprises. To support cooperation, one has to deal with the partners' privacy respect. In fact, cooperation needs a certain degree of inter-visibility in order to perform interactions and data exchange. Nevertheless, cooperation may be employed as a cover to internalize the know-how of partners. To preserve privacy and autonomy, one must reduce workflow inter-visibility as tiny as the cooperation needs. We present in this paper a novel reduction-based method to preserve partners' privacy. The principle of the algorithm is to get rid of all activities as well as control and data flows that don't play any direct role into cooperation and don't affect the behavior of the original workflow.

Keywords: cooperation, visibility, abstraction, inter-organizational worklows.

1 Introduction

In context of globalization, organizations are increasingly using process-aware information systems to perform automatically their business processes. Based on such systems, organizations focus on their core competencies and access other competencies through cooperation, moving towards a new form of network known as virtual organization. To support cooperation, one has to deal with the important issue which is the partners' privacy respect. In fact, cooperation needs a certain degree of inter-visibility in order to perform interactions and data exchange. Nevertheless, cooperation may be employed as a cover to internalize the know-how of partners. The question here is how to best preserve the know-how of each partner and capitalize on the accumulated experience and knowledge to allow cooperation and to improve productivity.

The study of the existing approaches shows that most of the proposed solutions does not respect the privacy of participants [7,8] often due to the limitation of a single overall process ownership. In order to preserve privacy and autonomy of workflow of participants, we propose to reduce workflow inter-visibility to be as little as the cooperations need.

M. Weske, M.-S. Hacid, C. Godart (Eds.): WISE 2007 Workshops, LNCS 4832, pp. 166–177, 2007.

To meet this objective, we have developed the *CoopFlow* approach [1,2,5] inspired by the Service-oriented Architecture that requires three operations: publish, find, and bind. Service providers publish services to a service broker. Service requesters find required services using a service broker and bind to them. Accordingly, our approach consists of three steps: workflow advertisement, interconnection, and cooperation. In fact, for building an inter-organizational workflow, each organization has to advertise, within a common registry, a description of its offered and required activities within their workflows. For workflow interconnection, each organization identifies its partners using a matching mechanism. Workflow cooperation consists of deployment and execution of an inter-organizational workflow involving different, may be heterogeneous, workflow management systems (WfMSs). In this paper we will focus on the first step of *CoopFlow*, especially on the workflows abstraction procedure to preserve the partners' privacy.

The rest of this paper is organized as follows. Section 2 summarizes our approach to inter-organizational workflow cooperation and presents a running example. Section 3 describes our novel method of workflow abstraction based on structural reduction process. Section 4 discusses related work for inter-organizational workflow cooperation. Conclusion and perspectives are presented in Section 5.

2   CoopFlow: An Approach for Workflow Cooperation

In line with our fundamental objective to support a virtual organization we have developed a novel bottom-up approach that consists in three steps: (1) advertisement of parts of organizations' workflows that could be exploited by other organizations, (2) interconnection of cooperative workflows and (3) workflow cooperation and monitoring according to cooperation policies (rules). In the following, we present this approach. Next sections detail the abstraction procedure to preserve the partners' privacy.

The approach we present here is inspired by the Service-oriented Architecture that requires three operations: publish, find, and bind. Service providers publish services to a service broker. Service requesters find required services using a service broker and bind to them. Accordingly, our approach consists of three steps: workflow advertisement, workflow interconnection, and workflow cooperation.

Step 1: Workflow Abstraction. For building an inter-organizational workflow, each partner has to advertise, using a common registry, its offered and required activities within its workflow. In order to preserve privacy and autonomy, one must reduce workflow inter-visibility to be as tiny as the cooperation needs. Therefore, we propose here to advertise an abstraction of workflow's behavior. The abstraction process is presented in section 3.

Step 2: Workflow Interconnection. To carry out a work that is not with the range of only one organization, a partner begins by searching organizations with complementary skills via the cooperative interfaces they published. Partners identification is based on an automated research of the new enterprises and potentially partners, looking for joining a virtual enterprise. Research will be based on the semantic description of information, which the enterprise requires, and the level of the cooperation that it wishes to establish (*i.e.* the profiles of the business processes to be interconnected). The various profiles published can be managed within an accessible registry on the Web. In order to construct a virtual organization, we have to match cooperative interfaces. Matching takes into account the flow control, the data flow and semantic descriptions of cooperation activities. Given two cooperative interfaces, the matching result can be (1) positive (*i.e.* interfaces match) (2) negative (*i.e.* interfaces do not match) or (3) conditional (*i.e.* interfaces match if a given condition holds). If the matching result is not negative, the cooperative interfaces are interconnected.

Step 3: Workflow Cooperation and Monitoring. The third and last step within our approach for workflow cooperation consists in the inter-organizational workflow deployment and execution. To do that, we have developed a workflow cooperation framework that allows different workflow management systems to interconnect and making their workflows cooperate. In addition to cooperation, this framework mainly enforces cooperation policies identified during the workflow interconnection step. Cooperation policies enforcement could be implemented by many styles, we can cite: (1) organizations trust each other sufficiently to interact directly with each other. Cooperation policies enforcement is completely distributed. (2) no direct trust between the organizations exist, so interactions take place through trusted third parties acting as intermediaries.

In the remaining of this paper we'll interest in the *CoopFlow*'s first step, especially we propose a workflow reduction-based abstraction procedure and demonstrate step by step how this abstraction preserve partners' privacy.

3   Workflow Abstraction

This section presents the workflows' abstraction procedure. We begin by giving some preliminary definitions useful for the comprehension of the rest of this paper. Then, we focus on the abstraction process and propose a set of reduction rules as well as an abstraction algorithm to reduce the visibiliy of partners' workflows as tiny as the cooperation needs.

3.1   Definitions

Definition 1. *A Workflow W(P,T,F) is determined by:*

- *a finite set $P=\{p_1, p_2,\ldots,p_n\}$ of places*
- *a finite set $T=\{t_1, t_2,\ldots, t_m\}$ of transitions $(P \cap T = \emptyset)$*
- *a set of arcs $F \subseteq (P \times T) \cup (T \times P)$*

The set of input (output) places for a transition t is denoted $\bullet t$ $(t\bullet)$. The set of transitions sharing a place p as output (input) place is denoted $\bullet p$ $(p\bullet)$.

The set $T = T_{coop} \cup T_{int}$ where T_{coop} is the set of cooperative activities and T_{int} is the set of internal activities.

Definition 2. *A petri net W(P, T, F) is a WF-net (Workflow net) if and only if [6]:*

1. *W has one source place $i \in P$ such that $\bullet i = \emptyset$ and one sink place $o \in P$ such that $o \bullet = \emptyset$, and*
2. *Every node $x \in P \cup T$ is on a path from i to o.*

Definition 3. *Let t_i, $t_j \in T$ and $p_{ij} \in P$. t_i and t_j are sequent and denoted $seq(t_i, t_j, p_{ij})$ if and only if : (1) $\bullet p_{ij} = \{t_i\}$, (2) $p_{ij} \bullet = \{t_j\}$, (3) $t_i \bullet = \{p_{ij}\}$ and (4) $\bullet t_j = \{p_{ij}\}$.*

Definition 4. *Let t_i, $t_j \in T$ and p_{ij}^i, $p_{ij}^o \in P$. t_i, t_j are alternative and denoted $alt(t_i, t_j, p_{ij}^i, p_{ij}^o)$ if and only if: (1) t_i, $t_j \in p_{ij}^i \bullet$, (2) t_i, $t_j \in \bullet p_{ij}^o$, (3) $\bullet t_i = \bullet t_j = \{p_{ij}^i\}$ and (4) $t_i \bullet = t_j \bullet = \{p_{ij}^o\}$.*

Definition 5. *Let t_i, t_j, t_k, $t_l \in T$ and p_{ki}, p_{il}, p_{kj}, $p_{jl} \in P$. t_i, t_j are synchronized and denoted $sync(t_i, t_j, t_k, t_l, p_{ki}, p_{il}, p_{kj}, p_{jl})$ if and only if: (1) $\bullet p_{ki} = \{t_k\}$, $p_{ki} \bullet = \{t_i\}$, (2) $\bullet p_{kj} = \{t_k\}$, $p_{kj} \bullet = \{t_j\}$, (3) $\bullet p_{il} = \{t_i\}$, $p_{il} \bullet = \{t_l\}$, (4) $\bullet p_{jl} = \{t_j\}$, $p_{jl} \bullet = \{t_l\}$, (5) $\bullet t_i = \{p_{ki}\}$, $t_i \bullet = \{p_{il}\}$, (6) $\bullet t_j = \{p_{kj}\}$, $t_j \bullet = \{p_{jl}\}$*

After having given the definitions of a workflow, a WF-Net and sequent, alternative and parallel activities, we will interest in the remainder of this section in the presentation of the set of abstraction rules allowing the reduction of a partner workflow.

3.2 Workflow Abstraction Rules

The objective of the abstraction is to preserve the privacy and to simplify workflows by removing all information that are not necessary to the cooperation description and hiding all the activities and control and data flows not playing a direct role in the cooperation nor in the initial workflow semantic retaining. Given an internal workflow W_i (P_i, T_i, F_i), the activities of W_i are reduced based on the following rules:

Sequent Activities

Rule 1 (Sequent internal activities). $\exists\ t_i$, $t_j \in T$, $p_{ij} \in P$ *where*
$P = \{p_1, \ldots, p_{ij}, \ldots, p_n\}$
$P' = P\text{-}\{p_{ij}\}$
$T = \{t_1, \ldots, t_i, t_j, \ldots, t_m\}$
$T' = T - \{t_i\}$
$F = \{f_1, \ldots, (t_i, p_{ij}), (p_{ij}, t_j), \ldots, f_k\}$
$F' = F - (\{(t_i, p_{ij}), (p_{ij}, t_j)\} \cup \{(p_k, t_i)/p_k \in \bullet t_i\}) \cup \{(p_k, t_j)/p_k \in \bullet t_i\}$
$C = seq(t_i, t_j, p_{ij})$, $t_i \in T_{int}$ et $t_j \in T_{int}$

Rule 1 shows that if we dispose of an internal activity t_i followed by another internal activity t_j and linked to it via p_{ij}, then we eliminate t_i and p_{ij} as well as all the flows between t_i and t_j. Moreover, we get rid of all the flows coming from $\bullet t_i$ to t_i and create new flows linking the places belonging to $\bullet t_i$ to t_j.

Rule 2 (Internal activity followed by a cooperative activity). $\exists\ t_i,\ t_j \in T,\ p_{ij} \in P$ *where*
$P = \{p_1, \ldots, p_{ij}, \ldots, p_n\}$
$P' = P\text{-}\{p_{ij}\}$
$T = \{t_1, \ldots, t_i, t_j, \ldots, t_m\}$
$T' = T - \{t_i\}$
$F = \{f_1, \ldots, (t_i, p_{ij}), (p_{ij}, t_j), \ldots, f_k\}$
$F' = F - (\{(t_i, p_{ij}), (p_{ij}, t_j)\} \cup \{(p_k, t_i)/p_k \in \bullet t_i\}) \cup \{(p_k, t_j)/p_k \in \bullet t_i\}$
$C = seq(t_i, t_j, p_{ij}),\ t_i \in T_{int}\ et\ t_j \in T_{coop}$

Rule 2 shows that if we dispose of an internal activity t_i followed by a cooperative activity t_j and linked to it via p_{ij}, then we eliminate t_i, p_{ij} and all the flows existing between t_i and t_j (*i.e.* (t_i, p_{ij}) and (p_{ij}, t_j)). In addition, we eliminate all the flows coming from $\bullet t_i$ to t_i and create new ones linking the places belonging to $\bullet t_i$ to t_j.

Rule 3 (A cooperative activity followed by an internal activity). $\exists\ t_i, t_j \in T,\ p_{ij} \in P$ *where*
$P = \{p_1, \ldots, p_{ij}, \ldots, p_n\}$
$P' = P\text{-}\{p_{ij}\}$
$T = \{t_1, \ldots, t_i, t_j, \ldots, t_m\}$
$T' = T - \{t_j\}$
$F = \{f_1, \ldots, (t_i, p_{ij}), (p_{ij}, t_j), \ldots, f_k\}$
$F' = F - (\{(t_i, p_{ij}), (p_{ij}, t_j)\} \cup \{(t_j, p_k)/p_k \in t_j\bullet\}) \cup \{(t_i, p_k)/p_k \in t_j\bullet\}$
$C = seq(t_i, t_j, p_{ij}),\ t_i \in T_{coop}\ et\ t_j \in T_{int}$

Rule 3 shows another case of sequent activities compaction where we dispose of a cooperative activity t_i followed by an internal one t_j and linked to it via p_{ij}. In this case, we remove the internal activity t_j, the place p_{ij} and all the flows belonging to $\bullet t_j$. Besides, we create new flows linking t_i to all the places belonging to $t_j\bullet$.

Alternative Activities

Rule 4 (Alternative internal activities). $\exists\ b_1,\ b_2,\ \ldots,\ b_n \in BS_{int},\ p_i,\ p_o \in P$ *where*
$T = \{b_1, b_2, \ldots, b_n\}$
$T' = T - \{b_2, \ldots, b_n\}$
$F = \{f_1, \ldots, (p_i, b_1), (b_1, p_o), \ldots, (p_i, b_n), (b_n, p_o), \ldots, f_k\}$
$F' = F - \{(p_i, b_2), (b_2, p_o), \ldots, (p_i, b_n), (b_n, p_o)\}$
$C = alt(b_1, b_2, \ldots, b_n, p_i, p_o)\ et\ b_1, b_2, \ldots, b_n \in BS_{int}$

In rule 4 we show that if we dispose of alternative activities $t_1, \ldots, t_n$ then we remove all activities t_i, $i \in \{2, 3, \ldots, n\}$, as well as all the flows that are directly linked to it.

Rule 5 (Internal activities alternative to blocks of activities). $\exists\, b_1, b_2,$ $\ldots, b_j \in BS$, $\exists\, b_{j+1}, \ldots b_n \in B/BS$ $p_i, p_o \in P$ *where*
$T = \{b_1, b_2, \ldots, b_m\}$
$T' = T - \{b_2, \ldots, b_j\}$
$F = \{f_1, \ldots, (p_i, b_1), (b_1, p_o), \ldots, (p_i, b_j), (b_j, p_o), \ldots, f_k\}$
$F' = F - \{(p_i, b_2), (b_2, p_o), \ldots, (p_i, b_j), (b_j, p_o)\}$
$C = alt(b_1, b_2, \ldots, b_n, p_i, p_o)$ *et* $b_1, b_2, \ldots, b_j \in BS_{int}$

Rule 5 shows that if we dispose of internal activities $b_1, \ldots, b_j$ alternative to blocks of activities $b_{j+1}, \ldots, b_n$, then we remove all internal activities b_i, $i \in \{2, 3, \ldots, j\}$ as well as all the control flows that are directly linked to them.

Rule 6 (Alternative internal and cooperative activities). $\exists\, b_1, b_2, \ldots b_j$ $\in BS_{int}$, $b_{j+1}, b_{j+2}, \ldots b_n \in BS_{coop}$, $p_i, p_o \in P$ *where*
$T = \{b_1, \ldots, b_j, b_{j+1}, \ldots, b_n\}$
$T' = T - \{b_1, b_2, \ldots, b_j\}$
$F = \{f_1, \ldots, (p_i, b_1), (b_1, p_o), \ldots, (p_i, b_j), (b_j, p_o), \ldots, f_k\}$
$F' = F - \{(p_i, b_1), (b_1, p_o), \ldots, (p_i, b_j), (b_j, p_o)\}$
$C = alt(b_1, b_2, \ldots, b_n, p_i, p_o)$ *et* $b_1, \ldots, b_j \in BS_{int}$

In Rule 6 we show that if we dispose of a set of internal activities t_i, $i \in \{1, \ldots, j\}$ alternative to a set of cooperative activities or blocks of activities t_k, $k \in \{j+1, \ldots, n\}$, then we remove all internal activities t_i, $i \in \{2, \ldots, n\}$ as well as all the control flow that are directly linked to them.

Rule 7 (A single internal activity followed by a set of alternative activities). $\exists\, b_i \in BS_{int}$, $b_j, \ldots b_k \in B$, p_i, p_j *et* $p_k \in P$ *where*
$P = \{p_1, p_2, \ldots, p_j, \ldots, p_n\}$
$P' = P - \{p_j\}$
$T = \{b_1, \ldots, b_i, \ldots, b_m\}$
$T' = T - \{b_i\}$
$F = \{f_1, \ldots, (p_i, b_i), (b_i, p_j), \ldots, f_k\}$
$F' = F - \{(p_i, b_i), (b_i, p_j)\} \cup \{(p_i, b_l),\ l \in \{j, \ldots, k\}\}$
$C = alt(b_j, \ldots, b_k, p_j, p_k)$

In Rule 7 we show that if we dispose of an internal activity followed by a set of alternative activities, then we remove the internal activity as well as the control flows that are directly linked to it.

Rule 8 (Alternative activities followed by a single internal activity). $\exists\, b_i \in BS_{int}$, $b_j, \ldots b_k \in B$, p_i, p_j *et* $p_k \in P$ *where*
$P = \{p_1, p_2, \ldots, p_k, \ldots, p_n\}$
$P' = P - \{p_j\}$
$T = \{\ldots, b_i, \ldots, b_m\}$

$T' = T - \{b_i\}$
$F = \{f_1, \ldots, (b_l, p_j), (p_j, b_i), (b_i, p_k), \ldots, f_k, \ l \in \{j, \ldots, k\}\}$
$F' = F - \{(p_j, b_i), (b_i, p_k), (b_l, p_j), \ l \in \{j, \ldots, k\}\} \cup \{(b_s, p_k), \ s \in \{j, \ldots, k\}\}$
$C = alt(b_j, \ldots, b_k, p_i, p_j)$

Rule 8 illsustrates the case of a set of alternative activities followed by a single internal activity. In this situation we remove the internal activitiy as well as all the control flows that are directly linked ti it.

Rule 9 (Renaming of internal activities). $\exists \ b_1 \in B_{int}, \ b_2, \ldots b_n \in B, \ p_i,$ $p_o \in P \ where$
$T = \{b_1, \ldots, b_n\}$
$T' = (T - b_1) \cup INTERNAL$
$F = \{f_1, \ldots, (p_i, b_1), (b_1, p_o), \ldots, (p_i, b_n), (b_n, p_o), \ldots, f_k\}$
$F' = (F - \{(p_i, b_1), \ (b_1, p_o)\}) \cup \{(p_i, INTERNAL), \ (INTERNAL, p_o)\}$
$C = alt(b_1, b_2, \ldots, b_n, p_i, p_o)$

In Rule 9, we show that if we dispose of an internal activity alternative to a set of cooperative activities, then it is renamed to *INTERNAL*. The objective of the transformation is to hide the activity details without affecting the semantic of the workflow.

Parallel Activities

Rule 10 (Parallel internal activities). $\exists \ b_0, \ b_1, \ldots, \ b_{n-1}, \ b_n \in BS \ et \ p_{0i},$ $p_{i(n+1)} \in P, \ i \in \{1, \ 2, \ldots, \ n\} \ where$
$P = \{p_1, \ldots, p_{ij}, \ldots, p_n\}, \ j \in \{1, \ 2, \ldots, \ 2k\}$
$P' = P - \{p_{ij}\}, \ j \in \{2, \ldots, \ n-1\}$
$T = \{b_0, \ldots, b_{n-1}, \ldots, b_m\}$
$T' = T - \{b_i\}, \ i \in \{2, \ldots, \ n-1\}$
$F = \{f_1, \ldots, (b_0, p_{ij}), (p_{ij}, b_j), (b_j, p_{i(k+j)}), (p_{i(k+j)}, b_n), \ldots, f_k\}, \ j \in \{1, \ldots, \ k\}$
$F' = F - \{(b_0, p_{ij}), (p_{ij}, b_i), (b_i, p_{i(k+j)}), (p_{i(k+j)}, b_n)\}, \ j \in \{2, \ldots, \ k\}$
$C = sync(b_l, \ldots, b_n, p_{ij}) \ et \ b_l \in BS_{int}, \ l \in \{1, \ldots, \ k\}, \ j \in \{1, \ldots, \ 2k\}$

In Rule 10 we show that if we dispose of a set of parallal activities, then we retain only one activity and we remove the others.

Rule 11 (Internal activities parallal to blocks of activities). $\exists \ b_0, \ b_1,$ $\ldots, \ b_{n-1}, \ b_n \in B \ et \ p_{il} \in P, \ l \in \{1, \ 2, \ldots, \ 2k\} \ where$
$P = \{p_1, \ldots, p_{ij}, \ldots, p_n\}, \ j \in \{1, \ 2, \ldots, \ 2k\}$
$P' = P - \{p_{il}\}, \ l \in \{1, \ldots, \ j, \ k+1, \ldots, \ k+j+1\}$
$T = \{\ldots, b_1, \ldots, b_j, \ldots, b_m\}$
$T' = T - \{b_l\}, \ l \in \{1, \ldots, \ j\}$
$F = \{f_1, \ldots, (b_0, p_{il}), (p_{il}, b_l), (b_l, p_{i(k+l)}), (p_{i(k+l)}, b_n), \ldots, f_k\}, \ l \in \{1, \ldots, \ j\}$
$F' = F - \{(b_0, p_{il}), (p_{il}, b_l), (b_l, p_{i(k+l)}), (p_{i(k+l)}, b_n)\}, \ l \in \{1, \ldots, \ j\}$
$C = sync(b_0, \ldots, b_n, p_{il}) \ et \ b_x \in BS_{int}, \ l \in \{1, \ldots, \ 2k\}, \ x \in \{1, \ldots, \ j\}, \ b_y \in$
B/BS

In Rule 11, we show that if we dispose of a set of internal activities parallal to blocks of activities, then we remove all the internal activities.

Rule 12 (Parallel internal and cooperative activities). $\exists\ t_0,\ t_1,\ \ldots,\ t_n,$ $t_{n+1} \in T \cup B$ et p_{0i}, $p_{i(n+1)} \in P$, $i \in \{1,\ 2,\ \ldots,\ n\}$ where

$P = \{p_1, \ldots, p_{0j}, p_{j(n+1)}, \ldots, p_n\}$, $j \in \{1,\ 2,\ \ldots,\ n\}$
$P' = P\text{-}\{p_{0j}, p_{j(n+1)}\}$, $i \in \{1,\ \ldots,\ i\}$
$T = \{t_0, \ldots, t_n, \ldots, t_m\}$
$T' = T - \{t_j\}$, $j \in \{1,\ \ldots,\ i\}$
$F = \{f_1, \ldots, (t_0, p_{0j}), (p_{0j}, t_j), (t_j, p_{j(n+1)}), (p_{j(n+1)}, t_{n+1}), \ldots, f_k\}$, $j \in \{1,\ \ldots,$ $n\}$
$F' = F - \{(t_0, p_{0j}), (p_{0j}, t_j), (t_j, p_{j(n+1)}), (p_{j(n+1)}, t_{n+1})\}$, $j \in \{1,\ \ldots,\ i\}$
$C = sync(t_0,\ \ldots,\ t_n,\ p_{0j},\ p_{j(n+1)})$, $t_j \in T_{int}$ pour tout $j \in \{1,\ \ldots,\ n\}$ et $t_j \in$ $T_{coop} \cup B$ pour tout $j \in \{i+1,\ \ldots,\ n\}$

In Rule 12, we show that if we dispose of a set of internal activities parallal to a set of cooperative of block of activities, then we remove all the internal activities as well as all the control flows that are directly linked to them.

Iteration

Rule 13 (Iterations). Let p be an iteration and $t_i \in T_{int}$ s.t $t_i\bullet = \bullet t_i = p$ then

$\quad P = \{p_1, \ldots, p_n\}$
$P' = P$
$T = \{t_1, \ldots, t_i, \ldots, t_n\}$
$T' = T - \{t_i\}$
$F = \{f_1, \ldots, (p, t_i), (t_i, p), \ldots, f_m\}$
$F' = F - \{(p, t_i), (t_i, p)\}$
$C = iter(p)$

Rule 13 shows that if we dispose of an iteration formed by a single internal activity then this later is removed.

After having presented the diffrent reduction rules, in the next Section, we will show how to use these rules to transform an internal workflow into a cooperative one.

3.3   Abstraction Algorithm

In this Section we give the algorithm of abstraction of workflows. Given an initial workflow, the principle of the algorithm consists in removing all internal activities that don't play any direct role in the cooperation and whose elimination doesn't affect the visible behavior of the initial workflows.

Thus we start by identifying the various reduction patterns and then applying them the suitable reduction rules. This procedure is repeated until the removing of all internal activities that don't play a direct role in the cooperation.

The methods used in this algorithms are *existsSEQ()*, *existsPAR()*, *existsALT()*, *existsIT()*, *existsInternal-Then-ALT()*, *existsALT-Then-Internal()* that allow to identify workflow patterns and *applyRules()* that allow to apply a rule to a certain pattern.

Algorithm 1. Abstraction d'un workflow

1: **Abstract(workflow w)**;
2: **workflow** w';
3: **workflow*** patron = **null**;
4: **if** ((patron =existsSEQ(w)) $\neq$ null) **then**
5:     w' $\leftarrow$ applyRules(w, patron, $[R_1, R_2, R_3]$);
6: **else if** ((patron =existsPAR(w)) $\neq$ null) **then**
7:     w' $\leftarrow$ applyRules(w, patron, $[R_{10}, R_{11}, R_{12}]$);
8: **else if** ((patron =existsInternel-Then-ALT(w)) $\neq$ null) **then**
9:     w' $\leftarrow$ applyRules(w, patron, $[R_7]$);
10: **else if** ((patron =existsALT-Then-Internel(w)) $\neq$ null) **then**
11:     w' $\leftarrow$ applyRules(w, patron, $[R_8]$);
12: **else if** ((patron =existsALT(w)) $\neq$ null) **then**
13:     w' $\leftarrow$ applyRules(w, patron, $[R_4, R_5, R_6]$);
14: **else if** ((patron =existsIT(w)) $\neq$ null) **then**
15:     w' $\leftarrow$ applyRules(w, patron, $[R_{13}]$);
16: **end if**
17: **if** (patron $\neq$ null) **then**
18:     Abstract(w');
19: **else**
20:     renameINT(w);
21:     **return**;
22: **end if**

The termination of the abstraction algorithm consists in providing a well founded order on the algorithm parameters. Let W be the set of workflows and order relation $\prec$ defined on the set of workflows as follows: for two workflows $w(P, T, F)$ and $w'(P', You, F')$ of W, $w \prec w'$ if and only if $card(P) + card(T) + card(F) < card(P') + card(You) + card(F')$ such that $card(E)$ denotes the number of elements of a given set E and $<$ represents the usual order on the natural numbers. The order *prec* is well-founded since it does not contain infinite descendant elements. Let $w(P, T, F) \in W$. We note W_1 the set of $w'(P', You, F')$ such that Abstract(W) calls Abstract (w'). $forall\ w'(P', You, F') \in W_1$, $card(P') + card(You) + card(F') < card(P) + card(T) + card(F)$ since the size of an abstracted workflow is always less than the size the initial workflow. Indeed, this is valid for the whole of the abstraction rules. By consequent, the algorithm of abstraction terminates.

In the graph theory, a plain graph is a graph with the particularity of being able to be represented on a plan such that none of the arcs crosses the others. It is known, in the field of the grpah theory that finding a pattern in a plain graph has a linear complexity [3]. However it is evident that a structured workflow is a plain graph (by construction). Thus, the application of reduction rules has a linear complexity and consequently the algorithm complexity is polynomial in the worst case.

3.4   Running Example

For illustration, consider the workflow of a product provider presented in Figure 1-(a) using Petri nets as specification language [6]. First, the provider waits for an order request. Then he notifies the client that her order was taken into account and he assembles the components of the product. After that two cases can happen: the client is a subscriber (she often orders products) or she is not. In the first case, the provider sends the product and the invoice and waits for the payment. In the second case, the provider sends the invoice, waits for the payment and then sends the product. Filled activities, in Figure 1, are the ones that cooperate with the partner. The application of the abstraction algorithm to the internal workflow of the provider results in the abstracted workflow of Figure 1.

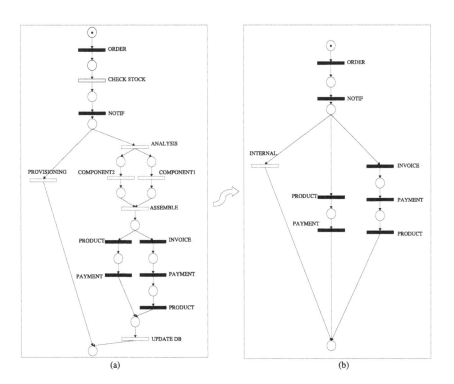

Fig. 1. The provider's abstraction

4   Related Work

Research on workflow management has focused on inter-organizational issues and much has been achieved so far. In [7], the author presents some forms of workflow-interoperability and focuses on capacity sharing, chained execution, subcontracting, (extended) case transfer, and loosely coupled [8]. Problems to be encountered on the way to workflow interoperability include mainly autonomy of local workflow processing, confidentiality that prevents complete view of local workflow [10], and especially flexibility that needs no definition of a global workflow that describes cooperation between local workflows.

In the public-to-private approach, a common public workflow is specified and partitioned according to the organizations involved by private refinement of the parts based on a notion of inheritance. Each partner has a copy of the workflow process description. The public-to-private approach consists of three steps. Firstly, the organizations involved agree on a common public workflow, which serves as a contract between these organizations. Secondly, each task of the public workflow is mapped onto one of the domains (*i.e.*, organization). Each domain is responsible for a part of the public workflow, referred to as its public part. Thirdly, each domain can now make use of its autonomy to create a private workflow. To satisfy the correctness of the overall inter-organizational workflow, however, each domain may only choose a private workflow which is a subclass of its public part [8]. A drawback of the approach public-to-private is the lack of the preservation of established workflows. In fact, in this approach, one has to look for which rules, in what order and how many times one has to apply them in order to match the established workflow with the public part which is deduced from partitioning of the public workflow. If not impossible, this is hard to do. Moreover, there is no defined procedure to do that.

The inter-organizational cooperation problem has also been addressed by the *CrossFlow* approach [4,9] that uses the notion of contracts to define the business relationships between organizations. This approach does not support arbitrary public processes.

5   Conclusion and Perspectives

In this paper we presented important steps to provide support for inter-organizational workflow cooperation. They consist in the advertisement, interconnection and cooperation of workflows. To preserve partners' privacy, we proposed a polynomial algorithm for workflows' behaviors abstraction based on a reduction procedure. The principle of the abstraction is to reduce the workflow inter-visibiliy as tiny as cooperation needs by removing all activities as well as the set of the control and data flows that don't play any direct role in the cooperation. The abstraction concept provides a high degree of flexibility for participating organizations, since internal structures of cooperative workflows may be adapted without changes in the inter-organizational workflows.

References

1. Chebbi, I., Dustdar, S., Tata, S.: The view-based approach to dynamic inter-organizational workflow cooperation. Data and Knowledge Engineering Journal 56(2), 139–173 (2006)

2. Chebbi, I., Tata, S.: CoopFlow: a framework for inter-organizational workflow co-operation. In: Proceedings of International Conference on Cooperative Information Systems, Agia Napa, Cyprus, pp. 112–129 (31October-4 November 2005)
3. Eppstein, D.: Subgraph isomorphism in planar graphs and related problems. In: SODA 1995: Proceedings of the sixth annual ACM-SIAM symposium on Discrete algorithms, pp. 632–640, Philadelphia, PA, USA, Society for Industrial and Applied Mathematics (1995)
4. Grefen, P., Aberer, K., Hoffner, Y., Ludwig, H.: Crossflow: Cross-organizational workflow management in dynamic virtual enterprises. International Journal of Computer Systems Science & Engineering 15(5), 277–290 (2000)
5. Tata, S., Chebbi, I.: A bottom-up approach to inter-enterprise business processes. In: In Distributed and Mobile Collaboration, 13th IEEE International Workshops on Enabling Technologies, Infrastructure for Collaborative Enterprises, Modena, Italy (June 2004)
6. van der Aalst, W.M.P.: Interorganizational workflows: An approach based on message sequence charts and petri nets. Systems Analysis - Modelling - Simulation 34(3), 335–367 (1999)
7. van der Aalst, W.M.P.: Loosely coupled interorganizational workflows: Modeling and analyzing workflows crossing organizational boundaries. Information and Management 37, 67–75 (2000)
8. van der Aalst, W.M.P., Weske, M.: The p2p approach to interorganizational workflows. In: Proceedings of the 13th International Conference on Advanced Information Systems Engineering, pp. 140–156. Springer, Heidelberg (2001)
9. Van Dijk, A.: Contracting workflows and protocol patterns. In: Proceedings BPM, Eindhoven, The Netherlands (June 2003)
10. Zhao, J.-L.: Workflow management in the age of e-business. In: Tutorial at the 35th Hawaii International Conference on System Sciences, Waikoloa, Hawaii (January 2002)

Using Control Patterns in Business Processes Compliance

Kioumars Namiri[1] and Nenad Stojanovic[2]

[1] SAP Research Center CEC Karlsruhe, SAP AG, Vincenz-Prießnitz-Str.1
76131 Karlsruhe, Germany
Kioumars.Namiri@sap.com
[2] FZI Karlsruhe, Haid-und-Neu-Str. 10-14
76131 Karlsruhe, Germany
Nenad.Stojanovic@fzi.de

Abstract. The realization and documentation of an effective Internal Controls System is required by regulations such as Sarbanes Oxley Act (SOX). In this paper we introduce a pattern based approach for modeling of the Internal Controls in Business Processes. They can be captured as declarative rules and checked during execution time of processes. The approach supports the definition of the controls outside of the operative Business Processes run by e-Business Systems in order to enable the reuse of process models and controls in different business and compliance environments. A detailed discussion on the domain model of Internal Controls and the system architecture necessary for realizing the approach is provided.

Keywords: BPM, Regulatory Compliance, Internal Controls, Patterns.

1 Introduction

Regulations such as Sarbanes Oxley Act (SOX) [1] require the implementation of an effective Internal Controls system and define it as a management responsibility in enterprises. Internal Controls system is a well know technique to ensure among others reliable financial statements in enterprises and the transaction in their e-Business Systems. COSO (Committee of Sponsoring Organizations of the Treadway Commission) has already proposed in early 90s an integrated framework [2], which is recognized by regulation bodies and auditors as a de facto standard for realizing the Internal Controls System.

Following the COSO recommendations for the realization of Internal Controls Process a set of effective **Controls** have to be designed in order to *prevent* or *detect* the occurrence of the identified risks based on a defined set of Control Objectives for Business Processes. The controls must be tested and used in daily operations. The Internal Controls Compliance is not a one-time task; it is rather a continuous process.

This paper introduces an abstraction layer above a Business Process, in which the controls are formally modeled and evaluated against existing process models and instances. It describes a novel, model-driven approach for the automation of Internal Controls in an enterprise, based on their conceptual separation from Business Process Management (BPM). The approach advocates the use of control patterns in the

M. Weske, M.-S. Hacid, C. Godart (Eds.): WISE 2007 Workshops, LNCS 4832, pp. 178–190, 2007.

proposed abstraction layer. Two sets of patterns are introduced: i) a set of empirically determined high level control patterns, which represent the way Compliance/Business Process experts communicate about the compliance domain and ii) a set of system level control patterns based on the property specification pattern system proposed originally by Dwyer et al [5], in which the system developers speak about the domain. Each pattern set should give software/system and compliance/business practitioners access to specify and design the compliance requirements.

We are mostly concerned with automation of the so called Application Controls (AC), which control Business Processes to support financial control objectives and to prevent or detect unauthorized transactions. However, the approach provides a general framework that can be applied with respect to any other compliance domain using BPM technology.

We start by a motivating scenario. In section 3 we introduce our domain model of the Internal Controls followed by the controls patterns that we propose in that domain in Section 4. In section 5 we introduce our approach for designing the controls and to ensure their effectiveness during runtime of BPs from a system architecture perspective. Concluding remarks are given in section 6.

2 Motivating Scenario

We use the Purchase-To-Pay (P2P) Process delivered by an ERP product as an example. The process starts by creating the request for a Purchase Order (PO) and ends when the payment of that PO is recorded in Accounting.

The Internal Controls compliance of such a Business Process depends on each enterprise specific risk assessment. Table 1 shows an excerpt of the risk assessment carried out by Compliance experts of two different enterprises. It shows their different control objectives, risks, and controls on the same standard P2P Process.

Table 1. Risk assessment on Purchase-To-Pay (P2P) Process for two different enterprises

Control Objective	Risk	Control
Enterprise A: Prevent unauthorized use	Unauthorized creation of POs and payments for not existing suppliers	1) POs for material types which have not been ordered during last year and an amount higher than 5000 $ must be double approved by two different purchasing clerks (Second Set of Eyes Control - SSE).
Enterprise B: React flexible on changes in the supplier market	Dependancy on one single supplier in the market	Minimum Number of Suppliers is 2 for material type 5: Keep at least two contracted Suppliers in your Supplier Relationship Management (SRM) System for the given material type.

We have identified that there exist frequently defined patterns of controls on Business Processes at different enterprises. Taking the perspective of an ERP vendor, providing this set of patterns in a repository where a certain pattern can be selected, instantiated to a real control, and applied on Business Processes by their customers brings a higher level of system and component reusability for the ERP/BP products. Taking the perspective of a customer company building their compliance on top of such a pattern repository can reduce the required domain specific knowledge in compliance projects. Therefore, the process models become nowadays too complicated, not readable and manageable when they are directly, i.e. manually enriched with the necessary compliance controls. In the rest of the text we present an approach that copes with this kind of complexity.

3 Domain Model for Internal Controls Compliance

3.1 Roles Involved

We distinguish three roles involved in Business Process Compliance with the specific interests/expertise, as follows:

Business Process Expert. A Business Process expert knows how to configure and maintain the processes having business objectives (goals) in mind. The business objective for, e.g. a purchasing process, is simply to set up a process in which internal orders created can be processed and sent to suppliers, to receive the ordered goods, and to pay the supplier invoices. This group of persons in an enterprise has no or little knowledge about regulations and compliance requirements, but very detailed knowledge on how a process is implemented.

Compliance Expert. The auditing consultants are Compliance/SOX experts and have detailed knowledge about the regulatory requirements. They have no or little knowledge about the realization of Business Processes in an enterprise. Their main task is rather to define and monitor the necessary controls according to the risk assessment and to notify other entities in the enterprise in case of control violations. They do not define how to bring a process in a compliant state because this is the task of Business Process experts. They are not involved in design and operation of operative Business Processes such as Purchase, Sales, Production etc.

External Auditors. External auditors are regulation bodies or official firms who certify the **design** and **effectiveness** of the Internal Controls system in an enterprise on a periodical basis. The external auditors are out of scope in this work.

3.2 Interplay of the Entities in the Domain Model

In the following we enrich the entities resulted from our analysis of (mainly not IT-related) COSO by additional entities. These additional entities will enable the model to serve to us as an operational basis for our approach later on. We discuss those parts of the model that are relevant for our approach in detail: *Business Process* and *Application Control* (we simply use the term Control).

3.2.1 Control - Business Process Model

Below we further detail the relationship between a *Business Process* and a *Control* as shown in Figure 1. We interpret a business process according to [6], where it is defined as a set of logically related tasks (or activities) to achieve a defined business outcome. An *Activity* can be aggregated by other activities. A special kind of activity is the *Coordinator* Activity (such as switch/fork/join etc.), which defines the behavior of the flow in a Business Process known from workflow modeling. An *Activity* consumes and produces *Business Documents* (such as *Purchase Order, Invoice etc.*). And finally an Activity is performed by a User (human or computer).

For each *Control* at least one *Recovery Action* must have been designed, which reacts on the violation of a control. The nature of the recovery action depends on the current role of the person involved in the Business Process compliance: The Compliance expert or the Business Process expert. We detail this in the next subsection.

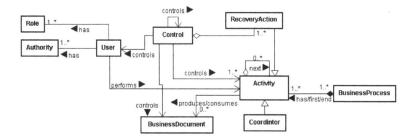

Fig. 1. Relationship between a Control and a Business Process

3.2.2 Role Based Recovery Action Model

Our model for Business Process compliance recognizes the fact that there are different roles. Every role has a specific view involved in the Business Process Compliance, introduced by a role based recovery action model. In the following we first explain the different types of possible recovery actions in case of a control violation:

> *Ignore:* The control violation is ignored.
>
> *Block:* The current instance of the BP, which generated a control violation, is blocked.
>
> *Notify (User, Message):* A notification message for the specified user *User* is created with the given message *Message*.
>
> *Retry:* The activity that generated the violation is retried again.
>
> *Rollback (Activity):* The current instance of the BP that generated the control violation is rolled back to the given activity *Activity*.
>
> *Recover (RecoveryProcess):* A previously designed recovery process *RecoveryProcess* is instantiated parallel to the current instance of the original BP that generated the control violation. The recovery process itself is an autonomous Business Process.

Please note that a combination of the above listed recovery actions is also possible such as *Retry & Notify* etc.

During the Control design a Compliance expert defines the recovery actions as a *minimal* set of actions regarding the Business Process logic. The decision on recovery action selection in a certain control design is up to the Compliance expert based on the enterprise specific risk assessment. Please remark that this leads to the fact that even for the same control violation different recovery actions may be designed in two different enterprises.

A Business Process Expert can modify and extend the recovery actions of a control, in order to avoid permanent blocking of process instances in case of a control violation. The valid combination of recovery actions set by the Compliance expert and Business Process expert follows these basic rules:

- A control violation always requires a reaction, particularly a single *Ignore* is never allowed.
- The recovery action designed by a Business Process expert is never allowed to "weaken" the original recovery action designed by the Compliance Expert. For instance if an Compliance expert requires a *Block & Notify* on a Business Process instance in case of a certain control violation, the Business Process expert is not allowed to redesign the recovery of a control to only *Notify*.

To clarify the Role-Based Recovery Action model we give below an example:

Scenario Revisited: Recall the required control "Minimum Numbers of Suppliers" specified for the Enterprise B in the "Motivating Scenario". The Compliance expert in that enterprise designs the control according to the risk assessment of enterprise B and decides to select the ***Block & Notify*** recovery action in case of the control violation. The Compliance expert at this stage is not concerned about all the possibly blocked P2P instances having material type 5 in their *PO* if the number of valid contracts to possible suppliers of this material type becomes lower than 2.

When the control is stored in the control repository, the business process expert having detailed knowledge about P2P process gets notified and checks the recovery action of the control. Since the business process expert has the Business Objective "Purchase Goods" in mind, he is aware that some process instances may be blocked completely by that control. Further he is aware of a Business Process *RfQProcessing* which creates a so called *Request for Quotation (RfQ)* for a supplier. The business objective of *RfQProcessing* is to contract the selected Supplier in the SRM system of enterprise B.

The Business Process expert extends the recovery action model of the control by adding the recovery action *Recover (RfQProcessing) & Retry* to the control design. In case of the control violation the *RfQProcessing* is enacted in parallel additionally to the current P2P Process instance. The process step is retried again and in case that the control violation does not exist anymore (*RfQProcessing* has increased the number of contracted suppliers in backend system SRM to 2 or more), the process instance can continue.

3.2.3 Controlled Entities

We can see that in the domain model of Business Process compliance for Internal Controls, we have four different types of first class entities: activities, business documents, users, and the controls. We refer to these four different entities as Controlled Entities (CE) in a Business Process.

We consider a control as a controlled entity in a Business Process because the effectiveness of a control should impact the execution of a Business Process. This means basically that if a control is not effective, i.e. its violation has no implications on a Business Process, the enterprise runs the risk of not being compliant. Thus the main tasks of Compliance experts include not only to design the controls but also to assure their effectiveness.

CEs have dependent artifacts in common in their structural composition as visualized in Figure 2. The concept of these artifacts will serve us as a basis for implementing the controls in Business Processes.

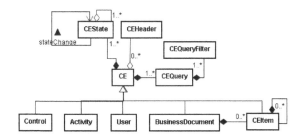

Fig. 2. Composition of Controlled Entity - CE

A CE may have additional Meta data information (*CEHeader*) specifying an instance of that CE in more detail. Each instance of a CE has a current state (*CEState*) and a set of valid state changes, which are caused by activities executed on an instance of that CE. Interestingly a *Control* itself has also a set of states and can be treated in the same way as a business document.

The item (*CEItem*) of a CE Business Document represents all sub parts of that entity (For instance a PurchaseOrder PO may contain several items for different material types as sub orders). The item can be a CE itself and it may consist of other sub items. The query of a CE (*CEQuery*) determines the number of all instances of that CE according to a given filter (*CEQueryFilter*). An example:

- A Query for all POs *POQuery* with a filter *POQueryFilter approved* POs in the *period of last quarter* and for a certain *supplier "XYZ"*

will return the number of all PO instances satisfying the given filter criteria.

4   Control Patterns

In the following we introduce two different sets of patterns, which we call high level and system level control patterns. They basically represent the same thing on different abstraction levels in a domain, namely frequently recurring/defined patterns of controls on Business Processes. The high level control patterns provide the basis for the terminology in which the Compliance experts communicate about the domain. We have determined the presented set of high level control patterns empirically by analyzing different kinds of typical ERP Scenarios (Purchasing, Sales and Human Resource Management and all belonging side and sub processes such as Goods Return, Payment, Dunning, etc.). Here we have grouped typical control categories that are defined on those Business Processes at different enterprises built on top of a provided set of process reference models inside an ERP Product.

The system level control patterns represent a more technical view on the controls and their introduction is aimed to facilitate the use of formal methods by system developers/technical personnel having the task of implementing the controls in ERP/BPM Systems. The system level patterns themselves are generic in their nature in that way that they are not bound to the usage of certain formal logics. Each development team can select its favorite and suitable technical representation of the system level control patterns which can vary from database-oriented/SQL to different temporal logics such as LTL or CTL (see Figure 3).

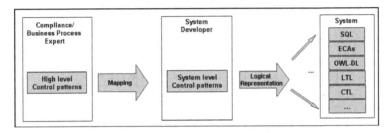

Fig. 3. From a High level Control pattern to its technical Representation

4.1 High Level Control Patterns

In Figure 4 we expose different categories of control patterns on Business Processes and give a brief description for each pattern category type without going into details of its sub categories:

SSE patterns: We already mentioned this kind of control patterns briefly in the scenario section, which basically requires the double eyes- principle on certain transactions. Here we add the comment that a control demanding a "higher number of eyes" would also be possible and would fall into this category as well.

Business Document control patterns: Here the syntax and semantics in and between different business documents are subject to the controls.

Inter Activity control patterns: The controls satisfying these patterns require that certain activities occur (or are absent) if certain set of other activities occur in a Business Process (or a side process).

Report patterns: Reports are collected based on attributes on certain types of activities and business documents in an enterprise during a certain period, e.g. monthly turnover reports. The purpose of report control patterns is not the definition of a report, but rather to control that a report has been generated respectively reports are compared to each other as required in the control.

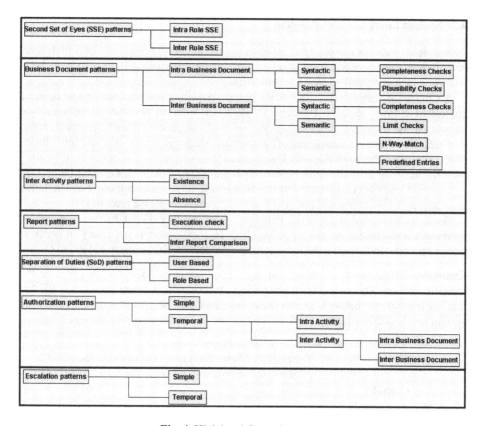

Fig. 4. High level Control Patterns

SoD patterns: In order to minimize fraud or misusage it is required that an activity is divided into sub activities and each sub activity is executed by different users or roles.

Authorization patterns: These controls limit users/roles access to CEs.

Escalation patterns: In case that detected controls are ignored by the responsible users, this fact can/has to be escalated to responsible entities in the enterprise.

Each pattern is specified by a set of attributes. Below we give an example for the specification of the pattern "N-Way-Match" including its description:

- **Description:** certain fields in header and items of different business document types belonging to the same Business Process instance must match each other
- **Subjected CE:** Business Document
- **Objected CE:** Business Document, Activity
- **Related to:** -
- **Control Trigger:** State change of an Activity or Business Document
- **Example:** 3-way match control on PO, Invoice, and Delivery of Business Documents of a P2P process instance if the supplier identification is identical.

4.2 System Level Control Patterns

System level patterns are used to represent the technical representation of a high level compliance pattern. Each high level control pattern corresponds to a system level pattern, which is described by a Control Strategy:

A *Control Strategy* defines the way a control monitors the behavior of one or more controlled entities inside a Business Process. In order to become active a *control* requires to be triggered according to the *state* of the process parameters in a *scope*. We defined the two elements of a control strategy *scope* and *pattern* based conceptually on the work done by Dwyer et al [5]. Although their patterns are mainly used for defining formal requirements on program specifications, they can be applied to Internal Controls compliance and the monitoring requirements there. For a detailed description of the scopes and patterns and their semantics please refer to [5].

We have extended the Dwyer patterns by an entity called *CECondition*, which represents a constraint on one or more CEs. This extension is necessary in order to reflect special conditions in the subjected and objected CEs, as discussed in [11].

Example: Recall the first control on the P2P Process of enterprise A given in the scenario section. This is an "Intra Role SSE" Pattern, which means that it is sufficient that each approver belongs to the same role and can be mapped to the following system level control strategy:

- *ControlTrigger* = Activity *"SelectSupplier"*
- *Scope = Between* the activity *"SelectSupplier"* and activity *"SendPO"*
- *Control Pattern = Bounded Existence* of $n=2$ on CE *"ApprovePO"*-Activity
- *CEConditions:*
  - *POHeader.amount > 5000$*
  - *$ApprovePO_1.User.Role$ = "Purchasing Clerk"*
  - *$ApprovePO_2.User.Role$ = "Purchasing Clerk"*
  - *$ApprovePO_1.User.Id \neq ApprovePO_2.User.Id$*
  - *$\forall t_i, \forall POItem_i \in \{PO.POItems\}, POQueryFilter_i = POItem_i.lastOrderDate$ $t_i = POQuery(POQueryFiliter) \mid t_i > 1\ year$*

5 The Approach

In order to realize the separation of the business and control objectives, our approach introduces another layer above the Business Process model. This layer is called "SemanticMirror". According to the assessed risks, a set of Controls is defined on that layer. Finally, by executing a Business Process, the semantic process layer will be continually updated with information needed for the evaluation of defined controls in order to ensure that compliance tests will pass. The approach spans over three phases, described below. The first two phases have a sub phase, which we call Business Process Model Adaptation.

5.1 Phase 1 - Control Design Phase

Before this phase, the process models may be non-compliant in terms of they do not contain the required controls according the risk assessment of the enterprise. During this phase, a Compliance expert goes through the relevant Business Process model, as it may be delivered by an ERP vendor. First, the Compliance expert selects an activity contained in the process model. Then he selects a certain control pattern from the control pattern repository. He instantiates the selected pattern by configuring it according to the enterprise's specific requirements. He then stores the control: a) the control is stored in the SemanticMirror and b) the currently selected activity in the process model is extended by the control (Business Process model adaptation).

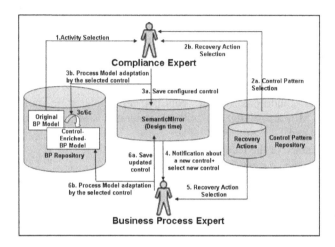

Fig. 5. Phase 1 and phase 2

5.2 Phase 2 - Recovery Action Design Phase

After a new control is created in the SemanticMirror, the according Business Process expert is notified about this fact. He checks the recovery action part of the control and, if necessary, he modifies/extends the recovery action model of the control. After this phase, the control in the SemanticMirror and the process model in the BP repository are updated with necessary modifications done by the Business Process expert (Business Process model adaptation).

The cooperative interactions of the actors and the systems during phase 1 and 2 are summarized in Figure 5.

5.3 Phase 3 - Business Process Execution Phase

This phase enables the bidirectional interaction between BPM and Internal Controls management: The SemanticMirror will be updated by information about the current instance of the Business Process enacted and if a control is violated, the recovery action defined in the control will be executed.

In order to enable the automated generation of the SemanticMirror during execution time, it has to be continuously updated when an activity is performed in the given Business Process instance.

In the following we describe the validation of control c during execution time of Business Process p with a recovery action on violation of c *Retry & Notify & Recover(r)*. The nodes *cd* and *cn* and their transitions exist after the Business Process Model Adaptation phase (See step 3c/6c in previous Figure 5). *cd* is a decision-node (Coordinator) and *cn* is an activity that generates a notification message. All steps are visualized in Figure 6:

- 1. The process context is written to SemanticMirror.
- 2a. As the state of the SemanticMirror changes in terms of adding/updating CE facts to it, the trigger of control c gets activated. The condition of c is determined by the values of the CE facts in the SemanticMirror itself or optionally by
- 2b. querying the necessary backend systems using the CEQuery of a subjected CE.
- 3. If the conditions of the controls are violated, a new fact in the SemanticMirror (*cViolation*) will be generated signaling that control c has been violated.
- 4. An instance of the recovery process r is generated.
- 5. The instance p steps into the decision node cd.
- 6a. cd, being a decision node, is a coordinator activity. The activity of cd queries the SemanticMirror for a fact instance called *cViolation*.
- 6b. In case of existence of a *cViolation* in the SemanticMirror, cd sets the transition to "*ok*", otherwise to "*notOK*".

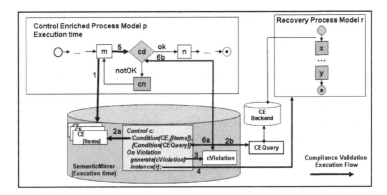

Fig. 6. Phase 3

Please notice that the approach described above will still detect a control violation in the SemanticMirror, even if a Business Process expert/technical consultant will remove the control from the process model being not aware of the necessity of that control: the process context is always written to the SemanticMirror during step 1 and the controls exist independently in the SemanticMirror. Further, the described approach enables dynamical application of the controls during the execution phase of a Business Process. There is a minimum overlap between Business Process design and compliance design.

6 Conclusion

In this paper we introduced a pattern based approach for modeling Internal Controls required by regulations such as SOX. They can be captured as declarative rules and checked during execution-time on Business Processes. We built the model based on the de facto Internal Controls standard called COSO. The approach supports the definition of the controls outside of the workflow in order to enable the reuse of process models and controls in different business environments.

Currently our approach requires the manual selection of a concrete control pattern and its specific design on a Business Process according to the enterprise-specific compliance needs. A higher level of automation can be brought to the approach by building a "Risk Repository" as a starting point of the approach.

Another issue that must be addressed is the inter-control dependency: in order to become effective, a "well-designed" control may depend on existence, effective design, and operation of other controls. This issue is also mentioned directly by law in [10]. We currently recognize this fact by introducing the "related to" attribute in a pattern specification. On a similar note, different designed controls can contradict, subsume or block each other in a Business Process. We have to extend the Control Design phase by concepts to detect and avoid such situations.

References

1. Pub. L. 107-204. 116 Stat. 754, Sarbanes Oxley Act (2002)
2. Committee of Sponsoring Organizations of the Treadway Commission (COSO), Internal Control – Integrated Framework (1992)
3. Hartman, T., Foley & Lardner LLP.: The Cost of Being Public in the Era of Sarbanes-Oxley (June 2005)
4. zur Muehlen, M., Rosemann, M.: Integrating Risks in Business Process Models. In: Proceedings of the 2005 Australasian Conference on Information Systems (ACIS 2005), Manly, Sydney, Australia, November 30-December 2 (2005)
5. Dwyer, M., Avrunin, G., Corbett, J.: Patterns in Property Specification for Finite-State Verification. In: Proceedings of the 21st International Conference on Software Engineering, pp. 411–420 (May 1999)
6. Davenport, T.H., Short, J.E.: The New Industrial Engineering: Information Technology and Business Process Redesign. Sloan Management Review 31, 11–27 (1990)
7. Governatori, G., Milosevic, Z., Sadiq, S.: Compliance checking between business processes and business contracts 10th International Enterprise Distributed Object Computing Conference (EDOC 2006), pp. 221–232. IEEE Press, Los Alamitos (2006)
8. Agrawal, R., Johnson, Ch., Kiernan, J., Leymann, F.: Taming Compliance with Sarbanes-Oxley Internal Controls Using Database Technology. In: Proc. 22nd Int'l. Conf. on Data Engineering (ICDE 2006), April 3 – 7, 2006, Altanta, GA, USA (2006)
9. Reichert, M., Dadam, P.: ADEPTflex – Supporting Dynamic Changes of Workflows Without Losing Control. Journal of Intelligent Information Systems 10(2) (1998)
10. Public Company Accounting Oversight Board (PCAOB), PCAOB Accounting Standard No. 2, paragraph 12.

11. Namiri, K., Stojanovic, N.: A Formal Approach for Internal Controls Compliance in Business Processes, 8th Workshop on Business Process Modeling, Development, and Support (BPMDS 2007), In conjunction with CAiSE 2007.
12. Giblin, C., Muller, S.: Brigit Pfitzmann (2006) from regulatory policies to event monitoring rules: Towards model driven compliance automation. IBM Research Report. Zurich Research Laboratory (Oct 2006)
13. Casati, F., Castano, S., Fugini, M., Mirbel, I., Pernici, B.: Using Patterns to Design Rules in Workflows. IEEE Transactions on Software Engineering 26(8) (August 2000)

A Framework for Evidence Lifecycle Management

Andreas Schaad

SAP Research
Vincenz Priessnitz Str. 1
Karlsruhe, Germany
andreas.schaad@sap.com

Abstract. Organisational control principles, such as those expressed in the separation of duties, delegation of obligations, supervision and review, support the main business goals and activities of an organisation. One specific type of an obligation is that of a workflow task. The delegation of a task from one principal to another will result in a review task on the side of the delegating principal, allowing for the meaningful tracking of work. The target of this review task is the delegated task and a defined evidence needs to be negotiated, generated by the delegatee and reviewed by the delegator. This paper expands on earlier work we have done on delegation and revocation of tasks and the resulting review relationships. The concept of evidence had not yet been treated in sufficient detail and accordingly, a more detailed discussion around a framework for evidence lifecycle management is the subject of this paper.

Keywords: Delegation, review, evidence, organisational control.

1 Introduction

Organisations are determined by their goal-orientation [7], [8]. High-level goals are decomposed until they become operational and executable. Workflow systems provide some of the technical means to support such goal decomposition and operationalisation. Workflow models allow to arrange tasks in a manner such that the execution of a workflow will result in the achievement of a goal.

Within workflow systems research we observe a tendency moving away from strict enforcement approaches towards supporting exceptions that are difficult to foresee when modeling a workflow [9]. Along those lines one specific set of mechanisms is to provide delegation features that allow for the exception-based delegation of a task at workflow execution time between principals. The ability to demonstrate the tracking of the distribution of work is imperative.

When tasks are delegated between principals, the delegating principal may become subject to a new task, that is, to review that the task he delegated will eventually be executed in a satisfiable manner. In fact, at the time of delegation the delegating principal will negotiate with the delegatee what kind of evidence needs to be provided such that the delegator can fulfill his review task.

M. Weske, M.-S. Hacid, C. Godart (Eds.): WISE 2007 Workshops, LNCS 4832, pp. 191–200, 2007.
© Springer-Verlag Berlin Heidelberg 2007

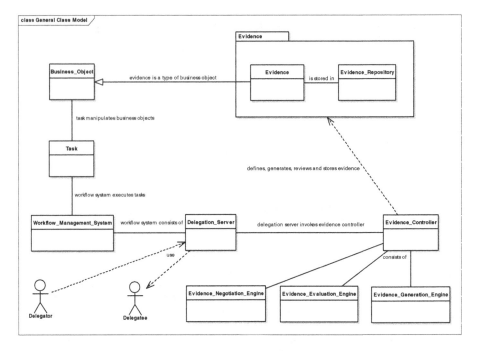

Fig. 1. Overall architecture

We propose a framework to support the negotiation, generation and evaluation of evidence in the context of delegated tasks. This framework entails a description of a proposed Evidence Controller and its relationship to a workflow system and business object repository. A set of protocols for the negotiation, generation and evaluation of evidence as well as their inter-dependencies is presented based on the evidence controller.

As it can be seen in figure 1, principals interact with applications and the underlying business objects through a workflow management system. A workflow management system will control the execution of tasks as defined in a workflow model. The execution of a task may result in a change of state of a business object. For example, such a workflow management system could support the processing of a loan application, supporting clerks in accessing credit check databases, customer files as well as automated calculation of the credit conditions of a client.

When an exception occurs, a principal has the ability to delegate the task he has to perform to another principal (the delegatee). The delegation server, a component of the workflow management system, will coordinate such delegation activities and keep track of which task has been delegated to whom. As part of a delegation, the delegator may request from the delegatee to provide him with evidence that the delegated task has been performed at some stage. As such, the workflow system will invoke the services of an evidence controller, a component that supports the negotiation, generation and evaluation of evidence.

2 Basic Concepts

2.1 Delegation and Review

One specific type of an obligation may be that of a workflow task. Obligations are continuously created, delegated, revoked or discharged according to the overall goals of an organisation and the general principle of distributing work. Ideally, there should never be any uncertainty about who currently holds an obligation, whether somebody has discharged his obligations, the effect of such a discharge, and who has to ultimately ensure that the tasks of an obligation are performed. For this reason it is necessary to hold to account persons who delegate obligations. This also includes the ability to trace back any delegated obligations to the initial delegator. In order for them to be able to give an account of the obligation that they have delegated, they must review it. We propose that this may be done by creating a review obligation referring to the delegated obligation. In this context, review is understood as an obligation referring to a previously delegated obligation which has to examine the results of the discharge of this delegated obligation. The holder of such a review obligation has then to make sure that the obligation he delegated has been carried out satisfactorily based on some evidence. We point to [13] for detailed formal discussion and examples.

A review does not act as a direct enforcement mechanisms for the delegated obligation, but as a post-hoc control and detective mechanism. If the review fails because the delegated obligation has not been discharged at all or satisfactorily, this may trigger corrective measures to be taken by the reviewer himself or some other principal.

In [7] a distinction between procedural and output controls is made as the components of administrative controls. We argue that review conforms to both these definitions of control. On the one hand the creation of a review is part of the delegation procedure for certain obligations, while on the other hand this review then controls the output of the discharge of a delegated obligation.

A set of formal models has been presented in [1],[2],[4] supporting our concepts of delegation and review. The Alloy [3] constraint analyis tool was used for formal verification (http://alloy.mit.edu/).

2.2 Evidence

Evidence is itself a specific type of business object that can be manipulated by a task. That means that execution of a task may generate evidence referring to some review task on the delegators side symbolizing that his prior delegation now requires him to check the results.

Being a business object, Evidence has some defined attributes. While these attributes are application specific, they can be grouped into primitive data types (such as integers and strings), the state of a business object (such as an loan application object which is in the state pending) and free types that are defined and evaluated by human user (e.g. textual definition of what determines the quality of some piece of work).

Being a business object, Evidence has itself states (Figure 4) which can be categorised as "partial, adduced, rejected and accepted". These states and their respective transitions are based on several events. The states "accepted" and "rejected" are determined by the execution of the review task. The states "partial" and "adduced" are determined by the actions performed by the delegatee and refer to the generation of evidence.

Evidence specifications are stored within an evidence repository. This allows principals to search this repository during the evidence negotiation phase for already existing evidence specifications that can be instantiated depending on the delegation context. Ad-hoc negotiated evidence can then also be stored in the repository for future reference.

3 General Scenarios

Delegation may have several reasons, ranging from simple automated workload management to supporting decomposition of tasks. Note that when speaking about principals, this can mean both, human actors as well as automated agents.

The following describe a set of delegation scenarios that outline the working of our proposed framework. Note that the evidence to be supplied by a principal subject to a delegation is characterised by being small enough so that the later review causes significantly less overhead on the delegator's side than having performed the delegated task himself. Examples 1 - 3 describe the delegation of tasks where the evidence is already specified within the evidence repository. Examples 4 and 5 describe the negotiation between two principals where there is no evidence proposed by the repository. Example 6 is an example of a delegation activity between two automated principals (e.g. applications or services).

- Example 1: We consider a principal delegating the task of preparing the quarterly results. When this task is delegated, the delegating principal would find that the evidence repository proposes him the numeric value of the final credit and debit sheet as possible evidence that the quarterly results have been prepared. This is in fact also a checksum, as the expected value would be a zero. This type of evidence is a set of numeric values for later Boolean comparison.
- Example 2: We consider a principal delegating the task of preparing the shipment for a customer. The evidence repository proposes him to use the final delivery note listing the shipment details as the evidence to be provided. This type of evidence is based on the business object "Shipment" and a query on its attributes.
- Example 3: We consider a principal delegating the task of following up on a sales opportunity. The evidence repository proposes him to check on the follow-up sheet (e.g. phone calls towards customer's number) as provided by the Customer Relationship Management. This type of evidence is based on the business object "Sales Opportunity" and a query on its attributes.
- Example 4: We consider a principal delegating the task of implementing a software component. In this case there is no predefined evidence suggested by

the evidence repository. The two principals agree that the evidence to be delivered by the delegatee consists of a set of unit tests and test scenarios that the delegatee needs to deliver. This type of evidence is specific to the delegated task but is still something that can be automatically verified and validated, e.g. unit tests succeeds or fails or test scenario shows that a specific property such as required time for computation is below defined threshold.

- Example 5: We consider a principal delegating the task of preparing the answer to a tender. This is a one-time project like activity, and no evidence template can be provided by the evidence repository. The delegating principal and the delegatee agree that the delegatee will prepare the tender and two types of evidence are agreed upon. The first is that of providing a draft document with some baseline calculations on the feasibility of answering the tender. The second is that of the final proposal that then needs to be signed-off by the delegating principal. The first evidence is an example of a numeric value, e.g. that the projected profit when entering the tender is within a set general range. The second type of evidence is in fact the entire business object which was the target of the delegated task, evaluation of this evidence is subjective though certain properties may be defined as part of general corporate style guide or existing best-practice templates.

- Example 6: The credit range checking application of a bank usually handles all bank internal queries as well as external queries by customers through a defined interface. Due to an increase in queries as well as down-time of a server, the application decides to delegate handling of all internal and external queries for small customers to a 3rd party offering the same services. The two applications agree to base the review of this delegation on the evidence of every credit range check returning a succeed or fail plus the integer value of the range. The delegating application will use these data for comparison of random samples with its historic data to detect any deviations.

4   Detailed Discussion

4.1   Overall Structure

Figure 6 details the overall architecture from a state-based perspective. We can observe that the top layer (Main Phases) shows the three states of evidence negotiation, generation and checking and how these states relate to each other.

Based on this overall state-machine, three other state machines were defined: A negotiation supporting state machine (Figure 3), an evidence generation supporting state machine (Figure 4) and an evidence review and check state machine (Figure 5). Since individual states in these machines do show some dependencies, we decided to model these by dotted lines, symbolising that some state in some state machine may either act as direct trigger or be a required state to cause a state change.

The phase of evidence negotiation was also modeled in form of a sequence diagram to emphasise the notion of "evidence as a business object", as one key distinguishing and novel aspect of our proposal.

4.2 Evidence Negotiation

Figure 3 shows that when two principals start negotiation of evidence for a task to be delegated, the first step is to check for an already existing evidence specification within the evidence repository. If this is the case, the principals do agree on this specification and the negotiation is finished.

If there is no specification or the existing specification needs to be modified, one of the two principals will propose a specification. Either this proposal is accepted by the other principal and the negotiation is finished, or the other principal makes a counterproposal which will eventually be subject to acceptance or the negotiation needs to be terminated and with it the delegation of a task.

Figure 2 expands on this state machine from a message exchange perspective. In particular, the delegatee may utter some preference in case there already is a set of existing predefined evidence specifications (Step 6).

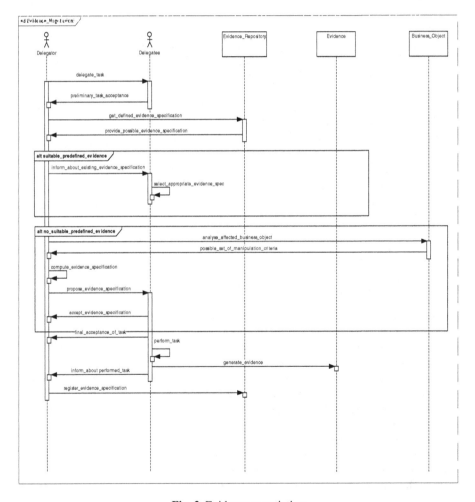

Fig. 2. Evidence negotiation

In case no predefined evidence specification can be obtained, the delegating principal will analyse the business object(s) that are likely to be affected by the delegation of a task (Step 7). This analysis will yield a set of possible manipulation criteria (Step 8). Based on these criteria, the delegating principal will compute the evidence specification (Step 9) which will then be accepted by the delegatee (Steps 10 and 11).

Only then will the preliminary task acceptance (Step 2) by the delegatee become final (Step 12) and once the delegatee successfully performed the task (Step 13), generated evidence (Step 14) and informed the delegator (Step 15), the delegator can register the negotiated evidence specification for future reuse (Step 16).

4.3 Evidence Generation

Evidence is then generated when performing a delegated task, according to the agreed evidence specification. Evidence may be partial, evolving with along the completion of the task. Once the task has been completed, full evidence will have been adduced which may then be accepted or rejected as part of the final review and checking phase. If the evidence has been rejected, the delegatee may be asked to deliver supplementary evidence expressed by setting the state of the currently adduced evidence back to partial.

4.4 Evidence Check and Review

When checking the evidence adduced by the delegatee as part of performing a delegated task, the first step is to determine whether an automated check is possible. If this is the case and the check succeeds, the delegated task has been accepted as performed. If an automated check is possible but fails, the reason for this will be determined resulting in either a choice of revocation, re-delegation and compensation or in a manual check.

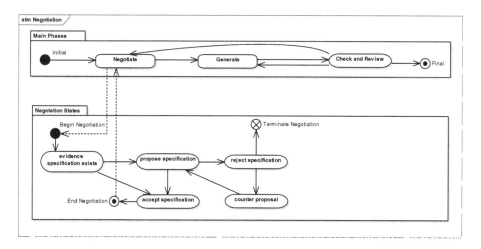

Fig. 3. State machine for negotiation with respect to overall protocol

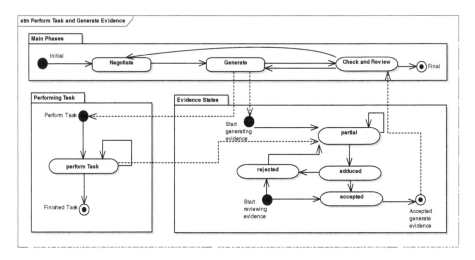

Fig. 4. State machine for evidence generation with respect to overall protocol

This manual check could have also been triggered in case of an automated check not being possible. In the simplest case, the manual check succeeds and the delegated task may be accepted as performed. If the manual check does not succeed, it may be the case that some secondary evidence is required which may eventually lead to a successful check. If neither automatic-, semi-automatic nor fully manual checks succeed, the check and review phase will terminate unsuccessfully.

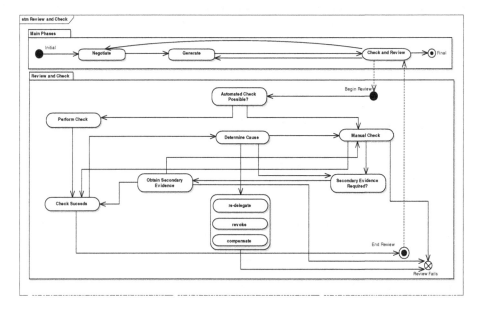

Fig. 5. State machine for evidence check and review with respect to overall protocol

5 Conclusion and Summary

In [1] we have presented our initial investigations into the delegation of obligations and the concept of review as one kind of organisational principle to control such delegation activities. This initial work led us to a more detailed and refined analysis of organisational controls [2], [4] with a particular emphasis on the notion of general and specific obligations (e.g workflow tasks).

However, we had not yet discussed the notion of evidence at a level of detail that would allow for its implementation in a workflow system supporting delegation activities and controlled distribution of work. This paper tried to fill this gap by providing a first framework for evidence lifecycle management, focusing on the stages of evidence negotiation, evidence generation and evidence checking. We chose a state-machine-based approach but did not yet formally verify its correctness (as we did in our previous work using the Alloy constraint analysis tool [3]). This verification is subject to future work.

Another future line of work will be to further investigate into the semantics of business object repositories to support evidence negotiation. Status and Action management tools and models for business objects such as those presented in [10], [11] may be a starting point.

As a last item we intend to investigate the relationship of evidence to the notion of revocation of delegated tasks [12]. More specifically, we will clarify how revocation actions (that may not necessarily stem from the original delegator) influence the generation of evidence or how they already take partially generated evidence into account.

References

1. Schaad, A., Moffett, J.: Delegation of Obligations. In: 3rd International Workshop on Policies for Distributed Systems and Networks (POLICY 2002), Monterey (2002)
2. Schaad, A.: A Framework for Organisational Control Principles, PhD Thesis. Department of Computer Science, University of York (2003)
3. Jackson, D.: A Micromodularity Mechanism. In: 8th Joint Software Engineering Conference, Vienna, Austria (2001)
4. Schaad, A., Moffett, J.: Separation, Review and Supervision Controls in the Context of a Credit Application Process – A Case Study of Organisational Control Principles. In: Schaad, A. (ed.) ACM Symposium of Applied Computing (2004)
5. Muller, J.: Delegation and Management. British Journal of Administrative Management 31(7), 218–224 (1981)
6. Moffett, J.D.: Delegation of Authority Using Domain Based Access Rules, in Dept of Computing. Imperial College, University of London (1990)
7. Mintzberg, H.: The structuring of organizations. Prentice-Hall, Englewood Cliffs, NJ (1979)
8. Pugh, D.: Organization Theory: Selected Readings. 4th ed. Penguin Business, Penguin Books (1997)
9. Rinderle, S., Reichert, M., Dadam, P.: Flexible Support of Team Processes by Adaptive Workflow Systems. Distributed and Parallel Databases - An International Journal 16(1), 91–116 (2004)
10. http://wi.wu-wien.ac.at/home/mendling/bpmdemo/paper5.pdf

11. http://www.sap.info/index.php4?ACTION=noframe&url=http://www.sap.info/public/INT/int/prnt/PrintEdition-1032042665dae07838-int/-1/articleContainer-295204268dc1a1890c
12. Schaad, A.: Revocation of Obligation and Authorisation Policy Objects. DBSec, 28–39 (2005)
13. Schaad, A.: An Extended Analysis of Delegating Obligations. DBSec, pp. 49–64 (2004)

Appendix

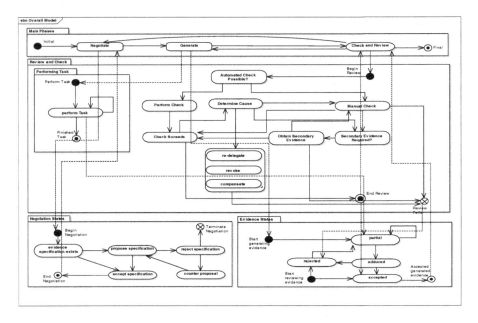

Fig. 6. Summary of general state machine (Figures 3 -5)

Collaboration for Human-Centric eGovernment Workflows

Khaled Gaaloul[1], François Charoy[2], Andreas Schaad[1], and Hannah Lee[3]

[1] SAP CEC Karlsruhe, Security & Trust Group
Vincenz-Priessnitz-Strasse 1, 76131 Karlsruhe, Germany
[2] LORIA - INRIA - CNRS - UMR 7503
BP 239, F-54506 Vandœuvre-lès-Nancy Cedex, France
[3] University Hamburg Vogt-Koelln-Str. 30 D-22527 Hamburg, Germany
khaled.gaaloul@sap.com, charoy@loria.fr, andreas.schaad@sap.com,
lee@informatik.uni-hamburg.de

Abstract. The execution of cross-domain eGovernment processes is a challenging topic. In earlier work, we presented an approach based on collaborative workflows to support eGovernment interoperability. However, such collaborative workflows often appear to be lacking transparency and control supporting concepts and mechanisms. These are needed as eGovernment workflows appear to be heavily human-centric. What is in many cases described as collaboration appears to be a mere coordination and synchronization of processes, often ignoring human-centric interactions. One type of transparency and control supporting mechanism in human-centric collaboration is that of task delegation.

In this paper we aim to analyse the gap between coordination and collaboration in the context of workflow management for eGovernment. First, we present a real case study to identify the key distinguishing factors regarding collaboration as opposed to coordination. Based on this, we present our approach to support cross-organisational collaboration. In particular, we will focus on the concept of delegation in the context of heavily human-centric collaborative workflows. Finally, we propose a delegation extension and structured set of future requirements regarding a coordination architecture presented in earlier work.

Keywords: eGovernment, R4eGov, workflow coordination, workflow collaboration, delegation.

1 Introduction

Electronic government (eGovernment) is the civil and political conduct of government, including services provision, using information and communication technologies. The concept of eGovernment has been gaining ground from initial isolated to extensive research and applications. The prerequisites for an e-Government enactment strategy are the achievement of a technological interoperability of platforms and a deeper cooperation and security at the organisational

M. Weske, M.-S. Hacid, C. Godart (Eds.): WISE 2007 Workshops, LNCS 4832, pp. 201–212, 2007.

level. Those requirements are related with the environment in which the public agencies operate, strictly constrained by norms, regulations, and result-oriented at the same time [1]. Actually, most governmental organisations offer electronic services within a collaborative environment. However, inter-organisational collaboration, especially by means of workflows, is not as widespread.

The R4eGov project consists of inter-organisational collaboration between European administrations [2]. An example domain for such collaboration is Europol[1] (European Police Office) and Eurojust[2] (European Judicial Cooperation Unit). It describes an interagency collaboration within the areas of law enforcement and justice. One of the objectives is to establish a collaboration, including information exchange between both parties based on legal constraints, such as European laws, to which they have to comply to, but sustain effective degrees of freedom for each department to solve their issues in the way they think is the most efficient and effective [3]. Those objectives can be achieved using collaborative workflows [4,5]. This is a novel approach supporting interoperability between organisations without the burden of centralized workflow management systems. The perspective is to enable a particular workflow model to be executed collaboratively by different workflow engines located on the private network perimeters of their respective owners.

However, recent works [6,7] presented new requirements such as control and transparency in collaborative workflows. What is in many cases described as collaboration appears to be coordination and synchronization of processes by ignoring human-centric interactions. Actually, we need to consider all the relevant participating systems and workflows even if they are not directly involved in the current control-flow sequence of the workflow.

This paper expands on earlier work we have done in R4eGov to support inter-organisational collaboration between European administrations based-workflow. We aim to elicit the collaborative requirements between Europol and Eurojust and the definition of methods and tools to support such an human-centric collaboration. This collaboration requires transparency and control supporting concepts and mechanisms. The concept of delegation had not yet been treated in sufficient detail in the context of heavily human-centric collaborative workflows, and is the subject of this paper to foster transparency and control mechanisms in collaborative workflows according to global policies and European law regulations in R4eGov.

The remainder of this paper is organized as follows. Section 2 presents a workflow example inspired from an R4eGov scenario and shows the difference between workflow coordination and collaboration. We motivate in section 3 the use of collaborative workflow management for eGovernment and present our approach to support cross-organisational collaboration. In particular, we will focus on the concept of delegation. Section 4 extends our approach and presents some future requirements. Section 5 presents some related work regarding delegation. Section 6 concludes and presents some future works.

---

[1] http://www.europol.eu.int/
[2] http://www.eurojust.europa.eu/

2   eGovernmental Workflow Scenario

We introduce in this section an R4eGov workflow scenario related to the European administrations collaboration. Europol and Eurojust are two key elements of the European system of international collaboration within the areas of law enforcement and justice. They carry out very specific tasks in the context of dialogues, mutual assistance, joint efforts and cooperation between the police, customs, immigration services and justice departments of the EU member states [3]. During their collaboration, Eurojust and Europol are involved and a number of legal instruments are used. A Specific scenario for this collaboration is the Mutual Legal Assistance (MLA)[3].

2.1   Mutual Legal Assistance (MLA)

Figure 1 depicts a global workflow scenario called Mutual Legal Assistance (MLA) involving two national authorities of different European countries regarding the execution of measures for protection of a witness in a criminal proceeding. This simplified collaborative workflow is inspired by and in parts derived from the case studies delivered in the European research project R4eGov [3].

The workflow shows a member of the Europol National Unit in country A asking for an MLA request. The rules of procedure on the processing and protection of personal data at Eurojust refer to a "Case Management System" (CMS). The measure is to be executed in country B. Europol National Unit makes a written request of assistance (witness protection) to Eurojust National Member (EJNM) A. Then, EJNM A opens a Temporary Work File in the CMS, and contacts EJNM B forwarding the request of assistance. The EJNM B contacts the responsible national authority of country B. Finally, steps will be taken by the responsible national authority to provide the requested assistance.

As shown in the example, four different parties are involved in this collaboration. After the Europol National Unit A sends the request to the corresponding contact point it waits until the follow-up is established by the other authorities. What is going on between the two EJNMs A and B is not mediated to the Europol National Unit A. Hence, this appears to be more than coordination and synchronization of processes between governmental organisations. Actually, several of the depicted tasks involve human interactions and are possibly time consuming. For instance, steps taken by the responsible national authority to provide the requested assistance might involve several stakeholders that decide on this subject, hereby considering different aspects like the severeness of the case or concurrent investigations. With respect to transparency, the current process status needs to be communicated to EJNM B. In addition, unexpected events can happen during any task (unexpected in the sense that it cannot be modeled beforehand, e.g. changing bilateral agreements between countries A an B) that need to be propagated to every other participant of the collaborative

---

[3] This case study has been performed in joint collaboration between Eurpol, Eurojust and Unisys in the context of the EU FP6 IST Integrated Project R4eGov.

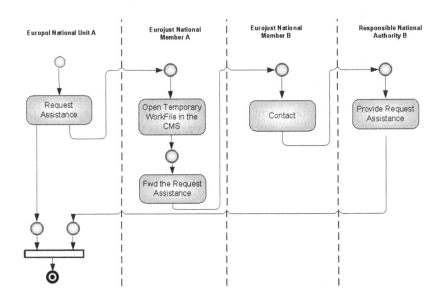

Fig. 1. Mutual Legal Assistance scenario

workflow (especially to the Europol National Unit A, the initiator of the collaborative workflow) even if they do not interact directly with each other in terms of control-flow.

2.2 Problem Statement

In the common understanding a collaborative workflow consists of one global workflow model composed of public views of each collaborative partner, with each public view abstracting a concrete private workflow behind it [5]. Existing choreography and collaboration approaches support the control-flow related message exchange that is part of the workflow model itself. Apparently this message exchange in case of status notification or cancellation is not part of the collaborative workflow model [6]. What is in many cases described as collaboration or collaborative mechanisms appears to be coordination and synchronization of processes by ignoring human-centric interactions. The requirements for interactions and monitoring can be summarized as transparency and control [8]. Transparency addresses the revelation of collaborative dependencies. This allows to react accordingly to exceptions and compensations implied by law regulations. Control fosters the behaviour of partners according to the collaborative policies (e.g. European laws).

This scenario depicts that we need to consider all the relevant participating systems and workflows even if they are not directly involved in the current control-flow sequence. Moreover, emergency situations can necessitate delegation of some activities intra and inter-organisations. For instance, EJNM B can delegate its part of the work to the last party. This delegation has to be legal and compliant with the R4eGov laws regulations policies. In the next section, we motivate the concept of delegation as a support for transparency and control within an human-centric collaboration between Europol and Eurojust.

3   An Extended Analysis of Collaboration for eGovernmental Workflows

As stated before, control and transparency are important for a successful eGovernment collaboration. Hence, we follow a decentralized approach, combining the local workflows to form a collaborative workflow, integrating the existing systems of the involved partners, and adding a decentralized collaborative administration architecture to support human interactions (e.g. during a delegation request).

3.1   Workflow Engine Encapsulation

One re-occurring requirement is to enable collaborative workflows across different organisations without changes to the existing IT landscape of each organisation. A solution that enables collaborative workflows therefore needs to be built on top of existing solutions. Considering that assumption, the purpose of the workflow engine encapsulation is to offer a common interface to collaborative components (that would need to be deployed on each participants system) independent of the underlying workflow engine in place.

Such an interface needs to work in both directions: The collaborative components need to access engine and process specific functionalities. The process, during its execution, needs to publish events or performs requests to the collaborative components. Dealing with those requirements, we propose to set up a workflow to workflow collaboration by realizing a layer which we call an Administrative Communication Layer (ACL).

3.2   Administrative Communication Layer (ACL)

Wolter and *al.* [6] proposed an abstract modular infrastructure for collaborative workflow management and identified key components to leverage an existing workflow system onto decentralized collaboration. As indicated by Figure 2 the proposed architecture is divided into a control-flow layer and an administration layer.

In [7], we developed a prototype extended collaborative workflow tool to support the collaboration between the MLA partners. The term Administrative Communication Layer refers to the distinction of administrative events (e.g. starting/completion of one task) with control events (e.g. triggering a workflow instance) in each workflow engine. Our collaborative communication is event-based, on demand, or a combination of both. Therefore, ACL enables administrative information exchange by mediating information to the collaborative event management and process management components of the collaborative

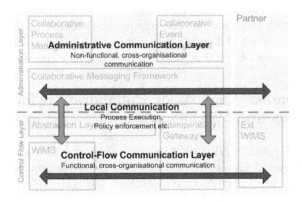

Fig. 2. Administrative Communication Layer

partners. The following aspects of administrative communication between the local process engine and the collaborative partners are supported:

- Status management to represent the overall status of the collaborative workflow (displaying the local process of the executing participant together with the overall workflow of all involved parties).
- Exceptions handling and execution of alternative scenarios which cannot be handled as part of the regular process model (e.g. EJNM delegates a part of his work to another authority due to legal changes).

3.3 Delegation Scenarios in MLA Request

Delegation is an important factor for secure distributed computing environments. It consists of delegating a part of a work to another partner according to laws regulations policies. Delegation can be motivated by many factors (e.g. lack of resources, organisational policies, etc.) [9] and can take place depending on the delegator/delegatee agreement. This agreement is closely related to the delegation criteria. In the following, we identify two different criteria of delegation from the MLA scenario:

Role-Based Delegation: The basic idea behind a role-based delegation is that users themselves may delegate role authorities to others to carry out some functions authorized to the former. Our interest is in the Eurojust's side, the main actors involved in Eurojust's information workflow are: Eurojust National Members and National Correspondents (NCs). Eurojust National Member of country A can play a role of a senior, an experienced prosecutor, or a judge. As a leader of the Eurojust organisation, EJNM, confronts problems that are particularly perplexing, collaborations are necessary for information sharing with members from the same organisation. Since EJNM believes in delegating responsibility, he would like to delegate certain responsibilities to the NC member where the

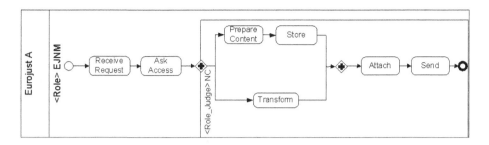

Fig. 3. Role-based delegation

former define the delegation condition based on the organisation Role Hierarchy (RH) (see figure 3). The delegation request is inspired from RDM2000 proposed by Zhang and al. [10].

Task-Based Delegation: Our interest is in the collaboration between the EJNM and the responsible of national authority in country B (see figure 1). What do we mean by task-based delegation is the delegation of a set of tasks to the delegatee. The condition of delegation depends on the organisational task alignment. Actually, a task Ti can be delegated if and only if the delegatee has a task Tj where Tj can give the same feedback/output than Ti and can substitute it.

Due to emergency situations, EJNM B needs to delegate to an external party. The main task of EJNM B consists of determining the Judicial Authority (JA). Since the latter organisation role is JA, we assume that a delegation request can be motivated by the factor of specialisation. Nevertheless, the condition of delegation doesn't depend on the delegation factor (the specialisation of the JA member). Actually, a delegation request can be done if and only if the latter organisation offers the same service (set of tasks) to ensure the well and coherent deployment of the process. Coherency is closely related to the global policies of the MLA request. Here, our concern is the output of the task and not the "how" of the task. Since, the delegatee (JA member) provides tools, platforms and solutions to handle this delegation, there is no need to go deeper in his process attributes and then disclose his privacy. The agreement can be defined according to tasks alignment between organisations. This alignment is defined in the global policies of the process.

3.4 Extended Architecture

The scope of our approach is to address user-to-user delegation supporting human-centric collaborative workflows. We propose an abstract modular architecture to extend the architecture presented in [6].

Once a delegator needs to delegate, ACL communicates a delegation request to the delegatee where the former invites the latter to accept to be in charge of one or more tasks of his local process.

A preliminary prototype implementation of the delegation mechanisms has been developed. It is based on mail request where a delegator sends a delegation

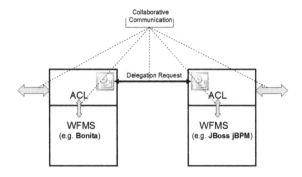

Fig. 4. Extended architecture supporting delegation

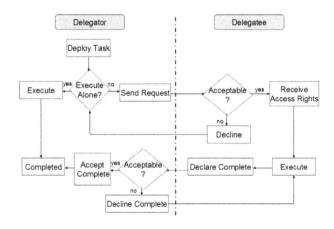

Fig. 5. Dialogue states and actions

request and wait for the approval of the receiver (the delegatee). The delegation dialogue between the two actors is depicted in figure 5. The acceptance/decline of the request and the delegating access rights (e.g. credentials, resources, etc.) will depend on the MLA global policies. Actually, *"Receive Access Rights"* step is more complicated: it depends on the delegation criteria (e.g. role, permission) and the availability of the resources to the delegatee. This step needs more investigations and time and will be a part of our future works.

4   Additional Requirements for Delegation

Delegation offers a suitable solution to support both ad-hoc and process-based interactions in an eGovernment context. In this highly dynamic environment, delegation is closely related to other concepts and mechanisms such as authorizations policies and revocation to allow the compliance and on-the-fly shift of responsibilities with respect to an ongoing collaboration both on a (atomic) task level and on a (global) process level.

4.1 Authorization Policy Model from the MLA Scenario

Authorization policy model can be used to specify public roles and their privileges and provides means to specify role mapping to internal and external roles of participating organisations. This specification will be used to identify the delegating access rights requirements afterwards. The Mutual Legal Assistance (MLA) scenario is being used as an example to illustrate the requirements of the authorization policy specification. Some of the related authorization policies rules are listed below:

1. **Europol National Unit A:** Only national units, liaison officers, and the Director, Deputy Directors or duly empowered Europol officials shall have the right to input data directly into the information system and retrieve it therefrom.
2. **Eurojust National Member A:** The case management system shall allow National Members to define the specific items of personal and non-personal data to which they wish to give access to other National Member(s), Assistant(s) or authorized staff members that are involved in the handling of the case.
3. **Eurojust National Member B:** When a National Member gives access to a temporary work file or a part of it to one or more involved National member(s), the case management system shall ensure that the concerned users have access to the relevant parts of the file but that they cannot modify the data introduced by the original author.
4. **Responsible National Authority B:** The case management system shall mark such data in a way that will remind the person who has introduced the data in the system of the obligation to keep these data for a limited period of time.

The involved participants are highly heterogeneous, and they intend to stay autonomous in terms of controlling their resources and executing tasks responsible for them. Moreover, fairly fixed, globally known roles for the collaboration are already established, and lastly, their privileges are often derived from European laws and regulations that must be followed by all participants. The last two characteristics make the Role Based Access Control (RBAC) model an extremely attractive choice to specify public roles and their privileges [11].

4.2 Revocation

Revocation is an important process that must accompany the delegation. It is the subsequent withdrawal of previously delegated objects such as a role or a task. A vast amount of different views on the topic can be found in literature [12,13] where each author having their own assumptions and opinions on how to model revocation. For simplification, our model of revocation is closely related to the delegation model based user-to-user. Actually, the decision of revocation is issued from the delegator in order to take away the delegated privileges, or the desire to go back to the state before privileges were delegated. The privileges consist of the delegating access right provided to the delegatee. Basically, delegating access rights issued from the delegator describes the permission given to access to the task resources such as rogatory letters or legal requests in the MLA scenario.

5 Related Work

In this section, we present a literature review related to the delegation requirements. Basically, we aim to come up with a delegation classification of models, policies, and technologies that will fit with the motivation criteria for delegation in our future works.

The eXtensible Access Control Markup Language is an XML-based, declarative access control policy language that lets policy editors to specify the rules about who can do what and when. As an OASIS standard, its greatest strength lies in interoperability [14]. Unlike other application-specific, proprietary access-control mechanisms, this standard can be specified once and deployed beyond the boundaries of organisations and countries. In [15], Rissanen and Firozabadi add new structured data-types to express chains of delegation and constraints on delegation. The main result of their research is an administrative delegation. It is about creating new long-term access control policies by means of delegation in a decentralised organisation. However, this approach does not cover ad-hoc interactions and seems to not support decentralized delegation in the context of MLA.

In [16,17], they tackle new requirements for delegation such as delegating in a dynamic and light-weight manner, performing single sign-on, and reusing existing protocols and software with minimal modifications. Welch and al. define *Proxy Certificates* allowing an entity holding a standard X.509 public key certificate to delegate some or all of its privileges to another entity. This delegation can be performed dynamically, without the assistance of a third party. However the problem with the X.509 proxy certificates is that commercial tooling for Web Services does not necessarily recognize and properly process these certificates [18]. Wang and Del Vecchio try to leverage and extend existing Web Services standards, without breaking the existing tooling by exploiting the Security Assertion Markup Language (SAML) inherent extensibility to create a delegation framework. They develop a set of verification rules for delegation tokens that rely on WSSecurity X.509 signatures, but do not force any trust relationship between the delegatee and the target service. However, this approach support heavily computing and is time consuming that may slack the MLA deployment during a delegation request.

Role-based access control (RBAC) is recognized as an efficient access control model for large organisations. Most organisations have some business rules related to access control policy. Delegation of authority is among these rules

[11]. In [19,20], authors extend the RBAC96 model by defining some delegations rules. Barka and Sandhu proposed a role-based delegation model. They
deal with user-to-user delegation. The unit of delegation in them is a role. However, users may want to delegate a piece of permission from a role [20]. Zhang
and *al.* propose a flexible delegation model named Permission-based Delegation
Model (PBDM). PBDM supports user-to-user and role-to-role delegations with
features of multi-step delegation and multi-option revocation. It also supports
both role and permission level delegation, which provides great flexibility in authority management. However, neither RBAC nor PBDM support the task-based
delegation criteria described in the MLA delegation scenario.

6   Conclusion and Future Directions

In this paper we presented a novel approach to support an human centric collaborative workflow for eGovernment. Our primary concern is to analyse the
gap between coordination and collaboration by distinguishing factors regarding
collaboration as opposed to coordination. Actually, transparency and control
supporting concepts and mechanisms are not taken into account in the context
of heavily human-centric collaborative workflows. To satisfy this need we propose an extended architecture supporting task delegation as a mechanism in
human-centric collaboration. Further, we discussed future requirements regarding a coordination architecture presented in earlier work.

We consider this paper as a primer for future related work in the areas of collaboration and security. Our concern will be to come up with a secure delegation
mechanism supporting privacy and dynamic human interactions by addressing
the delegating access rights issue. Moreover, we plan to further investigate the
area of compliancy accordingly to the R4eGov laws regulations policies.

References

1. Traunmüller, R. (ed.): EGOV 2004. LNCS, vol. 3183. Springer, Heidelberg (2004)
2. R4eGov Technical Annex 1. Towards e-Administration in the large(March 2006).
   http://www.r4egov.info
3. Eurojust / Europol collaboration (2006). SIXTH FRAMEWORK PROGRAMME,
   Information Society Technologies, R4eGov.
4. Schulz, K.A., Orlowska, M.E.: Facilitating cross-organisational workflows with a
   workflow view approach. Data Knowl. Eng. 51(1), 109–147 (2004)
5. Contenti, M., Mecella, M., Termini, A., Baldoni, R.: A Distributed Architecture
   for Supporting e-Government Cooperative Processes. In: Böhlen, M.H., Gamper,
   J., Polasek, W., Wimmer, M.A. (eds.) TCGOV 2005. LNCS (LNAI), vol. 3416, pp.
   181–192. Springer, Heidelberg (2005)
6. Wolter, C., Plate, H., Herbert, C.: Collaborative Workflow Management for eGovernment (September 2007). Accepted in the 1st international workshop on Enterprise Information Systems Engineering (WEISE)
7. Indrakanti, S., Gaaloul, K., Rahaman, M., Plate, H.: Prototype extended collaborative workflow tool (March 2007). Deliverable WP5-D3, SIXTH FRAMEWORK
   PROGRAMME, Information Society Technologies, R4eGov

8. Jensen, C., Scacchi, W.: Collaboration, Leadership, Control, and Conflict Negotiation in the NetBeans.org Community. In: 26th International Software Engineering Conference (2004)
9. Atluri, V., Warner, J.: Supporting conditional delegation in secure workflow management systems. In: SACMAT 2005 Proceedings of the tenth ACM symposium on Access control models and technologies, Stockholm, Sweden, pp. 49–58. ACM Press, New York (2005)
10. Zhang, L., Ahn, G.-J., Chu, B.-T.: A rule-based framework for role-based delegation and revocation. ACM Trans. Inf. Syst. Secur. 6(3), 404–441 (2003)
11. Belokosztolszki, A., Eyers, D.M., Moody, K.: Policy Contexts: Controlling Information Flow in Parameterised RBAC. Policy 00, 99 (2003)
12. Wainer, J., Kumar, A., Barthelmess, P.: DW-RBAC: A formal security model of delegation and revocation in workflow systems. Inf. Syst. 32(3), 365–384 (2007)
13. Hagstrom, A., Jajodia, S., Parisi-Presicce, F., Wijesekera, D.: Revocations-A Classification. In: CSFW 2001: Proceedings of the 14th IEEE workshop on Computer Security Foundations, Washington, DC, USA, p. 44. IEEE Computer Society Press, Los Alamitos (2001)
14. Moses, T. (ed.): eXtensible Access Control Markup Language (XACML) Version 2.0, OASIS. Last viewed on Mar. 28, 2007
15. Rissanen, E., Firozabadi, B.S.: Administrative Delegation in XACML. Swedish Institute of Computer Science, Kista-Sweden
16. Kesselman, I., Mulmo, C., Pearlman, O., Tuecke, L., Gawor, S., Meder, J., Siebenlist Welch, S., Foster, V.: X.509 Proxy Certificates for Dynamic Delegation (2004). 3rd Annual PKI R&D Workshop.
17. Rits, M., Schaad, A., Crosta, S., Pazzaglia, J.-C.: A Secure Public Sector Workflow Management System. In: 21st Annual Computer Security Applications Conference, Tucson, Arizona (December 2005)
18. Humphrey, M., Wang, J.: Extending the Security Assertion Markup Language to Support Delegation for Web Services and Grid Services. In: IEEE International Conference on Web Services (ICWS 2005), vol. 1, IEEE Computer Society Press, Los Alamitos (2005)
19. Barka, E., Sandhu, R.: Framework for role-based delegation models. In: ACSAC 2000: Proceedings of the 16th Annual Computer Security Applications Conference, Washington, DC, USA, p. 168. IEEE Computer Society Press, Los Alamitos (2000)
20. Zhang, X., Oh, S., Sandhu, R.: PBDM: a flexible delegation model in RBAC. In: SACMAT 2003: Proceedings of the eighth ACM symposium on Access control models and technologies, pp. 149–157. ACM Press, New York (2003)

International Workshop on Human-Friendly Service Description, Discovery and Matchmaking (Hf-SDDM)

Workshop PC Chairs' Message

Dominik Kuropka

Hasso Plattner Instiut
University of Potsdam
Prof.-Dr.-Helmert-Str. 2-3
14482 Potsdam, Germany
dominik.kuropka@hpi.uni-potsdam.de

Motivation

The rise of service-oriented architectures will boost the amount of available Web services in the future. This will put the question of proper Web service discovery and description on the agenda. Current technologies such as UDDI and ongoing research efforts on semantic service technology are usually focused on technical, theoretical and correctness issues rather than on the ease of use. This results in high requirements on both the service requesters who are searching for services, and the service providers who have to contribute proper descriptions of their services.

This workshop aims at being a platform for the discussion on how semantic service technology can be made easier to use and handle to make it applicable in "real-world" usage scenarios which involves "normal" users instead of hard-core logic and semantic experts.

Overview of Workshop Papers

Gennady Agre and Ivan Dilov present the INFRAWEBS Designer in their paper titled "How to Create a WSMO-based Semantic Service without Knowing WSML". This tool provides a user-friendly and easy-to-use way to graphically model semantic descriptions of Web services and goals according to the WSMO Framework.

In their paper on "User-friendly Semantic Annotation in Business Process Modeling" Matthias Born, Florian Dörr, and Ingo Weber presents an approach and their proof-of-concept prototype tool for the semantic annotation of business processes. Their main idea is to increase the precision of business process models by adding semantic information and thus ease the Web service discovery and composition when it comes to the implementation of business processes.

Ivan Markovic and Mario Karrenbrock take up the topic of semantic business process modeling in their paper on "Semantic Web Service Discovery for Business Process Models". They present an approach and a prototypical implementation for the discovery of Semantic Web services for process task implementation based on semantically enriched business process models.

An extension of the Web Service Modeling Toolkit (WSMT) is presented by Michael Stollberg and Mick Kerrigan in their paper on "Goal-based Visualization and Browsing for Semantic Web Services". This allows clients to browse and search for

M. Weske, M.-S. Hacid, C. Godart (Eds.): WISE 2007 Workshops, LNCS 4832, pp. 215–216, 2007.
© Springer-Verlag Berlin Heidelberg 2007

Web services on the level of goals that can be solved by them, which allows to better comprehend the available Web services from a problem-oriented perspective and significantly eases their handling.

David Lambert, Stefania Galizia, and John Domingue diagnose in their paper on "Agile elicitation of semantic goals by wiki" that there are three, usually disjoint parties involved when it comes to Semantic Web services: people who need a specific functionality (users), people who provide functionality by exposing Web services (service providers) and people who usually do the semantic annotation of services. In their paper the authors present a wiki-based technique which eases the communication between these three groups.

Web service searching can be performed by various stakeholders, in different situations, making different forms of queries appropriate. Therefore Jianguo Lu and Yijun Yu argue in their paper "Web service search: who, when, what, and how" that all those combinations of the stakeholders should result in different ways of implementation of the Web service search. Using a real web service composition example, the authors describe in their paper when, what, and how to search web services from service assemblers' point of view.

How to Create a WSMO-Based Semantic Service Without Knowing WSML

Gennady Agre[1] and Ivan Dilov[2]

[1] Institute of Information Technologies – Bulgarian Academy of Sciences
agre@iinf.bas.bg
[2] Faculty of Mathematics and Informatics - Sofia University
idilov@gmail.com

Abstract. In order to make accessible new Semantic Web Services technologies to the end users, the level of tools supporting these technologies should be significantly raised. The paper presents such a tool - an INFRAWEBS Designer – a graphical ontology-driven development environment for creating semantic descriptions of Web services and goals according to WSMO Framework. The tool is oriented to the end users – providers of Web services and semantic Web services applications, who would like to convert their services into WSMO based semantic Web services. The most character features of the tool – intensive use of ontologies, automatic generation of logical description of a semantic service from graphical models and the use of similarity-based reasoning for finding similar service descriptions to be reused as initial templates for designing new services are discussed.

Keywords: Semantic Web Services, Web Service Modeling Ontology, Graphical Modeling, Case-based Reasoning.

1 Introduction

At the moment the practical application of Semantic Web Service (SWS) technologies is still rather restricted due to several reasons, some of which are the high complexity of both OWL-S 10] and WSMO [12] Frameworks, the lack of standard domain ontologies, unavailability of mature tools supporting WSMO or OWL-S, and the absence of pilot applications focusing on every-day needs of consumers, citizens, industry etc., which can demonstrate the benefits of using semantics [15]. The IST research project INFRAWEBS [1] proposed a technology-oriented step for overcoming some of the above-mentioned problems. It focused on developing a Semantic Service Engineering Framework enabling creation, maintenance and execution of WSMO-based SWS, and supporting SWS applications within their life-cycle. Being strongly conformant to the current specification of various elements of WSMO, the INFRAWEBS Integration Framework (IIF) hides the complexity of creation of such elements by identifying different types of users of Semantic Web Service Technologies; clarifying different phases of the Semantic Service Engineering process, and developing a specialised software toolset oriented to the identified user types and intended for usage in all phases of the SWS Engineering process.

M. Weske, M.-S. Hacid, C. Godart (Eds.): WISE 2007 Workshops, LNCS 4832, pp. 217–235, 2007.
© Springer-Verlag Berlin Heidelberg 2007

The necessity of intensive development of tools supporting SWS technology and especially WSMO initiative has been clearly recognized in WSMO community and several such tools have been developed or still under development (e.g. WSMO Studio [17], Ontology Editing and Browsing Tool [16], WSML Editor [7]). However, the analysis of these tools for creating WSMO objects has shown that they are oriented mainly to the researchers and developers, working in the area of SWS technology, rather than to the real users of such a technology – Web service providers, semantic service brokers and semantic application providers. The creation of SWS is done by using structural text editors, which of course simplify the process for creating WSML description of a service, but still request strong knowledge of WSML – the logical language used for describing SWS and goals in WSMO [6].

A different approach was proposed in the INFRAWEBS Designer - a graphical, ontology-based, integrated development environment for designing WSMO-based SWS and goals. The INFRAWEBS Designer is oriented to Web service providers and Web service application providers, and does not require any preliminary knowledge of WSML. The most important characteristics of the Designer are:

- *User-friendliness:* it proposes an intuitive graphical way for constructing and editing service and goal descriptions, abstracting away as much as possible from the concrete syntax of the logical language used for implementing it. The WSML description of the semantic object under construction is automatically generated from the graphical models created by the user.
- *Intensive use of ontologies:* the process of constructing logical descriptions of semantic objects is ontology-driven - in each step of this process the user may select only those elements of the used ontologies that are consistent with the already constructed part of the object description.
- *Reusability:* creation of semantic descriptions is a complex and time-consuming process, which can be facilitated by providing the designer with an opportunity to reuse existing, similar descriptions, created by the designer herself or by other users. The INFRAWEBS Designer provides the user with such an opportunity by applying the case-based reasoning approach.

The detailed description of the INFRAWEBS Designer conceptual and functional architecture can be found in [2]. The main objective of this paper is to show how this tool can be used by service providers to create a WSMO-based semantic Web service and to present some conclusions on the tool applicability based on our experience in creating two pilot INFRAWEBS applications - the first uses a travel agency scenario [8] and the second is based on an eGovernment scenario [11].

The structure of the paper is as follows - the next section introduces a general structure of a WSMO-based SWS and describes how this structure is represented in the INFRAWEBS Designer. Section 3 considers a 6-steps procedure for creating and publishing the semantic (WSML) description of a Web service. Section 4 briefly presents two INFRAWEBS test beds and discusses some results of using the INFRAWEBS Designer for their development. The last section is a conclusion.

2 WSMO-Based Semantic Service and Its Graphical Representation

A WSMO-based semantic description of a Web service contains three types of information [12]: *the service top entity* is the information related to service identification (service non-functional properties, name spaces etc.), *service advertisement* (a logical description of service capability that is used for semantic service discovery), and a service choreography (a logical description of service behaviour that is used for communication with and execution of the service). Although such information is very heterogeneous it is displayed in the Designer in a uniform way – as a service tree in which internal (functional) nodes represent the roles of each specific portion of the information and are used for editing the tree structure, while the tree leaves serve as pointers to the content of each information portion (Fig. 1).

Two of the service tree sections – "Namespaces" describing the correspondences between short (so called SQ-names) and full names (IRI) of ontologies used for description of the service and "Imported ontologies" describing which ontologies are needed for service description are created by the Designer automatically. The list of imported ontologies is created as a result of analysis of the WSML text of logical axioms used for describing the service capability and choreography. The namespaces for these ontologies are created by means of a special algorithm assigning unique SQ-names for ontologies used in the service. The fully automatic way for creating these parts of WSMO service description guarantees the correctness of

Fig. 1. An example of service tree

the description and solves the problem with possible repetition of the same SQ-names used for describing different ontologies by different users. And, what is the most important, it allows the user of the Designer to create correctly these parts of WSMO SWS description without any knowledge of WSML syntax. "Non-functional properties" section of a service tree visualizes all non-functional properties (NFP) of the service entered by the service designer. The INFRAWEBS Designer provides a special structural text editor for creating and editing NFP of a WSMO service – the NFP Editor.

The most important part of the WSMO service identification information, is the service IRI, which uniquely identifies the service. Following the basic design principles of the INFRAWEBS Designer this record is also created fully automatically.

The service capabilities are represented in the service tree by five functional nodes - "Shared variables", "Preconditions", "Assumptions", "Post-conditions" and "Effects", which visualize the corresponding parts of a SWS description. The last four nodes can have leaves which are the named pointers to the corresponding WSML logical expressions (axioms) represented by their graphical models.

The last part of the service tree shows the WSMO service choreography which according to the WSMO specification consists of the service state signature and state transition (choreography) rules. The service state signature is represented as a sub-tree, which nodes correspond to different ontology concepts playing the specified roles ("in", "out", "shared" or "controlled"). Some of them can have successors pointing to the WSDL operations grounded to those concepts. The context-sensitive right-click menus associated with the nodes of the state signature tree allow creating and editing this part of the service tree.

The state transition rules are shown as sub-trees, which roots provide information of the rule type and concrete values of the rule parameters. Double-clicking the rule roots leads to visualizing and editing the graphical model corresponding to the logical WSML axiom describing the condition part of the selected rule.

The leaves of each transition rule sub-tree show the conclusion part of the corresponding rule (so called "update rules"). Each update rule is represented by a special graphical symbol corresponding to the rule type ("add", "delete" or "update") and a string corresponding to the WSML content of the rule (so called "fact"). Double-clicking the update rule leads to visualizing and editing the graphical model corresponding to the logical WSML axiom describing the "factual" part of selected rule. A set of context-sensitive right-click menus associated with each node of the choreography part of the service tree allows creating and editing the corresponding elements of the tree.

3 Designing a WSMO-Based Semantic Web Service

In the INFRAWEBS Framework the SWS design process is considered as a complex activity for creating a semantic (WSML) description of a given non-semantic Web service (described by its WSDL file), based on a set of appropriate WSML ontologies and (optionally) facilitated by a set of similar descriptions of other semantic services. In order to facilitate this very complex activity it is split into several steps, which are described in the next subsections in more details.

3.1 Step 1 - Finding a WSDL Description of a Semantic Web Service

The description of a WSMO-based semantic Web service can be seen as a semantic extension (annotation) of a real Web service (represented by its WSDL file) allowing automatic service discovery, enactment and execution. We assume that the selection or finding the appropriate WSDL file is, usually, a first step in a long process of converting an ordinary Web service into semantic (WSMO-based) service. That is why, the INFRAWEBS Designer provides the user not only with conventional means for selecting a WSDL file from her local computer store based on the file name, but with a specially designed tool for finding and selecting the desired WSDL file from the remote repository (the SIR component of the INFRAWEBS Framework [14]) based on the similarity of this file with its description represented by a special WSDL query form.

The query form allows specifying terms the user expects to find in the WSDL service metadata created during the service annotation in the SIR as well as words she expects to find in the service content, i.e. in the names of service operations, messages etc. Moreover, the user can select a category (or several categories), to which the desired WSDL file is expected to belong, using for this purpose a graphical representation of the service categorization scheme as a special WSML ontology. The filled query form is converted to the XML query and sent to the Organizational Memory (OM) – a special IIF components playing a role of the case-based memory in the INFRAWEBS Framework [5], which translates it to a SPARQL query and sends it to the SIR. After receiving the results, the OM ranks them according to the similarity coefficient and parameters specified by the user (the minimum similarity threshold and the maximum number of the results) and returns the ordered list back to the Designer along with some textual annotations. The user can browse the list of matched WSDL services and select the desired one, which will be downloaded from the SIR to the special in-memory store of the Designer - WSDL Store. The downloaded WSDL file can be saved manually to the local store of the user's computer or be saved automatically during saving the semantic service if the WSDL file will be grounded to the description of this semantic service.

3.2 Step 2 – Finding Appropriate WSML Ontologies

A process of converting a non-semantic Web service into a WSMO-based semantic Web service may be seen as an interactive ontology-driven process of Web service annotation. Creation of such an annotation crucially depends on the availability of proper ontologies. In practice finding the appropriate set of ontologies is one of the initial steps the user has to do before creating a semantic description of a Web service.

The IIF assumes that all ontologies are stored in the Remote Repository of WSMO Objects (DSWS-R [9]). Thus, the process of finding ontologies is implemented in the Designer as a complex procedure for communicating with the IIF, in which the user describes the expected content of the required ontology, the Designer translates this description into a XML query and sends it to the OM component, which plays a role of an indexer of the DSWS-R in the IIF. The OM matches the query against its internal representation of ontologies stored in the DSWS-R (cases) and returns a set of ontology identifiers, which are the most similar to the query. After the user has inspected and selected the desired set of ontologies, they are downloaded to the Designer's Ontology in-memory Store from the remote DSWS-R.

Fig. 2. Searching ontologies

The ontology query form allows the user to specify words she expects to be used for describing concepts or relations in the desired ontology. As an example of such words it is also possible to use the WSDL file, describing the Web service that the user wants to convert into semantic one. As usual, the user can set the minimal threshold for lexical similarity between the searching ontology and the query as well as the maximum number of matched ontologies she would like to be presented in the result section of the query form (Fig. 2).

Ontologies can be also loaded from the Designer Temporary Store (workspace) located at the user computer. Since the process of designing a semantic service is rather time consuming, the Designer allows the user to locally store the ontologies found in the remote repository and temporary loaded into the in-memory Ontology Store of the Designer. However, such manual saving of ontologies may be avoided since the Designer automatically saves all ontologies used for describing a semantic service when the service is saved into the workspace.

It should be mentioned that all WSMO objects (ontologies, services and goals) are identified in the Designer and DSWS-R by their IRI (not by names of files where they are stored). The specially designed file manager supporting the correspondence between a unique IRI of a WSMO object and a name of a file where the object is stored is used both for downloading and loading all WSMO objects.

Ontologies describe inheritance between concepts. A concept usually has one or more super-concepts that can be defined in other "imported" ontologies mentioned in the corresponding section of the ontology description. Normally the imported ontologies are not loaded into the Ontology Store and as a result, all concepts having their super-concepts defined in such ontologies can not inherit any of the super-concepts' attributes. In order to avoid this deficiency the INFRAWEBS Designer provides a mechanism for on-demand loading of imported ontologies when a concept, which super-concepts are defined in the imported ontologies, is selected.

It should be noted, that the Designer does not know in advance where the ontology that has to be loaded on-demand is stored. That is why it initially attempts to load the ontology from the local workspace and then, if the Store does not contain such ontology, tries to download the ontology from the remote store (DSWS-R).

All ontologies are loaded into the Ontology Store, which is a global structure accessible for all semantic services loaded into the Designer. Such an organization allows to design in parallel several new semantic services using the same set of ontologies from the Ontology Store. It is very convenient in cases, when the user is going to design several semantic services from a single complex Web service (e.g. Amazon.com), using different sets of the Web service operations.

3.3 Step 3 - Creation of a Service Choreography

The semantic service choreography describes how the service works. The choreography description of a WSMO service consists mainly of three parts [13]:

- *Top level choreography elements*, which include a set of imported ontologies used for describing the service choreography and a set of non-functional properties of the choreography. In the Designer the imported ontology set is constructed *automatically* by analyzing the content of the choreography state signature and rules. The non-functional properties of the service choreography can be created and edited in a uniform way by means of the NFP Editor (see Section 2).

- *State signature,* which contains main ontological concepts describing the service behaviour and how some of these concepts are related to input/output operations of the Web service (service grounding).
- *State transition rules,* which describe how the user can communicate with the semantic service in order to execute the Web (WSDL) service grounded to the semantic service description.

The creation of a state signature is implemented as a two steps procedure – specifying the mode (role) an ontology concept or relation will play in the state signature, and then selecting a proper concept for this role. The mode of the desired concept is selected from the right-click menu associated with the state signature node of the service tree while the selection of the concrete ontology concept (relation) for the specified node is realized as a selection from the window listing all ontology concepts (relations) available in the in-memory Ontology Store.

When a concept is selected to be used as "in", "out" or "shared" mode, an additional step (or several steps) are needed in order to specify to which WSDL operation the concept is grounded.

The WSMO service grounding plays a very important role to specify the connections between the ontological concepts used for semantic description of the Web service and input/output messages of the operations of the WSDL description of the service. *The service grounding process* is implemented as a two-steps procedure in which the user initially selects a node (ontology concept or relation) from the service state signature tree (with the mode "in", "out" or "shared") and then selects a proper operation message from the context-sensitive menu

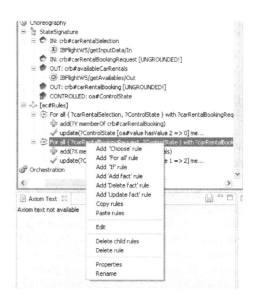

Fig. 3. Representation of a service choreography

graphically presenting the corresponding (in" or "out") WSDL operations. The grounding procedure is guided by a set of checks, which guarantee the semantic correctness of the created WSML description of the grounding. All groundings can be easily edited or deleted by means of the corresponding operations proposed as options in the context-sensitive right-click menus associated with the nodes of the state signature tree (Fig. 3).

In the WSMO choreography specification transition rules express changes of states by changing the set of instances (adding, removing and updating instances to the signature ontology). The different transition rules take the following form: *if* Condition *then* Rules *endIf, forall* Variables *with* Condition *do* Rules *endForall* and *choose* Variables *with* Condition *do* Rules *endChoose*. The *Condition* under which a rule is applied is an arbitrary logical expression as defined by WSML. The Rules may take the form of *Updates*, whose execution is to be understood as changing (or defining, if there was none) instances in an ontology. The update rules take the form of add, delete, and update, hence allowing to add/remove instances to/from concepts and relations, and also add/remove attribute values for particular instances. More complex transition rules can be defined recursively by if-then, for-all and choose rules. Since the syntax of *Condition* and *Rules* are different the creation of the transition rules in the Designer is split into two separate steps – creation of the conditional parts and creation of the conclusion part (update rules). An empty skeleton of a transition rule is created by specifying the desired type of the rule. These types are presented as options in the right-click menu associated with "Rules" node of the service tree. In order to guarantee the syntactical correctness the rule is always created as a pair – condition – update rule, where both parts are empty. The Double-clicking on the conditional part of the transition rule leads to opening the modeling area, where the user can create the graphical model of the conditional part of the rules.

3.4   Step 4: - Creation of a Service Capability

According to WSMO Framework a capability of a semantic service describes what the service can do. Usually, a capability of a WSMO-based semantic service is represented in terms of its inputs (pre-conditions), outputs (post-conditions), assumptions and effects. All these parts as well as conditional and conclusion parts of transition rules describing the service choreography are written as logical expressions (also called "axioms") in a special language – WSML. The conceptual syntax for WSML has a frame-like style - the information about a class and its attributes, a relation and its parameters and an instance and its attribute values is specified in a large syntactic construct, instead of being divided into a number of atomic chunks. WSML allows using of variables that may occur in place of concepts, attributes, instances, relation arguments or attribute values [6]. Although the machines can easily handle such axioms, creating and comprehending the axioms are very difficult for humans. That is why we have developed an approach in which the text (WSML) representation of such expressions is automatically generated from their graphical models. The approach has been implemented as a special software component called Axiom Editor. A detailed description of the functionality of this component as a stand-along software component is presented in [3]. It was also included as an Eclipse third party plug-in in the last version of WSMO Studio[1]. The present section contains brief description of the main ideas of the approach as well as a description of an additional functionality of this component caused by its usage as a main tool for graphical creating the semantic service capability and choreography descriptions in the INFRAWEBS Designer.

3.4.1   Representation of WSML Logical Expressions
A WSML logical expression is graphically modeled as a labeled directed acyclic graph (LDAG), which can contain four types of nodes:

- A *root* node which may have only outgoing arcs. The node is used for marking the beginning of the graphical model corresponding to the WSML logical expression.

---

[1] The latest release can be downloaded from http://www.wsmostudio.org/download.html

Graphically the root node is represented as a yellow oval with the name corresponding to the role, played by the graphical model in the description of the WSMO service - "Precondition", "Postcondition", "Assumption" or "Effect" (for axioms forming the service capability); "if-then", "for all" or "choose" (for axioms forming conditional parts of the transition rules), and "add", "delete" or "update" – for axioms forming "factual" parts of the update rules. In the case of using this node for describing the service capability, the node corresponds to the WSML statement *defineBy*.

- Intermediate nodes called *variables*. Such nodes can have several incoming and outgoing arcs. Each variable has a unique name and poses a frame-like structure consisting of slots represented by attribute–value pairs. Such a variable corresponds to a notion of compound molecule in WSML [6] consisting of an a-molecule of type Var_i memberOf Γ and conjunction of b-molecules of type $Var_i[p_l$ hasValue $Var_{jl}]$ and $Var_i[p_k$ hasValue $Var_{kl}]$ respectively, where $Var_i, Var_{jl}, Var_{kl}$ are WSML variables and Γ is a concept from a given WSML ontology. Graphically each variable is represented as a rectangle with a header containing variable name and type (i.e. the name of concept, which has been used for crating the variable), and a row of named slots.

- Intermediate nodes called *relations*. A relation node corresponds to a WSML statement $r(Par_1, ..., Par_n)$, where r is a relation from a given ontology, and Par_1, ..., Par_n are WSML variables – relation parameters. Graphically each relation node is represented as a rectangle with a header containing relation name and a row of relation parameters.

- Intermediate nodes called *operators* that correspond to WSML logical operators *AND, OR, IF-THEN*[2], *NOT* and *Old-New*[3]. Each node can have only one incoming arcs and one (for *NOT*), two (for *IF-THEN* and *Old-New*) or more (for *AND* and *OR*) outgoing arcs. Graphically each operator is represented as a blue oval, containing the name of the corresponding operation.

- Terminal nodes (leaves) called *instances* that can not have any outgoing arcs. An instance corresponds to the WSML statement *Var* hasValue *Instance*, where *Var* is a WSML variable and *Instance* is an instance of a concept from a given ontology. Graphically an instance is represented by a pink rectangle with header containing the instance name and type.

Directed arcs of a graph are called *connections*. A connection outgoing from a variable or relation has the meaning of refining the variable (or relation parameter) value and corresponds to WSML logical operator *AND*. A connection outgoing from an operator has the meaning of a pointer to the operator operand.

The proposed model allows considering the process of axiom creation as a formal process of LDAG expanding (and editing) and to formulate some formal rules for checking syntactic and semantic (in relation to given ontologies) correctness of the constructed logical expressions.

---

[2] This operator corresponds to Implies and ImpliedBy operators in WSML.

[3] This operator can be used only for creating the update rules and corresponds to the WSML "=>" operator.

3.4.2 A Model for Constructing Logical Expressions

Constructing a logical expression is considered as a repetitive process consisting of combination of three main logical steps – definition, refinement and logical development. The *definition* step is used for defining some general concepts needed for describing the meaning of axioms. During this step the nature of a main variable defining the axiom is specified. Such a step is equivalent to creating a WSML statement *?Concept* memberOf *Concept*, which means that the WSML variable *?Concept* copying the structure of the *Concept* from a given WSML ontology is created. Attributes of the concept, which are "inherited" by the axiom model variable, are named *variable attributes.* By default the values of such attributes are set to free WSML variables with type defined by the definition of such attributes in the corresponding ontology.

The refinement step is used for more concrete specification of the desired properties of such concepts and may be seen as a specialization of too general concepts introduced earlier. This step is implemented as a recursive procedure of refining values of some attributes (or relation parameters) defined in previous step(s). In terms of our model each cycle in such a step means an expansion of an existing non-terminal node – variable (or relation). More precisely that means a selection of an attribute of an existing model variable and binding its value (which in this moment is a free WSML variable) to another (new or existing) node of the axiom model. The main problem is to ensure semantic correctness of the resulted (extended) logical expression. Such correctness is achieved by applying a set of context-sensitive rules determining permitted expansion of a given node.

The logical development step consists of elaborating logical structure of the axioms, which is achieved by combination of general concepts by means of logical operators AND, OR, IF-THEN and NOT. Such operators may be added to connect two independently constructed logical expressions or be inserted directly into already constructed expressions. The operation is controlled by context-dependent semantic and syntactic checks that analyze the whole context of the axiom.

It should be underlined that during this step the user is constructing the axiom by logical combination of main axiom objects defined in the previous steps. In other words, the logical operators are used not for refining or clarifying the meaning of some parameters of already defined objects, but for complicating the axiom by specifying the logical connections between some axiom parts which are independent in their meaning.

3.4.3 Creation of Service Capability Axioms

Creation of a service capability axiom is a complex process, which can be conceptually split into two tasks – the definition of a role the axiom should play in the service capability description, and the construction of a graphical model of the desired axiom. The first task is solved via using a context-sensitive right-click menus associated with different nodes of the service capability tree. They allow the user to create new or to delete, edit and rename already created service capability axioms and/or axiom definitions, represented as nodes in the service tree.

The graphical creating or editing the content of an axiom is performed in the graphical model area by means of operations provided by the Axiom Editor module of the INFRAWEBS Designer and is guided by the model of this process described in the previous subsection. At each step of her work the user is assisted by a set of context-sensitive menus proposing only that operations or ontology elements that are allowable for using in the current moment. In such a way the Designer guarantees both semantic and syntactic correctness of the axiom under construction (Fig. 4).

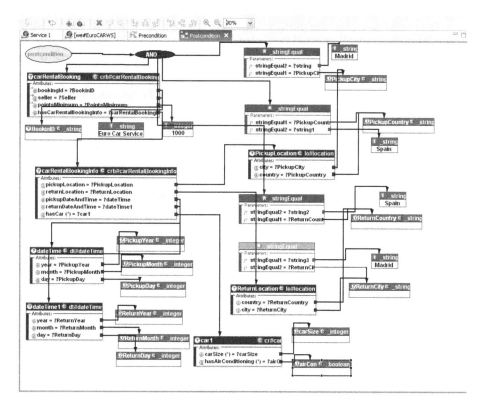

Fig. 4. An example of graphical representation of an axiom

The complete description of all menus and operations that the user can use and perform during the process of graphical creating the service capability axioms can be found in [3] as well as in "Axiom Editor User's Guide" (http://www.infrawebs-eu.org/opensoftware/Axiomeditor-1.5.0.pdf).

3.5 Step 5: Loading a Service

Most of the existing WSMO tools for creating a service description save such a description in a text file with extension .wsml. The tools provide specialized structural text editors facilitating the creation of WSML text of the service, which are integrated with WSMO Validator for checking the syntactical correctness of the description. In order to be compatible with such tools the INFRAWEBS Designer also stores the automatically generated WSML text of the service description in such text file. However, using only this single text file is not enough to guarantee the *semantic correctness* of the service description. That is why in the Designer we consider a WSMO service as a complex object consisting of:

- a text file containing the WSML description of the semantic service
- a set of text files containing the WSML descriptions of all ontologies mentioned in the "imported ontology" section of the service description
- a set of graphical files containing the graphical models of all WSML axioms used for describing service capability and choreography
- a text file containing the WSDL description of a Web service grounded to the semantic service

From the user point of view a service (or another WSMO object) is represented in the Designer as a single text file with extension .wsdl, which can be found and opened by means of a standard Open Dialog Window. However, the Designer is also equipped with a special "Navigator" module, which aims at browsing a special area of the Temporary Store – the Designer Workspace. The selecting a text file from this workspace leads to:

- Loading the service description into the memory and its visualization in the WSMO Navigator Window as a service tree.
- Loading all WSML ontologies mentioned as imported ontologies in the service description into the Ontology Store of the Designer.
- Loading the WSDL file grounded to the service into the WSDL Store.

The graphical models corresponding to different axioms forming the service descriptions are loaded to the Designer modeling area on-demand, when the graphical pointer to a concrete model (a leaf of the service tree) is selected in the service tree. The correspondences between graphical models and different parts of the WSML description of a service are maintained by a special internal module of the Designer - Graphical Model Manager.

3.5.1 Reusing Descriptions of Existing Semantic Services

Although the usage of graphical models makes easier the process of designing the WSML description of a semantic service, it still remains a rather complex and time-consuming activity. The next step towards facilitating this process is to provide the service designer with an opportunity to reuse the description of the existing semantic services. More precisely, the idea is to provide the user with graphical models of service description parts (e.g. capability axioms or transition rules) similar to what she wants to construct, which can be farther adapted by graphical means provided by the INFRAWEBS Designer.

In order to be reused, semantic descriptions of Web services and other objects (goals, ontologies, mediators) are considered in the INFRAWEBS Framework not only as logical (WSML) representations of these objects but also as a special class of text documents, containing natural language and ontology-based words. Such special "text" representation is extracted by the OM from each object stored in the DSWS-R and serves as a basis for constructing case representation of such an object. An INFRAWEBS case is a triple *{T, P, S}*, where *T* is the type of the WSMO object stored (service, goal or ontology), which determines the structure of object representation P; P is the special representation of a WSMO object as a structured text, and S is the service IRI, which is used for locating the service in the remote repository (DSWS-R) where the WSML (and graphical) description of the object is stored.

The user describes a semantic service to be found by filling a standard request form, which is sent to the OM playing a role of a case-based memory in the INFRAWEBS Framework. The form consists of three sections allowing constructing different queries based on the amount of information the user has in the current moment (Fig 5).

The first section ("Text") allows the user to describe the desired functionality of a service to be found by means of natural language keywords. All non-functional properties (of type "string") occurred in the WSML descriptions of semantic services stored in the DSWS-R will be matched against these keywords, and services with the best match will be returned. The second

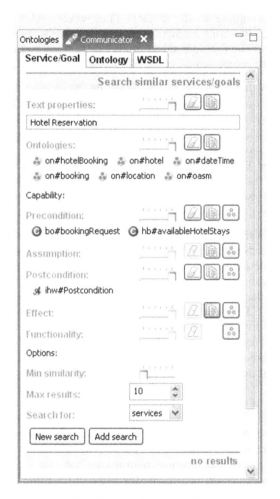

Fig. 5. The service request form

section ("Ontologies") allows the user to find services using a set of ontologies similar to that specified in the request form. Since in the INFRAWEBS Designer the user can work with several services in parallel, the filling of this section is done by pressing the corresponding button when the service containing the desired set of ontologies is active. By switching among descriptions of services loaded into the Designer, the user can construct the desired set of ontologies, which combines ontologies from different services. During processing the request form the Designer analyses each of the specified services and extracts from their descriptions a list of names of ontologies imported by each service. These names are used by the OM as keywords in the process of finding the existing semantic services using the same (or similar) set of ontologies The third section ("Capability") is devoted to the ontological description of the capability of the desired service. It is split into five subsections – the first four of them correspond to the sections of the service capability description according to the WSMO Framework, and the last one ("Functionality") allows constructing a general description of the capability of a service to be found.

The first four sub-sections can be filled by the following two ways (which can not be combined):

- Semiautomatic - by selecting the ontological elements (concepts, instances or relations) from the ontologies shown in the Ontology Store. In such a way the user can specify that she expects that the desired service should contain similar set of ontology keywords in the WSML (logical) description of the corresponding section of its capability.

- Fully automatic – by specifying the name of a service, whose corresponding capability description section will be used as an example of what the desired service should have in its capability section description.

When the user has no clear idea about the content of a concrete section of the desired service capability description or no "example" service descriptions exist, she can express her general idea of what ontology keywords the desired service capability description should have as a whole. Such general description is the content of the subsection "Functionality". The subsection can be filled by selecting ontological elements from all available ontologies shown in the Ontology Store.

The filled request form is translated into a complex XML query, in which each form subsection determines a separate (local) criterion for evaluating the similarity. The overall aggregated similarity is calculated based on a weighted sum of all local criteria [4]. The user can set the weight (from 0 to 1) for each criterion by means of a slider placed on the right from the corresponding form subsection (or section).

At the bottom of the form there is an additional section "Options", which allows the user to adjust two global parameters of the search. The first one, implemented as a slider, determines the minimum similarity threshold, which should be exceeded by the service in order to be returned as a search result. The second one sets a maximum number of elements that may be returned.

The query results returned by the OM are represented as an annotated list of services IRI ordered by the value of their ontological similarity coefficient or by the value of the lexical similarity coefficient if only the text section of the query form has been filled by the user. The Designer allows the user to send several queries to the OM without losing the results of previous query implementing in such a way the effects of searching alternatives.

The INFRAWEBS Designer allows working simultaneously with the descriptions of several services and/or goals. Such WSML objects can be downloaded from the remote repository (the DSWS-R) or loaded from the Temporary Store. In both cases the reuse of the object descriptions is organized as a procedure for copying and pasting different part of these descriptions. Following the basic design principles of the INFRAWEBS Designer, the user never works with WSML description of the object under construction. The Designer allows her to copy and paste graphical model of object capability axioms and/or axiom definitions, as well as the set of such models forming the whole capability description of the object (including the shared variables). The same is true for the choreography description of the object – it is possible to copy and paste the state signature, some state transition rules or the whole choreography of the service (or goal). Moreover, the Designer allows constructing a service description from the parts of several service and goal descriptions and vice versa. In all cases the Designer guarantees that the newly created description of a WSMO object is syntactically correct and consistent with the existing WSML ontologies.

3.6   Step 6: Uploading a Service to a Remote Repository

The main objective for creating the semantic description of a Web service is to allow the service to be discovered and used by other users. This is achieved by publication of the semantic service description in a remote repository (in the case of the IIF, it is the DSWS-R component). It is assumed that only valid WSML definitions are stored in the DSWS-R, which means that components providing functionality for creation or modification of ontologies, goals or Web services are expected first to validate them and only store them if they are valid. That is why the process of storing a WSMO object created in the Designer is implemented as a two-steps procedure – validating the object description and then uploading the created WSMO object to the DSWS-R.

3.6.1   Validating the Description of a WSMO Object Created in the Designer
As it has been mentioned several times, the Designer does not allow creating syntactically incorrect descriptions of WSMO services and goals. Moreover, all created descriptions are semantically consistent with the WSML ontologies used for creating these descriptions. That is why, only the semantic completeness of the created objects in respect to the WSMO specification for the corresponding objects should be verified. Since the Designer can be used for creating two types of WSMO objects – WSMO services and WSMO goals, two separate validation procedures (for each type of the objects) are developed. Both of them consist of a set of different tests that can issue two types of messages – error messages and warnings. The presence of error messages means that the description of the current object does not contain some elements, which are crucial for further use of the object, and that is why, such an object can not be uploaded to the DSWS-R. The presence of warnings reminds the user that some parts of the object description are missing but even in its present condition the object is usable. The user should decide whether to upload the object as it is or to complete the missing parts of the object description and only then to upload it. In order to facilitate the work of the user for removing the discovered incompleteness in the object description, the validation procedure is stopped only after performing all tests.

3.6.2   Uploading a WSMO Service (Goal)
After passing the validation phase a process for uploading the object to the remote local repository is started. The Designer considers (internally) a WSMO object (service or a goal) as a complex object consisting of object description, files containing the WSML descriptions of all ontologies mentioned as "imported ontologies" in the object description as well as a set of all graphical models corresponding to the logical expressions used for defining the object. That is why, all these elements of a WSMO object should be uploaded along with the WSML

description of the object. However, in order to avoid the problem with overwriting the already published ontologies, only imported ontologies, which have been loaded into the Ontology Store and not published in the DSWS-R, are uploaded. The graphical models are always uploaded along with the WSML description of an object and overwrite "old" models of the object if such exist.

4 Lessons Learnt

The INFRAWEBS Designer has been used for creating two Demonstrators. The first one is STREAM Flows! System (SFS) in which the customers can create and reuse travel packages [8]. The application is built upon a Service Oriented Architecture, accessing, discovering, composing and invoking Semantic Web Services for the management of the Travel Packages. The second Demonstrator is based on the eGovernment scenario, which illustrates some interactions carried out by semantic Web services in the scope of public administrations, and interactions among these and citizens and enterprises with emphasis on the E2A (Enterprise-to-Administration) integration [11].

The results of such test applications of the Designer can be summarized as follows:

- As a tool for automatic generation of programs (descriptions of WSMO services and goals written in WSML language), the Designer always produces syntactically correct programs that pass all checks applied by WSMO Validator and WSML Parser. Moreover, the auto-generated WSML programs are semantically consistent both with WSML ontologies used for their creation and with the last WSMO specification for WSMO objects described by these programs. The evidence for this conclusion is the ability to use the produced WSML files both for service discovery and execution.

- As a tool oriented towards end-users of semantic Web service technology, who are not familiarized with WSML language, it can be concluded that the proposed approach for graphical creating the WSMO objects is understandable and usable. Guided by the Designer the users are able to create rather complex (from logical point of view) axioms and transition rules without any knowledge of WSML syntax. The developed GUI is sufficiently intuitive and easy learnable by the users. The available set of primitives (the Designer operations) is adequate for constructing and editing descriptions of goals and services used in the test beds.

- The developed model of the process for graphical creation of WSML expressions is flexible enough and allows using alternative ways for constructing the same expression, depending on the experience and the way of thinking of the concrete user of the Designer.

- The developed mechanism for reusing the existing descriptions of WSMO objects via "copy-and-paste" of graphical models is understandable to the users and significantly reduces time and efforts for creating new objects.

- The integration of the Designer with facilities for creating WSMO goals has made it a powerful instrument for creating composite goals and significantly simplifies the work of the service application Designer.

The use of the INFRAWEBS Designer for preparation of the test bed demonstrators has shown some additional advantageous features of the tools:

- The Designer can be effectively used for validation of the semantic correctness and completeness of WSML descriptions of WSMO services and goals created *without* using the Designer. The ontology-driven approach used for creating logical expressions in the Designer, allows identification of all erroneous usages of names of ontologies, ontology concepts and concept attributes, as well as errors caused by improper use of the concept types. Moreover, the Designer is able to find errors and incompleteness in the description of imported ontologies used by the object as well as in description of the service interface. The problems discovered in a "third-party" WSML file can be easily identified based on error messages issued by the Designer and the graphical representation of logical expression, which is automatically produced by the tool. For identifying the problems, users who are more experienced in WSML can also analyze the difference between original WSML text written without using of the Designer and the WSML text automatically generated by the Designer from the graphical model of the corresponding logical expression.
- The Designer can be effectively used even for validation of the semantic correctness and completeness of WSML descriptions of WSMO ontologies. For example, the Designer built-in mechanism for on-demand loading of ontologies enables identifying the incorrect use of ontology names, imported by the concrete ontology, the lack of the cited ontology in the workspace as well as an incorrect use of concept attributes inconsistent with attribute inheritance rules. All these features of the Designer make it a powerful tool not only for novices in WSML but for WSML experts as well.

As main shortcoming of the INFRAWEBS Designer we can mention the insufficient support that the Designer provides to an "industrial" Web service provider for filling the *conceptual gap* between the descriptions of "normal" and semantic (WSMO-based) Web service. The WSMO representation of a semantic service is too complex and more logic oriented, which is significantly different from more functionally oriented non-semantic description of Web services. As a future work in this direction we foresee more intensive use of WSDL description of the service, for example for automatic generation of initial, rough WSML description of the semantic service state signature and choreography rules. Another promising step in this direction, in our opinion, is a more intensive use (and reuse) of existing descriptions of Web service business logic. For example, it will be fruitful to develop some methods for automatic generation of an initial WSMO-based choreography (or orchestration) from the available BPEL files.

Another weakness of the current version of the Designer, as a complex tool for creating WSMO-based services, is a lack of means for automatic creation of schema mapping, which is needed for finding (building) correspondence between data types presented in the WSDL file and WSML ontology concepts that describe the service domain. The WSMO community has not found yet a commonly accepted solution for this problem. If the using of XSLT transformations will be accepted as a WSMO standard, in the future versions of the Designer we are going to integrate it (or to develop) with some tools for automatic generation of such transformation given WSML ontologies and the WSDL file of the service.

5 Conclusions

The paper has presented the INFRAWEBS Designer – a graphical ontology-driven development environment for creating semantic descriptions of Web services and goals according to WSMO Framework. This tool is oriented to the end users – providers of Web services and semantic Web services applications. The Designer allows a user to graphically compose a WSMO-based semantic description of a given Web service based on existing WSDL description of this service and a set of WSML ontologies. To ease this process the graphical descriptions of look-alike semantic Web services can be found and used. Such descriptions can serve as templates for the WSMO object under construction. In fact, the reuse of available semantic Web services makes the otherwise tedious process of composing a WSMO object (requiring expert knowledge of the WSMO model and WSML language) a task doable for the ordinary Web service developers. Easing the hurdle of WSMO object construction is an achievement of the INFRAWEBS project that has a potential impact on the adoption of semantic Web services on a larger scale.

The INFRAWEBS Designer is a standalone desktop Java application, which is based on Eclipse Rich Client Platform (RCP). It is distributed as a simple archive and does not require any complex installation – after the archive is expanded, the Designer is ready for use.

The Designer is also available as open source under LGPL license (http://www.infrawebs.eu/index.html?menue=dissemination&site=open_software).

References

1. Agre, G., Pariente, T., Marinova, Z., Nern, H.-J., Lopez, A., Micsik, A., Boyanov, A., Saarela, J., Atanasova, T., Scicluna, J., Lopez, O., Tzafestas, E., Dilov, I.: INFRAWEBS – A Framework for Semantic Web Service Engineering. In: di Nitto, E., Sassen, A.-M., Traverso, P., Zwegers, A. (eds.) At your service: An overview of results of projects in the field of service engineering of the IST programme. MIT Press Series on Information Systems (in print)
2. Agre, G.: INFRAWEBS Designer – A Graphical Tool for Designing Semantic Web Services. In: Euzenat, J., Domingue, J. (eds.) AIMSA 2006. LNCS (LNAI), vol. 4183, pp. 275–289. Springer, Heidelberg (2006)
3. Agre, G., Kormushev, P., Dilov, I.: INFRAWEBS Axiom Editor - A Graphical Ontology-Driven Tool for Creating Complex Logical Expressions. International Journal Information Theories and Applications 13(2), 169–178 (2006)
4. Agre, G.: Using Case-based Reasoning for Creating Semantic Web Services: an INFRAWEBS Approach. In: Proc. of EUROMEDIA 2006, Athens, Greece, pp. 130–137 (April 17-19, 2006)
5. Andonova, G., Agre, G., Nern, H.-j., Boyanov, A.: Fuzzy Concept Set Based Organizational Memory as a Quasi Non-Semantic Component within the INFRAWEBS Framework. In: Proc. of the 11th IPMU International Conference, Paris, France, pp. 2268–2275 (July 2-7, 2006)

6. de Bruijn, J., Lausen, H., Krummenacher, R., Polleres, A., Predoiu, L., Kifer, M., Fensel, D.: D16.1 – The Web Services Modeling Language (WSML). WSML Draft (2005)
7. Kerrigan, M.: Developers Tool Working Group Status. Version 1, Revision 4, http://wiki.wsmx.org/index.php?title=Developer_Tools
8. López-Cobo, J.-M., López-Pérez, A., Scicluna, J.: A semantic choreography-driven Frequent Flyer Program. In: Proc. of FRCSS 2006 -1st International EASST-EU Workshop on Future Research Challenges for Software and Services, Vienna, Austria (April 1st 2006)
9. Marinova, Z., Agre, G., Ognyanov, D. Final Dynamic DSWS-R and integration in the IIF. INFRAWEBS Deliverable D4.4.3 (February 2007)
10. Last accessed (May 2007), http://www.daml.org/services/owl-s/
11. Riceputi, E.: Requirement Profile 2 & Knowledge Objects. INFRAWEBS Deliverable D10.5-6-7.2 (September 2006)
12. Roman, D., Lausen, H., Keller, U. (eds.): D2v1.3 Web Service Modeling Ontology, WSMO Final Draft (21 October 2006)
13. Roman, D., Scicluna, J., Nitzsche, J. (eds.): D14v1.0. Ontology-based Choreography. WSMO Final Draft (15 February 2007)
14. Saarela, J.: Specification & General SIR, Full Model & Coupling to SWS Design activity. INFRAWEBS Deliverable D3.2-3-4.2 (July 2006)
15. Software, Services and Complexity Research in the IST Programme: An overview. Software Technologies Unit, Information Society and Media DG, European Commission, Ver. 1.0, draft (September 2006)
16. http://sourceforge.net/project/g.org
17. http://www.wsmostudio.org

Goal-Based Visualization and Browsing for Semantic Web Services

Michael Stollberg and Mick Kerrigan

Digital Enterprise Research Institute (DERI),
University of Innsbruck, Austria
{firstname.lastname}@deri.at

Abstract. We present a goal-based approach for visualizing and browsing the search space of available Web services. A goal describes an objective that a client wants to solve by using Web services, abstracting
from the technical details. Our visualization technique is based on a
graph structure that organizes goal templates – i.e. generic and reusable
objective descriptions – with respect to their semantic similarity, and
keeps the relevant knowledge on the available Web services for solving
them. This graph is generated automatically from the results of semantically enabled Web service discovery. In contrast to existing tools that
categorize the available Web services on the basis of certain description elements, our tool allows clients to browse available Web services
on the level of problems that can be solved by them and therewith to
better understand the structure as well as the available resources in a
domain. This paper explains the theoretic foundations of the approach
and presents the prototypical implementation within the Web Service
Modeling Toolkit WSMT, an Integrated Development Environment for
Semantic Web services.

1 Introduction

The provision of suitable search facilities for Web services is one of the major
challenges for realizing sophisticated SOA technologies. One desirable feature are
browsing facilities that support Web service application developers in the search
and inspection of potential candidate services for a specific problem. Existing
tools for this mostly follow the registry approach already defined in UDDI [3]:
Web services are categorized with respect to certain description elements, and
graphical tools support the search and browsing of these registries.

We take a different approach for visualizing the search space of available
Web services. Our technique is based on goal templates as generic and reusable
descriptions of objectives that clients want to achieve by using Web services. We
organize them in a graph structure with respect to their semantic similarity, and
we keep knowledge on the suitability of the available Web services for solving
the goals that is obtained from Web service discovery runs. Our graphical user
interface visualizes this graph and provides browsing facilities for this. This is a
novel approach that allows clients to browse and understand the available Web
services on the level of the problems that can be solved by them.

M. Weske, M.-S. Hacid, C. Godart (Eds.): WISE 2007 Workshops, LNCS 4832, pp. 236–247, 2007.
© Springer-Verlag Berlin Heidelberg 2007

The basis for our search space visualization is the *Semantic Discovery Caching* technique (short: SDC), a caching mechanism for enhancing the computational performance of automated Web service discovery engines [14]. Its heart is the so-called SDC graph that organizes goal templates in a subsumption hierarchy and captures knowledge on the suitability of the available Web services from Web service discovery results. The SDC graph is generated automatically by semantic matchmaking of sufficiently rich formal descriptions of goals and Web services. It provides an index structure for the efficient search of suitable Web services, and thus serves as the data structure of our search space visualization. The graphical user interface is implemented in the Web Service Modelling Toolkit WSMT [9], an Integrated Development Environment for the Semantic Web service technology developed around the WSMO framework [5].

This paper explains the theoretic foundations of our visualization technique and presents the prototype implementation. Section 2 recalls the idea of the goal-based approach for Semantic Web services. Section 3 explains the structure, definition, and properties of the SDC graph, and Section 4 presents the visualization and browsing support developed in WSMT. Section 5 discusses the approach and positions it within related work, and Section 6 concludes the paper. For illustration and demonstration, we use the shipment scenario from the SWS Challenge, a widely recognized initiative for the demonstration of SWS techniques (see `www.sws-challenge.org`).

2   The Goal-Based Approach for Semantic Web Services

In order to introduce into the overall context, the following explains the aim of the goal based approach for Semantic Web services as promoted by the WSMO framework and recalls the foundations of our approach from previous works.

While the initial Web service technology stack as well as most approaches in the field of Semantic Web services (SWS) only pay little attention to the client side of SOA technology, the WSMO framework promotes a goal driven approach for this [5]. Therein, a goal is the formal description of an objective that a client wants to achieve by using Web services. Goals focus on the problem to be solved, abstracting from technical details on how to invoke a Web service. The overall aim is to facilitate problem-oriented Web service usage: the client merely specifies the objective to be achieved as a goal, and the system automatically discovers, composes, and executes the necessary Web services for solving this. Therewith, goals shall allow to lift the client-system interaction to the knowledge level in the tradition of previous AI technologies for automated problem solving.

Figure 1 provides an overview of the goal-based SWS framework that has been developed throughout several works around WSMO [7,16,15,18,14]. This distinguishes *goal templates* as generic and reusable objective descriptions which are stored in the system, and *goal instances* that denote concrete client requests and are defined by instantiating a goal template with concrete inputs. On this basis, we can separate *design time* and *runtime* operations. At design time, suitable Web services for goal templates are detected. For simplification, we here use

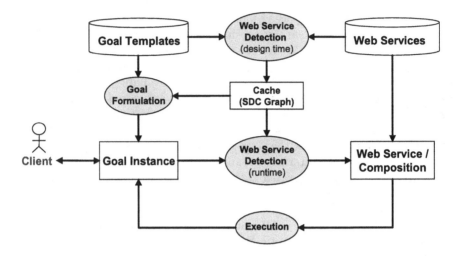

Fig. 1. Overview of Goal-based SWS Framework

'Web service detection' as a general term for SWS discovery, selection, ranking, and composition techniques. The result is captured in the SDC graph which provides an index for the efficient search of goal templates and Web services (see Section 3). At runtime, a client defines a concrete objective in terms of a goal instance, and the captured knowledge is used to detect the suitable Web services in an efficient manner. Finally, the detected Web service or composition is executed and the result is reported to the client.

For illustration, let us consider the following setting from the shipment scenario as described in the SWS Challenge. A client wants to ship a package of 40 lb weight from New York City to Bristol in the UK, and there are several available Web services that offer package shipment. In our framework, a goal template describes the objective of package shipment on the schema level, e.g. that the sender is located in a US city and the receiver in a European city, and the maximal weight class (e.g. from 0 - 1.5 lb, 1.5 - 10 lb, etc.). Let the client create a goal instance for this goal template by defining the concrete input data. To determine the suitable Web services for solving the goal instance, the system can make use of the knowledge on design time Web service detection results. This allows to develop efficient and workable SWS technologies. For example, in this specific use case there are only 4 Web services out of all available ones that are suitable to solve the goal template. One of these must be chosen to be executed for solving the goal instance: this requires a detailed investigation at runtime of only 4 Web services instead of all the available ones.

Summarizing, goal instances are the primary element for clients to interact with the system, i.e. end-users who want to use Web services to solve a certain task. The aim of the visualization and browsing technique presented in this paper is to aid clients in the goal instance formulation process as well as to provide graphical support for better understanding the available resources.

3   The Semantic Discovery Caching Technique

The following resumes the *Semantic Discovery Caching* technique (short: SDC), a caching mechanism for enhancing the computational efficiency of Web service discovery [14]. Its heart is the SDC graph that provides an automatically generated index structure for efficient search of goals and Web services, which serves as the basis for the search space visualization presented in Section 4.

The main purpose of the SDC technique is to increase computational performance of the Web service discovery task, which is considered as the first processing step for detecting the Web services that are suitable for solving a goal [12]. It captures design time discovery results for goal templates in the SDC graph, and effectively uses this knowledge in order to minimize the computational costs of Web service discovery for goal instances at runtime. The SDC graph organizes goal templates in a subsumption hierarchy, and the leaf nodes represent the available Web services whose suitability for solving goal templates is explicated by directed arcs. Therewith, the SDC graph provides a directed graph structure that describes the usability of Web services in a problem domain with respect to the goals that can be solved by them. The following explains the structure of the SDC graph and the underlying semantic matchmaking techniques.

3.1   Formal Functional Descriptions and Semantic Matchmaking

The SDC technique is based on the Web service discovery approach presented in [15] that performs semantic matchmaking on the goal template and the goal instance level on the basis of rich functional descriptions.

A functional description $\mathcal{D} = (\Sigma, \Omega, IN, \phi^{pre}, \phi^{eff})$ is defined over a signature Σ with respect to a domain ontology Ω; IN is the set of input variables, ϕ^{pre} is the precondition, and ϕ^{eff} is the effect wherein the predicate $out()$ denotes the outputs. Such a functional description precisely describes the possible executions $\{\tau\}_W$ of Web services, respectively the possible solutions $\{\tau\}_G$ of goal templates with respect to the start- and the end-states. We denote the functional usability of a Web service W for a goal template G by the matchmaking degrees shown in Table 1; therein, $\phi^{\mathcal{D}}$ is a FOL-formula of the form $\phi^{pre} \Rightarrow \phi^{eff}$ that defines the formal semantics of $\mathcal{D}$ as an implication between the possible start- and the possible end-states. The first four degrees distinguish different situations wherein W is usable to solve G, and the *disjoint* degree states that this is not given.

A goal instance is defines as a pair $GI(G) = (G, \beta)$ with G as the corresponding goal template and β as the input binding for the input variables IN defined in the functional description $\mathcal{D}_G$. A goal instance must be a valid instantiation of the corresponding goal template in order to be a consistent objective description. This is given if $\mathcal{D}_G$ is satisfiable under the input binding β. Then it holds that $\{\tau\}_{GI(G)} \subset \{\tau\}_G$, meaning that the solutions of the goal instance are a subset of those for its corresponding goal template. In consequence, only those Web services that are suitable for solving the goal template G are potential candidates for every of its goal instances $GI(G)$ and no others can be. This allows to effectively distinguish design- and runtime operations as explained above.

Table 1. Definition of Matching Degrees for $\mathcal{D}_G$, $\mathcal{D}_W$

Denotation	Definition	Meaning
$\mathbf{exact}(\mathcal{D}_G, \mathcal{D}_W)$	$\Omega \models \forall\beta.\ \phi^{\mathcal{D}_G} \Leftrightarrow \phi^{\mathcal{D}_W}$	$\{\tau\}_G = \{\tau\}_W$
$\mathbf{plugin}(\mathcal{D}_G, \mathcal{D}_W)$	$\Omega \models \forall\beta.\ \phi^{\mathcal{D}_G} \Rightarrow \phi^{\mathcal{D}_W}$	$\{\tau\}_G \supset \{\tau\}_W$
$\mathbf{subsume}(\mathcal{D}_G, \mathcal{D}_W)$	$\Omega \models \forall\beta.\ \phi^{\mathcal{D}_G} \Leftarrow \phi^{\mathcal{D}_W}$	$\{\tau\}_G \subset \{\tau\}_W$
$\mathbf{intersect}(\mathcal{D}_G, \mathcal{D}_W)$	$\Omega \models \exists\beta.\ \phi^{\mathcal{D}_G} \wedge \phi^{\mathcal{D}_W}$	$\{\tau\}_G \cap \{\tau\}_W \neq \emptyset$
$\mathbf{disjoint}(\mathcal{D}_G, \mathcal{D}_W)$	$\Omega \models \neg\exists\beta.\ \phi^{\mathcal{D}_G} \wedge \phi^{\mathcal{D}_W}$	$\{\tau\}_G \cap \{\tau\}_W = \emptyset$

3.2   The SDC Graph – Structure and Definition

The SDC graph is automatically generated from the results of design time Web service discovery on goal templates. It organizes goal templates in a subsumption hierarchy with respect to their semantic similarity, which constitutes the indexing structure of the available Web services. The leaf nodes represent the Web services that are functionally usable for solving the goal templates.

We consider two goal templates G_i and G_j to be similar if they have at least one common solution. Then, mostly the same Web services are usable for them. We express this in terms of *similarity degrees* $d(G_i, G_j)$; defined analog to Table 1, they denote the matching degree between the functional descriptions $\mathcal{D}_{G_i}$ and $\mathcal{D}_{G_j}$. The SDC graph is defined such that the only occurring similarity degree is $subsume(G_i, G_j)$. This allows efficient search because then (1) the solutions for the child G_j are a subset of those for the parent G_i, and thus (2) the Web services that are usable for G_j are a subset of those usable for G_i.

In consequence, the SDC graph consists of two layers. The upper one is the *goal graph* that defines the subsumption hierarchy of goal templates by directed arcs. This constitutes the index structure of the available Web services with respect to the goals that can be solved by them, whereby the subsumption hierarchy represents the specialization in a problem domain. The lower layer is the *usability cache* that explicates the usability of each available Web service W for every goal template G by directed arcs that are annotated with the usability degree $d(G, W)$. This is generated from the results of Web service discovery on the goal template level that is performed at design time. The discovery operations use this knowledge structure by inference rules of the form $d(G_i, G_j) \wedge d(G_i, W) \Rightarrow d(G_j, W)$ that result from the formal definitions.

Figure 2 illustrates the SDC graph for our running example along with the most relevant inference rules. There are three goal templates: G_1 for package shipment in Europe, G_2 for Switzerland, and G_3 for Germany. Their similarity degrees are $subsume(G_1, G_2)$ and $subsume(G_1, G_3)$, which is explicated in the goal graph. Consider some Web services, e.g. W_1 for package shipment in Europe, W_2 in the whole world, W_3 in the European Union, and W_4 in the Commonwealth. Their usability degree for each goal template is explicated in the usability cache, whereby redundant arcs are omitted: the usability degree of W_1 and W_2 for G_2 and G_3 can be directly inferred, thus the arcs are omitted.

Structure of SDC Graph	Inference Rules for $subsume(G_i, G_j)$
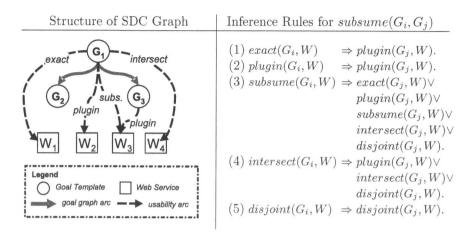	(1) $exact(G_i, W) \Rightarrow plugin(G_j, W)$. (2) $plugin(G_i, W) \Rightarrow plugin(G_j, W)$. (3) $subsume(G_i, W) \Rightarrow exact(G_j, W) \vee$ $plugin(G_j, W) \vee$ $subsume(G_j, W) \vee$ $intersect(G_j, W) \vee$ $disjoint(G_j, W)$. (4) $intersect(G_i, W) \Rightarrow plugin(G_j, W) \vee$ $intersect(G_j, W) \vee$ $disjoint(G_j, W)$. (5) $disjoint(G_i, W) \Rightarrow disjoint(G_j, W)$.

Fig. 2. Example of a SDC Graph and Inference Rules

The SDC graph provides an index structure of the search space of the available Web services with respect to the goals that can be solved by them. It can be automatically generated, and – in contrast to most existing clustering techniques for Web services – it is based on semantic matchmaking that ensures a high accuracy of the obtained graph structure. The generation and maintenance is supported by algorithms that ensure that the SDC graph exposes its properties at all times. We refer to [14] for details on this, in particular for the resolution of intersection matches in the goal graph which is necessary to ensure that the only similarity degree is *subsume, cf.* clause (i) in the following definition.[1]

Definition 1. *Let $d(G_i, G_j)$ denote the similarity degree of goal templates G_i and G_j, and let $d(G, W)$ denote the usability degree of a Web service W for a goal template G. Given a set $\mathcal{G}$ of goal templates and a set $\mathcal{W}$ of Web services, the SDC graph is a directed acyclic graph $(V_{\mathcal{G}} \cup V_{\mathcal{W}}, E_{sim} \cup E_{use})$ such that:*

(i) $V_{\mathcal{G}} := \mathcal{G} \cup \mathcal{G}^I$ is the set of inner vertices where:
- *$\mathcal{G} = \{G_1, \ldots, G_n\}$ are the goal templates; and*
- *$\mathcal{G}^I := \{G^I \mid G_i, G_j \in \mathcal{G}, d(G_i, G_j) = \mathsf{intersect}, G^I = G_i \cap G_j\}$ is the set of intersected goal templates from $\mathcal{G}$*

(ii) $V_{\mathcal{W}} := \{W_1, \ldots, W_m\}$ is the set of leaf vertices representing Web services

(iii) $E_{sim} := \{(G_i, G_j) \mid G_i, G_j \in V_{\mathcal{G}}\}$ is the set of directed arcs where:
- *$d(G_i, G_j) = \mathsf{subsume}$; and*
- *not exists $G \in V_{\mathcal{G}}$ s.t. $d(G_i, G) = \mathsf{subsume}, d(G, G_j) = \mathsf{subsume}$.*

(iv) $E_{use} := \{(G, W) \mid G \in V_{\mathcal{G}}, W \in V_{\mathcal{W}}\}$ is set of directed arcs where:
- *$d(G, W) \in \{\mathsf{exact}, \mathsf{plugin}, \mathsf{subsume}, \mathsf{intersect}\}$; and*
- *not exists $G_i \in V_{\mathcal{G}}$ s.t. $d(G_i, G) = \mathsf{subsume}, d(G_i, W) \in \{\mathsf{exact}, \mathsf{plugin}\}$.*

---

[1] The SDC prototype is open source software available from the SDC homepage at `members.deri.at/~michaels/software/sdc/`. It is realized as a discovery component in the WSMX system (the WSMO reference implementation, `www.wsmx.org`). We use VAMPIRE for matchmaking, a FOL automated theorem prover.

4   Visualization and Browsing

We now turn towards the visualization and browsing support for the search space
of Web services. The aim is to provide a graphical representation that allows
clients to better understand the available Web services as well as the problems
that can be solved by them. This occurs to be desirable in order to determine
which tasks in a client application can be solved by the Web services. Moreover,
the graphical user interface supports clients in the selection of an appropriate
goal template for expressing a specific objective or request, therewith providing
graphical support for the goal instance formulation process (see Section 2).

The search space visualization uses a SDC graph as the data structure. As
explained above, this provides an index structure of Web services on the level
of goal templates that is obtained by semantic matchmaking and thus exposes
a high accuracy of the relevant relationships of goals and Web services. The
SDC graph visualization is implemented as a new plug-in for the Web Service
Modeling Toolkit WSMT [9], an Integrated Development Environment (IDE)
implemented in the Eclipse framework that provides the graphical user- and
developer tools for the Semantic Web service technologies developed around
the WSMO framework. In particular, we extend the WSMT Visualizer [8] that
provides a graph-based editor and browser for ontologies.[2]

For illustration of the visualization and browsing facilities, we have created
the SDC graph for the original data set of the shipment scenario as defined in the
SWS Challenge. This defines 5 Web services that offer package shipment from
the USA to different destinations in the world, and a collection of exemplary
client requests. These correspond to goal instances in our framework, while goal
templates define the generic and reusable objective descriptions.[3]

The following presents the technical realization of the SDC graph visualization
and browsing in WSMT, and explains the obtained surplus value for clients as
well as the integration with SWS environments for automated goal solving by
the discovery, composition, and execution of Web services.

4.1   SDC Graph Visualization in WSMT

As explained above, a SDC graph consists of two layers: the upper one is the
goal graph wherein the existing goal templates are organized in a subsumption
hierarchy, and the lower layer is the *usability cache* that explicates the suitability of the available Web services for each goal template (*cf.* Definition 1). The
visualization of an SDC graph that is initially presented to the user displays the
subsumption hierarchy of the goal graph along with a cluster of the usable Web
services for each goal template. This allows to get a complete overview of the

---

[2] The Web Service Modeling Toolkit is open source software available for download
from the sourceforge web site at: http://wsmt.sourceforge.net.

[3] The complete resources for this use case are available at http://members.deri.at/~
michaels/software/sdc/resourcesSWSC.zip; the scenario description is given at
http://sws-challenge.org/wiki/index.php/Scenario:_Shipment_Discovery.

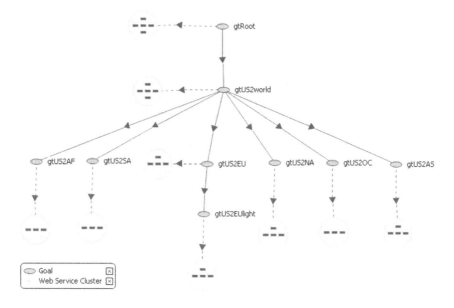

Fig. 3. SDC Graph Visualization in WSMT

available resources. The user can then browse and navigate through the search space via more detailed perspectives as we shall explain below.

Figure 3 shows the visualization of the SDC graph for the shipment scenario in WSMT. The root node of the SDC graph `gtRoot` is a goal template that describes the objective of shipping a package of any weight from anywhere in the world to anywhere in the world. This denotes the most general objective description in the problem domain of package shipment. The goal template `gtUS2world` is concerned with package shipment from the USA to anywhere in the world; this is a child node of `gtRoot` in the goal graph because every solution for `gtUS2world` is also a solution for `gtRoot` but not vice versa (see Section 3.2). Analogously, the child nodes of `gtUS2world` describe the objective of package shipment from the USA to specific continents, the goal templates on the subsequent levels of the goal graph specify local regions and cities, and finally the lowest levels differentiate the weight classes for the packages.

Two interesting properties result from the structure and definition of the SDC graph. At first, the Web service cluster of a child node contains a subset of the Web service cluster of its parent. For example, from the 5 Web services for `gtUS2world` only 3 are usable for the goal template `gtUS2AS` that defines package shipment from the USA to Asia. This reflects the nature of most problem domains: the deeper a goal template is allocated in the goal graph, the more specialized is the described objective, and the fewer Web services are usable for it. Secondly, a SDC graph may contain disconnected subgraphs when there are two goal templates that do not have any common solution. One can understand each connected subgraph to cover a particular problem domain: apart from the one for the shipment scenario, there could be a set of goal templates for the problem domain of ticket booking, and another one that is concerned with purchase order management. Furthermore, it is possible to generate the goal templates for a problem domain from a given goal description and the underlying domain ontologies. For example, in the shipment scenario we can generate gtUS2world from gtRoot by restricting the molecule for the sender location in the goal description from *world* to *USA*. Eventually, we can generate all goal templates that can be expressed on the basis of the domain ontologies and therewith maximize the granularity of the SDC graph [14].

The SDC graph visualization plug-in reuses the JPowerGraph[4] graphing library developed for the visualization of ontologies and other WSMO elements in WSMT [8]. This provides a powerful framework with a number of layout algorithms for user-facing, interactive graphs with any type of interconnected data. The prototype employs a simple vertical-tree layout; however, a spring-layout algorithm where the nodes in the graph repel each other while the edges between nodes draw them back together can be employed to display larger and more complex SDC graphs. The display of Web service clusters extends the technique for the clustering of ontology instances in larger knowledge bases.

4.2   Browsing Facilities

The SDC graph visualization as explained above is complemented with browsing facilities that allow to inspect goal and Web service descriptions on more fine-grained levels. By double-clicking on a goal template in the SDC graph, the user can step down to the next level wherein the visualization is focused on the selected goal template. This view shows the relevant branch of the goal graph, and the suitable Web services for the selected goal template along with the concrete usability degrees as the disaggregation of respective Web service cluster. Figure 4 below illustrates this for the goal template gtUS2EU that corresponds to the introductory example discussed in Section 2. As the next level of detail, the user can browse the specification of individual goal and Web service descriptions by double-clicking on the element in the SDC graph; these facilities already exist in WSMT, and we omit further screen shots here with respect to space limitations. The segmentation of the browsing support into multiple levels allows to manage the complexity of the SDC graph while presenting all relevant information to the user in a browsable fashion.

In addition, the SDC graph visualization is equipped with a number of features for adjusting the visual presentation. Zoom and rotate functionalities allow to position and size the graph. The graph display can also be adjusted by dragging and dropping nodes into other positions. Individual node types can be filtered to increase the understandability, e.g. filtering out the Web service clusters allows to see the goal graph more clearly. In the case of multiple sub-graphs – i.e. when several problem domains are captured in the SDC graph as explained above – the root nodes can be used to filter out entire subgraphs so that the user can focus on the relevant problem domain.

---

[4] The JPowerGraph is a Java library available under an LGPL open source license and can be downloaded from http://jpowergraph.sourceforge.net.

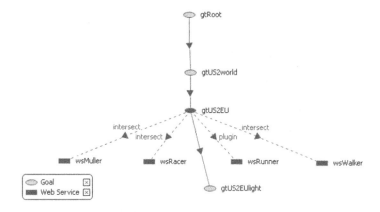

Fig. 4. SDC Graph Browsing for a Single Goal Template

4.3   Integration with SWS Environments

The Web Service Modeling Toolkit WSMT is an IDE for the WSMO framework that aims at supporting developers and end-users through all activities in the software life-cycle of Semantic Web services. It currently encompasses text- and graph-based editing facilities for WSMO elements, a graphical tool for the creation of ontology mappings, and the graphical editing and browsing tool that we have extended for visualizing and browsing of SDC graphs.

Alongside the textual and graph-based editing support, the WSMT provides full syntactic and semantic validation of the entities created by the developer in order to ensure that modeling mistakes are caught early in the life cycle and do not propagate to later activities. Beyond the validation, the WSMT embeds reasoning support and discovery engines that allows the automatic and manual testing and verification of ontologies, goals, and Web service descriptions. Furthermore, the WSMT deployment functionalities allow to register WSMO elements within a Semantic Execution Environment (SEE) like WSMX [6] or the IRS system [2], which provide the facilities for the automated discovery, composition, and execution of Web services in order to solve a given goal.

Thus, the development of the SDC graph visualization as a new plug-in for the WSMT ensures a tight integration and reusability of already existing editing facilities as well as a seamless interaction with SWS environments for the automated solving of goals by the detection and execution of Web services. In fact, the SDC graph along with the presented visualization can be seen as a registry technique for goal-based SWS applications.

5   Related Work

We are not aware of any other existing approach or available tool that provides visualized browsing of Web services in a similar way. Goal-based approaches for Semantic Web services are rare in literature; however, some works that are not based on the WSMO framework also adopt the idea of managing Web services on the level of semantically described client requests, e.g. [11].

Most existing graphical user interfaces for the search and inspection of available Web services follow the registry approach already supported by UDDI, i.e. to organize Web services with respect to certain description elements that often are based on domain ontologies (e.g. [13,17,1]). Other tools provide Web service search engines in a google-like style, e.g. [10]. Our approach is based on a graph structure that is automatically created by semantic matchmaking of sufficiently rich semantic descriptions, which ensures a high precision and accuracy of the relevant relationships and does not require any additional descriptions apart from those required by the SWS techniques for automated goal solving.

Existing graphical user interfaces for managing WSMO elements provide form-based registry views that simply enlist the available goals and Web services (e.g. in the IRS system [2] or in WSMO Studio [4]). Our visualization tool overcomes the deficiencies in the comprehensibility and scalability by providing multi-leveled browsing facilities that allow clients to navigate from the complete overview of the search space down to detailed views on individual resources.

6   Conclusions and Future Work

This paper has presented a novel approach for the visualization and browsing of Web services that allows clients to comprehend and inspect available Web services on the level of problems that can be solved by them, abstracting from the technical details which are the focus of most existing tools. We use goals as the constituting element, and SDC graphs that define subsumption hierarchies of goal templates as the indexing structure and capture knowledge on the suitability of the available Web services from discovery results. The multi-leveled browsing facilities allow to navigate from the overall view of the search space down to more detailed views, and the realization as a plugin-in for the WSMT ensures a tight integration with SWS environments for automated goal solving.

For the future, we plan to extend the SDC graph visualization with graphical user interface and API for the creation and management of goal instances, and to further integrate the SDC technique within SWS environments.

Acknowledgements. This material is based upon works supported by EU funding under the projects SUPER (FP6-026850) and SEEMP (IST-4-027347-STP).

References

1. Abramowicz, W., Haniewicz, K., Kaczmarek, M., Zyskowski, D.: Architecture for Web services Filtering and Clustering. In: Proc. of the 2nd International Conference on Internet and Web Applications and Services (ICIW 2007), Mauritius (2006)
2. Cabral, L., Domingue, J., Galizia, S., Gugliotta, A., Norton, B., Tanasescu, V., Pedrinaci, C.: IRS-III – A Broker for Semantic Web Services based Applications. In: Cruz, I., Decker, S., Allemang, D., Preist, C., Schwabe, D., Mika, P., Uschold, M., Aroyo, L. (eds.) ISWC 2006. LNCS, vol. 4273, Springer, Heidelberg (2006)

3. Clement, L., Hately, A., von Riegen, C., Rogers T. (eds.): UDDI Version 3.0.2. Approved Standard, OASIS (2004), online: http://uddi.org/pubs/uddi_v3.htm

4. Dimitrov, M., Simov, A., Momtchev, V., Konstantinov, M.: WSMO Studio - A Semantic Web Services Modelling Environment for WSMO (System Description). In: ESWC. LNCS, vol. 4519, Springer, Heidelberg (2007)

5. Fensel, D., Lausen, H., Polleres, A., de Bruijn, J., Stollberg, M., Roman, D., Domigue, J.: Enabling Semantic Web Services. The Web Service Modeling Ontology. Springer, Heidelberg (2006)

6. Haller, A., Cimpian, E., Mocan, A., Oren, E., Bussler, C.: WSMX - A Semantic Service-Oriented Architecture. In: Proceedings of the International Conference on Web Service (ICWS 2005), Orlando, Florida (2005)

7. Keller, U., Lara, R., Lausen, H., Polleres, A., Fensel, D.: Automatic Location of Services. In: Gómez-Pérez, A., Euzenat, J. (eds.) ESWC 2005. LNCS, vol. 3532, Springer, Heidelberg (2005)

8. Kerrigan, M.: WSMOViz: An Ontology Visualization Approach for WSMO. In: Proc. of the 10th International Conference on Information Visualization (IV 2006), London, England, July 2006 (2006)

9. Kerrigan, M., Mocan, A., Tanler, M., Fensel, D.: The Web Service Modeling Toolkit - An Integrated Development Environment for Semantic Web Services (System Description). In: ESWC. LNCS, vol. 4519, Springer, Heidelberg (2007)

10. Lausen, H., Haselwanter, T.: Finding Web Services. In: Proc. of the 1st European Semantic Technology Conference (ESTC), Vienna, Austria (2007)

11. Maximilien, E.M.: Human-Based Semantic Web Services. In: Proc. of the 1st SWS Challenge Workshop, Stanford, California, USA (March 2006)

12. Preist, C.: A Conceptual Architecture for Semantic Web Services. In: McIlraith, S.A., Plexousakis, D., van Harmelen, F. (eds.) ISWC 2004. LNCS, vol. 3298, Springer, Heidelberg (2004)

13. Srinivasan, N., Paolucci, M., Sycara, K.: Adding OWL-S to UDDI - Implementation and Throughput. In: Proc. of the 1st International Workshop on Semantic Web Services and Web Process Composition at the ICWS, San Diego, USA (2004)

14. Stollberg, M., Hepp, M., Hoffmann, J.: A Caching Mechanism for Semantic Web Service Discovery. In: Aberer, K., et al. (eds.) ISWC 2007. LNCS, vol. 4825, pp. 480–493. Springer, Heidelberg (2007)

15. Stollberg, M., Keller, U., Lausen, H., Heymans, S.: Two-phase Web Service Discovery based on Rich Functional Descriptions. In: ESWC 2007. LNCS, vol. 4519, Springer, Heidelberg (2007)

16. Stollberg, M., Norton, B.: A Refined Goal Model for Semantic Web Services. In: Proc. of the 2nd International Conference on Internet and Web Applications and Services (ICIW 2007), Mauritius (2007)

17. Verma, K., Sivashanmugam, K., Sheth, A., Patil, A., Oundhakar, S., Miller, J.: METEOR-S WSDI: A Scalable P2P Infrastructure of Registries for Semantic Publication and Discovery of Web Services. Journal of Information Technology and Management 6(1), 17–39 (2005)

18. Vitvar, T., Zaremba, M., Moran, M.: Dynamic Service Discovery through Meta-Interactions with Service Providers. In: ESWC 2007. LNCS, vol. 4519, Springer, Heidelberg (2007)

Agile Elicitation of Semantic Goals by Wiki

David Lambert, Stefania Galizia, and John Domingue

Knowledge Media Institute, The Open University, Milton Keynes, UK
{d.j.lambert,s.galizia,j.b.domingue}@open.ac.uk

Abstract. Formal goal and service descriptions are the shibboleth of the semantic web services approach, yet the people responsible for creating them are neither machines nor logicians, and rarely even knowledge engineers: the people who need and specify functionality are not those who provide it, and both may be distinct from the semantic annotators. The gap between users' informal conceptualisations of problems and formal descriptions is one which must be effectively bridged for semantic web services to be widely adopted. We show how a simple technique—using a wiki to collect user requirements and mediate a progressive, iterative refinement and formalisation of user goals by domain experts and their knowledge engineer colleagues—can achieve this. Further, we suggest how the process could be extended, so as to itself benefit from semantic technologies.

1 Introduction

Service oriented computing (SOC) offers a promising new approach to programming, resource sharing, and organisational collaboration. Semantic web services address several of the problems SOC faces as the number and complexity of services grows, such as finding appropriate services, composing, and invoking them correctly. But the mechanisms used to enable this magic require formal, logical specifications of user goals and the web services that can satisfy them.

We are currently working with biomechanics researchers who have chosen semantic web services as the best platform to support their work. In this context, we faced the problem of capturing the users' notions of their goals, and translating them to formal representations. These formalisations, for the static Semantic Web as well as Semantic Web Services, are far from intuitive. In our case, we have tried to bridge the chasm with a methodology where domain experts can express their requirements in natural language and, through interaction with a semantic web expert mediated by a wiki, progressively refine their goal into one expressible in a formalism suitable for use by semantic web services.

We review the context of the work in the next section, then examine the problem of goal conception and description for users in section 3. In sections 4 and 5 we present our solution and a worked example of the method, respectively. Section 6 outlines the future direction of the work. Related research is discussed in section 7, and we conclude in section 8.

M. Weske, M.-S. Hacid, C. Godart (Eds.): WISE 2007 Workshops, LNCS 4832, pp. 248–259, 2007.
© Springer-Verlag Berlin Heidelberg 2007

2   Background

In this section, we recount a short history of the two sides of our problem, as well as the source of our solution. First, we introduce our application domain, an on-going programme to develop web services for use in a biomechanics application. Section 2.2 reviews semantic web services, noting why they have been selected as the most promising solution for our application. Finally, in section 2.3 we look at the existing software process for LHDL, with which they were comfortable and wished to use to develop semantic web services goals.

2.1   The Living Human Digital Library

The creation of in-silico models of entire organisms has been identified as a 'Grand Challenge' problem for informatics, and several projects have begun working towards the construction of multi-domain, multi-scale models. Our work concerns one such project, the 'Living Human Digital Library' (LHDL) [1], which intends to lay a technical foundation for virtual physiomes by first developing techniques and infrastructure for distributed modelling and analysis of the human musculoskeletal system.

For the immediate purposes of supporting LHDL, web services are appropriate: they address the need for distributed, autonomous provision and invocation of computational services and data storage facilities that the web services approach provides. Longer term, simulations of entire physiomes will require integration across scales and between disciplines (e.g. chemistry, biomechanics, clinical) and sub-systems (e.g. neurological, renal, cardiac). These programmes are about co-ordination: the intention is not to create a single federation of services that define a single virtual physiome, but rather a framework to enable the integration of services to suit particular requirements—even to the point of modelling individuals for clinical purposes. As the number of services available for use, and the number engaged in any one simulation, increase, it will become infeasible to manage them manually. With the future in mind, LHDL is investigating semantic web services as the most promising technological solution.

2.2   Web Services and Semantics

Service-oriented computing [2], and especially web services [3], have forced a paradigm shift in computing provision. They enable computation to be distributed, and easily invoked over the internet. 'Virtual organisations' of services can be constructed for tasks the component services were not designed for. However, as services become more complex, and their numbers increase, it becomes more difficult to comprehend and manage their use. Tasks such as service discovery, composition, invocation, process monitoring and fault repair cannot be successfully automated for web services, because the descriptions involved are only syntactic, and require human engineers to interpret them. Semantic web services [4] add rich, formal semantics to enable this automation. By modelling the purpose and interfaces of the services in logical formalisms such as description logics [5] or abstract state machines, we allow machines to reason in powerful ways about the services in ways that otherwise must be done by humans, or are simply too expensive to be done at all.

The Web Services Modelling Ontology (WSMO) [6] is a leading framework for semantic web services. Its four key concepts of domain ontologies, goals, web services, and mediators evidence its commitment to separation of concerns. WSMO insists on a clear distinction between user goals and their realisation by web services, thus enabling capability-based invocation. The user's needs and context are given first-class status in the modelling process, while intelligent middleware can determine how to satisfy a user's goal with the services available to it. Similarly, the necessary loose-coupling of services, goals, and ontologies is handled by the systematics use of mediators, which intervene in several places where otherwise heterogeneity would cause incompatibility. Between ontologies, *OO-mediators* perform ontology mapping wherever necessary; *WW-mediators* allow web services to interact correctly, primarily addressing choreography mismatches; user goals are mapped to web services by *WG-mediators*; and *GG-mediators* allow the creation of new goals by composing others.

Our WSMO implementation is the Internet Reasoning Service (IRS) [7], a general-purpose semantic services platform which has been used in several domains including business process management, e-learning, and e-government. In its current implementation, it adopts and extends the epistemological commitments of WSMO. Its internal representation format is OCML [8], a frame based knowledge modelling language. The IRS can invoke web services exposed via SOAP or XML-RPC, and export legacy Java and Common Lisp code as web services by automatically generating wrappers. Goals can be executed by sending SOAP messages or making HTTP GET requests, thus supporting the REST paradigm. A process of 'elevation' deals with mapping the XML messages of services to internal ontological representations expressed in OCML.

2.3   LHDL's Existing Software Development Process

Even as LHDL moves towards a web-based infrastructure, the project must continue to support the development of the legacy client software. For some time the LHDL members responsible for the LhpBuilder software (covered in section 3.1) had been successfully using agile development methods, and wanted to retain them.

Agile development [9] is a software development philosophy which emphasises people and communication over (usually heavy-weight) processes. There are several flavours of agile development, but they agree on the following 'agile manifesto' (http://agilemanifesto.org/):

- individuals and interactions over processes and tools
- working software over comprehensive documentation
- customer collaboration over contract negotiation
- responding to change over following a plan

These principles are typically realised in the following ways:

- the writing of use-case 'stories' which capture a facet of functionality that the customer describes in their own terms, and that become specifications for the software developers
- rapid turnaround, where users see their requirements implemented within weeks, fostering trust between customer and engineers
- emphasis on working, executable code instead of design documents
- simple solutions, which should never be more complicated than the current requirements necessitate
- continuous improvement, including refactoring, lessens the cost of future development
- test-driven development, applying automated tests to code

In this paper, we are particularly interested in the first two points, since these are the aspects of agile development most concerned with requirements specification. In LHDL, domain experts and software developers used wikis to develop and record the use-cases. Wikis [10] are websites where the content is user-editable. Wikis lower the bar for generating web content by both providing a simplified language for data entry, and sidestepping bureaucratic control of websites. The wiki engines which drive them often provide additional functionality such as versioning and notification. They are frequently used to support community websites, like BiomedTown, since they support a very collaborative workflow. Users can add their own material and edit the work of others, and the iterative, distributed efforts of many users—often experts—can quickly lead to impressive content.

3   What Is Involved in Creating Goals?

Having established that semantic web services are an appropriate way to attack the problems LHDL has set out to tackle, we face a new inconvenience: how can users who are not IT-experts construct the formal goal definitions? In this section, we examine the user's and then the middleware's perspectives on semantic web services, and then present criteria for reconciling the two in the context of LHDL project.

3.1   The User's View

The user experience in LHDL is mediated by the LhpBuilder and a community website, BiomedTown (`www.biomedtown.org`). The community services include forums, wikis, mailing lists and file storage, and are accessed via a web browser. The principle desktop tool is LhpBuilder [11], a legacy application which enables a user to create, store, and manipulate Virtual Medical Entities (VMEs). VMEs are collections of data such as MRI images, gait analysis data or finite element analysis results. LhpBuilder can perform operations such as extracting two-dimensional slices from volume data, virtual palpations, or combining motion-capture data with bone images.

Some of the tasks a user may wish to carry out include: registering as a member of BiomedTown (for any of several projects hosted there); searching and retrieving data resources; using data resources within LhpBuilder; creating new data resources by editing existing ones, or by defining processing pipelines on existing data; importing and exporting data resources from LhpBuilder; uploading data objects to the repository; and adding meta-data to stored data objects. These tasks are defined as 'stories', written by the users, and stored at BiomedTown.

There are different classes of users, who have different relationships with the goal generation processes. Most users will simply use existing goals, often without realising that they are goals: for example, by submitting a normal web form, or by invoking some functionality through LhpBuilder which is implemented through semantic services. Another class of users will go to the lengths of suggesting or requesting new goals, but will not take part in seeing them through the specification process. Those who actively participate in the generation of goals will be a small minority. Even these practitioners, who are technically savvy and familiar with particular computational tools of their trade, do not typically write Perl programs, as may bioinformaticians working in genetics or proteomics, nor are they familiar with the logical languages used on the semantic web.

3.2   The Machine's View

Semantic web services require several components, which in the case of the WSMO framework, include the following:

- *user goal description*
- *domain ontologies*
- *web service description* description of web services
- *mapping goals to web services* either directly or using composition
- *identifying mediator requirements* mismatches between ontologies, goals, and
  web services identified and dealt with

of which only the first two should be of interest to the typical user, and we will only consider the first here. WSMO, and hence IRS, impose a strict division between goal and service. This allows us to explicitly model the user's needs, without regard to how it might be implemented. This allows the middleware to better understand the context of a goal invocation, and flexibility in how to satisfy it. An IRS goal consists of several components:

name which identifies the goal
superclasses which may anchor the goal in a goal taxonomy
inputs the parameters passed to the goal
output the returned value
capability which is a context in which the goal is applicable

Goals may have several superclasses, so the taxonomy is a graph, not a tree. Inputs and outputs are named parameters, and each is typed by association with a concept from an appropriate domain ontology. The capability in turn is expressed by four kinds of axioms:

- *preconditions* and *postconditions* conditions on the inputs and outputs that must be met for the goal to execute
- *assumptions* and *effects* conditions in the world which should hold true and after invoking the goal, respectively

Preconditions and postconditions can be verified at invocation time by the middleware or the services themselves. Assumptions and effects are predicates on a world state which cannot be easily verified by the middleware or services at run time, and which may be unverifiable in principle. All four are sentences in restricted predicate logic, and all are optional (or true by default, whichever interpretation suits).

A goal definition in IRS's internal representation language of OCML, and a corresponding graphical representation are shown in figures 2 and 3 respectively.

3.3   Requirements for a Goal Formalisation Process

Given the discrepancy between users who can describe their goals informally and perhaps imprecisely, and the representation required by semantic middleware, we required a process that meets the following criteria:

1. *Perform requirements capture* We are concerned not just with generating the formal goal, but with the very act of discovering what the user wants.
2. *Generate formal goal descriptions* Identification and description of semantic goals using requirements documents. Necessary domain ontologies created or reused.
3. *Generate natural-language documentation* Not only are formal descriptions hard to write for non-specialists: they are not much easier to read.
4. *Easy to use* Users must be comfortable with the process itself.
5. *Fit well with current practice.* The users have a methodology which worked well for the non-web services version of the software and which they intend to use as they move to web services. They are happy with the results, and comfortable with the process.
6. *Support distributed development.* The teams responsible for LhpBuilder and the semantics are geographically separated, so collaboration must work at a distance. This will often be the case in SOC environments, since one of SOC's key features is its distributed nature.

4   A Lifecycle for Agile Goal Specification

Our solution is iterative collaborative refinement of goals, mediated by a wiki. Just as wikis simplify the HTML notation of websites, so we use a wiki to simplify the entry of goals. Where the wiki engine turns simplified markup into HTML, we use the intervention of ontology engineers to refine the informally stated, natural language requirements into OCML ones. The lifecycle then looks like this:

1. User conceives task and develops story
2. User enters natural language goal definition in wiki

3. Knowledge engineer clarifies the natural language
4. User agrees or refines this new definition
5. Knowledge engineer creates the formal goals, retaining the natural language
   as documentation

In actual use, the process will involve more iteration, sometimes a substantial amount, depending on circumstances. The user's initial goal descriptions are lodged in terms of natural language descriptions. For instance, a user might say that they want to search for VMEs. We use a template to structure the definition (see figure 1 for a completed example). The distinction between precondition/-postcondition versus assumption/effect is not only often subtle and difficult for domain experts to comprehend, it can also be an arbitrary distinction, since it depends on how the interface develops. This requires input from the engineer as well as the user, and emerges in the process. Initially, we just ask for 'before' and 'after' conditions.

Following submission, a semantic web services expert reviews the goal, refining it by making the types and conditions more concrete (i.e. aligning it with the current ontology). The goal may suggest a class of goals which are best separated, in which case the engineer can split the goal into several pages and proceed with each. The domain ontology (or ontologies) may also require extension or revision in the light of the developing goal. The domain ontology can usefully be inspected in a graphical format by the domain expert, to ensure the correct terms are being used. At this point, the engineer has essentially formalised the goal, but checks with the user via the formalised natural language. If this is correct, the engineer proceeds to a fully formal representation but retains the natural language definitions as comments. This provides documentation, which can be hyperlinked to other pages in the wiki. This can also be used as a 'cookbook' by semantic engineers when they construct other goals.

5   Example

In this section, we illustrate the process of requirements elicitation and goal formalisation for an LHDL project goal. We use the example of a user requirement to find the URLs of VMEs which match given search criteria.

The user begins by filling the template form: figure 1 is a goal showing the use-case story. The ontology engineer begins by creating a new goal class, `search-goal`. The user seems to want several kinds of goal, searching by one of several criteria such as donor attributes, data type, or VME attributes, or creation attributes. The engineer divides them out into separate pages, linked from the general `search-goal` superclass's page. Common to all, however, is that every goal returns a list of URLs: this can be recorded on the top-level goal's page. We will focus here on searching by acquisition attributes.

The user's story for this particular goal type says the following: "Example: Find all scans with slice spacing smaller than 2mm, generated with an axial scan."

Fig. 1. Wiki page with a goal in development. Note that some parameters have been given types and are hyperlinked to the relevant pages.

```
(defclass search-goal (lhdl-goal) ?goal
   (output-role :type (list-of vme-url)))

(defclass search-by-acquisition (search-goal) ?goal
   ((input-role acquisition-filter :type acquisition-filter)
    (has-postcondition
     (kappa (?goal)
            (and (has-value ?goal acquisition-filter ?filter)
                 (has-value ?filter slice-spacing-max
                                    ?slice-spacing-min)
                 (has-value ?filter slice-spacing-min
                                    ?slice-spacing-max)
                 (has-value ?filter scan-type ?scan-type)
                 (has-value ?goal output-role ?urls)
                 (forall ?url ?urls
                         (and (<= (value ?url slice-spacing)
                                  ?slice-spacing-max)
                              (>= (value ?url slice-spacing)
                                  ?slice-spacing-min)
                              (= (value ?url scan-type)
                                 ?scan-type)))))))))
```

Fig. 2. The search goal in OCML

The search is expressed by a list of criteria which must be true of each URL returned. Again, another page is built where the engineer can develop and explain the a search filter for acquisition data:

```
(defclass acquisition-filter ()
  ((slice-spacing-max :type float)
   (slice-spacing-min :type float)
   (scan-type :type scan-type)))
```

But this is explained to the user in the following terms: "The user creates a search filter object with field which reflect maximum or minimum values that are acceptable for VMEs."

At this point, or perhaps after some iterations in which the user and engineer reach agreement via English, the OCML descriptions are in place. The result is fully formalised (figure 2).

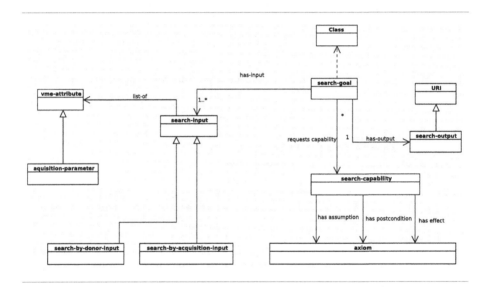

Fig. 3. An intermediate depiction of the search goal as UML

6   Future Development

We have used this method successfully to produce real goal definitions and built services to support them, but there is obviously scope for enhancement. Most conspicuous is the absence of semantics, the use of which would open several options. The process could be partially automated and brought within the semantic web services umbrella. An obvious integration would be with semantic wikis, in which the final ontological goal descriptions are stored in a knowledge base and intelligently extracted into the wiki as required, instead of being merely presented as text in the wiki [12]. Goals could be categorised simultaneously goals at the wiki and semantic levels.

If we reexamine the agile manifesto in section 2, we see that we have not addressed all the points. Without pushing the analogy too far, we can ask what it would mean to have 'working code': this might correspond to having the formal definitions stored in a reasoner which would continually check for consistency (and refactoring could be partially addressed by checking for redundancy).

Similar problems confront those creating service descriptions. Although service builders are likely to be software engineers and therefore might be expected to be more familiar with formal notations, they may still need help with particular formalisms like WSMO. Wikis provide a convenient meeting place for software engineers and semantic web services 'consultants'.

7 Related Work

The semantic services literature is replete with work on service descriptions and useful, machine-reasonable semantics [13] and how to attach them to directory services but the question of where the semantics themselves come from is largely ignored. Most of the talk is of describing services, or discovering them, not defining users' intent.

The IRS was previously used in MiAKT [14], brokering the invocation of services for medical imaging. The two best-known bioinformatics projects using web services are ^{my}Grid and BioMOBY. ^{my}Grid [15] is an on-going project which provides bioinformaticians with workflow tools which can alleviate the chores of manually discovering genome-related web services and data stores, and the subsequent programming to invoke them. They essentially worked backwards from already implemented services, annotating them and then using the annotations to constrain (by reasoning over input/output types) and suggest workflow construction (services were also (coarsely) categorised by task type). In the ^{my}Grid project, DAML+OIL was initially used [16], but moved to using an extended RDF [17]. In particular, they note that DAML-S does not intrinsically support task typing. This is a disadvantage, because users think more along the lines of tasks they must complete, and not about the inputs and outputs to them. They have also looked at the question of workflow discovery [18]. Where ^{my}Grid has generated third party annotations of existing, non-semantic web services, the BioMOBY [19] project set out to create a unified ontology, with services strictly adhering to the standard terminology and XML message structures. Despite the ontology itself being developed collaboratively, in an 'open source' way, this approach precludes incorporation of legacy services and third-party annotation.

^{my}Grid and BioMOBY are targeted at the genetics and molecular biology communities where practitioners had long used scripting languages to call web services. They are thus not addressing the goal formulation problem to the same extent, since the users have already mostly formulated them, and have practice in refining them to an executable form, as well as being more conscious of what services are available. Even then, in both projects, familiarity of the

practitioners with the ontology languages was considered more important than their expressivity. LHDL has a commitment to applying a comprehensive semantic web services framework in a domain where there has previously been little use of web or grid technologies. The goals and practises for the new computational environment are naturally less developed, and requirements elicitation plays a more prominent role.

8   Conclusions

The LHDL project is driven by researchers in biomechanics who have opted to use semantic web services technologies to simplify the provision and use of their computational and data services. They must specify semantic web services goals, but are not experts in the relevant formalisms. This problem has been largely ignored, but threatens to be a bottleneck as demand for semantic web services increases from the small number currently built by semantic web researchers.

In our approach, we closed the gap by using a wiki to mediate communication between domain experts and knowledge engineers, allowing the progressive formalisation of goals initially expressed in natural language. Since the Biomed-Town citizens were already using the wiki to record use-cases for their agile development process, it was a natural step to adopt the wiki for goal requirements recording, and then further to perform the 'agile development' *in* the wiki. The wiki's normal function as a communal blackboard means the final definitions can be annotated by the users.

The point of the semantic web, of course, is to give the machine a greater understanding so that it can reason about our problems and provide intelligent assistance. We plan to implement this technique as a workflow within our web services platform, and offer more hints from the middleware, both to the domain experts and engineers.

Acknowledgements. This work was supported by the Living Human Digital Library project, European Union programme FP6-026932.

References

1. Viceconti, M., Taddei, F., Van Sint Jan, S., Leardini, A., Clapworthy, G., Galizia, S., Quadrani, P.: Towards the multiscale modelling of musculoskeletal system. In: Bioengineering modeling and computer simulation (2007)
2. Papazoglou, M.P.: Service-Oriented Computing: Concepts, Characteristics and Directions. In: WISE 2003. Proceedings of the Fourth International Conference on Web Information Systems Engineering (2003)
3. Christensen, E., Curbera, F., Meredith, G., Weerawarana, S.: Web Services Description Language (WSDL) 1.1 (2001)
4. McIlraith, S., Son, T., Zeng, H.: Semantic web services. IEEE Intelligent Systems (2001)
5. Martin, D., Burstein, M., Hobbs, J., Lassila, O., McDermott, D., McIlraith, S., Narayanan, S., Paolucci, M., Parsia, B., Payne, T., Sirin, E., Srinivasan, N., Sycara, K.: OWL-S: Semantic markup for web services (2004)

6. Fensel, D., Lausen, H., Polleres, A., de Bruijn, J., Stollberg, M., Roman, D., Domingue, J.: Enabling Semantic Web Services. Springer, Heidelberg (2006)
7. Cabral, L., Domingue, J., Galizia, S., Gugliotta, A., Norton, B., Tanasescu, V., Pedrinaci, C.: IRS-III: A Broker for Semantic Web Services based Applications. In: Cruz, I., Decker, S., Allemang, D., Preist, C., Schwabe, D., Mika, P., Uschold, M., Aroyo, L. (eds.) ISWC 2006. LNCS, vol. 4273, Springer, Heidelberg (2006)
8. Motta, E.: An Overview of the OCML Modelling Language. In: 8th Workshop on Knowledge Engineering: Methods & Languages KEML 1998 (1998)
9. Highsmith, J., Cockburn, A.: Agile software development: the business of innovation. Computer 34, 120–127 (2001)
10. Leuf, B., Cunningham, W.: The Wiki Way: Quick Collaboration on the Web. Addison-Wesley Longmann, Reading (2001)
11. Van Sint Jan, S., Viceconti, M., Clapworthy, G.: Modern visualisation tools for research and education in biomechanics. In: IV 2004. Eighth International Conference on Information Visualisation, vol. 00, pp. 9–14. IEEE Computer Society, Los Alamitos (2004)
12. Fischer, J., Ganter, Z., Rendle, S., Stritt, M., Schmidt-Thieme, L.: Ideas and Improvements for Semantic Wikis (2006)
13. Paolucci, M., Soudry, J., Srinivasan, N., Sycara, K.: A Broker for OWL-S Services. In: Proceedings of the 2004 AAAI Spring Symposium on Semantic Web Services (2004)
14. Shadbolt, N., Lewis, P., Dasmahapatra, S., Dupplaw, D., Hu, B., Lewis, H.: Mi AKT: Combining Grid and Web Services for Collaborative Medical Decision Making. In: Proceedings of The UK e-Science All Hands Meeting 2004 (2004)
15. Stevens, R.D., Robinson, A.J., Goble, C.A.: myGrid: personalised bioinformatics on the information grid. Bioinformatics 19(1), i302–i304 (2003)
16. Wroe, C., Stevens, R., Goble, C.A.R., Greenwood, M.: A Suite of DAML+OIL Ontologies to Describe Bioinformatics Web Services and Data. International Journal of Cooperative Information Systems 12(2), 197–224 (2003)
17. Lord, P., Bechhofer, S., Wilkinson, M., Schiltz, G., Gessler, D., Hull, D., Stein, C.G.L.: Applying semantic web services to bioinformatics: Experiences gained, lessons learned. In: McIlraith, S.A., Plexousakis, D., van Harmelen, F. (eds.) ISWC 2004. LNCS, vol. 3298, pp. 350–364. Springer, Heidelberg (2004)
18. Goderis, A., Li, P., Goble, C.: Workflow discovery: the problem, a case study from e-science and a graph-based solution. In: International Conference on Web Services (ICWS 2006), pp. 312–319 (September 2006)
19. Wilkinson, M.D., Links, M.: BioMOBY: An open source biological web services proposal. Briefings in bioinformatics 3(4), 331–341 (2002)

User-Friendly Semantic Annotation in Business Process Modeling

Matthias Born, Florian Dörr, and Ingo Weber

SAP Research, Karlsruhe, Germany
{Mat.Born,F.Doerr,Ingo.Weber}@sap.com

Abstract. Current problems in Business Process Management consist of terminology mismatches and unstructured and isolated knowledge representation in process models. Semantic Business Process Management aims at overcoming many of those weaknesses of Business Processes Management through the use of explicit semantic descriptions of process artifacts. However, this vision has a prerequisite: semantic annotations need to be added to the process models. In this paper, we present an approach that allows flexibly annotating semantics in a user-friendly way, by exposing ontological knowledge to the business user in appropriate forms and by employing matchmaking and filtering techniques to display options with high relevance only. By adding semantic information the precision of process models increases, ultimately supporting Web Service discovery and composition. As a proof-of-concept, the work has been implemented prototypically in a process modeling tool.

1 Introduction

A core aspect of Business Process Management (BPM) is creating models of business processes. These models are used in various contexts: communication, documentation, implementation, and automated execution. Companies try to establish a common basis of business terminologies using process repositories. However, this approach only helps to a certain degree and does not address the overall business process lifecycle. General issues within a business process lifecycle are that business consultants and IT experts do not speak the same language, do not share the same concepts of processes, or use the same tools. Semantically annotated process models could enable support for the modeler in various associated tasks: reusing parts of process models when creating new models; making process models executable; detecting cross-process relations; facilitating change management; and providing a structured basis for knowledge transfer. Semantic Web Service technology, like Web Service Modeling Ontology (WSMO) [16] or OWL-S [14,2] provide methods and tools for creating these machine-accessible representations of knowledge. The Semantic BPM (SBPM) approach attempts to take BPM to the next level by integrating and utilizing semantics to improve the modeling and management of business processes.[7]

Within this context, our paper presents an approach for the integration of semantics in modeling tools to support the graphical modeling of business processes with information derived from domain ontologies. For this purpose, we

M. Weske, M.-S. Hacid, C. Godart (Eds.): WISE 2007 Workshops, LNCS 4832, pp. 260–271, 2007.
© Springer-Verlag Berlin Heidelberg 2007

identify suitable semantic information to specify business process models more precisely. We then present a concept for the integration of the identified semantic information in modeling tools utilizing ontological descriptions of the business process models and the domain world to augment and annotate process models. We exploit the particular nature of business process models, e.g., their control and data flow, and suggest a specific structure of the domain ontology. This structure defines business objects in the domain of discourse, alongside with their lifecycle. If the domain ontology is not expressed in this structure, our techniques for string-based matchmaking are still applicable, but the additional precision in filtering with regard to the objects and their states is lost.

Finally, we present matchmaking functionalities for supporting users in modeling semantically annotated process models. This is achieved by matching elements of the graphical business process model with elements of domain ontologies. We demonstrate a way how textual fragments are used to match semantic annotations to elements of a process model based on the Business Process Modeling Notation (BPMN)[1,2]. In short, this is achieved by comparing the context and any given textual descriptions to the applicable instances in a domain ontology.

Our approach enables that partial process fragments can be used to support the modeling tasks given an underlying ontological description of the domain. Further, the augmented business process models facilitate the discovery of appropriate Web Services. The semantic annotations can be used when querying an enterprise-wide process model repository, by allowing for more informed search techniques and fuzzy results.

As a proof-of-concept, the SAP Research modeling tool "Maestro for BPMN" has been extended with a prototypical implementation of this approach.

The remainder of the document is structured as follows. The requirements for our solution are described in Section 2. Based on this, we present an approach for user-friendly semantic annotation in Section 3, followed by a description of the related prototypical implementation in Section 4. Subsequently, related work is discussed in Section 5. Finally, Section 6 concludes.

2   Requirements

In the scope of our approach the additional information, i.e. the semantic annotations, sometimes referred to as tags or markups, for semantically enriched process models, should comply with the following requirements. First of all, the information should be definable by business experts during modeling time. Users of modeling tools should be able to understand the semantic information they deal with. Therefore, the semantic information may not be too IT specific or low-level. The ease of use is essential regarding user acceptance. Second, the additional information should facilitate the realization of the processes and support querying the process space. It should allow users to specify process models more

---

[1] OMG, BPMN Information, http://www.bpmn.org/

[2] Note that our conceptual solution is independent of BPMN as a concrete graphical business process modeling notations.

precisely, couple these process models tighter to existing domain concepts, and help to find and compose according Web Services for the activities of the process models (Web Service discovery and composition). Third, in order to support the semantic augmentation of the process models, information should be derived from appropriate domain ontologies. In order to realize this kind of support the following competency questions should be taken into account:

1. *What are the objects, states & actions of a domain and what are their names?*
2. *What are possible states of a certain object?*
3. *What are possible predecessor and successor states of a certain state?*
4. *Which actions are possibly relevant for a certain object?*
5. *Which objects are manipulated (state changes) by a certain action?*
6. *Which state changes (transitions) for which objects can be caused by which actions?*

Based on these requirements the next section shows the generic approach which we developed for user-friendly annotation of process models.

3   Solution Approach

To provide the information described above, different BPMN elements are extended and utilized. Figure 1 illustrates the main components of our semantic extension for business process modeling tools. Especially, BPMN Data Objects and Associations are used to describe the activities of a process model more precisely by defining associated objects and their state transitions: Data Objects identify the objects an activity deals with and Associations link the Data Objects to the corresponding activities in the process diagram. The user modeling the process may name and define the activities, may specify pre- and post conditions for the activities in natural language, and may define the objects as well as the objects' states before and after an associated activity has been executed within the graphical model. Through our extensions for business process modeling tools, business experts are supported in specifying this additional information during the graphical modeling phase of business processes. To help users annotating their process models, different matchmaking functionalities utilize parsed-in domain ontologies, which in turn describe available elements like objects and their states. The matchmaking functionalities thus help to link the model elements to available domain concepts.

The following subsections show how we used ontologies and describe the different matchmaking functionalities in more detail.

3.1   Ontology Engineering

We identified the following additional semantic information for business process models:

- The objects relevant for each activity in the process model can be specified and, if applicable, get directly linked to according objects of a domain.

- The states of these objects, before and after the according activity on the objects has been performed, can be specified.
- Natural language definitions of pre- and post conditions for the activities within a process model can be provided.

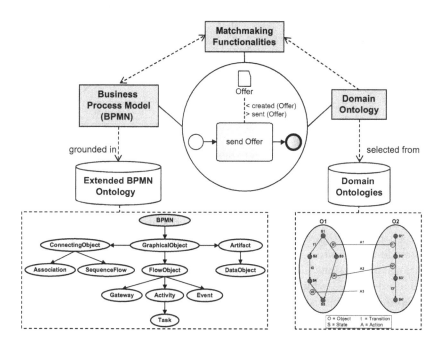

Fig. 1. Realization Overview

All this information helps to ease the realization of the processes and supports querying the process space by answering questions regarding which objects are manipulated within which business process models.

Two kinds of ontologies are used to enable the semantic support of modeling activities. First, the evolving *sBPMN* ontology [1] of the SUPER[3] project as a format for representing BPMN process models, featuring basic concepts and attributes for standard BPMN elements, has been extended. Our ontology provides possibilities to define states of a Data Objects before and after corresponding activities have been executed, to link objects, states, and activities to elements of domain ontologies describing them, and to capture natural language pre- and post conditions for activities. With these extensions, the *sBPMN* ontology can be used as an internal and external format for semantically augmented BPMN process models in the scope of our approach. Second, we will define a possible structure of domain ontologies along with a short concrete example. Our domain ontology covers information concerning domain objects and states which helps

---

[3] Integrated Project SUPER (Semantics Utilised for Process Management within and between Enterprises), http://www.ip-super.org/

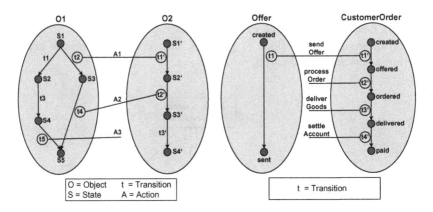

Fig. 2. Domain Ontology Visualization

to model business processes more precisely. This kind of domain ontology may then be used within process modeling tools to support the user in finding and defining appropriate activities, data objects, and states.

The left part of Figure 2 illustrates the information which our domain ontology provides. The ontology contains information about domain objects, states, transitions, and actions. For each domain object, possible states and state transitions are described which together form the lifecycle of a domain object. Actions represent activities in the domain and can cause multiple state transitions on different objects. States are described by more fine-granular definitions (e.g. constraints over attributes). The right part of Figure 2 shows an extract from an example of a domain ontology. Domain ontologies that provide this kind of information support our semantic business process modeling approach. During modeling, these domain ontologies are queried and utilized to help the user specifying model elements and states by proposing appropriate domain concepts or instances.

Not only states but also objects can be defined, described, and considered. The objects defined in the ontology represent the nouns used within this language, actions may be regarded as verbs and states virtually are adjectives. The structure relates objects (nouns), actions (verbs), and states (adjectives) to each other and thus defines a normalized modeling language in the scope of specific domains. This correlation and interrelation together with ontology-specific descriptions and definition potentials (e.g. hierarchies, inheritance, subsumption relationships, etc.), which may be exploited for reasoning tasks, is a major contribution and benefit of the approach developed within this paper.

3.2   Name-Based Matchmaking

Different matchmaking functionalities that are required to bridge the gap between the business process model and the domain world are incorporated into our semantic extension approach, in order to allow matching elements of the graphical business process model with elements of domain ontologies and to support the user in semantically specifying or refining the process. Utilizing appropriate domain ontologies, the matchmaking functionalities address the problem of deriving a list of proposals for a selected model element (Data Object, Activity, Association/State) that a user has chosen for semantic refinement. To solve this problem, we use a combination of different text and name matching methods and utilize process diagram context information as well as domain ontology knowledge. For name matching tasks, we use a combination of heuristic comparison methods on the strings of characters, well-known string distance metrics [4,3], and matching methods considering synonyms and homonyms. The additional utilization of the diagram context information of selected model elements and domain knowledge to match model fragments with domain instances leads to even better results. Domain element proposals can be derived with the help of elements already specified in the process model and the information covered by domain ontologies.

3.3 Process Context-Based Matching

Furthermore, the actual control and data flow in the process model can be leveraged in our approach.

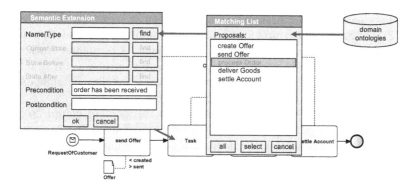

Fig. 3. Simple Ordering Example

- With respect to the control flow, the lifecycle of a domain object can be used to suggest next activities during modeling. E.g., in Figure 3, the modeling tool can suggest *"process Order"* as the next activity in the process, because it follows *"send Offer"* in the CustomerOrder's lifecycle.
- Also, the lifecycle can be employed for making targeted suggestions for the semantic refinement of activities, e.g., by not suggesting an already used action again in a process model.

Analogously, the data flow in process models can be supported. Note that consistency checks between process control, data flow, and object lifecycles can be provided as well. Furthermore, the semantic business process models contribute to mechanize the realization of the processes. The additional information about how objects are changed by the activities of a process model can help to derive pre- and post conditions for goal descriptions. By considering the states – which may for example be defined in terms of object-attribute-constraints in domain ontologies – of all associated objects before and after a particular activity has been carried out, high-level goal definitions for the activities of a business process model can be derived. These goal definitions can then be used as an input to business process composition approaches [19] or directly to functional discovery engines, like the prototype integrated in Web Service Execution Environment (WSMX)[4]. By comparing goal descriptions, such discovery engines try to find appropriate Web Services to achieve the goals. On this way, Web Services implementing the desired functionality can automatically be discovered for the activities of a business process model. If no single Web Service satisfies the goal, a composition of Web Services can be searched, which together then achieve the goal.

4 Prototypical Implementation

The solution approach has been successfully implemented using the SAP Research modeling tool, namely "Maestro for BPMN". Our assumption of the implementation is that a domain ontology was created before a business experts starts modeling the process. The application is based on the Tensegrity Graph Framework[5], which provides basic functionalities like rendering, editing of diagram elements, event propagation mechanisms, a command stack, and a persistency service for diagrams. To give an impression how context information can be utilized for matchmaking tasks, Figure 4 sketches a simple example scenario concerning the definition of a Data Object.

The *Activity* "send Offer" has been linked to the *domain action* "send Offer". This action causes two transitions, one of them affecting the *domain object* "Offer" the other one affecting the *domain object* "Customer Order". When the user wants to define the *Data Object* more precisely, these two *domain objects* are proposed as possible elements because the *Data Object* is associated to the *Activity* "send Offer". If the *Activity* "send Offer" is not linked to the *domain action* "send Offer" the list of proposals would only contain the *domain object* "Offer", which can be found via name matching with "send Offer". Similar matchmaking capabilities regarding *Data Objects*, *States*, and *Activities* are also incorporated into our extension approach. A whole business process diagram could be described in this manner.

Figure 5 shows an example for an enriched business process model. The Data Object states before the execution of the associated activity are indicated by a "less than" sign ($<$), the states afterwards by a "greater than" sign ($>$). The Data Objects and states identified with the help of domain ontologies could be linked to appropriate elements within these ontologies. Not all objects or states

---

[4] http://www.wsmx.org/
[5] http://www.tensegrity-software.com

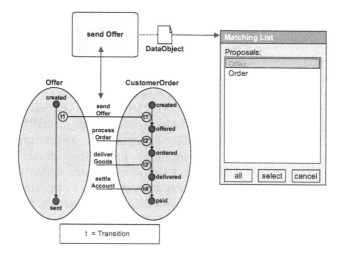

Fig. 4. Data Object Definition Example

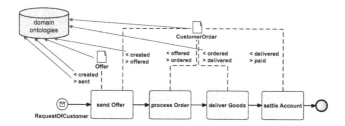

Fig. 5. Use Case Example

may be found in domain ontologies, there may also be unknown or new objects and states in the business process diagrams modeled. These unlinked objects and states may indicate that such new elements should be created and realized respectively within the domain described by the ontologies in order to implement the business process model.

The semantically augmented process models can be used to enhance querying the process space, by answering questions regarding which objects are manipulated within which business processes. If, for example, the EU enacts a new law concerning the expiration date of dairy products, the augmented process models and the knowledge of the domain ontologies can help to find out, which business processes are affected by the new law.

5   Related Work

Since this paper envisions an improvement regarding process modeling activities, the work is related to the field of business process modeling tools. There is a plethora of different BPMN modeling tools like for example Intalio Designer[6], the Eclipse STP BPMN Modeler[7], and ITP-Commerce's Process Modeler[8]. To our knowledge, no commercial tool enables the semantic annotation of business process models.

The working group concerned with semantic business process modeling and analysis tools within the SUPER project is currently working on a prototype of a Business Process Modeling Ontology (BPMO) Editor for modeling semantic business processes in SUPER. The first version of the BPMO modeling environment provides basic functionality for enriching existing process models with semantic annotations (e.g. by assigning a pre-defined semantic goal to an activity).[5,18] In contrast, the approach introduced within this paper allows for a flexible and fine-granular semantic annotation of processes by directly accessing single ontology elements from the business domains of interest. Regarding process ontologies our work is closely related to the *sBPMN* ontology [1] developed in the SUPER project. The *sBPMN* ontology acts as the basis for our prototypical implementation. In a different context, Lin and Strasunskas [12,13] present an approach of a General Process Ontology and discuss a semantic annotation framework.

Other related work can be found in [8,9,10], which is also based on semantic business process modeling. In [10], a class of Petri Nets, so-called Pt/T-Nets are modeled in OWL, and elements can be brought in relation to (ontology-based) data items. These relations are specified in terms of attributes and values, and how these are inserted or deleted as the process executes. There seems to be a clear focus on data instances, and there is a prototypical implementation in a Petri Net modeling tool. [8] presents, among others, a slightly updated model of the Petri Net ontology together with an approach to process decomposition, which is based on the "linguistic specificity" of the terms used in the labels of the elements in a process model. One of the goals is to normalize the terminology in a process model, but rather in terms of the abstraction level used in the labels. The focus in this work, in contrast, is on specifying the semantics of a process model by making use of a given domain ontology. Also, the data objects in our domain ontology structure can be easily used for capturing business objects, not only data.

The domain ontologies in this paper describe – among other information – data objects including their life-cycles. Hence, any suitable formalism may be used instead which is based on Finite State Machines [6]. However, our ontology-based approach allows capturing intentional human language and machine interpretable information jointly in a standardized language. With the help of these domain descriptions, object state information can be added to the process model. With a different focus, the idea of relating object state information to business process models is also presented in [17]. The approach taken there is to check the consistency of business process models and object lifecycles by adding object

---

[6] Intalio Designer, Available: `http://www.intalio.com/products/designer/`

[7] STP BPMN Modeler, Available: `http://www.eclipse.org/stp/bpmn`

[8] Process Modeler for Microsoft Visio, Available: `http://www.itp-commerce.com/`

state information to specific points in a process model and comparing the process model's usage of the objects with the objects' lifecycles. The approach presented in this paper has a quite different target, as an object's lifecycle is not perceived as an authoritarian model. In [11], an approach for automatic "synthesis" of processes is presented, which means calculating the optimal combination of them. This work describes how annotated reference process models using UML syntax may be used for the synthesis task and describe synthesis algorithms.

The work concerning name-based matchmaking (see Section 3.2) is related to schema matching efforts. A survey of Rahm and Bernstein [15] presents a nice overview of this research area. However, the fundamental difference is that schema matching tries to map between two formalized schemas (e.g. defined in XML), whereas the matching problem addressed in this paper is to find appropriate domain ontology elements given a free-text name entered by the user and context information derived from the process model, concerning the selected element. Nevertheless, the name matching tasks are similar for the two approaches. A comparison of Cohen et al. [4] describes different string distance metrics, some of which are utilized for name matching tasks in this work. Therefore, these established, general approaches for name matching tasks are reused to some extent and are enhanced with new (context-related) matching functionalities.

6   Conclusion

In this paper, we presented an approach for the integration of semantics in modeling tools to support the graphical modeling of business processes with information derived from domain ontologies. The goal is to make these business process models more precise, to ease their (semi-automatic) realization, and to enable querying large process model repositories. For this purpose, we defined suitable semantic information, ontology-based descriptions for business process models and domains, and matchmaking functionalities to support users in modeling semantically annotated process models. Our current implementation assumes the existence of suitable domain ontologies which potentially are created by knowledge engineers. An example approach is to specify the concepts of such a domain ontology manually and derive the instance data from other (non-semantic) models, e.g. from MDD[9]/MDA[10] artifacts or other (structured) knowledge bases. An existing domain ontology does not replace a process model as the purpose of both is quite different. A domain ontology can be seen as a knowledge base for the actual process model, whereas a business process model is used by business experts to capture process knowledge graphically. A remaining open research question is whether a business expert should be able to change the structure of a domain ontology and, if so, what are the appropriate methods and techniques.

We also outlined how Web Service discovery and composition can be facilitated without forcing business process modelers to use unfamiliar, technical terminology, or even formal (ontology) languages. At the same time our approach is

---

[9] Model Driven Development.
[10] Model Driven Architecture.

more flexible and generic then just attaching predefined goals to the activities of a process model. Besides the ontology-based matching functionalities supporting the process modeling activities with semantic techniques is a major benefit of our approach as opposed to the capabilities of current modeling tools. To prove our concept the SAP research modeling tool "Maestro for BPMN" has been used for a prototypical implementation. In addition, a smart technological concept is needed to utilize the semantically annotated process models and find appropriate services to implement these processes. The areas of potential future work comprise the utilization and enhancement of our approach for the implementation of Web Service discovery and for querying the process space as well as for composition activities in order to (semi-automatically) realize business process models and make them executable. We took first steps towards this goal, which may serve as a basis for further enhancements and new extensions.

Acknowledgement. The work published in this paper was partially conducted within the EU project SUPER (www.ip-super.org) under the EU 6th Framework.

References

1. Abramowicz, W., Filipowska, A., Kaczmarek, M., Kaczmarek, T.: Semantically enhanced business process modelling notation. In: Hepp, M., Hinkelmann, K., Karagiannis, D., Klein, R., Stojanovic, N. (eds.) Proceedings of the Workshop on Semantic Business Process and Product Lifecycle Management (SBPM 2007) in conjunction with the 3rd European Semantic Web Conference (ESWC 2007). CEUR Workshop Proceedings, Innsbruck, Austria (June 7, 2007)
2. Burstein, M., Hobbs, J., Lassila, O., Mcdermott, D., Mcilraith, S., Narayanan, S., Paolucci, M., Parsia, B., Payne, T., Sirin, E., Srinivasan, N., Sycara, K.: OWL-S: Semantic Markup for Web Services. In: W3C Member Submission (November 2004)
3. Chapman, S., Norton, B., Ciravegna, F.: Armadillo: Integrating knowledge for the semantic web. In: Proceedings of the Dagstuhl Seminar in Machine Learning for the Semantic Web (February 2005)
4. Cohen, W.W., Ravikumar, P., Fienberg, S.E.: A comparison of string distance metrics for name-matching tasks. In: Proceedings of the IJCAI-2003 Workshop on, pp. 73–78, Acapulco, Mexico (August 2003)
5. Dimitrov, M., Simov, A., Stein, S., Konstantinov, M.: A BPMO based semantic business process modelling environment. In: Hepp, M., Hinkelmann, K., Karagiannis, D., Klein, R., Stojanovic, N. (eds.) Proceedings of the Workshop on Semantic Business Process and Product Lifecycle Management (SBPM 2007) in conjunction with the 3rd European Semantic Web Conference (ESWC 2007). CEUR Workshop Proceedings, Innsbruck, Austria, June 7, 2007 (June 2007)
6. Gill, A.: Introduction to Theory of Finite-state Machines. McGraw-Hill Education, New York (1962)
7. Hepp, M., Leymann, F., Domingue, J., Wahler, A., Fensel, D.: Semantic Business Process Management: A Vision Towards Using Semantic Web Services for Business Process Management, pp. 535–540 (2005)
8. Koschmider, A., Blanchard, E.: User Assistance for Business Process Model Decomposition. In: Proc. 1st IEEE Intl. Conf. on Research Challenges in Information Science, Ouarzazate, Marokko, pp. 445–454 (April 2007)

9. Koschmider, A., Oberweis, A.: Ontology based Business Process Description. In: Castro, J., Teniente, E. (eds.) CAiSE 2005. LNCS, vol. 3520, pp. 321–333. Springer, Heidelberg (2005)

10. Koschmider, A., Ried, D.: Semantische Annotation von Petri-Netzen. In: Proceedings des 12. Workshops Algorithmen und Werkzeuge für Petrinetze (AWPN 2005), Berlin, Germany, pp. 66–71 (September 2005)

11. Lautenbacher, F., Bauer, B.: Semantic reference and business process modeling enables an automatic synthesis. In: Proceedings of Semantics for Business Process Management Workshop (SBPM) of the European Semantic Web Conference (ESWC), Budva, Montenegro (2006)

12. Lin, Y., Strasunskas, D.: Ontology-based semantic annotation of process templates for reuse. In: Proc. of 10th Intl. workshop EMMSAD 2005, Porto, Portugal (June 2005)

13. Lin, Y., Strasunskas, D., Hakkarainen, S., Krogstie, J., Slvberg, A.: Semantic annotation framework to manage semantic heterogeneity of process models. In: Dubois, E., Pohl, K. (eds.) CAiSE 2006. LNCS, vol. 4001, pp. 433–446. Springer, Heidelberg (2006)

14. Mcguinness, D.L., van Harmelen, F.: OWL Web Ontology Language Overview (February 2004)

15. Rahm, E., Bernstein, P.A.: A survey of approaches to automatic schema matching. The VLDB Journal 10(4), 334–350 (2001)

16. Roman, D., Keller, U., Lausen, H., de Bruijn, J., Lara, R., Stollberg, M., Polleres, A., Feier, C., Bussler, C., Fensel, D.: WSMO - Web Service Modeling Ontology. Applied Ontology 1(1), 77–106 (2005)

17. Ryndina, K., Küster, J.M., Gall, H.: Consistency of business process models and object life cycles. In: Kühne, T. (ed.) MoDELS 2006. LNCS, vol. 4364, pp. 80–90. Springer, Heidelberg (2007)

18. Stollberg, M., Norton, B.: A refined goal model for semantic web services. In: Proc. of 2nd International Conference on Internet and Web Applications and Services (ICIW 2007) (May 2007)

19. Weber, I., Markovic, I., Drumm, C.: A Conceptual Framework for Composition in Business Process Management. In: BIS 2007: Proceedings of the 10th International Conference on Business Information Systems, Poznan, Poland (April 2007)

Semantic Web Service Discovery for Business Process Models

Ivan Markovic and Mario Karrenbrock

SAP Research
Karlsruhe, Germany
{ivan.markovic,mario.karrenbrock}@sap.com

Abstract. Information technology is seen as a critical tool to increase the level of automation when incorporating new business requirements due to changing needs. Therefore, one of the key challenges in the organizations today is to ensure the alignment between the business goals and the flexibility and responsiveness of IT systems to meet those goals. In this work we make a step forward in bridging the eternal gap between business and IT. Contribution of this work is two-fold. First, we present an intuitive way of specifying expressive user requests for the implementation of tasks in business process models. Second, we design a comprehensive approach to discovery of Semantic Web Services for process task implementation. The work has been prototypically implemented in a process modeling tool and can be used in different scenarios.

1   Introduction

Modern businesses constantly look for ways to shorten the time to act on business opportunities and differentiate themselves from competitors. Information technology (IT) is seen as a critical tool to increase the level of automation when incorporating new business requirements. Therefore, one of the key challenges in the organizations today is to ensure the alignment between the business goals and the flexibility and responsiveness of IT systems to meet those goals.

Business process (BP) modeling is used to capture business requirements. BP models are created by business experts to enable a better understanding of business processes, facilitate communication between business people and IT experts, identify process improvement options and serve as a basis for derivation of executable business processes.

In this work, we make a step forward to deriving executable processes out of BP models, i.e. mapping the view of a process model from the business perspective to the IT perspective (Web services) in an organization. In particular, we apply semantic technologies to represent the business goals set by business experts and services at the IT level and design an approach for translating these goals to their corresponding implementation in the IT system.

There are two major initiatives working on developing a standard for the semantic description of Web services: OWL-S [1] and Web Service Modeling Ontology (WSMO) [2]. These two initiatives aim to enable effective exploitation

M. Weske, M.-S. Hacid, C. Godart (Eds.): WISE 2007 Workshops, LNCS 4832, pp. 272–283, 2007.

of semantic annotations in discovery, composition, execution and interoperability of Web services. WSMO provides several advantages when compared to OWL-S: it's conceptual model has a better separation of the requester and provider point of view, it provides better language layering and it describes user requirements in a more natural fashion [3]. The Web Service Modeling Language (WSML) [4] provides a formal syntax and semantics for WSMO. In this work, we use WSML as a representation language for the domain ontologies as well as Semantic Web Service (SWS) descriptions.

This paper is structured as follows. In Section 2 we present a motivating scenario. Section 3 gives a list of requirements for the designed solution. In Section 4 we present our discovery framework. Section 5 discusses some related work. We conclude and give an outlook on future research in Section 6.

2 Motivating Scenario

To motivate our approach, we use a business travel scenario in the domain of e-tourism. This scenario describes the process of organizing business trips for the employees in a company. This process comprises of booking flights, hotels, trains, cars, shuttles etc. The organization of business trips for employees is not the core business of most companies and therefore companies constantly try to decrease travel costs. Thus, companies are interested in using price discounts and serving their employees by booking a convenient trip.

To give an illustrative example, let's imagine that one of a company's sales representatives must participate at a conference in Vienna. The company resides in Cologne and the sales representative wants to fly from an airport nearby. The conference will last three days and the employee will therefore stay in a hotel. Taking this description into account, several tasks must be processed by the company to organize the trip of the sales representative: i) book a flight from Cologne to Vienna, ii) book a hotel for the stay, iii) rent a car in Vienna. The figure below depicts the process model for our business travel scenario:

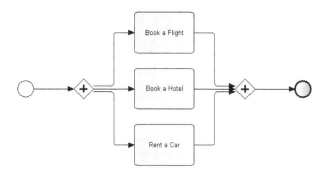

Fig. 1. Business travel scenario

Note that we used the Business Process Modeling Notation (BPMN)[1] for specifying the BP model. The tool used for this purpose is the SAP Research modeling tool "Maestro", which has also been used as a basis for building the Discovery System described in Section 4.

In this work we want to support the business expert (refered to as "user" in the sequel) in deriving executable business processes which implement BP models as the one shown in Figure 1. As a first step in achieving this, we address the following problems: i) supporting the user in creating machine processable representations of his intentions (goals) for individual tasks in the BP model, ii) designing automated support for mapping the specified user goals to the IT artifacts that implement the desired functionality. This will reduce the amount of work for the user significantly as he doesn't have to search the Internet for services to book flights, hotels and cars or contact a tourist agency. In the following section we present a list of requirements that a designed solution to this problem must fulfill.

3   Requirements Analysis

- *Req. 1: Support for user-friendly specification of the user goals* The users of the system are business people, therefore they must be able to specify their intentions in an easy and convenient way.
- *Req. 2: Querying support* There needs to be a mechanism that will perform matching of user requests against SWS descriptions to get a list of services that provide the required functionality. The matchmaking algorithm should return a ranked list of SWS that match the functional request.
- *Req. 3: User preferences specification* Provide support for an intuitive specification of user preferences. It is not enough to search for SWS based on functional requirements, but also non-functional properties such as price, speed, availability, etc. The system must be able to evaluate these requests against SWS descriptions and rank the list of services based on the degree of match.
- *Req. 4: Flexibility* The designed solution must provide support for relaxation and refinement of user queries. In the case that we get too less results matching the query, we can relax the query, i.e. incrementally abstract elements of the query, e.g. by using subsumption hierarchy. Similarly, if we get too many results, the user needs to be provided with a possibility to refine his request.

In the next section, we describe how our solution meets the listed requirements.

4   Discovery Framework

In our previous work [5], we presented a general framework for derivation of executable service orchestrations from business process models. Here we take a step further and design a comprehensive approach to discovery of SWS for process task implementation.

---

[1] http://www.bpmn.org/

4.1   Process Task Annotation

The first step in the discovery lifecycle is to annotate a process task with a WSMO goal. The WSMO goal describes the capability which a desired WSMO Web service must provide. It formaly captures the user's desired functionality of a process task.

The user must be able to specify his requirements in a natural way. In WSMO, goals are used to represent the user's intention in a formalized way and define the desired capability of a Web service. Each goal is represented as a set of WSML logical expressions and consists of the same elements as the capability description of a WSMO Web service [2]. It is unrealistic to assume that i) the business expert can express his request by writing logical expressions, ii) it is possible to automatically translate arbitrary user request to a WSMO goal.

To overcome this problem, we propose using lightweight ontologies for describing the user domain. These ontologies can be used for generating the user request templates which can be instantiated with concrete parameters by the user. Consider our example scenario described in Section 2 and suppose that the user wishes to specify the intention "I want to book a flight from Cologne to Vienna" as a WSMO goal. In this case, the user domain is flight booking with *flightTicket* as a domain concept used for annotation. Note that we consider the approach for annotating the process task in terms of "Object to be delivered" presented in [6] as the most intuitive way of describing user goals. Our assumption is that the lightweight ontologies for capturing the user domain can either be reused or created by an IT expert without much effort. In the process of generating the goal template, not every concept from domain ontology is considered relevant for task annotation. There are three conditions to be fulfilled to consider a concept relevant for annotation:

1. Concept has attributes
2. At least one attribute type is based on a complex type
3. Concrete instances are available for an attribute

To emphasize this, we give an illustrating example using the concept *flightTicket* from the flight domain ontology.

In Figure 2 we can see that the concept *flightTicket* has four attributes, satisfying our first condition. The attribute *flightNumber* of the concept *flightTicket* is based on the simple data type "String" and the other three attributes namely *originAirport*, *destinationAirport*, and *planeType* are based on complex types. The attributes *originAirport* and *destinationAirport* are both of type *Airport* which presents another concept in the flight domain ontology. The same holds for the attribute *planeType* of type *Plane* which is also modeled in the domain ontology. Having identified these three complex attributes in the concept *flightTicket*, the second condition to consider a concept relevant for task annotation is satisfied. The rationale for the second condition is that advanced querying and reasoning can only operate with values that are linked to the context and the hierarchy of the domain ontology. Concepts with attributes of only simple types don't contain any domain ontology specific context information and are

```
concept flightTicket
  nonFunctionalProperties
    dc#description hasValue "Ticket for a specific flight"
  endNonFunctionalProperties
  flightNumber ofType _string
  originAirport impliesType airport
  destinationAirport impliesType airport
  planeType ofType plane
```

Fig. 2. Concept *flightTicket*

therefore not considered relevant for task annotation. The third condition to be satisfied by a concept is that there must be at least one defined instance for every complex attribute in the domain ontology. In the flight domain ontology an instance Cologne is defined for the concept *Airport* and an instance Airbus330 is defined for the concept *Plane*. Note that the flight domain ontology could not be presented due to space limitations.

A template for specifying user requests is generated from the selected domain ontology, as shown in Figure 3.

The figure shows a screenshot of the Maestro modeling tool that has been extended to support SWS discovery in BP models.

In the first step, the user selects a particular task that he wants to annotate (in our case *Book a Flight*). After choosing the option *Specify*, a screen for defining a user request, *Build Postcondition*, is shown. Using a drop-down box, the user can specify the domain ontology he wants to use for annotating the task. When a domain ontology is chosen, the rest of the GUI is built by reading the concepts and their attributes from the selected ontology. In our case, the user specified *flightDomainOntology* which has a concept *flightTicket* with attributes *originAirport, destinationAirport* and *planeType*. The default value of the attributes for all concepts in the template is *undefined*. Now the user can instantiate the attribute values according to his needs and respectively assign the dinamically built, parameterized goal template as a formalized WSMO goal description by selecting the option *Save*. Note that in the case shown in Figure 3 the user does not care for the specific type of plane to fly with and leaves the default value for this attribute. Following these steps, the user can select other domain ontologies (*Hotel booking, Car rental*) for annotating the remaining tasks in the process model. In this way, we support the user to intuitively define his request while hiding the complexity of logical expressions for goal specification.

4.2   Functional Discovery

To address the second requirement described in Section 3, we designed a matchmaking engine which evaluates the user-defined WSMO goals against a repository of SWS for each process task. In this step we match the functional part of the SWS description (capability) with the desired functionality from the user goal.

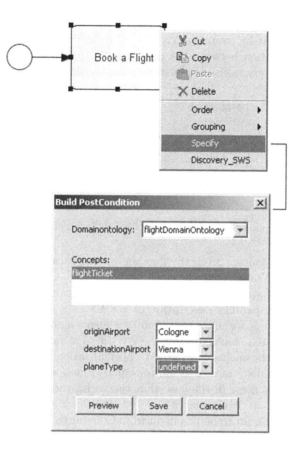

Fig. 3. Goal template

We follow a particular matching technique proposed in [7]. This technique is based on the intersection of ontological elements in service descriptions and rates two descriptions as relevant whenever they specify an overlapping functionality. We distinguish between various degrees of match introduced in [8] and [9]: the degree of match between concepts C_r and C_p in the user request and service description is rated "exact" if C_r and C_p are semantically equivalent, as "plugin" or "generalization match" if C_r is a subconcept of C_p, as "subsumes" or "specialization match" if C_p is a subconcept of C_r, as "intersect" if C_r and C_p are not disjoint, and as "fail" if they are. We further follow [6] and apply the degrees of match to semantic descriptions that capture the service functionality as a whole by a concept in an ontology. Consequently, we distinguish between a *concrete service* which contains all details about service parameters and an *abstract service* where an approximate description of functionality is given. An abstract service captures a set of concrete services. The intuition is that Web service descriptions should be filtered gradually, in several steps, to achieve the best discovery results as shown in Figure 8.

The presented idea of intersection-based matchmaking is realized within our discovery component by means of a standard reasoning task of *concept satisfiability* of a conjunction between the concepts C_r and C_p, taken from the user goal and service descriptions. To illustrate this, consider Figure 4.

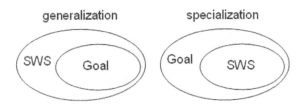

Fig. 4. SWS - goal degrees of match

We check if the concept from the domain ontology used for representing the functionality of a SWS matches the concept that describes the user goal specified during process task annotation. This verification is performed in two directions. First, we use the capability description of a SWS as a query against the user goal. In the next step, we use a user goal as a query against SWS descriptions. "Generalization match" means that the SWS offers a more abstract (or equal) functionality than the one user requested. "Specialization match" stands for having a SWS that fulfills a more specific (or equal) functionality compared to the one required by the user. In the case that the verification returns true in both directions, we have an "exact match".

Consider that we want to book a flight from Cologne to Vienna without having specified the type of plane we prefer. The object that should be delivered by the SWS is a flight ticket. In our example, the flight domain ontology is linked to the location domain ontology through a *locatedIn* attribute of the concept *Airport*. In the location ontology, concrete cites such as Vienna or Cologne are listed and also clasified in terms of concepts that group together cities in a certain geographic areas like regions and countries. Therefore, our knowledge base states that Cologne is a city in North Rhein-Westphalia (NRW), an example region which is a subconcept of Germany as it is located there. When we consider this kind of hierarchy, a "generalization match" for our request would be every SWS delivering a flight ticket and starting in Cologne, as well as the SWS describing the start location in a more abstract way, e.g. as a city in NRW. The same applies for the destination airport attribute. If we take into account that Vienna is in Austria we can state that also SWS offering flights to Austria can be considered relevant to satisfy the users intention.

After selecting a particular task, as in Figure 3, the user can initiate the matchmaking process by choosing *Discovery_SWS* option. Consequently, the list of matching SWSs is presented to the user as a result of the functional discovery process (Figure 5).

In this case, we get one match precisely corresponding to our request, one "specialization match" offering the flight with a particular type of a plane and

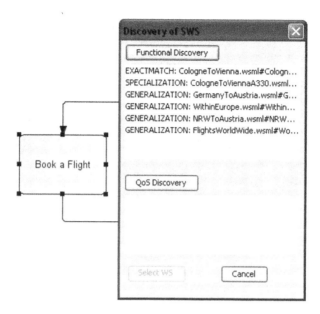

Fig. 5. Functional discovery process results

four "generalization matches" offering more abstract functionality, as derived from our knowledge base.

4.3   QoS-Based Discovery

The following step in the discovery sequence is discovery of SWS based on quality of service (QoS) criteria. For this action, a WSMO goal containing QoS requirements must be created. To support the QoS-based discovery of SWS we have adapted an available prototype[2] to our framework. A detailed description of the approach to QoS discovery for which the prototype was built can be found in [10], we will give a short overview of how we adapted the underlying ideas to our context. The core of the QoS discovery system consists of two ontologies: QoS Base and the ranking ontology. The QoS Base ontology defines the structure to be used by every domain QoS ontology, e.g. flight domain QoS ontology. This implies that for each domain ontology we need to define a QoS ontology which will extend the QoS Base structure to the specific domain. Based on the domain ontology selected (see Figure 3) and after functional discovery is performed (see Figure 5), selection of QoS discovery option will generate a template for QoS goal specification. This template is generated from the domain QoS ontology in a way similar to functional goal template generation described in Section 4.1. In Figure 6, we present the template generated from flight domain QoS ontology and used for specifying a WSMO goal consisting of QoS requirements.

---

[2] http://lsirpeople.epfl.ch/lhvu/download/qosdisc/

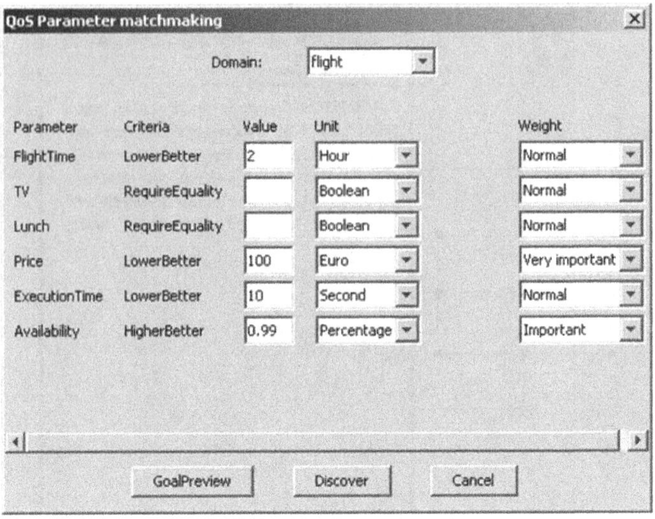

Fig. 6. Template for QoS goal specification

Rank 1: CologneToVienna.wsml#CologneVienna
Rank 2: GermanyToAustria.wsml#GermanyAus...
Rank 3: NRWToAustria.wsml#NRWAustria
Rank 4: WithinEurope.wsml#WithinEuropeFlights

Select WS   Cancel

Fig. 7. Selectable SWS

The user can define various parameters, such as *FlightTime*, *Price*, *Availability*, which are used for ranking purposes based on the information contained in the ranking ontology. As a feature, the QoS discovery template offers the option to assign weights to individual parameters which are used in the ranking algorithm. Note that the parameters whose weight is marked as *Very important* are eliminatory. This means that if the flight *Price* in Figure 6 is not less than 100 Euro the SWS is filtered out of the resulting set.

As a result of the QoS discovery process, the user gets a ranked list of SWS from which he can select the most suitable one (Figure 7).

4.4 SWS Selection

The essential concern of the discovery sequence is to filter the SWS suitable to provide the desired functionality and satisfying the required QoS parameters. To achieve this, two complementary discovery actions are performed in a sequence to decrease the amount of matching SWS at each step. In the last step of the discovery process, a single SWS is selected from the filtered set as depicted in the following figure:

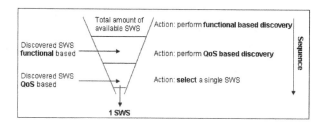

Fig. 8. Discovery Filter

After the user selects the SWS, a binding of the service to the process task is performed. This results in storing the service descriptor in the task properties and marking the process task as an executable service task.

4.5 Usage Scenarios

We disinguish two possible scenarios for using the discovery component:

- *Scenario 1: Design-time SWS discovery* In this scenario both types of discovery are performed during design-time, resulting in a SWS selection for each task during process modeling phase. Note that this is the scenario we presented in this work.
- *Scenario 2: Partial design-time SWS discovery* Discovery based on functional aspects takes place at design-time as it is a computationaly expensive operation, performing on a full set of SWS from the repository. The set of SWS filtered in this step is linked to each process task and the model is ready for being deployed to the execution engine. At runtime, the engine calls the QoS discovery component in order to do further filtering of pre-discovered SWS based on non-functional properties. The execution engine automatically selects the highest ranked SWS for process task implementation. This scenario is designed for the situations where e.g. the availability of services is low, thus ensuring the correct execution of process models during runtime.

5 Related Work

In [11], an approach to model Web services composition for business process model implementation is presented. The user is supported in modeling by filtering out the services that can be used in the next modeling step based on matching the pre- and postconditions. Main difference to our approach is that the modeling is done using a bottom-up approach, i.e. the user actually models service orchestration. We consider the top-down approach where the business expert starts from designing a business process model much more intuitive. Furthermore, the user can not specify his request in a user-friendly manner. He is only supported by suggestions for the next available service to use in the composition. In addition, discovery based on QoS parameters is not supported in this work.

The approach presented in [12] aims to apply Model Driven Development techniques to business process modeling. We consider semantic technologies as more powerful since they, apart from model transformation, bring the value of automated reasoning over resources. This is the main difference to our approach, as we are able to do automated matchmaking of user requests against the available service descriptions. In this way, we reduce the amount of manual work in the process of deriving an executable process model in comparison to [12].

So far, we have not seen other approaches addressing the problem of discovery of SWS for process task implementation in business process models.

6   Conclusion

In this work we presented an approach for an intuitive and expressive specification of user goals for process task annotation in BP models. Furthermore, we devised a solution for comparing user requests against the available SWS descriptions, resulting in a ranked list of SWS for task implementation. The complete approach has been prototypically implemented and can be used in different scenarios during design-time and runtime.

We are currently working on the extension of our matchmaking algorithm to support matching the dynamic (behavioral) part of SWS description. This will allow the user to pose behavioral constraints on the SWS that he wants to find. As part of our future work, we will devise an approach to validation of behavioral conformance for discovered SWS in a process model. In the long term, we plan to investigate techniques for composition of SWS to support situations when a process task can not be implemented by a single SWS, but rather an orchestration of several SWS.

Acknowledgments

This work has been partially supported by the European Union within the FP6 IST project SUPER[3] (FP6-026850).

References

1. Martin et al.: OWL-S: Semantic Markup for Web Services (2004),
   http://www.w3.org/Submission/OWL-S/
2. ESSI WSMO working group: Web Service Modeling Ontology (WSMO) – Final Draft (October 2006), http://www.wsmo.org/TR/d2

---

[3] www.ip-super.org

3. Lara et al.: A conceptual comparison between wsmo and owl-s (2005), http://www.wsmo.org/TR/d4/d4.1/v0.1/d4.1v0.1_20050412.pdf

4. ESSI WSML working group: Web Service Modeling Language (WSML) – Final Draft (October 2006), http://www.wsmo.org/TR/d16/d16.1/

5. Weber, I., Markovic, I., Drumm, C.: A conceptual framework for composition in business process management. In: Abramowicz, W. (ed.) Business Information Systems. LNCS, vol. 4439, pp. 54–66. Springer, Heidelberg (2007)

6. Preist, C.: A conceptual architecture for semantic web services. In: McIlraith, S.A., Plexousakis, D., van Harmelen, F. (eds.) ISWC 2004. LNCS, vol. 3298, pp. 395–409. Springer, Heidelberg (2004)

7. Trastour, D., Bartolini, C., Preist, C.: Semantic web support for the business-to-business e-commerce lifecycle. In: WWW, pp. 89–98 (2002)

8. Li, L., Horrocks, I.: A software framework for matchmaking based on semantic web technology. In: WWW, pp. 331–339 (2003)

9. Paolucci, M., Kawamura, T., Payne, T.R., Sycara, K.P.: Semantic matching of web services capabilities. In: Horrocks, I., Hendler, J. (eds.) ISWC 2002. LNCS, vol. 2342, pp. 333–347. Springer, Heidelberg (2002)

10. Vu, L.-H., Hauswirth, M., Porto, F., Aberer, K.: A search engine for qoS-enabled discovery of semantic web services. Int. J. of Business Process Integration and Management 1, 244–255 (2007)

11. Schaffner, J., Meyer, H., Weske, M.: A formal model for mixed initiative service composition. In: IEEE SCC, pp. 443–450. IEEE Computer Society Press, Los Alamitos (2007)

12. Koehler, J., Hauser, R., Küster, J., Ryndina, K., Vanhatalo, J., Wahler, M.: The role of visual modeling and model transformations in business-driven development. In: Fifth International Workshop on Graph Transformation and Visual Modeling Techniques (2006)

Web Service Search: Who, When, What, and How

Jianguo Lu[1] and Yijun Yu[2]

[1] School of Computer Science, University of Windsor
`jlu@cs.uwindsor.ca`
[2] Computing Department, The Open University
`y.yu@open.ac.uk`

Abstract. Web service search is an important problem in service oriented architecture that has attracted widespread attention from academia as well as industry. Web service searching can be performed by various stakeholders, in different situations, using different forms of queries. All those combinations result in radically different ways of implementation. Using a real world web service composition example, this paper describes when, what, and how to search web services from service assemblers' point of view, where the semantics of web services are not explicitly described. This example outlines the approach to implement a web service broker that can recommend useful services to service assemblers.

Keywords: Web service searching, web service composition, signature matching, XML Schema matching.

1 Introduction

Web service reuse is the number one drive for service oriented architecture. To reuse web services, it is paramount to develop web service repository architectures and searching methods. There have been tremendous researches on web service searching [2, 4, 7, 16]. However, in many cases, web service searching means different things for different people. Before implementing web service searching platforms and methods, we need to discuss who needs to search web services, when searches happen exactly, what are the queries to be sent out, and, once the queries are formulated, how to execute the queries.

This paper delineates various stakeholders in web service searching, and tries to give answers to the above questions using a real world web service composition example, where the semantics of web services are not explicitly described. In this example, we constructed a real web service from five atomic services. During the integration process, various searches are carried out in order to find relevant and reusable services in this scenario.

This paper details web service searching from a web service assembler's perspective. An assembler starts from an abstract description for the composite web service. From the description, an initial query is constructed in the form of service signature, i.e., the name of the operation and its input and output types. Based on the search results and the current service signature, subsequent queries are derived. As

M. Weske, M.-S. Hacid, C. Godart (Eds.): WISE 2007 Workshops, LNCS 4832, pp. 284–295, 2007.
© Springer-Verlag Berlin Heidelberg 2007

service signatures are well-structured in XML, such queries can be found using approximate XML Schema matching [8].

2 Cube of Web Service Searching

Unlike web pages that are presented for humans to read, web services are meant to be invoked by programs. Hence web services are usually searched by programmers, or sometimes by software agents that can automatically adapt their behavior by using new services. Either way, web services are consumed by programs.

Different stakeholders search for web services for different purposes, using different resources, and in different ways. Main stakeholders in web service searching can be categorized as follows:

1) *Web service end users:* End users are programmers who search for web services in order write a program to invoke them directly as is.

2) *Web service assemblers*: web service assemblers search for web services in order to compose them to perform some tasks that cannot be fulfilled by a single service. Once reusable atomic services are found, assemblers can use conventional programming languages to compose the services, either manually or supported by service composition tools.

3) *Web service brokers*: web service brokers are programs that assist web service assemblers by recommending relevant web services during the assembly process. Just the same as various code recommendation systems for conventional programming languages [24], web service brokers can watch over the shoulders of assemblers and are able to recommend services proactively according to existing code that has been written by service assembler.

4) *Web service agents:* They are intelligent programs that are able to automatically find relevant web services to use at system run time, when a new task occurs or when existing web services is not functioning properly and a replacement is called for.

The classification of various web service searches can be depicted in Figure 1. In addition to the main stakeholders in web service searching, there are a variety of forms of queries to search web services, including:

1) A set of keywords [5, 17];
2) Signature or part of the signature of the service [25, 17];
3) Context of the service to be used [24, 22, 23];
4) Semantic description of the service [16, 17].

These different kinds of queries form the Y axis in Figure 1.

Another dimension is *when* the searches are carried out. Roughly speaking, web services searching can happen at development time or run time. For web service end users and assemblers, web service searching happens at development time and is initiated by humans. Service brokers can recommend the services proactively while a web service is being developed. Service agents will search and consume the services

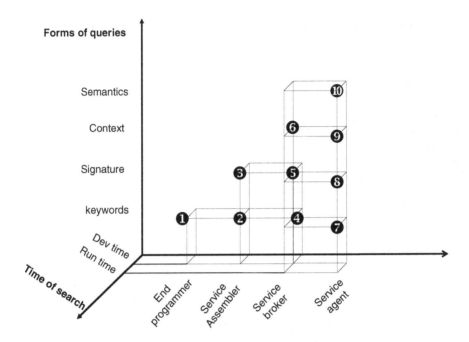

Fig. 1. Varieties of web service searches

dynamically at run time. In this case the service agent needs to have the complete semantic description of the service in order to conduct the correct search without human intervention.

Combinations of the three parameters (i.e., who, what, and when) constitute the variety of the searches. However, not every combination makes sense. For example, for end-user programmers or service assemblers, searches should happen in development time instead of run time, and usually keyword based search is more convenient (search type ❶). If the number of returns is large, maybe signature search can be performed to narrow down the results. Although semantic description is essential in determining equivalent web services automatically, for end user programmers it is neither necessary nor practical to write semantics such as ontology or functional specification to search for services.

On the other hand, if an intelligent agent wants to replace an existing service, semantics for the services must to provided in order to decide whether they are performing the same task (search type ❿ in Figure 1). In this case, search happens at run time instead of development time.

For service assemblers, keywords or signatures may be enough since programmers can judge whether the search results are good. Besides, it would be too cumbersome for programmers to spell out the semantics or the context of the web services in details. Hence the search type could be ❶ and ❸.

For a service broker, the program has the knowledge of the current code that service assembler has written, hence it has the context the service will be used (type

❻ search). This context, including other services already used and even the documentation, can be utilized to recommend the next service to be used [24, 1, 22].

Each combination determines how the search should be implemented. For example, search type ❶ is usually implemented by information retrieval methods such as vector space model. Type ❿ includes inferences on ontology and functional specifications [17, 16].

3 When Do We Need to Search for Web Service

Since web service searching can happen in many different situations, it is not possible in this paper to discuss all of them in details. In the following anatomy of web service searching, we will focus on search type ❸ in Figure 1, i.e., we suppose that web service assemblers will search for the services. We will discuss exactly when we need to search the services, and what the queries will be.

As a running example, let us start with the following task for service assemblers:

$$\textit{Given a zipcode, find its closest airport name.} \qquad (1)$$

Given the large number of web services that are available on the web, it is reasonable to assume that there should be a solution for such a problem. But how to solve the problem is by no means obvious. Before starting to write the program, service assemblers should first formalize the problem. Following the conventional program specification methods, the task could be formalized as follows by defining the concepts *zipcode*, *airport* and "*closest*":

Description 1. for all *Zipcode*, find *Airport*, such that

1) *isAirport(Airport) ∧ distance(Zipcode?, Airport ?, Distance)*, and for any other airport *Airport'*,
2) *isAirport(Airport') ∧ distance(Zipcode?, Airport'?, Distance')*
   →Distance' ≥ Distance. $\qquad (2)$

Here *distance* and *isAirport* are two predicates that need to be refined so that they can correspond to some web services. There are arguments in the predicates. For example, in *distance(X?, Y?, Z)*, X, Y, and Z are arguments in the predicate. An argument with a question mark adornment such as *X?* denote that the value of argument X needs to be provided. Arguments without a question mark such as Z denote the returned value after the service is executed.

If there is an existing service that implements (2), then the task has been fulfilled. Otherwise, which is true in this case, we need to refine (2) into subtasks (2.1) and (2.2) that may be implemented by existing Web services.

$$\textit{isAirport(Airport)} \qquad (2.1)$$
$$\textit{distance(Zipcode?, Airport?, Distance)} \qquad (2.2)$$

The task *isAirport(Airport)* returns a set of airports without any input. The other task *distance(Zipcode?, Airport?, Distance)* accepts a zipcode and an airport code, and returns the distance between them.

At this stage, web service assembler needs to search for those two services. Searching for a service for the *isAirport(Airport)* specification using signature

$$isAirport: \rightarrow Airports$$

doesn't return an exact match. However, a similar service (http://www.farequest.com/ FASTwebservice.asmx?WSDL[1]) can be found, whose signature (i.e., the name of the operation and its input and output types) is

$$stateAirport: stateAbbr \rightarrow Airports,$$

where *Airports* is the Schema for *airport(code, city, state, country, name)**. The predicate representation of the service is

$$stateAirport(StateAbbr?, Airports) \tag{2.1'}$$

As a service assembler, what is the next service to search for? Next section will give more cases as for when searches are carried out as development of the composite web service unfolds. Searching for this kind of similar services is also not a trivial task. Section 5 will discuss in more details regarding how to find this kind of related services.

4 What Are the Queries to Search for the Relevant Web Services

Even though now we are assuming using type 3 search and the queries are in the form of signatures, it is not always clear what the queries are exactly. In the running example, once we have the problem description, service assemblers know that we need to search for the predicates referred in the specification, such as *isAirport*. The query in the form of signature is *isAirport: $\rightarrow$ Airports*. In other cases queries to be issued may not be straightforward, as we will see in the next section.

4.1 Query Formulation

By issuing the query *isAirport: $\rightarrow$ Airports*, we can find the service *stateAirpor(StateAbbr?, Airports)*. At this stage we cannot invoke the *stateAirport* service yet in our composite service since it needs to use a *StateAbbr* as input in order to return the airport data. In order to obtain the state name, we need a service in the following signature:

$$\rightarrow <stateAbbr/> \tag{S1}$$

Or, we can utilize some known values from our existing input list. Currently, the only input is *ZipCode*. Hence we can search for a service of the following signature as an alternative:

$$<ZipCode/> \rightarrow <stateAbbr/> * \tag{S2}$$

From here and hereafter, we omit the service name in signature when it is not important.

---

[1] All the web services listed in this paper are active during the month of April 2007. As web services are volatile, some of them may not be functioning now.

Now using these two signatures S1 and S2 to search for web services, we found the following web service(http://www.farequest.com/FASTwebservice.asmx?WSDL):

> *zipState(ZipCode?, State),*

Whose signature is <ZipCode/> → <State/>*
Up to this stage, is*Airport(AirportCode)* is refined into

> *zipcodeState(ZipCode?, StateAbbr)*
> ∧ *stateAirport(State?, AirportCode)*

Generalizing from this example, the service assembler can use the following rule to form the query:

$$\frac{A \rightarrow B}{\rightarrow C} \qquad B \rightarrow C \qquad A \in \Sigma \qquad \text{(Rule 1)}$$

The meaning of the rule is that to derive a service of signature →C, suppose that we already have a service of signature B→C, and suppose we have A in the known list, we need to find a service of signature A→B.

Now Description 1 is refined as the following:

Description 2. for all *Zipcode*, find *Airport*, such that

2) *zipcodeState(Zipcode?, State)* ∧ *stateAirport(State?, AirportCode)* ∧ *distance(Zipcode?, AirportCode?, D)*, and

3) for any other AirportCode',
    zipcodeState(Zipcode?, State)
    ∧ *stateAirport(State?, AirportCode')*
    ∧ *distance(Zipcode?, AirportCode'?, D')*
    → D' ≥ D.   (3)

In Description 2, predicates *zipcodeSate* and *stateAirport* correspond to two real web services. Before Description 2 can be implemented, the predicate *distance* needs to be refined further into a real service. Similar to the previous steps, first we search for *distance(Zipcode?, Airport?, Distance)*. There is no exact match again. The closest match is the following service (http://ws.cdyne.com/psaddress/addresslookup.asmx?wsdl):

> *calculateDistanceMiles(latitute1?,longitude1?, latitude2?, longitude2?, DistanceInMiles)*

Whose signature is

> *calculateDistanceMiles: (latitute1,longitude1, latitude2, longitude2)*
> → *DistanceInMiles*

Since the inputs *latitude* and *longitude* in this service are not in the known list, we need to find a service that provides those parameters, i.e., we need to find a service that is compatible to the following signature:

<Zipcode | State/> → *<latitude/><longitude/>*

And

<AirportCode | State/> → *<latitude/><longitude/>*

We add <AirportCode>, <Zipcode> and <State> in the input type because those values are already available at this stage.

To formalize the process of generating the above query, we need Rule 2 as below:

$$\frac{A \rightarrow B}{A \rightarrow C} \quad B \rightarrow C \quad \text{(Rule 2)}$$

The meaning of the rule is that to derive a service of type A→C, suppose that we already have a service of type B→C, then we need to find a service of type A→B so that A→B and B→C can be composed into a service which is of type A→C.

In general, Rule 2 can seldom be applied directly. A general form would be the following:

$$\frac{A_1 |D_1|...|D_n \rightarrow B_1 \quad A_2 |D_1|...|D_n \rightarrow B_2}{(A_1, A_2) \rightarrow C} \quad (B_1, B_2) \rightarrow C \quad D_i \in \Sigma \quad \text{(Rule 3)}$$

Rule 3 means that to find a service of type (A1, A2)→C, and if we already have found a service of type (B1, B2)→C, what we need is to find a service of type A1|D1|...|Dn→B1, and A2|D1|...|Dn→B2, where Di is the type of available values.

Corresponding to Rule 3, we need to find a service of type

(Zipcode, Airport) → *Distance.*

And suppose that we have already found a service of type

(latitute1, longitude1, latitude2, longitude2) → *Distance*

Hence the services we need to search for should be compatible to the following types:

<Zipcode | State/> → *<latitude/><longitude/>*

And
<AirportCode | State/> → *<latitude/><longitude/>*

Using those two queries, the following two web services are found:

airportCoordinate(AirportCode?,
  LatitudeDegree,LatidudeMinute, LongitudeDegree, LatitudeMinute)

zipCodeCoordinate(ZipCode?, LatDegrees, LonDegrees).

Using those two web services, the *distance* predicate is refined into the following three services (predicates):

airportCoordinate(AirportCodeCode? LatitudeDegree, LatitudeMinute,
LongitudeDegree, LongitudeMinute)
∧ *zipCodeCoordinate(ZipCode?, LatDegrees, LonDegrees).*
    ∧ *calculateDistanceMiles(latitute1?,longitude1?, latitude2?,*
    longitude2?, Distance)

Using the above three web services found, Description 2 is derived into the following:

Description 3. for all *ZipCode*, find *Airport*, such that

1) *zipcodeState(Zipcode?, State)* ∧ *stateAirport(State?, AirportCode)*
    ∧ *airportCoordinate(AirportCode?LatitudeDegree, LatitudeMinute,*
    LongitudeDegree, LongitudeMinute)
    ∧ *zipCodeCoordinate(ZipCode?, LatDegrees, LonDegrees)*
    ∧ *calculateDistanceMiles(latitute1?, longitude1?, latitude2?,*
    longitude2?, Distance)
    and for any other *AirportCode'*,
2) *zipcodeState(Zipcode?, State)* ∧ *stateAirport(State?, AirportCode')*
    ∧ *airportCoordinate(AirportCode'?LatitudeDegree, LatitudeMinute,*
    LongitudeDegree, LongitudeMinute)
    ∧ *zipCodeCoordinate(ZipCode?, LatDegrees, LonDegrees)*
    ∧ *calculateDistanceMiles(latitute1?,longitude1?, latitude2?,*
    longitude2?, Distance')
    → *Distance' ≥ Distance*   (4)

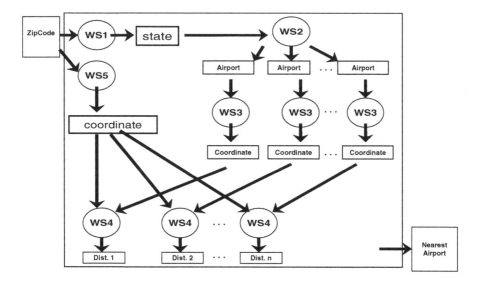

Fig. 2. The composite web service

Now that all the predicates in the composite web service definition refer to existing web services, we can generate the program to glue those web services. The integration program can be written in existing general purpose programming languages such as Java, or in languages for web service composition such as BPEL. The overall picture of the composition is depicted in Figure 2.

5   How to Search for Relevant Services

Now that we have described the ways to formulate the queries, the next task is to construct a query to locate relevant services. Given our first query for example, it is not a trivial task to run the query *"isAirport:* → *Airport"* in order to find the approximate matching

stateAirport: stateAbbr→ *airport(code, city, state, country, name)**

Note that there are at least two issues need to be tackled. One is the matching between tag names in XML Schema. For example, <state/> should be matched with <stateAbbr/>. The other is the matching between the structures of the schemas. For example, *Airport* in Description 1 needs to be matched with the structure *(airport(code, city, state, country, name))**.

In order to find approximate signature matchings, we need to construct a matching algorithm between XML Schemas, since input/output types in web services are described in XML Schema. Because XML Schemas are trees, we reduced schema matching problem to the classic tree matching problem [8], and developed Common Substructure algorithm to find the matching effectively. Instead of giving rigorous definitions for such matching problem, we use the following example to illustrate the problem and the solution.

Figure 2 shows two similar schemas. In order to find the matches, we locate the largest common substructures as described in Figure 3. In the system, we use Wordnet

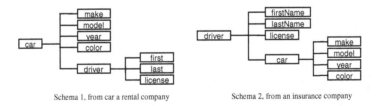

Schema 1, from car a rental company   Schema 2, from an insurance company

Fig. 3. Car-Driver Schemas

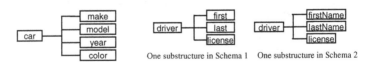

One substructure in Schema 1   One substructure in Schema 2

Fig. 4. A common substructure (left) and two similar substructures (middle and right figures)

to capture the synonyms. In addition, composite tag names such as *StateAbbr* is broken into a word list (*State, Abbr*).

6 Conclusions

There have been tremendous researches on web service searching. The notion of web service searching varies greatly. We classify various searches in terms of the stakeholder who will initiate the search. In the case of web service assembler, we described in detail as for when the searches are needed, what are the queries should be issued, and how the queries should be executed, by locating five real world web services that are needed in creating a new web service. In particular, we give the formal rules to derive the service queries in the process of service composition. This formalism can be used to implement a service broker that can recommend the services to programmers, i.e., the search type ❺ in Figure 1. If context information is included, the service broker can be expanded to search type ❻.

This paper outlines the plan for implementing a web service broker, illustrated using a concrete real word web service composition example. While service assembler refines the definition of a composite web service, service broker recommend the relevant atomic services that could be used, mainly rely on approximate signature matching between atomic web services and the tasks at hand. We have already implemented the XML Schema matching system [8] as our first step in implementing such a system. The query will be automatically generated using the rules outlined in this paper. In addition, context information will be used to increase the precision of recommendation.

Our early work on a generic matching system encompasses all kinds of queries ranging from keywords to signatures and ontology in the form of description logic [17]. While it is a comprehensive matching system involving various matching algorithms, yet we need to answer the questions such as who will use the system, and how the queries are formed. This paper is a step towards answering those questions in one particular scenario.

Acknowledgements. We would like to thank Debashis Roy and Deepa Saha for implementing the composite web service, and the anonymous reviewers for their helpful comments.

References

1. Agarwal, V., Dasgupta, K., Karnik, N., Kumar, A., Kundu, A., Mittal, S., Srivastava, B.: A service creation environment based on end to end composition of Web services. In: WWW 2005. Proceedings of the 14th international Conference on World Wide Web, Chiba, Japan, May 10 - 14, 2005, pp. 128–137. ACM Press, New York (2005)
2. Benatallah, B., Hacid, M., Leger, A., Rey, C., Toumani, F.: On automating Web services discovery. The VLDB Journal 14(1), 84–96 (2005)
3. Bultan, T., Su, J., Fu, X.: Analyzing Conversations of Web Services. IEEE Internet Computing 10(1), 18–25 (2006)

4. Caverlee, J., Liu, L., Rocco, D.: Discovering and ranking web services with BASIL: a personalized approach with biased focus. In: ICSOC 2004. Proceedings of the 2nd international Conference on Service Oriented Computing, November 15 - 19, 2004, pp. 153–162. ACM Press, New York (2004)

5. Dong, X., Halevy, A., Madhavan, J., Nemes, E., Zhang, J.: Similarity Search for Web Services. In: Proc. of VLDB (2004)

6. Dustdar, S., Schreiner, W.: A survey on web services composition. Int. J. Web and Grid Services 1, 1–30 (2005)

7. Elgedawy, I., Tari, Z., Winikoff, M.: Exact functional context matching for web services. In: ICSOC 2004. Proceedings of the 2nd international Conference on Service Oriented Computing, November 15 - 19, 2004, pp. 143–152. ACM Press, New York (2004)

8. Lu, J., Wang, J., Wang, S.: XML Schema Matching, IJSEKE, International Journal of Software Engineering and Knowledge Engineering (in Press)

9. Lu, J., Yu, Y., Mylopoulos, J.: A Lightweight Approach to Semantic Web Service Synthesis. In: ICDE Workshop, International Workshop on Challenges in Web Information Retrieval and Integration, Tokyo (2005)

10. Matskin, M., Rao, J.: Value-Added Web Services Composition Using Automatic Program Synthesis. In: Bussler, C.J., McIlraith, S.A., Orlowska, M.E., Pernici, B., Yang, J. (eds.) CAiSE 2002 and WES 2002. LNCS, vol. 2512, pp. 213–224. Springer, Heidelberg (2002)

11. McIlraith, S.A., Son, T.C.: Adapting golog for composition of semantic web services. In: Proc. of the 8th Int. Conf. on Principles and Knowledge Representation and Reasoning (KR 2002), Toulouse, France (2002)

12. Medjahed, B., Bouguettaya, A., Elmagarmid, A.K.: Composing web services on the semantic web. The VLDB Journal 12, 333–351 (2003)

13. ProgrammableWeb, http://www.programmableweb.com

14. Ponnekanti, S.R., Fox, A.: SWORD: A developer toolkit for web service composition. In: Proc. of the 11th Int. WWW Conf (WWW 2002, Honolulu, HI, USA (2002)

15. Rao, J., Su, X.: A survey of automated web service composition methods. In: Cardoso, J., Sheth, A.P. (eds.) SWSWPC 2004. LNCS, vol. 3387, Springer, Heidelberg (2005)

16. Sirin, E., Parsia, B., Hendler, J.: Composition-driven filtering and selection of semantic web services. In: AAAI Spring Symposium on Semantic Web Services (2004)

17. Sycara, K., Lu, J., Klusch, M.: Interoperability among Heterogeneous Software Agents on the Internet, Technical Report CMU-RI-TR-98-22, CMU, Pittsburgh, USA

18. Wong, J., Hong, J.I.: Making mashups with marmite: towards end-user programming for the web. In: CHI 2007. Proceedings of the SIGCHI Conference on Human Factors in Computing Systems, San Jose, California, USA, April 28 - May 03, 2007, pp. 1435–1444. ACM Press, New York (2007)

19. Wu, D., Parsia, B., Sirin, E., Hendler, J.A., Nau, D.S.: Automating DAML-S web services composition using SHOP2. In: Fensel, D., Sycara, K.P., Mylopoulos, J. (eds.) ISWC 2003. LNCS, vol. 2870, Springer, Heidelberg (2003)

20. Yu, Y., Lu, J., Fernandez-Ramil, J., Yuan, P.: Comparing Web Services with Other Software Components. In: International Conference on Web Services, ICWS 2007,

21. Zhang, L., Chao, T., Chang, H., Chung, J.: XML-Based Advanced UDDI Search Mechanism for B2B Integration. Electronic Commerce Research 3, 1-2, 25–42 (2003)

22. Mandelin, D., Xu, L., Bodík, R., Kimelman, D.: Jungloid mining: helping to navigate the API jungle. In: PLDI 2005. Proceedings of the 2005 ACM SIGPLAN Conference on Programming Language Design and Implementation, Chicago, IL, USA, June 12 - 15, 2005, pp. 48–61. ACM Press, New York (2005)

23. Wong, J., Hong, J.I.: Making mashups with marmite: towards end-user programming for the web. In: CHI 2007. Proceedings of the SIGCHI Conference on Human Factors in Computing Systems, April 28 - May 03, 2007, pp. 1435–1444. ACM Press, New York (2007)

24. Ye, Y., Fischer, G.: Supporting Reuse by Delivering Task-Relevant and Personalized Information. In: Proceedings of 2002 International Conference on Software Engineering (ICSE 2002), Buenos Aires, Argentina (May 19-25, 2002)

25. Zaremski, A.M., Wing, J.M.: Specification matching of software components. ACM Transactions on Software Engineering and Methodology (TOSEM) 6(4), 333–369 (1997)

International Workshop on Personalized Access to Web Information (PAWI)

Workshop PC Chairs' Message

Sylvie Calabretto[1] and Jérôme Gensel[2]

[1] LIRIS Laboratory, Bâtiment Blaise Pascal
Avenue Jean Capelle, 69621 Villeurbanne Cedex, France
[2] LIG Laboratory, STEAMER Team
681 rue de la Passerelle, 38402 Saint Martin D'Hères, France
Sylvie.Calabretto@insa-lyon.fr, Jerome.Gensel@imag.fr

Through the Web, huge amounts of information are now widely accessible. As a matter of fact, the Web can be seen as a large corpus accessed through Web-based Information Systems (WIS) which rely on classical client/server, n-tier or even services-oriented architectures. Due to recent advances in wireless and ubiquitous computing technologies, these WIS can now be used by nomadic as well as sedentary users, connecting from different kinds of access devices (workstation, laptop, PDA, mobile phones, etc.).

Considering the Web as a huge and growing corpus, one of the main challenges of WIS designers is to prevent users from experiencing the all too prevalent cognitive and informational overload. Meeting this goal guarantees to some extent the usability and the durability of a WIS. While in the field of Information Retrieval some solutions have been proposed to query specialized and limited corpora with good performances, access to information on the Web as a whole has still to cope with the limited capabilities of existing search engines.

Personalization is one of the keys towards a better access to information. The notion of personalization is closely related to the notion of *context*. The context (of use) of a Web-based Information System can be seen as an extensible set of meta-data. These meta-data can be of very different kinds: information defining the user's profile, hardware and software characteristics of the used access device, parameters of the physical environment surrounding the user and his/her access device, etc.

Three steps are essential in the management of the context. The context modeling phase consists of determining whether the considered context is extensible or not and what are its components. Once described and made operational by a given representation formalism, the context model has to be filled (context acquisition phase). The system can then exploit the context model for adaptation purposes. It consists of completing the query with context elements to make it more precise and/or to trigger an adaptation mechanism in order to adapt both the content and the information presentation, sent as a result of the interaction between the user and the WIS.

The PAWI workshop brings together researchers from various fields (Information Systems, Information Retrieval, and Adaptive Hypermedia) to present and discuss recent ideas and results about personalization and access to information on the Web.

M. Weske, M.-S. Hacid, C. Godart (Eds.): WISE 2007 Workshops, LNCS 4832, pp. 299–300, 2007.
© Springer-Verlag Berlin Heidelberg 2007

The nine selected papers cover several aspects linked to personalized access to Web information:

- *User Models* dedicated to Personalization are proposed by Santos *et al.* whose approach combines user modeling and machine learning techniques in order to produce adaptive interfaces in the field of Life Long Learning (LLL) in Higher Education. Targeting a meta-level approach, Abbas *et al.* presents a user's profile modeling framework whose goal is twofold: first, based on a meta description, profile and context can be described, second, profile rules corresponding to the current user's profile are selected in order to adapt information. Chevalier *and al.* present a generic and flexible user model which supports the interoperability of the information search tools (matching between profiles coming from various applications) and allows the system to model the user like any other element involved in the application. In Daoud *and al.*, the authors present a semantic representation of the user interests based on the ODP ontology, and a method for learning and maintaining the user's interests.
- More generic than user models, *Context models* are proposed by Jrad *et al.* who adopt the ontology formalism in order to describe three kinds of contexts (user, working, use) for the improvement of Web recommender systems. The work Carrillo *et al.* gives a formalization of three types of user preferences: Activity, Result and Display), a Contextual Matching Algorithm, for comparing the current context of use and the user preferences.
- New peripherals are explored by Windfeld *et al.* who propose an interesting discussion about the expected benefits of using large, high resolution screens. Such large displays ease interactive collaboration.
- Tracing the link paths followed by users in order to identify navigation patterns or classes of users is the purpose of the framework proposed by Monaco *et al.* for the structural modelling and statistical navigational analysis of Web applications. From these observations, information restructuring, pre-fetching and adaptation can be applied.
- Van *and al.* address the problem of document relevance estimation in digital libraries. These authors focus on co-citation based methods for document selection. They propose some similarity measures to be exploited within a personalized retrieval process.

We thank all the authors who have submitted a paper for their interest to PAWI Workshop, the program committee for their excellent reviews and the WISE conference organization.

User Modeling for Attending Functional Diversity for ALL in Higher Education

Olga C. Santos, Alejandro Rodriguez-Ascaso, Jesús G. Boticario, and Ludivine Martin

aDeNu Reserach Group. Artificial Intelligence Department.
Computer Science School, UNED.
C/Juan del Rosal, 16. Madrid 28040. Spain
{ocsantos,arascaso,jgb,ludivine.martin}@dia.uned.es
http://adenu.ia.uned.es

Abstract. In this paper we provide our general approach and discuss relevant issues in providing a dynamic user modelling approach for attending functional diversity for accessible lifelong learning (ALL) in Higher Education. Our approach to provide universal and personalized access lies on combining user modelling and machine learning techniques to cope with the needs for ALL with a pervasive support of standards and supporting the full life cycle of service adaptation. The modelling differs from others in i) coping with interactions and context of the user that can only be considered at runtime and ii) characterising interaction capabilities of different kinds of devices. Models are used to personalize and adapt learning materials, pedagogical settings and interactions in the environment to satisfy both the individual learning needs and the access preferences, taking into account the context at hand.

Keywords: Personalization, User-centered design, Context, Access devices, Machine learning, Multiagent systems, Knowledge modelling, Education, Distance Learning, Accessibility, Disability, Functional diversity.

1 Introduction

It is a fact that the introduction of Information and Communication Technologies (ICTs) in education has made possible moving the traditional teaching paradigm based on the one-size-fits-all maxim to a user-centered approach in which technology supports attending in a personalized way the learning needs of the students. Nowadays, these learning needs are becoming more and more required in current knowledge-based societies, in which there is an increasing demand a continuous updating of knowledge, known as the lifelong learning (LLL) paradigm. Works in instructional design, adaptive hypermedia, user modelling, artificial intelligence or human computer interaction, to name but a few, can be taken as a solid foundation for this approach. However, there are still open issues to be solved, which relay on the proper integration of the existing different techniques to effectively attend the learning needs in a personalized way. Moreover, user information has to be considered regarding the way they interact with the system due to their technological

M. Weske, M.-S. Hacid, C. Godart (Eds.): WISE 2007 Workshops, LNCS 4832, pp. 301–312, 2007.
© Springer-Verlag Berlin Heidelberg 2007

context, preferences or ability profile. This last factor is of major importance in this new paradigm. The LLL paradigm is to be addressed by higher education (HE) institutions, which have the infrastructure and experience to handle this challenge.

In line with the functional diversity paradigm [1] that we fully subscribe, some relevant stakeholders (e.g. ISO/IEC JTC1 SC36) consider that learners experience a disability when there is a mismatch between the learner's needs (or preferences) and the education or learning experience delivered. Disability is thus not viewed as a personal trait, but as a consequence of the relationship between a learner and a learning environment or resource delivery system. While pursuing accessibility, providers of e-learning services should adapt learning objects to both personal and context circumstances.

Our approach to provide universal and personalized access to LLL lies on combining user modelling and machine learning techniques to produce adaptive interfaces that cope with the needs for all users in HE with a pervasive support of standards. Some of the issues we are addressing are: 1) the semantic interoperability among the different data available, such as user data (personal, interaction preferences, learning styles, ability profile, learning needs, previous knowledge, background, course outcomes, ...), context (e.g. device, environment) and course (metadata of contents), 2) personalized content and service delivery taking into account the user, context and course data, and 3) the generation of context-sensitive feedback through the provision of dynamic support considering the current situation.

In this sense, aDeNu (Adaptive Dynamic on-line Educational systems based oN User modelling) is leading the development of a flexible, standard-based architecture to support the lifelong learning paradigm within the European Unified Approach for Accessible Lifelong Learning (EU4ALL Project - IST-2005-034778). This project, scientifically coordinated by aDeNu group, focuses on developing learning services for accessible lifelong learning (ALL) at HE institutions which are prepared both to assist learners and to support service providers. It is based on previous research of the group within aLFanet Project (IST-2001-33288), where we researched for three years on how to combine design and runtime adaptations to provide recommendations to learners [2]. However, when taking a comprehensive user-centred approach where functional diversity has to be addressed, open research questions appear, since neither the data to be gathered nor the way to present the information is the same. Some of the open questions are: a) acquisition, updating and management of user models, and very specially which user, usage and context data have to be considered, b)knowledge representation and reasoning for user models, following a standard based approach combined with machine learning techniques, c) adaptation to user, in order to enhance accessibility and usability in web-based system interfaces, d) context awareness and device modeling, considering how to integrate user and device profiles, e)recommender systems to promote dynamics for students with disabilities, f)evaluation of user experience when providing adapted responses to people with functional diversity.

In this section we support our objectives with institutional and operational reasons, review some related work in the field and briefly discuss the state of the art. Next we introduce the approach followed by the EU4ALL project and describe the scenario upon which this approach is based. We then present the user modelling for ALL, focusing on the way the accessibility needs can be attended. Finally, we make some conclusions and outline the future works.

1.1 Why Personalized and Accessible Services for ALL?

Personalized learning provides facilities for LLL to upgrade the skills of people with disabilities. However, this "student-centered approach" poses too many challenges to both traditional HE institutions and distance teaching universities. One key question to be answered is how to construct learning management systems (LMS) that support the HE user-centered scenario. Most courses on current LMS hardly offer any information about which didactical methods and models they use. As far as adaptation is concerned, they just offer predefined settings for a particular course that turn out to be the outcome of extensive customizations. Moreover, in HE institutions eLearning services are included in a wide variety of ICT services that cover many other relevant issues that could eventually affect user's behavior in a particular learning situation within a concrete course such as management of users (faculty staff, students, tutors, and administrative people), contents of varied nature (exams, study guides, calendars, bibliographic resources, videos, audios, etc.) and communication channels and means (e-mail, forums, news, radio, educational TV, IP telephone, etc.).

Moreover, two different types of reasons underpin the importance of attending accessibility issues at HE institutions throughout Europe. On the one hand, from the institutional viewpoint, different European action lines and initiatives relate to this issue. The "eEurope 2005" aims at bringing ICT applications and services to everyone, every home, every school and to all businesses. What is more, the European Lisbon strategy is not just about productivity and growth but also about employment and social cohesion. The purpose here is to put users at the centre. It will improve participation, open up opportunities for everyone and enhance skills. Moreover, on 25 October 2006, the European Parliament adopted the Commission's ambitious proposals for a new action programme in the field of education and training. For the first time, a single programme will cover learning opportunities from childhood to old age under the umbrella of the "Lifelong Learning Programme".

The Lifelong Learning Programme is actually an over-arching structure that is built on four pillars, or sub-programmes, one of them focused on addressing the teaching and learning needs in formal HE. In particular one of the key activities to be supported is "development of innovative ICT-based content, services, pedagogies and practice for lifelong learning". With respect to those from disadvantaged groups the LLL programme remarks that "there is a need to widen access for those from disadvantaged groups and to address actively the special learning needs of those with disabilities". The same orientation has been previously advocated in different European countries by means of their respective legislations and related actions plans. To name but a few, the 2001 legislation SENDA (Special Educational needs and Discrimination Act – which became the Disability Discrimination Act (DDA) Part 4) in UK, the "Plan Avanza" from the Spanish Ministry for Industry, Tourism and Commerce, and the German Barrier free Information Technology Regulation (BITV).

On the other hand, and in accordance with the 2006 European Ministerial Conference on "ICT for an inclusive society", from the operational viewpoint the main goal is to provide the required ICT services that enable people with disabilities as well as elderly citizens, while not excluding other minority groups, to get improved access to education. This enhances their potential and autonomy, because they get less dependent on others, can embark in new professional activities, and also can enjoy these "basic" services just as any regular citizen.

1.2 Overview of Related Work

The analysis done shows that although there are several research works on modelling users to take into account their accessibility needs, they just focus on adapting the user interface and work with static user models (user profiles) that are managed directly by the users. Users define their preferences, characteristics, abilities and needs and the system applies predefined rules to present the interface according to the user profile. Some of these works are briefly described next.

One of the first approaches for a universal framework in design and development of user interface systems targeted at users with functional limitations at university was defined in the CORES project [3]. They set out from the Slinky metamodel of user interface management systems, which split the functional core from the user interface component, and incorporated the user modelling to support the required degree of flexibility and intelligence to dynamically adapt the user interface to the users' needs. Particular modifications of usual modelling techniques were considered to build such models when the user has any impairment (i.e. the I/O device transparency). However, one limitation of their work is that they only consider the speech characteristics of the users, which helps them to make the system's usability evaluation one dimensional and thus, much simpler.

The IRIS project (IST-2000-26211) has developed a design support environment built on open source frameworks and web service technologies to help design web applications/services based on user modelling. The user modelling includes both usage and access device profiles based on P3P (Platform Privacy Preferences)[1] and CC/PP (Composite Capabilities/Preferences Profiles)[2] specifications and the deployment of user profile agents. The opposite approach is followed by [4]. Instead of a proactive approach oriented to the design of accessible web sites, they follow a reactive approach based on introducing web accessibility enhancements in existing web sites. They have developed the eAccessibilityEngine tool which employs adaptation techniques to automatically render web pages accessible to users with different types of disabilities by transforming web pages to attain conformance to W3C Web Accessibility Initiative[3] considering. The output depends on the specific user needs and the assistive software and hardware used and it is based on stereotypes directly kept by the users.

The limitation of these approaches is that the adaptation does not take into account the users' interactions during run-time in order to enrich the user model and adapt the interface. In this sense, [5] propose a model based architecture which uses both decision-making at runtime and the rule-based paradigm to enable the efficient use of all kinds of user information (static and dynamic). However, they do not explicitly consider accessibility needs in the user model.

The samples of the state of the art presented above show that the systems addressing the particular needs of people with functional diversity work on user

---

[1] P3P: http://www.w3.org/P3P/

[2] CC/PP: http://www.w3.org/Mobile/CCPP/

[3] W3C WCAG: http://www.w3.org/WAI/

profiles with static information of the users' features. There are systems that consider both static and dynamic data, but without paying special attention to the needs derived from the functional diversity. Moreover, they all promote adaptation at information presentation, processing and storage levels. However, the adaptations needed in the learning process go beyond the adaptation of contents and their presentation, and should take into account both the learning models and the psychological procedures.

These adaptations are not explicit and cannot be directly managed by the user. Specifications such as IMS Learning Design[4] can contribute to specify guidelines in the delivery of learning contents and services for people with disability considering pedagogical and psychological requirements. However, in practical situations that is not sufficient since it is not possible (in terms of time, effort and knowledge) to specify the whole design in advance. Moreover, it is not feasible either to collect directly from the users the values of the required attributes to build the model that the adaptation tasks will consider. There is a need for an intelligent support that analyses the learners' interactions at run time and process them to learn their usage preferences and thus, some attributes of the user model. In particular, artificial intelligence techniques, such as data mining (DM) to extract the knowledge from the interactions, and machine learning (ML) to classify user according to their behaviour, are used when dynamic support is required. In turn, collaborative filtering (CF) techniques are useful to offer recommendations to the users based on other users' experience and the similarity of user models (and thus, needs and preferences) among the users.

2 EU4ALL Approach

Students and professionals with disabilities have problems in accessing lifelong learning because of the diverse barriers that may exist in the various stages they must go through to realise their learning goals. They have to negotiate pre-established general procedures, which are usually more focused on the institutions needs than the students, and generally consider a single standard set of student needs. These are far from considering the individual needs and preferences of their student users. Learning ideally should be a personalised and adaptive process for ALL, which from the beginning till the end should consider the learner's specific needs. Students requiring "Accessible Lifelong Learning" (ALL) suffer from a lack of information about pre-established procedures and practices that meet their needs. In addition, there are many difficulties in providing the appropriate infrastructure to support them.

To address these issues, EU4ALL project,seeks to define and construct an extensible architecture of European-wide services to support lifelong learning in HE. In the next subsections we describe the objectives of the project and outline a scenario to present our approach.

2.1 Objectives

The concrete objectives of EU4ALL are as follows:

1. From an in-depth research, achieve a unified, agreed, shared and usable vision of the standards work, users' requirements, service definition, technologies

---

[4] IMS-LD: http://www.imsglobal.org/learningdesign/

2. Define practical specifications and implement in terms of standards an open and extensible architecture of services for ALL which is prepared both to assist learners and to support service providers
3. Provide user-centered services that consider individual user's needs and preferences, pedagogical guidelines and adaptive behaviour based on users' interactions
4. Bring together major service providers, like mega-universities to foster the awareness of best practices in providing educational services for ALL
5. Impact on major standardisation bodies, identifying where the creation of new standards or extension of existing ones supports the establishment of the EU4ALL framework and pursue this into the relevant standard bodies
6. Create a channel for the diffusion and benchmarking of these research results in all major distance training universities in Europe by means of an European-wide ALL repository, which facilitates a common understanding of learning methodologies, access needs, cognitive requirements, assessment procedures and LLL issues for people with disabilities.

One key issue in EU4ALL is to support the personalization of services to attend individual user's needs and their evolution over time. In this respect, services provided following objective 3, will be integrated in the architecture according to an iterative process along different prototypes, which includes technology users and other relevant stakeholders throughout the entire process of design and redesign.

2.2 Sample Scenario

Within this section we would like to outline some situations that illustrate the benefits that modelling users could bring to e-learning systems.

When Gabriella was enrolled in her university, she was asked to include her personal information in the e-portfolio of the university site. She is dyslexic, and during the process she was assisted by a consultant from the Disability Office. Due to her impairment, she has problems for organisation and time management, for remembering, concentrating, and understanding. Every time she connects to the system, she receives Post-It style messages through which she is reminded to start a new lesson. As the layout for a new lesson, the server provides her with the corresponding text. Should the lesson contain a graphical symbol, it is automatically substituted by vocal information based on familiar words that support comprehension.

Eva is a vision impaired student using her faculty computer lab to access a forum where a discussion is going on about gender issues in developing countries. Once she has logged into the forum, and according to her user profile, the computer checks that there is a screen reader installed, and then it is run to output vocally the WAI compliant web based content. Eva plugs the jack connector of her earphones into the Braille identified computer deck, to preserve her privacy, and also to avoid any annoyance for the colleagues working around her.

John drives every morning to work when traffic is very heavy. The full journey takes more than one hour. In such situations, it is very convenient for him to take a lesson of Psychology from his distance learning university. He uses his internet enabled mobile phone, and the hand-free set that is integrated with the Bluetooth and HiFi system of his car. The university server detects that John is connecting from his car terminal, and automatically streams the audio version of the lesson. John listens to a synthetic voice reading the text and summarizing the key content of lesson statistics. Would traffic get more fluid, and driving demand more concentration, he pronounces the voice command "Stop" to pause the lesson. He can resume it when desired.

Modelling user related information is essential to permit the automatic adaptation of content to user needs and preferences:

- User and context data. It deals with the needs of the person regarding how the interaction must be performed (input and output) because of impairments (dyslexia, visual disability) or because of context of use (driving a car). The information can be acquired by explicit or implicit procedures (e.g. filling the enrolment form, or detecting that the connection is performed from a car).
- Use of assistive technologies and enhancements. When a person with impairments is accessing the e-learning system, assistive technologies must be identified, and content must be provided accordingly. The content can be accessible just because of its universal design (WAI compliant content) or because the content can be generated/composed/enhanced (alternative vocal content substituting images). Assistive technologies must be seamlessly interoperable with generic consumer goods. Hands free kits are quite interoperable with car systems, and constitute an example of assistive technologies in use by people without disabilities.
- Content. Metadata about content must be available to identify its layout capabilities (textual, vocal content). Authoring tools should ease this to professionals.

3 User Modeling for ALL

The user model is a core element in an adaptive system since it stores the known data from the users' features and actions in a way that it is understood by the systems. Its purpose is to support the system in reasoning about the needs, preferences and future behaviors of the user in order to produce adaptation tasks or recommendations. Specifications can be useful to build explicit user models. However, as introduced before, they are not sufficient and a hybrid approach which combines explicit and implicit models is required. The later types of models are obtained from ML tasks on usage and interaction data. As introduced in section 1.2, ML techniques can be used to learn the user preferences from their behaviour in the system.

On the other hand, when learners have disabilities that affect their interaction with a computer system, not only the different interests and sets of needs have to be taken into account, but also the preferred access strategy for each user. The requirements, preferences and access strategies that assistive technologies and adaptations bring into play make this a very complex picture and provide a severe test for the design of representations of user information to support them. An open research issue that we are dealing with in EU4ALL project is to define how to store this information in a standardised way for learners with disabilities.

3.1 Supporting Functional Diversity in e-Learning by User Modelling

User modelling may constitute a significant driver for personalisation of e-learning, and thus improve notably the quality of user experience. This approach includes the provision of accessible e-learning services and contents, accordingly to user circumstances. These are composed by personal information (needs or preferences), user context, and the interaction capabilities of the device being used for the communication with the information system.

Providing appropriate delivery of contents to persons with functional diversity will require an in-depth characterisation of the interaction capabilities of user agents, including potential assistive technologies which support users with functional limitations to interact with the e-learning system. Comprehensive, standard based characterisation of devices will also be of use in situations where context imposes restrictions for the way in which content is delivered to people with or without disabilities, such as noisy environments, locations with inadequate lighting, etc. Consistently, formal definition of context is also required.

The learning resource also demands a metadata description. According to the draft standard produced by ISO/IEC JTC1 SC36, by learning resource we mean the learning content that is transmitted over the communications network. This resource could be characterised by its requirements/potentialities in terms of interaction (input/output) with the user, as well as the available alternatives, if any.

This on-going standardisation group proposes the consideration of two sets of information, related to characterisation of accessibility needs and e-learning resources in terms of human interaction:

- The description of a learner's accessibility needs and preferences including: a) how resources are to be displayed and structured, b) how resources are to be controlled and operated, and c) what supplementary or alternative resources are to be supplied.
- The description of the characteristics of the resource that affect how it can be perceived, understood or interacted with by a user, including: a) what sensory modalities are used in the resource, b) the ways in which the resource is adaptable (i.e. whether text can be transformed automatically), c) the methods of input the resource accepts and d) the available alternatives.

On the other hand, CC/PP was created by W3C as a framework to support the formal specification of characteristics describing users and client devices taking part in web based communications. This formal characterization enables the negotiation between web entities (namely clients and servers) with the aim to produce a suitable presentation for the exchanged content. CC/PP makes use of W3C RDF (Resource Description Framework) as the language to model the needed metadata.

Despite the CC/PP framework potentially supports the description of users and their agents, its industry implementation has mostly focused the latter so far. The User Agent Profile (UAProf), a CC/PP based standard developed by the Open Mobile Alliance (OMA), was designed to allow Internet enabled phones to send a profile of their capabilities to a server for obtaining an optimised layout of contents. Furthermore, emerging mobile communication services based on user location include this information by making use of the UAProf element used for storing terminal location.

Even though there is no commercial implementation available, the above mentioned research by the IRIS project focused on the definition of a user profile based on the CC/PP framework. This profile includes personal information, user preferences regarding input and output modes, as well as preferences about how to interact with the content (navigation, search and highlighting preferences). The delivery context is also described through the location component, complemented with biometrical and emotional data. Furthermore, the same work proposes an extension of the UAProf Hardware-Platform and Software-Platform components through the definition of input and output subcomponents. These subcomponents are used to characterise interaction capabilities of user agents, including assistive technologies. The project also issued an implementation for blending user and device profiles, based on an iterative process of translating user interaction needs into the interaction capabilities of the service [6].

3.2 Covering the Full Life-Cycle of Services' Adaptation

In order to effectively adapt the contents and services to ALL, the needs and preferences of the users (i.e. the user model) have to be managed along the full life cycle of eLearning. Therefore, EU4ALL architecture supports the full life cycle of adaptation for service provision, which covers the following phases: design, publication, use and auditing [7]. The management of this cycle is focused on individual users' needs, which are diverse and change over time. To cope with these issues, an approach based on standards for combining design and runtime adaptations is being developed considering related work performed in the aLFanet project.In EU4ALL architecture, the management of service adaptation consists on a loop of continuous interactions, where the following phases are identified:

- Design: provides the logic for the pre-design adaptations and provides the hooks and information upon which the runtime adaptation bases its reasoning. The services are to be defined in terms of IMS-LD (extended to fulfil the requirements).
- Publication: includes the storage and management of data to be retrieved by the different components. The usage of standards (e.g. IMS family) guarantees the required interoperability.
- Use: focuses on the environment while running the service. It deals with the delivery of contents taking into account the learning design specifications, the user profile and the accessibility requirements by properly adapting the user interface and by producing dynamic and contextual recommendations.
- Auditing: closes the cycle and provides reports on the actual use of services.

Although the needs and preferences of the users are being managed along the full life cycle of eLearning, there is still a need to implement accessibility and usability evaluations along the eLearning cycle to enrich the adaptation coverage. The goal is to ensure that usability and accessibility requirements can be assured and validated in the whole cycle. Methodologies are detailed elsewhere [8]. Therefore, results and impacts from the adaptation cycle are evaluated against accessibility and usability criteria, which are themselves user centered and therefore closely related to the user model.

3.3 Implementation

The strategies presented here are being developed within a flexible, extensible, open, standard-based architecture of services for ALL which is prepared to assist learners and support service providers [9]. Relevant components of the architecture regarding the personalization task are: 1) User Modelling for ALL (UM4ALL -see below), 2) Device Model Manager, aware of the CC/PP capabilities of the device, 3) the Learning Object Metadata Repository, which characterizes the contents that are to be delivered, 4) the Psycho-pedagogical module, delivering IMS-LD guidelines defined by the corresponding experts, and 5) the Content Personalization service, which selects the appropriate content to be delivered taking into account the user preferences and context, the device capabilities and the content's metadata. Other components are in charge of providing the user agent features and tracking the users' interactions.

In particular, at aDeNu we focus on the User Modelling for ALL system, or UM4ALL. It consist on three parts, the User Model itself (which stores the user personal data, needs and preferences in IMS Learner Information Profile –IMS LIP[5]-, including its extension for accessibility, i.e. IMS Access for All[6]), the User Modelling Engine, which includes the ML and DM algorithms to learn some of the attributes of the model, and the Recommending system, which copes with the needs for ALL via generating dynamic recommendations that can be provided to the users at runtime. In this way, it provides personalized services delivery in terms of dynamic recommendations on what to do. By analyzing users interactions, context-sensitive feedback is given to learners taking into account the current situation in the course and the device at hand. In order to build the recommendations, collaborative filtering techniques are applied.

Moreover, other modules can interact with the User Model (via web services) to get the corresponding information about the user so the user interface and the content can be modified according to the users' needs, pedagogical and psychological procedures, the context in which the user is accessing the information and the device used for that purpose.

The UM4ALL is a particularization of A2M [10] and follows the same approach as defined in aLFanet Adaptation Module [2], which includes a two level hierarchy of multi-agent architectures that work autonomously to solve adaptation tasks. The high level consists of a set of agents that interact to select the recommendations to be given to the learners (i.e. the Recommending system). In turn, the low level is used to learn the attributes of the models from the interaction data (i.e. the User Modelling Engine). A hybrid approach that combines knowledge-based methods and machine learning techniques is to be used, since both approaches are complementary.

4 Conclusions and Future Works

In this paper we have provided our general approach and discussed relevant issues in providing a dynamic user modelling for attending functional diversity for accessible lifelong learning in HE. The user model is the key element in the adaptation process since it is needed by the system to reason about the needs, preferences and future behaviors of the user in order to produce adaptation tasks or recommendations.

---

[5] IMS LIP: http://www.imsglobal.org/profiles/index.html
[6] IMS Access for All: http://www.imsglobal.org/accessibility/

When learners have disabilities that affect their interaction with a computer environment, not only the different interests and sets of needs that build the user model have to be taken into account, but also the preferred access strategy for each user. Providing appropriate delivery of contents to persons with functional diversity will require an in-depth characterisation of the interaction capabilities of user agents, including potential assistive technologies which support users with functional limitations to interact with the e-learning system.

An open research issue that we are dealing with in EU4ALL project is to define how to store this information in a standardised way for learners with disabilities. In this sense, we are making an intensive use of standards (IMS LD, IMS LIP, IMS Access for All and CC/PP) to allow the semantic interoperability of the different data available and provide personalized content and services delivery. Moreover, thanks to the dynamic analysis of users interactions, context-sensitive feedback is given to learners taking into account the current situation.

User modelling within EU4ALL relies on standards and the state of the art in this field to build a personalized e-learning platform. The project envisages the use of quantitative and qualitative methods for definition of user profiles [11]. Quantitative methods include eliciting of user information using questionnaires, observing users and stereotyping using statistical information. Qualitative methods provide objective data such as records of the tasks performed by the users and sensors to get information about the user activities and context of the user environment.

Specifically, our approach to provide universal and personalized access lies on user modelling and the generation of dynamic recommendations to cope with the needs for ALL with a pervasive support of standards covering the full life cycle of service provision. The modelling differs from others not only in i) taking into account the interactions and context of the user that take place at runtime and ii) characterising interaction capabilities of user agents, but in defining how we apply the models built. User models are used to personalize and adapt learning materials, pedagogical models and interactions in the environment to satisfy both the individual learning needs and the access preferences.

EU4ALL will ensure that user modelling solutions consider human factor recommendations which are applicable (ETSI EG 202 325). These guidelines propose that profile solutions 1) must be provided for the primary benefit of the user, 2) should give the user rights to modify profile contents affecting directly him/her and 3) should give the user the right to accept or reject proposed changes to the profile.

The project will design interfaces with the IMS e-Portfolio[7]. This specification constitutes the appropriate framework for providing users with control about the profile that characterizes them within the e-learning system.

Acknowledgements

EU4ALL project is funded by the EU 6th Framework Program under the grant FP6-IST-FP6-034778.

---

[7] IMS-ePortfolio: http://www.imsglobal.org/ep/index.html

References

1. Ondeck, D.M.: The Challenges of Functional Inability Home Health Care Management & Practice 15(4), 350–352 (June 2003)
2. Boticario, J.G., Santos, O.C.: An open IMS-based user modelling approach for developing adaptive learning management systems. Journal of Interactive Media in Education (in press)
3. Cudd, P.A., Freeman, M., Yates, R.B., Wilson, A.J., Cooke, M.P., Hawley, M.S.: The CORES project. In: Proceedings of the 4th international conference on Computers for handicapped persons table of contents, pp. 493–495. Springer, Heidelberg (1994)
4. Alexandraki, C., Paramythis, A., Maou, N., Stephanidis, C.: Web Accessibility through Adaptation. In: Miesenberger, K., Klaus, J., Zagler, W., Burger, D. (eds.) ICCHP 2004. LNCS, vol. 3118, pp. 302–309. Springer, Heidelberg (2004)
5. Makris, C., Tsakalidis, A.K., Vassiliadis, B., Bogonikolos, N.: Adapting Information Presentation and Retrieval through User Modelling. In: ITCC 2001, pp. 399–404.
6. Velasco, C.A., Mohamad, Y., Gilman, A.S., Viorres, N., Vlachogiannis, E., Arnellos, A., Darsenitas, J.S.: Universal access to information services-the need for user information and its relationship to device profiles. In: Universal Access in the Information Society, vol. 3(1), pp. 88–95. Springer, Heidelberg (2004)
7. Van Rosmalen, P., Boticario, J.G., Santos, O.C.: The Full Life Cycle of Adaptation in aLFanet eLearning Environment. Learning Technology newsletter 4, 59–61 (2004)
8. Martin, L., Gutierrez y Restrepo, E., Barrera, C., Rodriguez Ascaso, A., Santos, O.C., Boticario, J.G.: Usability and Accessibility Evaluations along the E-learning Cycle. In: Proceedings from International Workshop on Web Usability and Accessibility (in press)
9. Santos, O.C., Boticario, J.G.: European Unified Approach for Accessible Lifelong Learning. In: Mendez-Vilas, A. (ed.) Proceedings of the 4th Int. Conf. on Multimedia and ICTs in Education. Current Developments in Technology-Assisted Education (2006)
10. Santos, O.C.: A standard-based approach for modeling 'all' users in adaptive learning management systems with artificial intelligence techniques. In: Proceedings CAEPIA (in press)
11. Kadouche, R.: Towards strategies and methods for disabled users profile, Workshop on Personalisation of ICT products and services. ETSI (October 2004)

Exploiting Profile Modeling for Web-Based Information Systems

Karine Abbas[1], Christine Verdier[2], and André Flory[1]

[1] LIRIS-UMR 5205 CNRS, INSA de Lyon*, Claude Bernard Lyon1 University,
Lumière Lyon 2 University, Ecole Centrale de Lyon, Bât.Blaise Pascal, 7 avenue Jean Capelle,
69621 Villeurbanne Cedex
{karine.abbas,andre.flory}@liris.cnrs.fr
[2] LIG- UMR 5217 CNRS, équipe SIGMA, Joseph Fourier University, 681 rue de la Passerelle,
BP. 72, 38402 Saint Martin d'Hères Cedex, France
christine.verdier@imag.fr

Abstract. With the considerable amount of data and the diversity of the user's needs, the personalization of information becomes a real challenge in web-based information systems. This paper presents a personalized access technique, based on a profile modeling. This technique requires two steps: 1) building user's profile and 2) using profile content for filtering information. The first step consists in modeling a global profile which can be used for different and independent requirements of personalization. The profile modeling takes into account the user's context when defining user's profiles. We describe a global profile in three levels: meta-description level, description level for a specific domain, instances (data). The second step consists in selecting the profiles rules which correspond to the user and his current context and to filter information accordingly.

Keywords: personalized information access, profile modeling, web-based information system, context.

1 Introduction

Current web-based information systems try to improve continuously the access quality to information in different domains: economic, scientific, technical, etc. However the considerable amount of data increasing tremendously over time, and its large accessibility trough the web have made finding relevant information a quite challenging task. Indeed, a user is quickly submerged by a considerable number of information often unintelligible while searching information on the web, which leads to user disorientation. Then it becomes essential to take into account the needs and the heterogeneity of users in the information access process in order to adequately tailor user-system interaction. The personalization and user modeling process are the key elements to propose solutions to these challenges. Several works have been proposed in the literature such as recommendation system, adaptive hypermedia application, digital libraries, etc. However, the majority of them proposes actually specific solutions, suited only for predefined personalization requirements (usually device

M. Weske, M.-S. Hacid, C. Godart (Eds.): WISE 2007 Workshops, LNCS 4832, pp. 313–324, 2007.

characteristics and user preferences) and is hardly reusable when new personalization functionality needs to be added. Furthermore most of them do not take into account the current situation of users and their physical environment when it design personalized access tools. Certainly, the need for information can differ according to user's preferences and goals. But with the spread of mobile device providing access to the web everywhere and anytime, it is important to take into account the user context which modifies the personalization process accordingly.

In this context, we propose an approach which models user's profiles in order to personalize user-system interaction and to provide relevant information. This approach requires two steps: 1) building a generic profiles model and 2) using the profiles content to filter data before displaying it. This personalized access approach should guarantee a high level of flexibility in terms of 1) responsiveness to highly changing requirements of personalization and 2) suitable actions to be undertaken to meet these requirements.

This adaptation approach will be integrated in an Iconic Interface Management System (I^2Ms) [1]. This system aims to access data friendly and easily in a graphical interface in which data are represented with icons.

This article is organized as follows: section 2 deals with the general architecture of the system. Section 3 describes the principle of the profiles modeling. Section 4 presents a scenario of use of the profile model. Section 5 shows how to use these profiles in the data filtering process. Section 6 presents related works

2   General Approach

The approach we propose aims to facilitate access to relevant information, adapted to users' needs and context of use. This personalised process is based on the profile modeling. This leads to define characteristics which distinguish users. These characteristics are classified into two parts: user characteristics called user profile and context characteristics called context profile. The first part presents a collection of information concerning personal data (name address, etc.), professional data (activity, organisation, etc.), domain of interest, preferences, goals, competences, context, access right, etc. The second part concerns the description of parameters (such as location, task, etc.) which characterise the user contextual situation. The profiles modeling process depends strongly on the context parameters. We can have various profiles according to the user contextual situations. To the end, we propose to build a generic profile model which can be used for different and independent requirements of personalization, possibly not defined in advance. The model is flexible and reusable to guarantee that different requirements of personalization and different contextual situations can be taken into account in the personalization process.

This personalization approach requires two steps: the former consists in building the global profile in order to contain a collection of profiles for a set of contexts. The latter uses the profiles content for the filtering process applied to data in response to user query. The query result is represented in a XML graph. Thus the personalization approach provides information access personalised in the appropriate form at the right time and with minimal user effort.

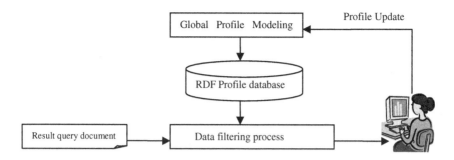

Fig. 1. General Architecture

3 Global Profile Modeling

We propose to model a generic profile in order to cover the different personalization goals for the different scenarios of use. These scenarios depend on user's needs and his environment. An orderly do not have the same preferences and knowledge if he works in hospital or he follows-up a patient with cancer at home. Then, the generic model, called global profile, has to take into account the heterogeneity of users and the different contextual situations. The global profile models two basic notions: the user profile and the context profile. The first profile describes information characterizing a user such as personal data (name, address, etc.), professional data, preferences, goals, skills, etc. the second profile represents a description of an autonomous aspect of the contextual situation in which a user interacts with the system. This situation can be described by a set of parameters such as the location, role, task, expertise domain and devise.

3.1 Profile Model Representation

The global profile is modeled in three levels:

- A meta-description level: this level corresponds to a high abstract description of concrete concepts characterizing the user and his context.
- A description level of concrete concepts: this level describes concepts adapted to a particular domain (medicine for example). These concepts are a specialisation of the abstract meta-description objects.
- Instances: the concepts described for a specific domain are instantiated.

The definition of the concepts describing a global profile is supported by well-estimated needs in terms of personalization and well-understand contextual situations.

The definition of user profile meta-description consists in supporting an abstract view of the concrete concepts describing user profile. This profile is structured in dimensions such as skills, preferences, etc. Then a *dimension* describes information that characterizes a user profile. A dimension is related to one or more attributes. An *attribute* describes information on a dimension. It can be simple or complex. A *simple attribute* is associated with a value, whereas a *complex attribute* expresses two forms: 1) an *association* relation between a set of (simple or complex) attributes or 2) *decomposition* relation. In the two cases, this attribute is called relation. The instances of the composition relation are ordered (or unordered) sequence of instances of other attributes. The attribute simple can be isolated information associated to a printable value (integer, boolean, string and so on) or information referring to a set of instances of a class associated with it.

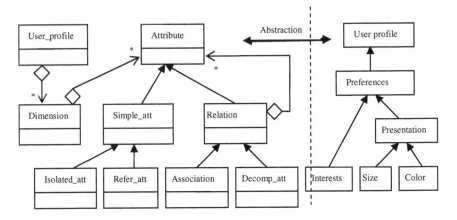

Fig. 2. User profile meta-description and instantiation

Figure 2 presents the user profile meta-description and an example of the profile description. This example is composed of the dimension "Preferences" which contains Interests (simple attribute) and Presentation (complex attribute). The attribute "Interests" refers to the class "Object". The attribute "Presentation" represents an association composed of two simple attributes: Color and Size.

The context modeling focuses on the mobile use of the personalized access system. A object-oriented representation of context is illustrated in the figure 3. The context is modeled by two basic classes *Context* and *C_parameter*. The first class is a superclass of the second one which is a superclass of the specialization classes characterizing a set of contexts. Moreover, we have semantic links between context parameters. For example, a process is composed of a set of tasks.

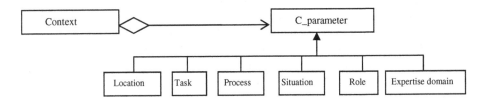

Fig. 3. Context modeling

The dependence relation between user profile and context profile is modeled by a semantic relation between class *Dimension* and class *C_parameter* (figure 4). This relationship defines a profile composed of one or more dimensions depending on one

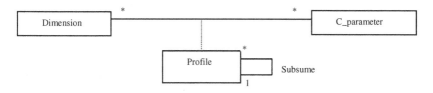

Fig. 4. Profile modeling

or more C_parameters. Then for each contextual situation (one or more C_parameter), we can have a profile with a set of dimensions.

Then, the global profile is a collection of generic profiles. A generic profile is a function which associate a set of context parameter C1,C2,.., Cj with $1 \leq j \leq n$ to a set of dimensions D1,D2,.., Dk with $1 \leq k \leq n$. the profile attributes are associated to values.

The profile model uses the subsumption relation which allows to classify profiles in order to generate inheritance relation between them. Intuitively, given two profiles P1 and P2 included in global profile. P2 contains dimension D1 and depends on context parameters C1 and C2 and P1 contains D2 and depends on C1. We say that P2 subsumes P1 if context parameters of P1 are included in the ones of P2. Then P2 inherits the dimensions of P1 (an example is gived in section 4).

Let us sum that, a global profile is a collection of profiles. Each profile is composed of dimensions. Each dimension is composed of attributes (simple and/or complex) and depends on one or more context parameters.

3.2 Profile Model Management

To implement the profile metadata and instances we use the RDF language [2] for several reasons: it is flexible (it supports schema refinement and description enrichment) and extensible (it allows the definition of new resources or properties

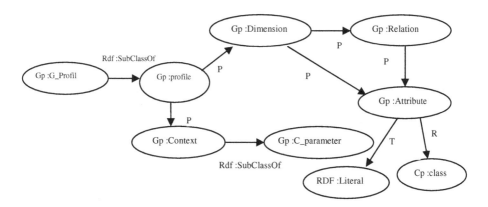

Fig. 5. RDF profile description

without updating the RDFS schema). The description of global profile concerns the structure and instances of both profiles: user profile and context profile.

RDF is a language based on resource-property-object triples designed for expressing statement about resources on the web. For example we can have a statement like (User, name, smith). User is a resource, name is a property of class and smith is an object. RDF schema is required for including concepts of class, property, subclass and sub-property relationships, a primitive type 'Literal', bag and sequence types, domain and range constraints on properties.

Figure 5 shows a RDF profile meta-model in which Ellipses denote class, P denote RDF:domain, R denote RDF:range and T denote RDF:type

We propose a tool for implementing a profile and managing it. A graphical interface is proposed to define the different objects and instances.

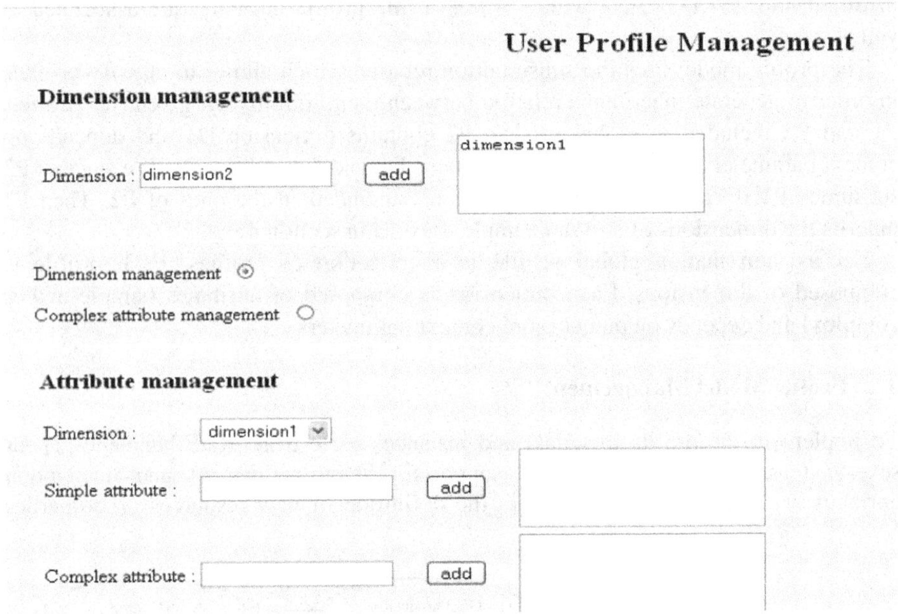

Fig. 6. Graphical interface

Figure 6 show a screen of the interface which allows to add new dimensions and attributes. This screen is composed of three parts: 1) the first part is dedicated to define dimensions, 2) the second one allows to specify if the dimension will be managed in the bottum part otherwise complex attribute and 3) the third part is reserved to manage attributes. If dimension management is selected, simple attributes and complex attributes are defined for a given dimension. Otherwise they are defined for a complex attribute.

The definition of a specific profile can not use all abstract profile objects. The expert builds his own profile with respect to the personalization needs.

4 Personalization Approach for I²Ms

This personalization approach will be implemented in an Iconic Interface Management System (I²Ms). This system aims mainly at accessing medical data friendly and easily in a graphical interface in which data are represented with icons. Then the goal of this approach is to personalize information before displaying it in the interface. Firstly the personalization approach has to define user profiles. Secondly these profiles can be used for filtering information. We use the generic profile model described above for building a profile structure specific to medical domain. The profile model is classified into three dimensions:

Profile (Access right, Skills, Preferences)

- *Access right* defines access rules to medical data in order to guarantee the confidentiality and the availability of data.
- *Skills* are represented by a combination of knowledge and expertise degree. They are primarily characterized by the capacity of a user to understand medical terms.
- *Preferences* mainly concern the user's interests and their choices for presentation of medical information. To define the preferences, a questionnaire is used and a user-system interaction is studied.

The personalization process is applied to medical objects that are disease, diagnosis, medical history, biological tests, etc. represented in a hierarchical structure, called Unique Medical Record Structure (UMRS). This structure is built by integrating heterogeneous medical record structures from legacy information systems into a unique medical record structure. A medical record integration technique has been proposed in [3].

Our profile model introduces the role for modeling profiles (figure 7). Indeed, users acting a common role can have the same goals, preferences, access right, etc. Then it is interesting to regroup them by role and assign profiles to roles. The profile is called role profile. Thus a given user acting a given role can inherit personalization rules defined in the role profile.

Now, we describe the composition of each dimension of role profile:

- The access right dimension is modeled by a complex attribute "Authorize" which is composed in three simple attributes: "Role, Object, and Action". Role and Object refer to class role and class Object. Action is an isolated attribute which has a value "R, W, D, S" for read, write, delete and send.
- The preferences dimension is composed of a complex attribute pref_obj composed of three simple attributes: Role, Object, Page_type. The attribute page_type identifies the type of the page like navigation page or search page.
- The skills dimensions define a simple attribute and a complex attribute. The simple attribute is identified by Level which designates one of the three levels: beginner, intermediary or expert. The complex attribute is Object which refers to one or more instances of the class Object.

As the I²Ms is used within the framework of health network, users can work in very different conditions, using different technologies and have different needs and priorities[4]. For example, an orderly can work in different places; he can both follow

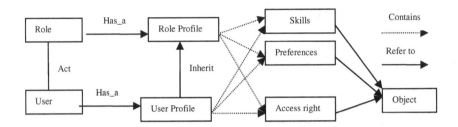

Fig. 7. User and role profile modeling for I2Ms

up a patient at home and work in hospital. Then we have to take into account the different contexts when modeling profiles.

We define three context parameters which can influence the user's behaviour. These parameters are Expertise domain, Location, Role. Then the modeling of dimensions depends on these parameters.

Table 1. Classification of profile according to contexts

Context Dimension	Role	Location	Expertise domain	Profile
Access right	Yes	No	No	**P1**
Preferences	Yes	Yes	No	**P2**
Skills	Yes	Yes	Yes	**P3**

Dimensions are classified in three profiles (P1, P2 and P3) according to context parameters. Each profile has an identifier. As the context parameters of P1 are included in ones of P2 then Profile1 is subsumed by P2. In the same manner, P2 is subsumed by P3.

There are exceptional situations or rules specific to a given user and not applied to a role acted by the user. For example, in common situations, a physician can read his medical record. But in emergency, he can have a direct access to every medical record. These exceptional situations are taken into account in a profile of the given user. Then if we have personalization rules specific to a user and not applied to a role, these rules are integrated in the user profile. The user profile is also classified in three dimensions: access right, skills and preferences.

The Access right dimension of the user profile is specified by two complex attributes: *Acess_user* composed by two simple attributes *ref_Play* and *constraint* and *Authorize_user* composed of *refPlay1, Action, Object, refPlay2, constraint*. This first complex attribute specifies that a user playing a given role (referred to an instance of class act) uses the system according to the defined constraint. For example, a physician is on duty on Saturday night in the hospital. Then he is authorized to access information on Saturday of 10 p.m to 8 a.m. The second complex attribute is applied on data access authorizations. The attribute ref_Play1 of this relation refers to a person that describes the constraint (an administrator or a user). The attribute refPlay2 represents a user on which the constraint is applied. For example, a physician (represented by ref_play1) can delegate his rights to another physician (represented by ref_play2).

The Skills dimension is modeled by defining two complex attributes: expertise and experience. The first attribute is composed of three attributes: Disease, User, and Degree. Disease refers to class disease which contains the disease classification. User refers to the class User. Degree is an isolated attribute which has an integer value. The second complex attribute is composed of three attributes: user (refer to class User), role (refer to class Role) and number_exp which is an isolated attribute indicating the number of month practiced in the role.

5   Use of the Profiles in the Personalization Process

The personalization approach is based on a filtering process in two steps. The first step selects the profiles rules which correspond to user and his current context from profile database. The second step applies the filtering rules on XML data. This data is extracted from various databases in response to a user query and structured in a XML document.

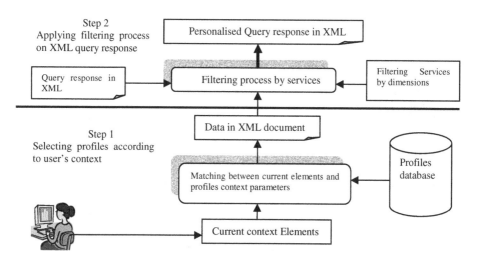

Fig. 8. The filtering process according the user's current context

5.1   Selecting Profiles According to the User's Context

This step selects from profile database profiles rules that are valid with respect to the user's current context. We remember that a user may have several profiles. Each profile contains dimensions (one or more) according to one or more context parameters. Then to obtain the dimensions content, we proceed in two steps: firstly we determine the context parameters of the user's current context situation; secondly we extract from profile database the dimensions which are valid with respect to the context parameters. The first step is performed by matching current situation parameters with ones of stored profiles. This operation is realized as follows:

- Let $C = \{C_1, ...,C_n\}$ a set of current situation parameters. We search all the possible combinations, called M, between the elements of C. For example, we have three parameters C_1, C_2 and C_3; the combinations proposed with the following elements are $\{C_1, C_2, C_3, C_1C_2, C_1C_3, C_2C_3, C_1C_2C_3\}$.
- We match each element of M with the context parameters of the stored profiles. We start by a single element (for example C_1, C_2 or C_3). If an element has the same set of parameters, we proceed as follow:

-We extract the part of data corresponding to the dimensions of the current profile and insert it in a XML document (dimension.xml) dedicated to filtering operation.
- We test if the current profile subsumes other profiles and their context parameters are among the graph elements. In this case, data of their dimensions are extracted and added to XML document.

Thus, extracted data (dimension.xml) will be used in the filtering process in order to personalize XML data.

5.2 Filtering Data According to Preferences

The filtering process is applied to XML data extracted from various databases in response to user's query. The process filters XML data according to a set of services. A service is a program developed for describing the steps necessary in the filtering data. In general, a service is proposed for each dimension. These services are based on the principle of matching process. It mainly consists in comparing between nodes of the XML graph and objects designated in each dimension. We use the XPATH tool for accessing XML nodes. Indeed some attributes of dimension refer to objects. These objects refer to metadata of system information database. Then there is a semantic link between dimension objects and XML graph nodes. If a node of XML graph corresponds to an object, a set of operations are performed in order to separate relevant information from irrelevant information. These operations have been described in each service for each dimension.

The filtering process starts by extracting data concerning each dimension from the file dimension.xml. Then a service dedicated for each dimension is performed. As a result, we have a modified XML document with data which will be displayed in the graphical interface.

6   Related Works

Accessing relevant information, adapted to user's needs and context of use is a real challenge in web-based information systems [5][6]. Many works have been developed in the literature to address the issue of personalization by presenting suitable models based on the user's profile. The main goal of these techniques [7][8][9] is to propose a global view of the user's characteristics, needs and preferences. These user models consists mainly in classifying the user characteristics into a set of categories such as demographic attributes (identity, personal data), professional attributes and behaviour attributes (traces of navigation on Web), goals, preferences, skills, Accessibility, Activity, Interest, Affiliation. The profile information are defined explicitly by a user through a questionnaire or implicitly by analyzing behavioural observation of the user navigation [7][8][9][10]. The majority of these works is actually specific solutions suited only for predefined adaptation requirements and are hardly reusable for a new adaptive application. Moreover these techniques have focused on the user and rarely considered the contextual situation in which a user interacts with the system. Taking into account the user's context is a key element to improve personalized information access process. A largest view is given by Dey[11], who defines context as "*any information that can be used to characterise the situation of an entity. An entity is a person, place or object that is considered relevant to the interaction between a user and an application*". Commonly, context aware systems limit the notion of context, referring to the situation in which the user is acting, to the concepts of user's location and device [12][13]. These approaches are focused on the system environment and ignored the user preferences. Nowadays some current models [14][15] try to take into account the user preferences and the context. But it is more interesting to have a flexible model which manages, in additional to the predefined categories, other user information such as skills, security, goals in order to provide a complete personalised system. Then, the combination of user characteristics and user context gives two important key elements to cover all facets of personalization process. Our work takes into account the challenge and proposes a personalization technique based on the both user and context modeling. Our technique builds a generic profile model in order to model a global profile which describes several categories characterizing a user. Thus, the flexible model can easily add new profiles, objects without modifying the global profile design.

7 Conclusion

In this paper, we have proposed a personalization approach based on the profile modeling for web-based information systems. This approach defines a global profile in order to take into account different contextual situations. Modeling a generic profile allows to propose a general model adapted to any application according to its personalization needs. This personalization technique is applied to a medical application in order to provide to user relevant data corresponding to his current context.

We expect to improve the profile modeling in order to take into account any contextual situation and user characteristics. We will work to adapt the user interface (content and presentation) according to data defined in the global profiles. For each profile dimension, a service is proposed to adapt the interface. A set of services will be developed and implemented to complete the data filter process and personalize user-system interaction anywhere and anytime.

References

1. Sassi, S., Verdier, C., Flory, A.: A new system for project management: the I2MS interface. In: TIGERA, Tunisie Hammamet (2007)
2. Brickley, D., Guha R.V.: RDF Vocabulary Description Language 1.0: RDF Schema. W3C Recommendation (February 2004)
3. Abbas, K., Verdier, C.: Design of a medical database transformation algorithm. In: International Conference on Enterprise Information Systems - ICEIS 2006, Paphos – Cyprus, pp. 23–27 (2006)
4. Bricon-Souf, N., Newman, C.: Context awareness in health care: A review. International Journal of Medical Informatics 76(1), 2–12 (2007)
5. Ceri, S., Daniel, F., Demald, V., Facca, F.M.: An approach to user-Behavior-Aware Web Applications. In: Lowe, D.G., Gaedke, M. (eds.) ICWE 2005. LNCS, vol. 3579, Springer, Heidelberg (2005)
6. Fiala, Z., Hinz, M., Meissner, K., Wehner, F.: A component based approach for adaptive dynamic web documents. Journal of web Engineering 2(1-2), 53–73 (2003)
7. Amato, G., Staraccia, U.: User profile modeling and applications to digital librairies. In: Proceedings of the 3rd European Conference on Research and avanced technology for digital libraries, pp. 184–187 (1999)
8. Razmerita, L.: User modeling and personalization of the Knowledge Management Systems, book chapter. In: Adaptable and Adaptive Hypermedia, to be published by Idea Group Publishing (2005)
9. Seo, Y.W., Zhang, T.: A Reinforcement Learning Agent for Personalized Information Filtering. In: Proceedings of the 2000 International Conference on Intelligent User Interfaces, New-Orleans, USA, pp. 248–251. ACM, New York (2000)
10. Shavlik, J., Goeks, J.: Learning Users' Interests by Unobtrusively Observing Their Normal Behaviour. In: Proceedings of the 2000 International Conference on Intelligent User Interfaces, New Orleans (2000)
11. Dey, A.: Understanding and using context. Personal and Ubiquitous Computing 5(1), 4–7 (2001)
12. Lemlouma, T., Layaida, N.: Context-aware Adaptation for Mobile Devises. In: IEEE Int. Conf. on Mobile Data Management, pp. 106–111. IEEE Computer Society, Los Alamitos (2004)
13. Burell, J., Gray, G.K., Kubo, K., Farina, N.: Context-aware computing: a text case. In: Borriello, G., Holmquist, L.E. (eds.) UbiComp 2002. LNCS, vol. 2498, pp. 34–38. Springer, Heidelberg (2002)
14. Kirsch-Pinheiro, M., Gensel, J., Martin, H.: Representing Context for an Adaptative Awareness Mechanism. In: de Vreede, G.-J., Guerrero, L.A., Marín Raventós, G. (eds.) CRIWG 2004. LNCS, vol. 3198, pp. 339–348. Springer, Heidelberg (2004)
15. Chaari, T., Ejigu, D., Laforest, F., Scuturici, V.M.: A Comprehensive Approach to model and use Context for adapting Applications in Pervasive Environments. Int. Journal of Systems and Software (2007)

Learning Implicit User Interests Using Ontology and Search History for Personalization

Mariam Daoud[1], Lynda Tamine[1], Mohand Boughanem[1], and Bilal Chebaro[2]

[1] IRIT team SIG-RI, IRIT, Toulouse, France
{daoud,tamine,bougha}@irit.fr
[2] Lebanese university, Faculty of Sciences, Beirut, Lebanon
bchebaro@ul.edu.lb

Abstract. The key for providing a robust context for personalized information retrieval is to build a library which gathers the long term and the short term user's interests and then using it in the retrieval process in order to deliver results that better meet the user's information needs. In this paper, we present an enhanced approach for learning a semantic representation of the underlying user's interests using the search history and a predefined ontology. The basic idea is to learn the user's interests by collecting evidence from his search history and represent them conceptually using the concept hierarchy of the ontology. We also involve a dynamic method which tracks changes of the short term user's interests using a correlation metric measure in order to learn and maintain the user's interests.

Keywords: user's interests, search history, concept hierarchy, personalized information retrieval.

1 Introduction

The explosion of the information available on the Internet and its heterogeneity present a challenge for keyword based search technologies to find useful information for users [2][11]. These technologies have a deterministic behavior in the sense that they return the same set of documents for all the users submitting the same query at a certain time. On the other hand, the effectiveness of these technologies is decreased by the ambiguity of the user's query, the wide spectrum of users and the diversity of their information needs. Recent studies [2] show that the main reason is that they do not take into account the user context in the retrieval process.

The development of relevance feedback [15] and word sense disambiguation techniques [16] aim to assist the user in the formulation of a targeted query, and have shown an improvement of the information retrieval (IR) performance. Effectively, relevance feedback techniques require that a user explicitly provides feedback information, such as marking a subset of retrieved documents as relevant documents. On the other hand, the word sense disambiguation techniques use generally an ontology-based clarification interface and require that the user specify explicitly the information need. However, since these techniques force

M. Weske, M.-S. Hacid, C. Godart (Eds.): WISE 2007 Workshops, LNCS 4832, pp. 325–336, 2007.
© Springer-Verlag Berlin Heidelberg 2007

the user to provide additional activities, a user may be reluctant to provide such feedback and the effectiveness of these techniques may be limited in real world applications [5].

The above situation gave rise to contextual IR which aims to personalize the IR process by integrating the user context into the retrieval process in order to return personalized results. In [1] contextual IR is defined as follows: *Combine search technologies and knowledge about query and user context into a single framework in order to provide the most appropriate answer for a user's information needs.*

It is common knowledge that several forms of context exist in the area of contextual IR. The cognitive context reflects the user's domains of interest and preferences about the quality of the results returned by the system such as freshness, credibility of the source of the information, etc. We cite also the physical context which reflects constraints on the materials and the user geographical locality, etc. Personalized IR is done when additional information about the user context defined previously is integrated into the IR process so as contextual IR takes place. We present some works within the scope of the personalized IR where the context is modeled as being a user profile representing the user's interests. These works explored various techniques to build the user profile using implicit feedback techniques [10][16][18][3].

In order to endow personalized IR systems with the capability to focus their knowledge on the user's domains of interests, we extend in this paper a related work [17] on building and learning the user's interests across past search sessions in order to enhance the keyword representation of the user's interests to a semantic representation one using a concept hierarchy. We use in our approach both of the search history and the concept hierarchy to learn and maintain the long term user's interests at the time the user conducts a search. We also involve a method which tracks the changes in the short term user's interests.

The paper is organized as follows: Section 2 reviews previous works on personalized IR that learn and maintain the user's interests. Section 3 presents our extended approach of representing and maintaining the user's interests during search sessions. Finally, some conclusions and future works are given in section 4.

2   Related Work

Traditional retrieval models and system design are based solely on the query and the document collection which leads to providing the same set of results for different users when the same query is submitted. The limitation of such systems is that the retrieval decision is made out of the search context while the IR takes place in context. Effectively, the IR process depends on time, place, history of interaction, task in hand, and a range of other factors that are not given explicitly but are implicit in the interaction and the ambient environment, namely *the context* [4]. The definition of context in IR is widely abused. While the wireless networks provide IR possibilities that the users are embedded in a physical environment, a physical context has to be considered in the IR models so as the contextual IR takes place. Personalized IR aims to enhance the retrieval process by integrating the user context or the user profile into the IR process. Works in this area have explored several techniques to build and maintain the user profile using implicit feedback techniques. User's interests are often represented by keyword vectors [7] [17], concept vectors [10] or a concept hierarchy [6][9].

A representation of the user's interests as a concept hierarchy is explained in [6]. An implicit user interest hierarchy (UIH) is learned from a set of web pages visited by the user. A clustering algorithm is applied to group words of the documents into a hierarchy where the high level nodes reflect a more general interest and the leaf nodes are considered more specific and reflect the short term interests.

Webpersonae [10] is a personalized web browsing system based on a user profile that reflects multiple domains of interest. Each one is represented by a cluster of weighted terms. These domains of interest are built by clustering the web pages visited by the user. The system involves the recognition of the current domain of interest used to rerank the search results by comparing the vector representation of recent pages consulted by the user to each of the long term domains of interest.

Recent works exploit ontology-based contextual information to get a semantic representation of the user's interest. Ontology is a concept hierarchy organized with "'is-a"' relationships between them. Many efforts are underway to construct domain specific ontologies that can be used by web content providers. Effectively, the information overload on the Web increases the attempts to provide conceptual search where the semantic web takes place. This research area implies the use of the knowledge representation language [19][14] in order to specify the meaning of the web content according to a concept taxonomy.

ARCH [16] is a personalized IR system that uses both of the user profile which contains several topics of interest and the yahoo concept hierarchy to enhance the user query. The system represents the long term user context as a set of pairs by encapsulating the selected concepts and the deselected concepts that are relevant to the user's information need across search sessions. The short term context is the pair of the selected and the deselected concepts in the current search session. When a long term user context exceeds a similarity threshold with the short term context, the system updates it by combining it with the short term context.

Moreover, Vallet et al. [18] exploit a semantic representation of the user interest based on weighted concept vectors derived from ontology. They build a dynamic semantic representation of the current context which reflects the ongoing user's retrieval tasks and use it to activate a long term user preference or a user's topic of interest. The current context is updated dynamically by using the user's query and feedback information. The personalization is achieved by re-ranking the search results where the original score is combined with the score yielded by the similarity between the current context and the document.

Challam et al.[3] build a short term user contextual profile as a weighted ontology and use the ODP as reference ontology [13]. The ODP is a Web directory where its purpose is to list and categorize web sites. In this study, the weight of the concept reflects the degree to which it represents the current user's activities. This weight is computed using a classifier to classify a web page into a concept of the ontology. The classification consists on a similarity measure between web page's vector visited by the user and each concept vector representation of the ontology. Thus, the concept's weight is the accumulated weights of all the pages that are classified into the concept and summed with the weights of all children's concepts weights. This user profile is used to re-rank the search results by combining the original rank of the document and the conceptual rank computed using a similarity between the document and the user contextual profile.

This paper presents a new technique for building and learning the user's interests across past search sessions. We exploit in our approach both of the search history and the ODP ontology to learn the long term user's interests at the time the user conducts a search.

Comparatively to previous work in the same area, our approach has the following features:

– A semantic representation of the user context as being a weighted portion of a global ontology with taking into account the short term and the long term user's interests.
– A robust method to detect dynamically related and unrelated user's interests using a statistical rank-order correlation operator between the semantic representation of the user's contexts.

3   Building and Maintaining a Semantic Representation of the User Interest

Our main goal is to learn and maintain implicitly the long term user's interests through the passive observation of his behavior. We exploit a cognitive context for our retrieval model where the user's interests are represented semantically. We extend the keyword representation of the user's interests to get an enhanced one using the ODP ontology. In the remainder of this paper we use the term user context as being a vector reflecting the user interest at a certain time.

3.1   Building the User Interest

Our method runs in two main steps that are presented in the subsections that follow:

– The first one consists on building the user's interests using an intermediate representation of the user context which is a keyword-based representation in order to get a concept-based representation using the ODP ontology.
– The second step consists on learning and maintaining the user's interests. The learning algorithm is based on a correlation measure used to estimate the level of changes in the semantic representation of the user context during a period of time.

The Term-Based Representation of the User Interest: An Overview.
We present in this section an overview of the term-based building process of
the user's interests developed in a previous work [17]. The user is modeled by
two related components: an aggregative representation of the user search history
and a library of user contexts reflecting his interests when seeking information.
More precisely, our approach uses the evidence collected across successive search
sessions in order to track potential changes in the user's interests. At time s,
the user is modeled by $U = (H^s, I^s)$ where H^s and I^s represent respectively the
search history and a set of user's interests at time s. A matrix representation is
used to represent the search history which is the aggregation of the search session
matrix. Let q^s be the query submitted by a specific user U at the retrieval session
performed at time s. We assume that a document retrieved by the search engine
with respect to q^s is relevant if it is explicitly judged relevant by the user or
else, some implicit measures of the user interest such as page dwell time, click
through and user activities like saving, printing etc, can be applied to assume
the relevancy of a document. Let D^s be the related set of assumed relevant
documents for the search session S^s, $R_u^s = \cup_{i=s_0..s} D^i$ represents the potential
space search of the user across the past search sessions. We use matrices to
represent both user search session and search history. The construction of the
search session matrix, described below, is based on the user's search record and
some features inferred from the user's relevancy point of view. The user search
session is represented by a Document-Term matrix S^s: $D^s * T^s$ where T^s is the set
of terms indexing D^s (T^s is a part of all the representative terms of the previous
relevant documents, denoted $T(R_u^s)$). Each row in the matrix S^s represents a
document $d \in D^s$, each column represents a term $t \in T^s$. In order to improve
the accuracy of document-term representation, the approach introduces in the
weighting scheme a factor that reflects the user's interests for specific terms. For
this purpose, it uses term dependencies as association rules checked among T^s
[8] in order to compute the user term relevance value of term t in document d
at time s denoted $RTV^s(t, d)$:

$$RTV^s(t, d) = \frac{w_{td}}{dl} * \sum_{t' \neq t, t' \in D^s} cooc(t, t') \tag{1}$$

w_{td} is the common Tf-Idf weight of the term t in the document d, dl is the
length of the document d, $cooc(t, t')$ is the confidence value of the rule $(t \rightarrow t')$,
$ccoc(t, t') = \frac{n_{tt'}}{n_t * n_{t'}}$, $n_{tt'}$ is the number of documents among D^s containing t and
t', n_t is the number of documents among D^s containing t and n_t' is the number
of documents among D^s containing t'. $S^s(d, t)$ is then determined as:

$$S^s = RTV^s(t, d) \tag{2}$$

The user search history is a $R_u^s * T(R_u^s)$ matrix, denoted H^s, built dynamically
by reporting document information from the matrix S^s and using an aggregative
operator combining for each term, its basic term weight and relevance term value

computed across the past search sessions as described above. More precisely, the matrix H^s is built as follows:

$$H^0(d, t) = S^0(d, t)$$

$$H^{s+1}(d, t) = H^s \oplus S^{s+1} = \begin{cases} \alpha * w_{t,d} + \beta * S^{s+1}(d, t) \\ if \ t \notin T(R_u^{(s)}) \\ \alpha * H^s(d, t) + \beta * S^{s+1}(d, t) \\ if \ t \in T(R_u^{(s)}) \\ H^s(d, t) \ otherwise \end{cases} \tag{3}$$

$$(\alpha + \beta = 1), s > s_0$$

After the representation of the search history, a weighted keyword representation of the user context K^s is extracted and reflects the user's interests at learning time s. The term's weight reflects the degree to which the term represents the user context. It is computed by summing for each term in $T(R_u^s)$ the columns in H^s as follows:

$$c^s(t) = \sum_{d \in R_u^s} H^s(d, t) \tag{4}$$

$K^s(t)$ is normalized as follows: $c^s(t) = \frac{K^s(t)}{\sum_{t \in T^s} K^s(t)}$. This original approach models the user context as a set of weighted keywords reflecting the user's interests in a search session. Given a user being interested in the military domain for a given search session, then we find terms of the military domain in the top of the term-based representation of the user context. The maintaining process of the user's interests between search sessions is accomplished using a rank order correlation measure applied on two consecutive keyword contexts of retrieval session. Change to a different domain interest contribute to add or change the keywords of the user context. This approach is faceted to a risk error due to the lack of not taking into account the semantic relation between words, so as the changes of interests depend on a distinctive difference in the rank order distribution of the keyword-based context representation, independently of their belonging to the same user's information need. We aim to enhance the keyword representation of the user interest to a semantic representation one that outcomes the limit of tracking changes in the user's contexts. Related and unrelated user's contexts are detected using the same measure but applied on a semantic representation of them. Effectively, enriching the keyword representation of the user's interests with concepts from the core ontology has two benefits: first, instead of the keyword representation, it provides a semantic meaning of the user's interests. Second, tracking the changes of the short term user interest is more reliable and accurate when they are represented semantically.

The Concept-Based Representation of the User Interest Using Ontology. We present in this section the method for a concept-based representation of the user interest in a semantic context to be stored in I^s. To get the semantic representation of the user context, we map the keyword vector representation of the user context described in the previous section on the concepts of the ODP ontology [13], thus we obtain a weighted concept hierarchy which is the semantic representation of the user context at a certain time.

– *Reference ontology and representation of domain knowledge* There are many subject hierarchies created manually and designed to organize web content for easy browsing by end users. We cite the online portals such as yahoo [1], Magellan[2], Lycos [3] and the open directory project [13]. Considering that the Open Directory Project (ODP) is the most widely distributed data base of Web content classified by humans, we use it in our profiling component as a fundamental source of a semantic knowledge to represent semantically the user's interests. We show in Fig.1. the concept hierarchy of the ODP ontology. Various methods can be utilized to represent the concept vector of the ODP ontology. In our approach, we use a term-vector based representation for the concepts developed in [3]. We are interested by the top three levels of the ontology to represent a set of general user's interests. Each concept of the hierarchy is associated to a set of related web pages. These documents are used to represent the term vector representation of the concept. The content of the pages associated to the concept j are merged together to create a super-document sd_j to obtain a collection of super-documents, one per concept, that are pre-processed to remove stop words and stemmed using the porter stemmer to remove common suffixes. Thus, each concept is treated

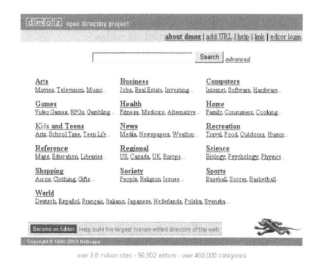

Fig. 1. The concepts in the ODP ontology

---

[1] http://www.yahoo.com
[2] http://www.mckinley.com
[3] http://point.lycos.com/categories/index.html

as a n-dimensional vector in which n represents the number of unique terms in the vocabulary. Each term's weight in the concept's vector is computed using the $tf * idf$ weighting scheme and normalized by their length. The term weight of the term i in concept j is computed as follows:

$$w_{ij} = tf_{ij} * idf_i \qquad (5)$$

Where
tf_{ij}=number of occurrences of t_i in sd_j
N=the number of super-documents in the collection
n_i=the number of super-documents containing t_i

– *Semantic representation of the user interest* The basic idea to get a semantic representation of the user context is to map the keyword representation of the user context on the concept hierarchy. Concepts and user context are represented in the vector space model as explained in the previous section. Thus the mapping consists on a cosine similarity between vectors and has as output a weighted concept vector which represents the semantic representation of the user context. The semantic vector c^s represents the short term user context at learning time s and includes his short term interests. The dimension of the semantic vector c^s at learning time s is equal to the dimension of the top three levels of the ODP's domain ontology θ. The weight of a concept in the ontology reflects the degree to which it represents the user's short term interest and beliefs at time learning s. Let K^s be the keyword representation of the user context computed as explained in the previous section and V_j the term vector representation of a concept j from the ontology. The concept's weight is then computed as follows:

$$P_j = cos(V_j, K^s) \qquad (6)$$

We then order the concept vector by decreasing weight where concepts with high weights reflect the short term user interest.

These semantic user's interests are then reused in the various phases of the personalized information access. Thus, the library of the user's interests can be used for the:

- Query reformulation
- Query to document matching
- Re-ranking the search results

As example, given a user being interested by the field of computers in a certain search session. *Computers* is categorized at the first level of the ODP's concept hierarchy. We take into account the top three levels of the concept hierarchy, we suppose a more specific user interest in the search session to be the *software* which is the subcategory of *computers* and *malicious software* which is the subcategory of *software* category as shown is Fig.2.

We assume that the keyword representation of the user context include terms related to the specific domain of interest cited above, and are extracted from the user search history. Thus, certainly the keyword-based user context, mapped to the nodes of the concept hierarchy exceeds a similarity threshold with the vector representation of the category *computers* and its subcategory *software*, especially the *malicious software*. In this way, we identify the best matching concepts for the keyword user context. The semantic vector representation of the user context is then generated on the basis of concept's weights and the categories cited above, having the high weights, are ordered in the top of the vector-based semantic user context.

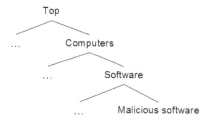

Fig. 2. A personal user's interests in a search session

3.2   Maintaining the User's Interests

As long as the user conducts a search, the system must tracks the changes in the short term user context. A new user context means a new user interest must be added to the library of the user's interests or a long term user interest may be reviewed. The key for providing an accurate profiling of the user's interests is to determine at what degree we shall update an existing user interest and when we shall add a new user interest to the library of the user's contexts. In our approach, we compare the current semantic context at time s noted cc^s and the previous one pc^s using Kendall rank-order correlation operator as showed in Fig.3. The Kendall rank correlation coefficient evaluates the degree of similarity between two sets of ranks given to a same set of objects. In our case the objects are the concepts of the ontology representing the user's interests. A change of domain of interest between search sessions contribute to change or to add keywords in the search history matrix, then to change the keyword based representation of the user context, we then conclude a significant change of the rank order of concepts being in the top of the concept-based representation of the current user's context. The Kendall rank correlation coefficient is given in the following formula:

$$\Delta I = (cc^s \circ pc^s) = \sum_{o \in \theta} (cc^s(o) - pc^s(o)) \tag{7}$$

where θ is the set of the top three levels of the ODP's concept hierarchy. The coefficient value ΔI is in the range [-1 1], where a value closer to -1 means that the semantic contexts are not similar and a value closer to 1 means that the semantic contexts are very related to each other. Based on this coefficient value, we apply the following strategy in order to learn the user's interests and so update the set of user interests in I^s:

1. $\Delta I > \sigma$ (σ represents a threshold correlation value). No potential changes in the user's contexts, no information available to update I^s;
2. $\Delta I < \sigma$. There is a change in the user's contexts. In this case we gauge the level of change, and two configurations may be presented: the change implies a refinement of a prior detected user's interest or else the occurrence of a novel one. In order to answer this question we do as follows:
   - select $c^* = argmax_{c \in I^s}(c \circ cc^s)$,
   - if $cc^s \circ c^* > \sigma$ then
     - refine the user's interest c^*: we define a refinement formula that combine the newly constructed semantic vector with the user interest c^* where the concepts weights computed in c^* are automatically reduced by a decay factor ζ, a real value in $[0, 1]$. The refinement of c^* is given as follows:
       $c^* = \zeta * c^* + (1 - \zeta) * cc$;
     - update the matrix H^s by dropping the rows representing the least recently documents updated, update consequently R_u^s,
   - if $cc^s \circ c^* < \sigma$ then add the new tracked interest in the library I^s, try to learn c^* a period of time by reinitializing the search history matrix to be equal to the current search session matrix as follows:
     set $H^{s+1} = S^s$, $s_0 = s$

An updating procedure of the library of the user's interests consists on managing their persistence. Indeed, it consists on removing some user's interests according to the updating frequency or the date of the last update. By this fact, we exclude the non recurrent contexts inserted in the library of the user's interests.

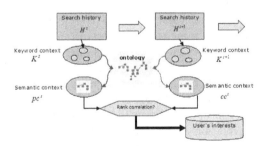

Fig. 3. Learning and maintaining process of the user's interests

4   Conclusion and Future Works

We proposed in this paper a new approach for building an ontology-based user's interests in the field of personalized IR. We improved a previous work for user modeling where the user interest consists on a keyword-based representation. In order to enhance the original approach, we exploit both the search history and a predefined ontology ODP to represent the user's interests conceptually. The basic idea consists on mapping the keyword representation of the user context to the concept hierarchy, and then each concept has a weight that reflects the degree to which it represents the user context at a certain time. The approach integrates the temporal dimension in the user's interests learning process. More precisely, we learn and maintain long term user's interests by updating the search history representation using the user relevancy point of view on familiar words from which we extract a short term user's interest.

A distinctive aspect in our approach is the use of the kendall rank order correlation measure between semantic representations of the user's interests. The benefit of using this measure is the gain of accuracy and reliability in tracking changes of the short term user's interests instead of applying it on a keyword-based representation of the user interests.

In future work, we plan to improve the maintaining process of the short term user's interests; Instead of tracking the changes of the user context by the user's queries submitted among search sessions, we aim to integrate a method for detecting session boundaries in order to activate the method for updating the library of the user's contexts. We define a session as a set of queries related to the same information need. We also tend to personalize the information retrieval process by using the user's interests in the query reformulation. We use the term of the concept representing the short term user interest in order to enhance the user query and personalize the search results to better meet the user's information needs in a search session.

In another hand, we plan to evaluate our approach experimentally using a large scale of quantitative data on the user search sessions and accurate contexts provided by the related queries during a reasonable period of testing a particular search engine.

References

1. Allan, J., et al.: Challenges in information retrieval and langage modelling. In: Workshop held at the center for intelligent information retrieval, Septembre (2002)
2. Budzik, J., Hammond, K.J.: Users interactions with everyday applications as context for just-in-time information access. In: Proceedings of the 5th international conference on intelligent user interfaces, pp. 41–51 (2000)
3. Challam, V., Gauch, S., Chandramouli, A.: Contextual Search Using Ontology-Based User Profiles. In: Proceedings of RIAO 2007, Pittsburgh USA 30 may - 1 june (2007)
4. Ingwersen, P., Jarvelin, K.: Information Retrieval in Context – IRiX. In: ACM SIGIR forum (2005)
5. Kelly, D., Teevan, J.: mplicit feedback for inferring user preference: A bibliography. In: SIGIR Forum (2003)
6. Kim, H.R., Chan, P.K.: Learning implicit user interest hierarchy for context in personalization. In: Proceedings of the 8th international Conference on intelligent User interfaces IUI 2003, Miami Florida USA January 12 - 15 (2003)

7. Lieberman, H.: Letizia: An agent thatassists web browsing. In: Proceedings of the International Joint Conference on Artificial Intelligence (IJCAI 2005), Montreal pp. 924–929 (August 1995)

8. Lin, S.H., Shih, C.S., Chen, M.C., Ho, J., Ko, M., Huang, Y.M.: Extracting classification knowledge of Internet documents with mining term-associations: A semantic approach. In: the 21th International SIGIR Conference on Research end Development in Information Retrieva (1998)

9. Liu, F., Yu, C., Meng, W.: Personalized Web Search For Improving Retrieval Effectiveness. IEEE Transactions on Knowledge and Data Engineering 16(1), 28–40 (2004)

10. Mc Gowan, J.P.: A multiple model approach to personalised information access. In: Master Thesis in computer science, Faculty of science, University College Dublin (February 2003)

11. Nunberg, G.: As Google goes, so goes the nation. New York times (2003)

12. Pazzani, M., Billsus, D.: Learning and revising user profiles. The identification of interesting Web sites, Machine learning 27, 313–331 (1997)

13. The Open Directory Project (ODP), http://www.dmoz.org

14. Fensel, D., Harmelen, F.V., Horrocks, I.D.: OIL: An Ontology Infrastructure for the Semantic Web. IEEE Intelligent Systems 16(2) (2001)

15. Rocchio, J.: Relevance feedback in information retrieval. In: Salton, G. (ed.) The SMART retrieval system - experiments in automated document processing, Prentice-Hall, Englewood Cliffs, NJ (1971)

16. Sieg, A., Mobasher, B., Burke, R., Prabu, G., Lytinen, S.: representing user information context with ontologies

17. Tamine, L., Boughanem, M., Zemirli, W.N.: Inferring the user's interests using the search history. In: Workshop on information retrieval, Learning, Knowledge and Adaptability (LWA 2006), ildesheim Germany November 9-11, pp. 108–110 (2006)

18. Vallet, D., Fernández, M., Castells, P., Mylonas, P., Avrithis, Y.: Personalized Information Retrieval in Context. In: Vallet, D. (ed.) 3rd International Workshop on Modeling and Retrieval of Context, Boston USA, pp. 16–17 (2006)

19. Lassila, O., Swick, R.: Resource Description Framework (RDF) Model and Syntax Specification. World Wide Web Consortium recommendation (February 22, 1999)

Contextual User Profile for Adapting Information in Nomadic Environments

Angela Carrillo-Ramos, Marlène Villanova-Oliver, Jérôme Gensel,
and Hervé Martin

LIG Laboratory, STEAMER Team
681 rue de la Passerelle, 38402 Saint Martin D'Hères, France
{carrillo,villanov,gensel,martin}@imag.fr

Abstract. In order to personalize information for a nomadic user, it is necessary to consider on the one hand, the *context of use* (which provides in particular a description of the conditions temporal, spatial, hardware, *etc.* under which users accesses the *Information Systems*), and on the other hand, *user preferences* which aim at expressing what the user would like to obtain from the system considering different aspects such as functionalities, content, display, *etc.*). We propose an approach generating a *Contextual User Profile* (*CUP*), profile which is made up only of the *user preferences* selected considering the current *context of use*. This approach proposes formalism for three different types of *user preferences* (*activity*, *result* and *display*). Additionally, this approach defines the *Contextual Matching Algorithm* which generates the *CUP* based on *user preferences* and on the *context of use*.

Keywords: preferences, adaptation, user profile, mobile device, context of use.

1 Introduction

The *personalization* of an information access process aims at adapting and delivering users information taking into consideration their profile, more precisely, the preferences included in their profile. The adaptation issue of applications executing on *Mobile Devices* (*MD*) can be considered from different points of view. One of them consists in defining *what* an application has to be adapted. In order to give a few examples, an application can be adapted considering the user's personal characteristics, preferences, culture, history in the system, current location, *etc.* and/or to the characteristics of the access device and network. These different criteria are generally (and sometimes in different ways) grouped together to build so-called *user profiles* and/or *context models* [4] [15]; both constitute the basis for adaptation of content, layout, *etc.* and to the user and/or *MD*.

When considering nomadic users accessing a *Web Information Systems* (*WIS*) through a *MD*, one of the challenges of the adaptation is that the value of some previously mentioned criteria, aggregated in a *context of use*, can evolve during a session [15]. According to Tamine *et al.* [15], the *context of use* is a set of both, elements such as location, connection time and current application and, the goals and intentions of the user during a session of information searching. Besides, an

M. Weske, M.-S. Hacid, C. Godart (Eds.): WISE 2007 Workshops, LNCS 4832, pp. 337–349, 2007.
© Springer-Verlag Berlin Heidelberg 2007

adaptation process can exploit *user preferences*, which can be expressed for a given session or for all sessions. *WIS* designers should provide users with tools that empower them to express their *preferences*: *i)* to choose and classify information that they want to obtain from the *WIS*; *ii)* to specify what they want to achieve on the *WIS* (*i.e.*, *activities* of a user in the system such as consultation, insertion, deletion or modification of data); *iii)* to visualize the system in the way in which they want information to be displayed on their *MD*. These tools would allow the system to know *user preferences* in order to adapt information presented to users. We define a "*user preference*" as a set of descriptions including: the *activities* that a user plans to achieve in the system (*e.g.*, consultation and data management) and the way in which this is achieved (*e.g.*, sequential, concurrent, conditional), the type and order of results of these *activities* (that we call "*content*"), and the way in which the user wants information to be displayed on the *MD* (specification of the expected format –image, video, text – with its characteristics). In this paper, we address the personalisation (based on *user preferences*) of *WIS* acceded using *MD*. Our work contributes in three senses: The *first contribution* consists in formalizing the notion of *user preference*. This formalism provides a support for the representation of three types of *user preferences*: *activity, result* and *display preferences* (see section 3).

The *second contribution* is a *Contextual Matching Algorithm* (*CMA*) which uses *user preferences* and the *context of use*[1] to define a *Contextual User Profile* (*CUP*) for a given user, during a given session. This algorithm examines a *user profile*[2] and analyzes each *user preference* to evaluate if this preference can be satisfied according to the *context of use*. This *context* allows the system to select only the *preferences* which are compatible with *user activities*. For example, the result of selection can be composed of *display preferences* which can be considered in function of characteristics of the *MD*. The *user preferences* retained are components of the *CUP*[3].

The *third contribution* of this work concerns conflict management which is not considered in the current version of the *CMA*. We have identified certain *conflict-types*, their causes and the way in which the system must react in order to solve them, or to inform the user about the existence of these conflicts.

This paper is structured in the following way. Firstly, we position our work according to our global objective: adaptation according to user and to *context of use* in an *IS*. We present in section 3, the concepts of *user profile* and of *user preference*, describing, for the latter, the classification that we have established. In section 4, we show how the *CUP* which is generated with the help of the *CMA*, considering the *user profile* and the current *context of use*. We then discuss the resolution of conflicts which can appear between *user preferences* belonging to the *CUP*. We present then in section 5 related works of our proposition before concluding in section 6.

---

[1] By "*context of use*", we refer to a set of data which allows characterization of the interaction between the user and the system. A *context of use* is composed of a representation of several elements such as the user activities, used device characteristics, and location or moment in which the user is connected. In our work, the *context of use* model is composed of information about user location, *MD* features, user access rights and user activities.

[2] The expression "*user profile*" has to be understood as indicating a set of *user preferences*. We are not interested in other elements which can enter in the constitution of a profile.

[3] The *contextual user profile* only covers *user preferences*.

2 Overview of Our Proposition

In this paper, we present the necessary elements to implement a *Contextual Profile Management System* (*CPMS*) (see Fig. 1). The *CPMS* is a component (integrable in the architecture of an *IS* or demanded by the *IS* as an external component[4]) which operates like a component dedicated to implement the adaptation in the *IS*. The role of the *CPMS* is providing the elements that allow proceeding to the adaptation (considering the user characteristics and his *context of use*) of the information managed by the *IS*. The *CPMS* takes as data input a *context of use* and a *user profile* (the latter being composed of the set of *user preferences*). Then, it proceeds to the *Preference Filter,* considering the *context of use,* by means of the application of a *Contextual Matching Algorithm* (see section 4.2). Only the preferences of the *user profile* which can be applied, given the *context of use,* are returned to constitute the *CUP*. This *CUP*, artefact produced by the *CPMS*, constitutes a realistic representation of those *user preferences* to which the information must be adapted, *i.e. i)* to the user considering the *context of use* and *ii)* to this same context. The realistic character of this representation reflects the fact that this *profile* is *contextual, i.e.* reconstructed for each session of the user based on the most relevant *preferences,* selected according to the current *context*. The notion of *user profile* is widely exploited as input data of the adaptation process. We see the contextualization of this *user profile* as means to refine this process, rendering it dynamic and evolutionary: the adaptation follows, indeed, the evolution of the context (*e.g.,* a change of the access *MD*).

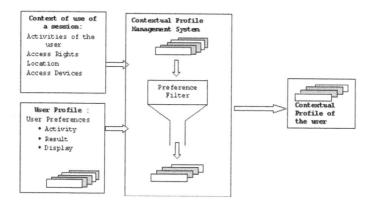

Fig. 1. Contextual Profile Management System (*CPMS*)

From the point of view of the adaptation to the user, we concentrate in this work on a *CUP* made up of *user preferences, i.e.,* expressions translating the desires of the user. We define *user preference* standing on user activities, or still on the nature and form of the delivered content. We focus on preferences expressed by the user. We do not present in detail the way in which information relative to the *user preferences* is

---

[4] The outsourcing of the *CPMS* can also allow the sharing of profiles between applications.

acquired. Considering the profile acquisition techniques presented in Kassab *et al.* [11], the *CPMS* is developed using information: *i)* provided by the user by means of dedicated interfaces; *ii)* defined like general user profiles; *iii)* deduced from the user's history, *i.e.*, from his previous sessions [3] [15].

Several works consider the *context of use* in order to personalize information [3] [15] [11] [4]. In our approach, we use a model of *context of use* composed essentially of four types of information: descriptions of activities that a user wants to execute during a session, access rights of a user to data in the *IS* which collaborates with the *CPMS*, user location (for example, *GPS* data) and characteristics of the access device used for connecting during the session (characteristics defined using *CC/PP* [10]). In this paper, we do not present in detail the model of *context of use* taken partially from Kirsch-Pinheiro [12]; a global level description of this model is sufficient for the understanding of the approach that we support.

In the following section, we present the notion of *user preference* and the different ways in which it is used in our approach to constitute a *CUP*.

3   User Preferences: Definition and Types

We propose three types of *user preferences*: **Activity preferences** concerning the activities that a user wants and can be achieved in the system. From the point of view of the system, an *activity* is consisted by a set of *functionalities*. We focus here on the *functionalities of consultation* which allow consulting information by using queries associated to these functionalities and, in a certain measure rend them operational. **Result preferences** concern the *content*: a user can choose results delivered to the user (among those obtained after the execution of the functionalities) and determine their order of presentation. **Display preferences** concern the way in which the user wants the information to be displayed on his *MD*. This includes on the one hand appearance, style, type of characters, *etc.*, and, on the other hand, characteristics of the display formats (*i.e.*, characteristics of the video, images, or sound).

Inside each type of *user preference*, we distinguish *general preferences* (which are applied for all sessions; this is the default value) and *specific preferences* (which apply to the current session). A *session* starts when a user connects to the system and executes one or several provided *functionalities*. We designate as F a *functionality* of name *fname* defined as a tuple composed of a list of input parameters ($<i_1, i_2 \ldots i_n>$) and a list of output parameters ($<o_1, o_2 \ldots o_k>$) :

$$F = fname(<i_1, i_2 \ldots i_n>, <o_1, o_2 \ldots o_k>)$$

In order to illustrate the notion of *activity*, let us consider a user who wants to receive on his cellular phone the weather predictions to illustrate our proposal. The *user profile* could, for example, specify when he wants to execute the *activity* "*Consultation of weather predictions*"; this user is only interested on a *content* composed of the predictions concerning the city where he is, for the current day, and with the temperature data in Fahrenheit degrees. His profile could indicate that he wants that the *display* to be realized as images rather than as text. Moreover, our proposal enables, in function of the results returned by an *activity*, defining that another *activity* must be committed. Therefore, if the weather predictions announce rain (*content* analysis), then the user wants to execute the *activity* "*Consult the cinema schedule*" which will be automatically committed and parameterized in function of the data of his profile (concerning location, hour, cinemas, genre of films, etc.).

A formalism of each one of the three types of *user preference* (*activity, result* and *display*) is presented below.

The *Activity preferences* describe the way in which a user aims at achieving his activities in the system. We define this type of *user preference* in the following way:

Activity_Preference (type, criteria, A)

Where **type** takes as values "*general*" or "*specific*". **criteria** is a set of adaptation criteria (*e.g.*, the location and the type of *MD*) considered for the execution of functionalities in a session. **A** is the *activity* that the user wants to execute in the system. This *activity* is expressed using a *string of functionalities* executed in a sequential, concurrent or conditional way. This string is expressed using a grammar defined in *BNF* ("*Backus Naur Form*") notation as follows:

```
Activity::= functionality [op functionality] |conditional | loop | nil ;
op::= sequence|concurrent ;
sequence ::= ";";
concurrent ::= "|" ;
conditional ::= "if" <condition> "then" Activity ["else" Activity] "end if" ;
loop ::= "while" <condition> "do" Activity "end while" ;
```

An *activity preference* for F_i, "*general*" and without other adaptation criterion is defined by:

Activity_Preference (General, (), F_i)

All functionality F_i can be associated to a *result preference* as follows:

Activity_Preference (General, (), (F_i, ResPrefF$_i$))

Where ResPrefFi is a ***Result preference*** which allows a user to choose and order the delivered contents that are result of the execution of a functionality.

We define a ***Result Preference*** as follows:

ResPrefF$_i$= Result_Preference (type, F_i, <(o_1,DisplayP$_1$),(o_2, DisplayP$_2$)...(o_k, DisplayP$_k$)>)

Where F_i is the name of the functionality and the last term is an ordered list of pairs (o_i, DisplayP$_i$) where o_i represents a result of F_i. The order of the list expresses the presentation order of the results. Finally, DisplayP$_i$ is the *display preference* applied to this result. A user can also choose, among the results delivered by the functionality, those which the user wants to obtain (*e.g.* only o_3, o_5, o_8). In this case, the *result preference* is defined as follows:

Result_Preference (type, F, <(o_3, DisplayP$_1$),(o_5,nil),(o_8, DisplayP$_3$)>)

Where *nil* means that the user did not define preferences for displaying this result. When a doctor is "*Consulting the medical tests*", he only wants to receive the results concerning the glycaemia curve (as images with the characteristics of the *display preference DisplayP_image$_1$*) and the *medical tests* associated with the *thyroid* (defined by text *T3*, *T4*, respectively associated with the *display preferences* `DisplayP_text₁` and `DisplayP_text₂`). This preference is defined as follows:

```
Result_Preference(General, "Consulting medical tests" <(glycaemia_
    curve,DisplayP_image₁),(T3,DisplayP_text₁),(T4,DisplayP_text₂)>)
```

The **Display preferences** describe the way in which the user wishes his *MD* to display the information (*e.g.*, in image format). These *user preferences* are defined for the tuples as follows:

```
Display_Preference (format, {characteristics}, substitution)
```

In these tuples, the **format** can take as values: *"video"*, *"text"*, *"image"* and *"audio"* and **characteristics** specify the values taken by the attributes which characterize the format. The term *substitution* corresponds to another *display preference* that the system will try to use instead of that defined if it cannot be satisfied (*substitution* can take as value *nil*). The *Display preference* P₁ corresponds to a preference for the display of a video, giving its dimensions and the file type (*e.g.*, width, height, type):

```
P₁ = Display_Preference (video, {200,300, AVI}, P₂)
```

Where P₂ is the *substitution preference* of P₁ and contains the characteristics for the text (police, size, color, file type):

```
P₂ = Display_Preference (text, {Arial, 10, bleu, .doc}, nil)
```

We can note that the *display preferences* can be referenced by the result preferences. In this case, they constitute preferences that will be applied to a format independently of a particular content (*e.g.*, in order to privilege text format for a session).

The *UML Class Diagram* which represents the relations between the three types of *user preferences* is shown in Fig. 2:

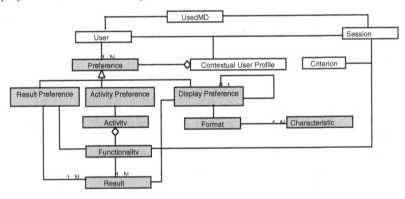

Fig. 2. UML Class Diagram of user preferences and their relations

4   The Contextual User Profile (*CUP*)

In this section, we show how the *CUP* is generated with help of the *Contextual Matching Algorithm*, using a *user profile* and the current *context of use*. In order to adapt information to *user characteristics* and those of his *MD*, we adopt a process with two steps (see Fig. 3):

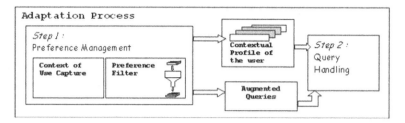

Fig. 3. Adaptation process

The *Preference Management* step (*step 1*) is achieved by the *Contextual Profile Management System* (*CPMS*). This step consists on the one hand, of the capture of the *context of use* of the session and on the other hand, of the selection of *user preferences* which can be applied considering the current session (see Fig. 3, *Preference Filter*). For example, for the location capture, it is possible to use a *GPS* device or methods such as the *SNMP* ("*Simple Network Management Protocol*") proposed in [14]. The result of *step 1* is a *CUP* (composed of the selected *user preferences*, see section 3) and one or several *queries* which are "*augmented*". An augmented query corresponds to the initial *query* associated to each consultation functionality (which is involved in the achievement of an *activity*, see section 3) to which the *CPMS* adds information about the *result* and *display preferences*. The *CUP* and the augmented queries are input parameters for the *Query Handling* step (step 2). This last step is not detailed here but it can be found in [6]. The following section details the *Preference Management* step, achieved by the *CPMS*.

4.1 Definition of the Contextual User Profile (*CUP*)

In our proposition, the *CUP* is composed of the *activity*, *result* and *display* preferences, which are applied considering the *context of use* of a session. For example, we only retain, from the *result preferences*, those which are not in contradiction with the user access rights. The *CUP* (u, s, *MD*) = {P_1, P_2, P_3...P_k} of a user "u", which connects through a device "*MD*" during a session "s" is the set of k *preferences* which are retained considering the *context of use*. We can note that the *CUP* is built by means of the analysis of the set P_u = {P_1, P_2,...,P_i, ..., P_n} constituted of *all* preferences defined by the user u, which include preferences defined independently of the *context of use* ("*general*" preferences) and those defined for the session in question ("*specific*" preferences). The analysis of the set P_u by the algorithm described in the following section relies on a preliminary phase of the organization of the *user preferences*. We do not describe here in a detailed way the priority system established for the organization of the preferences but we present its main principles: *i*) *activity preferences* are analyzed, then *result preferences* and finally *display preferences*. *ii*) In each type of *user preferences* (*activity*, *result*, *display*), *specific preferences* have the priority with regard to *general preferences*. If there are no *specific preferences* established, the system only considers the general ones. *iii*) If any *user preference* is defined, the system considers the *history of the user* in the *IS* and, in as last resort, the inherent constraints of the user's *MD* (in this case, the system builds itself the *preferences*).

4.2 The Contextual Matching Algorithm (*CMA*)

For each *preference P_i* of the complete set of *user preferences P*, we apply the following algorithm in order to verify if this P_i can be added to the *CUP* (see lines 2 to 4 of the algorithm): the system analyzes each P_i (for i ∈ [1, n]) and the set of *substitution preferences*[5] of P_i. This analysis finishes for each P_i if: P_i can be satisfied (see line 5) or, if a *substitution preference* of P_i can be satisfied (see lines 15 to 35) or finally if the *substitution preference* is *nil*. The analysis of the *substitution preference* of P_i starts if P_i can not be satisfied (see line 13). When the preference P_i (respectively, the *substitution preference* of P_i) can be satisfied, P_i (respectively, the *substitution preference* of P_i) is added into the *CUP* (see lines 7 and 15). The remainder of the *string of substitution preferences* is not analyzed (*i.e.*, they are added into the *Rejected Preference List*, see lines 11, 14, 24, 32).

In the algorithm bellow, we use the following abbreviations: **AP** is the *Analyzed Preference*, **RPL** is the *Rejected Preferences List*, **NAP** is the *New Analyzed Preference,* and **CUP** is the *Contextual User Profile*.

```
(1) i=1
(2) While (i <= n) do // For each Pi where i ∈ [1,n] :
(3)    AP = Pi     //Preference which is analyzed in this iteration
(4)    if (Pi∉CUP AND Pi∉RPL) then
(5)       if (Pi can be satisfied) then
(6)          Add Pi into CUP
(7)          AP = Substitution preference of Pi
(8)          While (AP <> nil) do
(9)             NAP = AP;
(10)            AP = Substitution preference of NAP
(11)            Add NAP into the RPL if and only if NAP∉CUP AND NAP∉RPL
(12)         end While
(13)      else
(14)         Add Pi into the RPL if and only if Pi∉CUP AND Pi∉RPL
(15)         AP = Substitution preference of Pi
(16)         While (AP <> nil) do
(17)            if (AP can be satisfied AND AP∉CUP AND AP∉RPL) then
(18)               NAP = AP;
(19)               AP = Substitution preference of NAP
(20)               Add NAP in CUP if and only if NAP∉CUP AND NAP∉RPL
(21)               While (AP <> nil) then
(22)                  NAP = AP;
(23)                  AP = Substitution preference of NAP
(24)                  Add NAP in RPL if and only if NAP∉CUP AND NAP∉RPL
(25)               end While
(26)            else
(27)               if (AP∈CUP OR AP∈RPL) then
(28)                  AP = Substitution preference of AP
(29)               else
(30)                  NAP = AP;
(31)                  AP = Substitution preference of NAP
(32)                  Add NAP in RPL if and only if NAP∉CUP AND NAP∉RPL
(33)               end if
(34)            end if
(35)         end While
(36)      end if
(37)   end if
(38)   Increment i. // i = i + 1
(39) End While
```

---

[5] The *substitution preferences* are only defined for the *display preferences*.

We use a *Rejected Preferences List (RPL)* to maintain *user preferences* which are not retained in the *CUP*, for two reasons. Firstly, some of these rejected preferences can still be useful. For example, let us suppose that a *user preference "a"* is the *substitution preference* of the preferences *"b"* and *"c"*. If *"b"* is satisfied, *"a"* will be added to the *RPL*. In the case where *"c"* can not be satisfied, *"a"* should be analyzed again in order to determine if it can be satisfied. Secondly, certain *user preferences* of the *RPL* can describe the characteristics of a format of media which must be displayed. For example, let us suppose that *"a"* is the *substitution preference* of *"b"*. *"a"* specifies the characteristics associated to the text and *"b"* specifies the characteristics associated to the video. If the video is supported by the *MD*, the system adds *"a"* (the characteristics of text) to the *RPL* privileging in this way the *user preference* concerning the video. If there is some information which cannot be displayed like text, the system is able to find again in the *RPL* characteristics preferred by the user for text (*"a"*) and can therefore be applied.

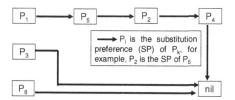

Fig. 4. User Preferences and their substitution preferences

In order to illustrate the *CMA*, we suppose that the set P_u of preferences of this user is the one presented in Fig. 4. If P_1 is satisfied, P_5, P_2 and P_4 are added into the *RPL*. The *CUP* generated by the algorithm are presented in Table 1.

Table 1. Profiles generated by the CMA

	CUP-1	CUP-2	CUP-3	CUP-4	CUP-5	CUP-6	CUP-7	CUP-8	CUP-9	CUP-10	CUP-11	CUP-12	CUP-13	CUP-14	CUP-15	CUP-16	CUP-17	CUP-18	CUP-19	CUP-20
P1	X	X	X	X																
P2										X	X	X	X							
P3	X		X		X	X				X	X			X	X		X	X		
P4														X		X	X			
P5							X	X												
P6	X	X			X		X		X		X		X		X		X			X

Each *CUP* corresponds to a given session and is generated in function of the *context of use* of the session. For example, let us suppose that one of the *MD* of the user only supports text. In this case, all *user preferences* associated to the display of images will not be satisfied during the sessions in which the user is connected through this *MD*. These preferences will not appear in the *CUP* generated for these sessions; they belong to the *RPL*. Table 1 presents for instance that *profile1* is composed of the *user preferences* P_1, P_3 and P_6 (P_2, P_4 and P_5 are in the *RPL*) then the *profile19* is "empty" (*i.e.*, it does not contain any *user preference*).

If there are no *strings of substitution preferences* (no *user preference* has *substitution preferences*) in a set of "*p*" *user preferences*, then the number of profiles (*P*) will be calculated in the following way:

$$P = f(p) = \sum_{i=1}^{p} C_p^i = \sum_{i=1}^{p} \binom{p}{i} = \sum_{i=1}^{p} \frac{p!}{(p-i)!\,i!} \qquad [1]$$

In order to calculate the number of *CUP* that the *CMA* generates considering the set of *user preferences* which contains the strings of *substitution preferences*, we calculate *P* using formula [1]. The number of *CUP* is obtained subtracting to *P* the number of combinations which contain a *user preference* with one or several of its *substitution preferences*. For example, let us suppose that we have 4 preferences P_1, P_2, P_3 and P_4 and that P_2 is the *substitution preference* of P_1. The total number of possible combinations is then 15 (4 combinations of 1 preference, 6 of 2, 4 of 3 and 1 of 4). If the total number of combinations where P_1 and P_2 appear together ((P_1, P_2) (P_1, P_2, P_3) (P_1, P_2, P_4) (P_1, P_2, P_3, P_4)) is 4, then the total number of *CUP* generated by the algorithm (for different sessions) is 11 (*i.e.*, 15-4). For the example shown in Fig. 4, the *CMA* can generate 20 different *CUP* (a *CUP* by session considering the contextual characteristics of this session). For a set of 6 preferences without strings of *substitution preferences*, we have 63 different *CUP* (6 combinations of 1 preference, 15 of 2, 20 of 3, 15 of 4, 6 of 5 and 1 of 6), but 43 of these combinations contain a preference with one or several of its *substitution preferences*. So, only 20 *CUP* are generated by the algorithm.

4.3 Management of Conflicts Between User Preferences

We can criticize of the *CMA* its relative simplicity in the measure that it does not evidence the way in which certain problems are managed when they occur during the *preference management* step[6]. In this section, we present some conflicts between preferences and other incompatibilities that we have identified and the way in which the system has to react if they occur. It is important to note that these ways of reacting are not always solutions to the conflicts but they allow users to at least be informed about the presence of conflicts. A representation of these conflicts can be found in [7].

Among the possible causes of conflicts, we consider: *i) Incompatibility between user preferences and access rights to the IS*. For example, a preference which includes an information request whose access is forbidden to the user. In this case, the *user preference* will not be satisfied and user will not obtain the desired information. *ii) Apparition of a conflict as consequence of the addition of a preference in the CUP*. For example, a preference can consist of obtaining information "*only*" in a format *k* then another preference, in the *CUP*, consists of obtaining the information "*only*" in a format *j*, with *k* different and incompatible with *j*. In the case of conflicting preferences having the same priority, a solution is to choose the preference which can be satisfied according to the formats supported by the *MD*. *iii) Incompatibility between the preferences and the technical constraints of the MD*. For example, the display format wanted by the user is not supported by his *MD*. In

---

[6] We can still note that the introduction of priorities between *user preferences* (see section 4.1) already establishes an answer for the conflict management.

this case, the system analyzes the eventual *substitution preferences* of the preference involved and verifies if there is a preference which can be satisfied. In the case in which information cannot be displayed in a format supported by the *MD*, the system must display a message which indicates the impossibility of displaying the information and the cause of this impossibility. *iv)* **Incompatibility between preferences of formats expressed by the user and availability of the required information** (the information is not available in the required format). The system displays the information using the format in which it is available and searches, among the *display preferences*, the characteristics that it must apply for this format.

Thus, in order to decide if a *user preference "can be satisfied"* (see section 4.2, lines (5) and (17)), a knowledge base registers the conflicts pre-identified during the design of the system [7]. The *CMA* exploits this knowledge in order to manage the conflicts. Since conflicts can occur between *user preferences*, the *CPMS* must integrate a mechanism for specifying conflicts and solving them. For their specification, we use the language proposed by Bell [2] which allows representation of conflicting events which occur simultaneously. This representation allows us on the one hand, to specify, for example, how the addition of a *user preference* can produce conflicts with other preferences belonging to the *CUP*, and on the other hand, to define the criteria in order to decide if a *preference* will be included in the *CUP*.

5 Related Works

Concerning the definition of the *user profile*, several works take into account the *interests of users* [3] [11] [15], their *history in the system* [3] [15], and their *preferences* [4] [15] in order to personalize information to users. Some works such as [4] and [15] define the *user profile* in a multidimensional way: the work presented in [15] exposes a profile of two dimensions, represented by the *history of the information requests* and the *recurrent information needs* of the user (based on the *user interests*). The work in [4] proposes a generic model of profiles composed of six dimensions: *i) personal data*; *ii) user interests*; *iii)* the *expected quality*; *iv)* the *delivery preferences*; *v) security*; *vi)* the *history* of user interactions. However, these propositions do not detail the mechanism of representation of the *context of use* and of the nomadic *user preferences* in order to adapt information.

Some works as *MADSUM* [8] and *AmbieAgents* [13] have explicit mechanisms to personalize the information for the user considering *user preferences,* in the case of *MADSUM*, and their *context of use*, in the case of *AmbieAgents*. However, these proposals do not specify a representation of the *context of use* nor of the *user preferences* in order to adapt information.

Adapting information according to characteristics of the user, needs, goals, tasks, knowledge or *user preferences* are some of the main aspects to bear in mind when designing and modelling *Adaptive Hypermedia systems* and *applications* [5], [11]. However, these works do not specify precisely how to adapt information to characteristics of nomadic environments, such as *location* (which can modify the information needs of users), and *MD features* (which constrain the information display on the access devices of users).

6 Conclusions and Future Work

In this paper, we have proposed formalism and mechanisms for the *preference management* of a nomadic user, in order to adapt her/him information delivered by an *Information System* (*IS*) according to her/his needs and to the current *context of use* (*i.e.*, location, user characteristics and those of her/his access *mobile device*). We have classified the *user preferences* in three types: *activity*, *result* and *display*. We have presented the *Contextual Matching Algorithm* (*CMA*) which defines the *Contextual User Profile* (*CUP*) taking into consideration the three types of *user preferences* expressed by the user. In order to generate the *CUP* for a session, the execution of this algorithm considers information about the *context of use* of this session and *user preferences*. We have also identified some conflicts which can occur and we have proposed some ways for solving them. Actually, we improve the *CMA* which generates the *CUP*. This version considers notably the eventual conflicts which could present in the moment of adding a preference to the *CUP* (conflicts between the preference and those belonging to the *CUP*). Concerning the conflict resolution, we actually study several methods (*e.g.*, *Argumentation* [1], *Branching rules* [9]) in order to implement them in the *CMA*.

References

1. Amgoud, L., Parson, S.: An Argumentation Framework for Merging Conflicting Knowledge Bases. In: Flesca, S., Greco, S., Leone, N., Ianni, G. (eds.) JELIA 2002. LNCS (LNAI), vol. 2424, pp. 27–37. Springer, Heidelberg (2002)
2. Bell, J.: Simultaneous Events: Conflicts and Preferences. In: Benferhat, S., Besnard, P. (eds.) ECSQARU 2001. LNCS (LNAI), vol. 2143, pp. 714–725. Springer, Heidelberg (2001)
3. Birukov, A., Blanzieri, E., Giorgini, P.: Implicit: An Agent-Based Recommendation System for Web Search. In: AAMAS 2005. Proc. of the 4th Int. Conf. on Autonomous Agent and MAS ACM Press, New York, pp. 618–624 (2005)
4. Bouzeghoub, M., Kostadinov, D.: Personnalisation de l'information: aperçu de l'état de l'art et définition d'un modèle flexible de profils. In: CORIA 2005. Proc of (Grenoble, France, March 9-11, 2005), pp. 201–218 (2005)
5. Brusilovsky, P.: Adaptive Hypermedia: From Intelligent Tutoring Systems to Web Based Education. In: Gauthier, G., VanLehn, K., Frasson, C. (eds.) ITS 2000. LNCS, vol. 1839, pp. 1–7. Springer, Heidelberg (2000)
6. Carrillo Ramos, A., Villanova-Oliver, M., Gensel, J., Martin, H.: Knowledge Management for Adapted Information Retrieval in Ubiquitous Environments. In: WEBIST 2005. LNBIP, vol. 1, pp. 84–96. Springer, Heidelberg (2007)
7. Carrillo Ramos, A.: Agents ubiquitaires pour un accès adapté aux systèmes d'information : Le Framework PUMAS. PhD Thesis of the University Joseph Fourier, France, March (2007)
8. Harvey, T., Decker, K., Carberry, S.: Multi-Agent Decision Support Via User Modeling. In: proc of AAMAS 2005, Utrecht, Netherlands, July, 25-29, pp. 222–229 (2005)
9. Herbstritt, M., Becker, B.: Conflict-Based Selection of Branching Rules. In: Giunchiglia, E., Tacchella, A. (eds.) SAT 2003. LNCS, vol. 2919, Springer, Heidelberg (2004)

10. Indulska, J., Robinson, R., Rakotonirainy, A., Henricksen, K.: Experiences in Using CC/PP in Context-Aware Systems. In: proc. of the 4th Int. Conf. MMD, Melbourne, Australia, January 21-24. LNCS, vol. 2574, pp. 247–261. Springer, Heidelberg (2003)

11. Kassab, R., Lamirel, J.C., Nauer, E.: Une nouvelle approche pour la modélisation du profil de l'utilisateur dans les systèmes de filtrage d'information basés sur le contenu: le modèle de filtre détecteur de nouveauté. In: proc. of CORIA 2005, Grenoble, France, March 9-11, pp. 185–200 (2005)

12. Kirsch-Pinheiro, M.: Adaptation Contextuelle et Personnalisée de l'information de Conscience de Groupe au sein des Systems d'Information Coopératifs. PhD Thesis, University Joseph Fourier, Grenoble, Septembre (In French) (2006)

13. Lech, T., Wienhofen, L.: A Scalable Infrastructure for Mobile and Context-Aware Information Services. In: proc of AAMAS 2005, Utrecht, Netherlands, July, 25-29, pp. 625–631 (2005)

14. Nieto-Carvajal, I., Botia, J.A., Ruiz, P.M., Gomez-Skarmeta, A.F.: Implementation and Evaluation of a Location-Aware Wireless Multi-Agent System. In: Yang, L.T., Guo, M., Gao, G.R., Jha, N.K. (eds.) EUC 2004. LNCS, vol. 3207, pp. 528–537. Springer, Heidelberg (2004)

15. Tamine, L., Bahsoun, W.: Définition d'un profil multidimensionnel de l'utilisateur. In: Proc of CORIA 2006, Lyon, France, 15-17 mars, pp. 225–236 (2006)

A Contextual User Model for Web Personalization

Zeina Jrad[1], Marie-Aude Aufaure[1,2], and Myriam Hadjouni[2,3]

[1] INRIA Paris-Rocquencourt, Domaine de Voluceau
78 153 Le Chesnay Cedex, France
{zeina.jrad,marie-aude.aufaure}@inria.fr
[2] Supélec - Plateau du Moulon - Service Informatique
91 192 Gif-sur-Yvette Cedex, France
marie-aude.aufaure@supelec.fr
[3] Riadi-Gdl Laboratory, ENSI Tunis, Campus la Manouba
La Manouba, 2010, Tunisie
myriam.hadjouni@riadi.rnu.tn

Abstract. Over the past years, information personalization has provided several valuable achievements on the improvement and optimization of Web searching and recommendation taking into account user's interests, preferences and contextual information. The main objective of a personalization system is to perform an information retrieval process taking into account the perception and the interest of the end-users. This paper focuses on how to model the user and his context in an extensible way that can be interpreted and used for personalization. We describe the architecture that provides personalization facilities based on the contextual user model for tourism usage[1].

Keywords: User modeling, context modeling, ontology-based modeling, Web personalization.

1 Introduction

Available data over the web became more and more complex and voluminous. Spatio-temporal aspects contribute to this complexity, as well as the lack of structure, the multidimensional nature, dynamicity and mass of data. The increasing interest for information retrieval on the web has led to the semantic web initiative of the World-Wide Web Consortium. The semantic web [1] is an extension of the current web which aims at enriching it so that it will be more comprehensible by computers. Adding semantic information is a necessary condition to the development of the semantic web. Metadata, annotations and ontologies [2][3][4] carry out this semantic information and constitute the foundation of the semantic web. They avoid ambiguities and allow the user to be provided with more relevant data. Another objective of the semantic web is to describe the semantic relationships between these results.

Over the past years, information personalization has provided several valuable achievements on the improvement and optimization of Web searching and

---

[1] This work is supported by the French National Research Agency through the Eiffel Project (semantic web and e-tourism).

M. Weske, M.-S. Hacid, C. Godart (Eds.): WISE 2007 Workshops, LNCS 4832, pp. 350–361, 2007.
© Springer-Verlag Berlin Heidelberg 2007

recommendation taking into account user's interests, preferences and contextual information. Personalization is also closely linked to navigation and visualization. Personalization can be defined as a correspondence between implicit and explicit users' needs, and the response given to these needs. In other words, and as stated by [5]: "the challenge in an information-rich world is not only to make information available to people at any time, at any place, and in any form, but specifically to say the right thing at the right time in the right way".

Amongst many successful examples, the domain of e-commerce has been a privileged personalization application domain on the Web, with many successful examples developed such as the well-known Amazon system [6]. Static user information refers to basic characteristics explicitly presented by the user during a registration procedure; while dynamic user information is collected through observing user's behaviours.

Recommender and Web personalization systems and information retrieval algorithms are some of the main background techniques used so far [7]. Semantic approaches use ontologies or personalization techniques to match users' needs. The main objective of personalization system is to perform an information retrieval process taking account the perception and the interest of the end-users [8]. Personalized content access intends to improve an information retrieval process by adding explicit user requests to implicit user preferences [9]. This is likely to better meet individual user needs and its overall satisfaction regarding the system outputs. Such request reformulations also disambiguates initial queries [10] [11]: for example, when a user ask for the term "conception", the query should be different when she/he is an architect or a computer science designer. Requests can be enriched with predefined terms derived from user's profile [12] [13]. User profiles can be also used to sort and organize query results. Multi-agent systems have been also suggested to provide an interaction layer between a user's profile, a personalization process and a web document [14]. Other approaches also consider social-based filtering and collaborative filtering [15], as well as rule filtering [16]. Rule filtering allows the definitions of rules based on static or dynamic profiles and are then used to improve the content delivered to a particular user. Web usage mining techniques can be used to extract information related to the navigations from logs files. The extracted knowledge can then be used to personalize a web site according to classes of users [18] [17].

Personalized user interactions can be done according to different steps [19]: modeling the user profile, acquiring user's data, designing reasoning and inference mechanisms, and generating personalized services. This paper presents an extensible user's context model that can be interpreted and used for personalization. This paper is organized as follows. Section 2 introduces user modeling approaches and related work in the area. Our framework for web personalization is described in details in section 3. We then conclude and give some perspectives.

2 User Modeling

The user model usually contains information on the goals, the needs, the preferences or the intentions of the users. More advanced user models can contain information related to psychic, emotional, physical, state, etc.

2.1 Modeling Approaches

The definition of the user model implies to represent and to store the most relevant characteristics of the user in the context of the application domain. There are many types of user models, two of them are particularly popular for personalized interaction:

- The simplest user model is the overlay model, wherein the user's knowledge is a subset of the system's knowledge [20]. In its simplest form the overlay model states if an item of the knowledge base is learned, it is not completely learned or is unknown. By comparing the user's knowledge with the expert's knowledge the system derives the user's lack of knowledge. The critical part of overlay modelling is to find the initial knowledge estimation. One of the main drawbacks of this approach is that it can't model the user's misconceptions of knowledge concepts, which is an important aspect within learning environments. More elaborated versions of overlay user models can differentiate between more detailed knowledge states.
- A stereotype user modelling approach classifies users into stereotypes. The users belonging to a certain class are assumed to have the same characteristics. Rich [21] has introduced the idea of stereotypes of users, models of groups of users sharing common interests or characteristics, in order to be used by a system, called Grundy, to recommend books that they might like. A simple self-description of the user enables the system to classify him/her, based on personality traits, in a certain category of users. Grundy uses stereotypes for users like: feminist, intellectual, sport-person, etc. Based on these categorizations the system is able to generate book recommendations. In order to improve the user model the system asks the user if he/she liked the recommendation and why. Using that feedback, it updates both the stereotypes and the model of the current user.

In this paper, we investigate the use of ontologies in modelling not only the user's preferences but also his global context (time, place, material, history, etc.). These aspects are developed in the following section.

2.2 User Context Ontology Based Modeling

Context can be thought of as the "extra", often implicit, information (i.e. associations, facts, assumptions), which makes it possible to fully understand an interaction, communication or knowledge representation.

Ontologies are a promising instrument to specify concepts and interrelations. They are particularly suitable to project parts of the information describing and being used in our daily life onto a data structure used by computers. Gu and al. [22] proposes a formal context model based on ontology using OWL to address issues including semantic context representation, context reasoning and knowledge sharing, context classification, context dependency and quality of context. The main benefit of this model is the ability to reason about various contexts. In [23], Mehta and al. propose the use of a common ontology based user context model as a basis for the exchange of user profiles between multiple systems and, thus, as a foundation for cross-system personalization.

In our work, the context needs to be more clearly defined as compared to the more general situation in knowledge representation. We therefore consider a contextual user model as a representation of combined information (implicit, explicit, etc.) that are directly related to one of the purposes of the system: the items recommendation or the interface adaptation. More details about the user profiles concepts and user model representation is given in the following sections.

3 A Context-Aware Model for Personalization

The capability to model user profile is at the heart of personalization systems. Therefore, one of our research issues is to design the profile structure adaptable to user interests' changes. A user profile can be either *static,* when the information contained is never or rarely altered (e.g., demographic information), or *dynamic* when the user profile's data change frequently.

3.1 Context-Aware Information Acquisition

The available techniques to collect information about users and the methods used to process such information to create user profiles and provide adapted content, presentation and/or structure, are varied. In the Eiffel project, our approach consists in automatically perform changes concerning the content or even the structure of a Web site, based on information concerning the user stored in the user's profile (fig 1). Such information is obtained either explicitly, using online registration forms and questionnaires resulting in static user profiles, or implicitly, by recording the navigational behaviour and/or the preferences of each user, resulting in dynamic user profiles. The advantage of explicit information is that the system is confident about information needs of its users and thus appropriate action can be taken such as offers and recommendations. In our tourism portal, this type of data is collected through:

- Questionnaire: this set of questions is necessary mainly for getting contextual information about the user and what he is searching for on this site. Thus, data concern personal context (age, gender, company, starting from, etc.), search purpose (explore, visit, learn, get experience, study, etc.), precisions (when, where, what budget, etc.), centers of interest (mountains, rivers, monuments, etc.) and ratings of priorities (a little, somewhat, very, too much, etc.).
- Registration: by registering, the user will be able to consult his calendar and selected items. In Eiffel we avoid to burden the user with boring questions; that's why we restricted registration data to only two parameters, which are the name and the email of the user.
- Feedback: in our interface, the system can get an idea of the attitude of the user through explicit feedback. For example, we let the user explicitly tells the system whether he/she likes or dislikes a place or an activity recommended for him/her by placing a heart sign or deleting the item. With this knowledge, the model of the user can be modified according to the user's preferences. A strong positive or negative feedback should result in a significant change of profile. If a feedback indicates that a user does not like a specific recommended place, a similar item should not appear in future recommendations.

However, a system whose purpose is to present personalized information needs to be able to infer the user's interests implicitly by observing user's actions and behaviour. In our system, this is done through:

- Cookies: which are small pieces of data sent by a Web site and stored on the client-browser computer and can be reused later on the server-Web site that sent the cookie as unique information concerning a user.
- Logs: a Web server log contains each access to a Web server with information such as the name of the client's computer, the date/time and the resource accessed.
- Feedback: we use some simplified approaches that enable inferring user interests through implicit feedback information, i.e. by observing user's actions and behaviours duration as we did in our team with the Broadway approach [24]. When the system determines a user's need, it recommends a list of items displayed with sample sentences from the summary of the item description (or an image according to the selected display method, see next section). In this way, users are able to predict roughly what the item or the place or the activity looks like. If a user selects, reads or bookmarks that item from the list, it can be inferred implicitly that in some degree he/she is interested within the content. Conversely, unread or unselected items recommended many times for the same user during a period of time or many sessions can be considered as uninteresting. The positive or negative implicit feedback changes interest level of the user in the contextual information.

3.2 User's Context Components

The user context can be described by a large set of facets. However, the user interacts with systems in different roles and is involved in different tasks in parallel, each of which is associated with a specific subset of the user context facets. To reflect this structuring the user context is divided into multiple working contexts (see Fig 1) grouping together user context facets that are related to and relevant for the same task and/or role of the user.

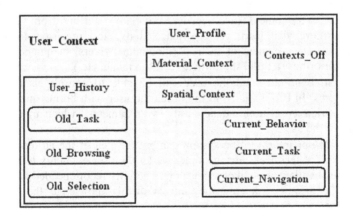

Fig. 1. Components of the context-aware model for personalization

According to the previously presented definitions, we decided to build a model upon the attributes from the following categories/components (see fig 1):

User_Context = User_Profile U User_History U Current_behavior U Contexts_Off

1. **User_Profile** = this category captures all the common information that can constitute a user profile. This includes:
 – General User Demographics: This category captures all the basic user information such as email address (required only for user identification and profile indexing), age, gender, as well as a series of optional information.
 – Travel preferences: This category captures user's preferences for travelling and building holidays planning. Typical preferences concern permanent preferences and interests (for example interested in museums, fishing activities, amount of money ready to spend for restaurant, favourite natural sites), temporary preferences and interests dependent of context (for example I like playing golf near the hotel, I like going on a ride when sun is shining), user situation (available time for walking activities, level of difficulties).
 – Interface preferences: This category captures user's preferences considering the means and the media in which user receives travelling instructions and information while browsing on the tourism portal, modality of items' presentation (only textual information, both textual and visual information, only visual information, etc.).
2. **User_History = Old_Task + Old_Browsing +Old_Selection**
Old_Task describes the reason why the user came to this site in the past i.e. the purpose of his old navigation. Old_Browsing points to what the user did in order to get answers to his queries whereas the Old_Selection designs the chosen items or the reservations he/she did while planning his travel or holidays. The User_History helps to keep track of completed browsing of the user including links, items and selections he/she did in planning his travel.
3. **Current_Behavior = Current_Task + Current_Navigation**. Current_Task describes the task the user is currently involved in, whereas the Current_Navigation points to a history of tasks completed so far within the current working context.
4. **Material_Context** = Terminal type used by the user while browsing the portal. The device used can be a laptop, a PC, a PDA, etc. This category is necessary in order to graphically adapt the interface according to the dimensions and graphical settings of the terminal.
5. **Contexts_Off** = User_Context of other users who already browsed the web site searching for tourism information.
6. **Spatial_Context =** This facet refers to the physical location of the user and to the time frame.

The context model proposed here is defined as an extension of the generic ontology user defined in [25], [26] and including/understanding various characteristics of a user, containing concepts, under concepts and relations between the various concepts.

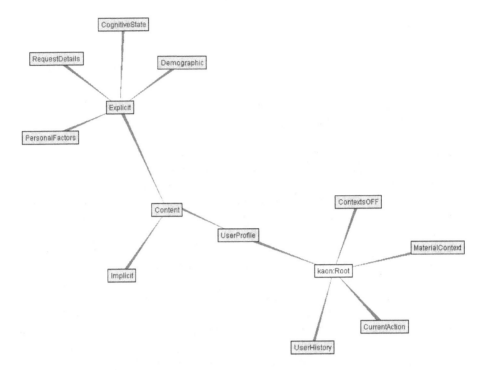

Fig. 2. A partial graph view of the user context ontology with OI Modeler

Our model, partially represented in fig. 2, has been implemented using OI-Modeler and KAON. KAON is a tool suitable for ontology management and for the development of ontology based applications [27]. It comprises a set of tools and APIs which are evolving continuously.

3.3   Basic Workflow of the Personalization System

In this section, we describe how the aforementioned model is incorporated in our system and we describe issues assisting the reader in understanding how our context model affects the browsing procedure. The main functionalities of the system (fig 3) can be summarized in the following steps:

1. Creation of a User profile: it associates the user with one category of users that are most close to his type of traveller. It is important in this module to make an evolutionary and not static detection of profile. Thus, this module contains the acquisition and observation functionalities in order to manage the user actions in the whole interaction. When new information about the user can be inferred, this module updates the user profile by adding the new information. The latter sends to the recommendation module the portion of user profile with the new preferences about the requested service.
2. Invocation of the tracing services when a user starts browsing on the considered web site.

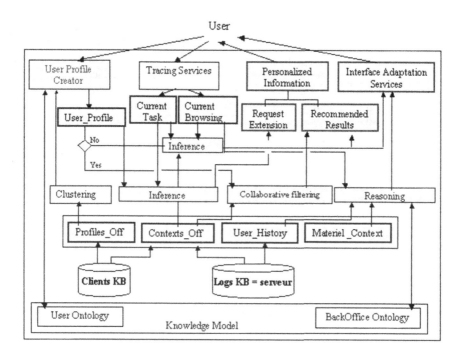

Fig. 3. Main functionalities of the system and basic workflow

3. Request extension: this is based on the User_Profile and the Current_Task in order to extend the user's query according to preferences indicated in his profile even if the user does not demand. For example, Mr. X formulates a query asking for a hotel whereas it is indicated somewhere in his profile that he prefers rooms with air-condition included. The purpose is to be able to add to his query the indicator "air-condition" even if Mr. X doesn't ask for it explicitly.

4. Item recommendation: this functionality is specialized in providing information about places of interest, such as a restaurant, a museum and so on. It provides the type of service/place it manages, and other features useful to better select them for the user. Each tourism object displayed to the user is an instance of a class described in ontology. This ontology is implemented in OWL language and is manually designed to fulfill the needs of the tourism application. It is imported in Mondeca ITM (Intelligent Topic Manager), a knowledge management tool [28]. The ontology can then be populated in two ways: either manually by tourism professionals or automatically by using knowledge-extraction techniques. Each of the objects contains structure of concepts and relations between them specific for that service. In addition they are provided with a part of users' profiles related to each service. For example the restaurant ontology will have all dishes proposed in menu with prices and ingredients, opening hours, number of available tables, average time spent in restaurant etc. Ontology of museums contains opening schedule, list of categories, etc.

5. Interface adaptation: with the new technologies, it is possible to create dynamic and very representative interfaces on the web. The interface designed in Eiffel is indented to be a new image-based search tool [29], an environment able to provide facilities for both browsing and displaying results. The objective is to present to the user an intuitive visual interface that may significantly reduce his cognitive load when doing a search on the web. Visualization is a promising technique that enables people to use a natural tool of observation and processing (their eyes as well as their brain) to extract knowledge more efficiently. Thus, the interface adaptation services intend to reconfigure the graphical design of the Eiffel portal according to user's profiles and browsing ways. This implies, for example, minimising links to some items which have never been chosen by the user and highlighting others that are most visited or chosen.

3.4 Inferring and Reasoning Algorithms

In our system, the recommendation of items or destinations is strongly related to the user's profile determined in the data-filtering process. Once the data concerning the users are collected (implicitly or explicitly or even in both ways), appropriate content is determined and delivered. This process is followed by information-filtering and Web usage filtering techniques as follows:

- The system tracks user behaviour and preferences and recommends items that are similar to items chosen in the past. For example, if a user shows an interest in castles or museums, or by a particular place, links to other related items will be presented.
- The system compares a user's tastes with those of other users in order to build up a picture of like-minded people. The choice of content is then based on the assumption that this particular user will value what like-minded people also enjoyed. The user's tastes are either inferred from previous choices or else measured directly by asking the user's opinion.
- We specify rules based on static or dynamic profiles that are then used to affect the content served to a particular user. For example, *association rules* could explicitly encode the fact that users who choose to visit place x and then place y may also be likely to be interested in visiting place z. More concretely, an interest in "Versailles castle" and "Arc de Triomphe" could potentially demonstrate a general interest in the monuments of the region.
- *Web usage mining* which relies on the application of statistical and data-mining methods to the Web server log data is then used in order to find a set of useful patterns that indicate users' navigational behaviours. Statistical analysis methods are applied to Web data to extract statistical information such as site activity, diagnostic, server, referrers and click stream analysis.

More advanced algorithms will be used such as clustering in order to group together users having similar characteristics and classification in order to map items into various classes such as different types of user profile. From the most used techniques for constructing the user model, analyzing user data and deriving new facts, we will use reasoning with uncertainty (e.g. Bayesian networks, fuzzy logic techniques) and machine learning techniques (e.g. neural networks algorithms).

We focused in this paper on the modeling aspects and don't want to go into further details concerning the reasoning features.

4 Conclusion and Future Work

In this paper, we have presented some background knowledge on modeling theory and personalization facilities which directly affect our user contextual model. We have also described our architecture that provides personalization facilities based on the contextual user model.

However, several issues remain open for further research in this area. One of the most interesting and important issue is the progressive and dynamic model creation. Specifications of rules that represent dependencies between model entities (derived from relevant theories) seem to be a promising solution, although hard to implement.

Another challenging issue concerns the combination of various filtering and reasoning algorithms in the process of personalisation. This work is actually in progress as a part of the Eiffel project.

References

1. Berners-Lee, T., Hendler, J., Lassila, O.: The Semantic Web. Scientific American 284(5), 34–43 (2001)
2. Gomez-Perez, A., Fernandez-Lopez, M., Corcho, O.: Ontological Engineering. Springer, New York (2003)
3. Gruber, T.: Toward Principles for the Design of Ontologies Used for Knowledge Sharing. In: Guarino, N., Poli, R. (eds.): International Journal of Human-Computer Studies, special issue on Formal Ontology in Conceptual Analysis and Knowledge Representation (1993)
4. Aufaure, M.A., Legrand, B., Soto, M., Bennacer, N.: Metadata- and Ontology- Based Semantic Web Mining. In: Taniar, D., Rahayu, J.W. (eds.) Web Semantics and Ontology, Idea Group Publishing (2006)
5. Fischer, G.: User Modeling in Human-computer Interaction. User Modeling and User-Adapted Interaction 11(1&2), 65–86 (2001)
6. Linden, G., Smith, B., York, J.: Amazon.com recommendations: item-to-item collaborative filtering. IEEE Internet Computing 7(1), 76–88 (2003)
7. Shahabi, C., Chen, Y.: Web Information Personalization: Challenges and Approaches. In: Bianchi-Berthouze, N. (ed.) DNIS 2003. LNCS, vol. 2822, pp. 5–15. Springer, Heidelberg (2003)
8. Vallet, D., Fernández, M., Castells, P., Mylonas, P., Avrithis, Y.: Personalized Information Retrieval in Context. In: 3rd International Workshop on Modeling and Retrieval of Context at the 21st National Conference on Artificial Intelligence, Boston, USA (July 2006)
9. Gauch, S., Chaffee, J., Pretschner, A.: Ontology based personalized search and browsing. Web Intelligence and Agent Systems 1(3-4), 219–234 (2003)

10. Pitkow, J., Schutze, H., Cass, T., Cooley, R., TurnBull, D., Edmonds, A., Adar, E., Breuel, T.: Personalized Search. Communications of the ACM 45(9) (2002)

11. Koutrika, G., Ioannidis, Y.: A Unified User-Profile Framework for Query Disambiguation and Personalization. In: Brusilovsky, P., Callaway, C., Nürnberger, A. (eds.) UM 2005. LNCS (LNAI), vol. 3538, Springer, Heidelberg (2005)

12. Glover, E.J., Lawrence, S., Birmingham, W.P., Giles, C.: Architecture of a Metasearch Engine that Supports User Information Needs. In: International Conference on Information and Knowledge Management (1999)

13. Liu, F., Yu, C., Meng, W.: Personalized Web Search For Improving Retrieval Effectiveness. IEEE Transactions on Knowledge and Data Engineering 16(1), 28–40 (2004)

14. Pogaènik, M., Tasiè., J.F.: Layered Agent System Architecture for Personalized Retrieval of Information from Internet. In: 10th European Signal Processing Conference. Intelligent Processing for Communication Terminals, September 4-8, Tampere, Finland (2000)

15. Van Setten, M., Moeleart, F.: Collaborative Search and Retrieval: Collaboration in Information Retrieval. In: GigaCE Report D1.1.2.3, TI/RS/2000/050. Telematica Institut, The Netherlands (2000)

16. Mobasher, B., Cooley, R., Srivastava, J.: Automatic Personalization Based on Web Usage Mining. Communications of the ACM 43(8), 142–151 (2000)

17. Da Silva, A., De Carvalho, F., Lechevallier, Y., Trousse, B.: Characterizing Visitor Groups from Web Data Streams. In: 2nd IEEE International Conference on Granular Computing (GrC 2006), Atlanta, USA, 10-12 May (2006), http://www-sop.inria.fr/axis/Publications/show.php?year=2006

18. Mobasher, B.: Web Usage Mining and Personalization. In: Singh, M.P. (ed.) Chapter in Practical Handbook of Internet Computing (2004)

19. Razmerita, L.: User modeling and personalization of the Knowledge Management Systems, in Adaptable and Adaptive Hypermedia, pp. 225–245. Idea Group Publishing, USA (2005)

20. Kobsa, A., Mller, D., Nill, A.K.-A.: An Adaptive Hypertext Client of the User Modeling System BGP-MS. In: Proceedings of the Fourth International Conference on User Modeling, Hyannis, MA, pp. 99–105 (1994)

21. Rich, E.: Stereotypes and User Modeling. In: Kobsa, A., Wahlster, W. (eds.) User Models in Dialog Systems, pp. 35–51. Springer, Heidelberg (1989)

22. Gu, T., Wang, X.H., Pung, H.K., Zhang, D.Q.: An Ontology-based Context Model in Intelligent Environments. In: Communication Networks and Distributed Systems Modeling and Simulation Conference (2004)

23. Mehta, B., Niederee, C., Stewart, A., Degemmis, M., Lops, P., Semeraro, G.: Ontologically-Enriched Unified User Modeling for Cross-System Personalization. In: Proceedings of User Modelling 2005, pp. 119–123. Springer, Heidelberg (2005)

24. Trousse, B., Jaczynski, M., Kanawati, R.: Using User Behavior Similarity for Recommendation Computation: The Broadway Approach. In: HCI 1999. proceedings of 8th international conference on human computer interaction, Munich (August 1999)

25. Razmerita, L., Angehrn, A., Maedche, A.: Ontology-based user modeling for Knowledge Management Systems. In: Brusilovsky, P., Corbett, A.T., de Rosis, F. (eds.) UM 2003. LNCS, vol. 2702, pp. 213–217. Springer, Heidelberg (2003)

26. Gavrilova, T., Brusilovsky, P., Yudelson, M., Puuronen, S.: Creating Ontology for User Modelling research. In: ECAI 2006. Workshop on Ubiquitous User Modeling in conjunction with the 17th European Conference on Artificial Intelligence (August 28, 2006)

27. Maedche, A., Motik, B., Stojanovic, L., Studer, R., Volz, R.: Ontologies for Enterprise Knowledge Management. IEEE Intelligent Systems (November/ December 2003)
28. Siorpaes K. and Prantner, K.: Evalution of Mondeca ITM. E-Tourism Working Draft (2004), http://138.232.65.141/deri_at/research/projects/e-tourism/2004/d12/v0.2/20041028/
29. Jrad, Z., Aufaure, M.A.: Personalized Interfaces for a Semantic Web Portal: Tourism Information Search. In. In: the 11th International Conference on Knowledge-Based and Intelligent Information & Engineering Systems, SWEA - KES2007 (2007)

Citation-Based Methods for Personalized Search in Digital Libraries

Thanh-Trung Van and Michel Beigbeder

Centre G2I/Département RIM
Ecole Nationale Supérieure des Mines de Saint Etienne
158 Cours Fauriel, 42023 Saint Etienne, France
{van,mbeig}@emse.fr

Abstract. In this paper we present our work about personalized search in digital libraries. Unlike other researches which use content-based methods, we focus on citation-based methods for this purpose. We propose a practical approach to estimate the co-citation relatedness between scientific papers using the Google search engine. We conducted some experiments to evaluate performance of different citation-based methods. The experimental results show that our approach is promising and applicable for personalized search in digital libraries.

1 Introduction

Nowadays, with the augmentation of a mass heterogeneous data available on the Web, the "information starvation" problem is no longer our main concern. However, another problem appeared, this is how to find relevant information from this huge source. We often use search engine to find relevant information, but popular Web search engines usually return a large number of results (thousands or millions) for a query and many of them are not relevant. An important reason is that the user query is usually short (less than 3 words on average [1]) and hence ambiguous. For example, with a short query like "java", without additional information, we can not know if the authors want to find information about an island, a kind of coffee, or a programming language. Even with a longer query like "java programming language", we still do not know which kind of document this user want to find. If she/he is a programmer, perhaps she/he is interested in technical documents about the Java language; however, if she/he is a teacher, perhaps she/he want to find tutorials about Java programming for her/his course.

The above problem could be solved by personalization techniques using user profiles. Generally, a user profile is a set of information that represent interests and/or preferences of a user. These information could be collected by implicitly monitoring user's activities [2,3] or by directly requesting users [4]. User profile could be used not only for personalized search [5], but also for different tasks like information filtering [6] or personalized visualization of search results [7]. In frame of digital libraries, user profiles could be collected from the papers

M. Weske, M.-S. Hacid, C. Godart (Eds.): WISE 2007 Workshops, LNCS 4832, pp. 362–373, 2007.

that users read in this library, from users search histories, from users browsing histories or explicitly specified by user etc.

Our work is about personalized search in digital libraries which consist of scientific papers. The originality of our approach is that, while most other researches about personalized search use content-based approaches to represent user profile and document and to compute the similarity between the user profile and search results for re-ranking, we focus on citation-based approaches for these purposes.

The rest of this paper is organized as follows. In the Sec. 2 we present principles of two famous citation-based methods to find similarity between scientific papers: bibliographic coupling method and co-citation method; we present in detail two approaches that we use to compute co-citation similarity: traditional method with Web of Science database and our Web co-citation method. In the Sec. 3 we describe the simulation of personalized search in a digital library using these methods. We conclude in Sec. 4.

2 Citation-Based Methods to Find Similarity Between Scientific Papers

An important characteristic of scientific papers in a digital library is the bibliographical relation between them. If an article appears in the bibliography section of another article, there must be at least a reason for the authors of the second article to cite the first article. In [8] Garfield stated 15 possible reasons for citing an article: paying homage to pioneers, giving credit for related work, identifying methodology/equipment etc. , providing background reading, correcting one's own work, correcting the work of others, criticizing previous work, substantiating claims, alerting to forthcoming work, etc. From the bibliographical relations between scientific papers, we can deduce the relatedness between them. However, in many cases, a simple direct relation "citing-cited" between scientific papers is not enough to represent their relatedness. Thus, new methods which can discover implicitly related papers were proposed. In 1963 Kessler [9] proposed the *bibliographic coupling* method. In this method, the similarity between two papers is computed based on the number of their *co-references*. He supposed that if two papers have common references in their bibliographies, they may focus (entirely or partially) on the same topic. In 1973 Marshakova [10] and Small [11] independently proposed another method called "co-citation". In this method, the relatedness between two papers is computed based on their *co-citation frequency*. The co-citation frequency is the frequency that two papers are *co-cited*. Two papers are said to be *co-cited* if they appear together in the bibliography section of a third paper. The two methods bibliographic coupling and co-citation are illustrated in Fig. 1 (arrows represent citations).

Both of these methods have their limits. In the bibliographic coupling method, the relatedness between two papers are fixed since their publication date because they are computed based on the number of their co-references which remains unchanged. In the co-citation methods, with the time two related papers may

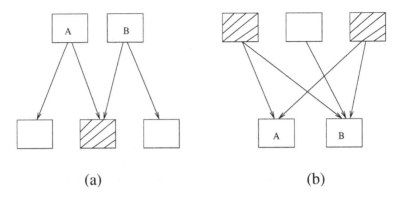

(a)   (b)

Fig. 1. Illustration of (a) bibliographic coupling and (b) co-citation

receive more and more citations and their co-citation frequency can increase. However if we want to know this citation information, we have to extract from the *citation graph* of the actual library or read from a *citation database*[1] which are usually limited; i.e. we can only know citing papers of a given paper if these citing papers exist within the same digital library or citation database.

The two methods bibliographic coupling and co-citation have been used widely for different purposes. The digital library CiteSeer uses these methods to find related papers. In [12] the co-citation method is used to create a patent classification system for conducting patent analysis and management. Recently, these methods are used in hyperlinked environment to find the relatedness between Web pages because of the similarity between the notion of "citations between scientific papers" and "links between Web pages". When applied to Web pages, the co-citation method seems to be better than bibliographic coupling therefore it is used more widely. In [13], Pitkow and Pirolli use co-citation method for clustering Web pages. In [14], Dean and Henzinger use co-citation and companion algorithms to find the related Web pages. Efron [15] uses this method to estimate *political orientation* in Web documents.

In [16] the authors used bibliographic coupling and co-citation methods for the classification of Brazilian Web pages. They also use these methods for the classification of scientific papers. They have found that co-citation method work very well in classification of Web pages. However this method has poor performance when it is used for classification of scientific papers in a digital library. This is because in a digital libray, we can only know citations from internal documents (i.e. documents inside the digital library) while citations from external documents are not available. In the case of Web page classification, the Web page collection is a subset of a search engine database which contains most of the link information available in Brazilian Web pages. That is why the link information is not limited like in the case of scientific papers. Because of this reason, the co-citation method work much better in Web pages collection than in scientific papers collection.

---

[1] A citation database is a system that can provide bibliographic information of papers.

The next part presents two approaches that we use to compute co-citation similarity between scientific papers: traditional method with Web of Science database and our Web co-citation method.

2.1 Using Web of Science as Citation Database

Actually there are many citation databases like Web of Science[2], Scopus[3] and digital libraries like CiteSeer, ACM Digital Library which provide citation information about scientific papers. After regarding in detail these sources, we decided to choose Web of Science as a citation database in our experiments. The Web of Science of Thomson ISI is an important database which is used widely for citation studies [17,18]. In 2005 there are about 35 million records in its database. The Web of Science provides access to current and retrospective multidisciplinary information from approximately 8,700 of the most prestigious, high impact research journals in the world. Besides, it also provides an API which facilitates the access to its database without using a Web browser. Another important reason for which we used Web of Science is that it contains most of journals and transactions used in our experiments (see Sec. 3.)

In Web of Science, an article is represented by a primary key called **UT**. Its API supports many operations on its database. Table 1 lists two important operations that are used in our work. More documentation about Web of Science search service could be found in [19].

Table 1. Some operations of Web of Science search service

Operation	Description
searchRetrieve	Performs a search and retrieves records
citingArticles	Searches the WOS database for records that cite a particular parent record and retrieves those records. The parent record is specified by its UT value

Thanks to the search service of Web of Science, if we know some information about a paper (like title, year of publication, journal etc.) we can use these informations to find the UT primary key of this paper in Web of Science database by calling the *searchRetrieve* function. Then using this UT primary key we can find all papers that cite this paper with the *citingArticles* function. From these information we can know the number of times that a paper is cited or the frequency that two papers are co-cited in Web of Science database.

2.2 Using Google as Citation Database

With the explosion of the World Wide Web, Web search engines have to become more and more complete in order to satisfy information needs of users and their indexes become bigger with time. For example, in 2005 Google claimed to index

---

[2] http://portal.isiknowledge.com
[3] http://www.scopus.com/scopus/home.url

over 8 billion Web documents. With their huge indexes, Web search engines could be a good source for many data mining tasks. For example, Turney et al. [20] used AltaVista search engine to find the semantic relation between words. They issued queries containing words to be examined to AltaVista search engine and noted the numbers of hits (matching documents) returned. Using these numbers they can deduce the relation between words.

Recently, a new method for citation analysis called Web citation analysis begins attracting the research community. Web citation analysis finds citations to a scientific paper on the Web by sending the query containing the title of this paper (as phrase search using quotation marks) to a Web search engine and analyze returned pages [21]. Because a Web search engine can index many kinds of document in many different formats, the notion of "citation" used here is a "relaxation" in comparison with the traditional definition. Vaughan and Shaw [21] used this method with Google search engine and compared with traditional bibliographic method using ISI database. Given an article, they classified Web documents that cite this article into 7 different categories: Journal (site of correspondence journal); Author (author, co-author, or one of their employers lists the articles in their pages); Service (a Web bibliographic service lists the article); Class (bibliography/reading list for a course); Paper (a scientific paper that is posted on the Web); Conference (conference announcement, report or summary/description); Other (cited in another way). Kousha and Thelwall [22] used a similar strategy called URL citations to find citations to articles of open access journals. However, in their work, the URL citation of a Web page is the mentions of its URL in the text of other Web page (and not its title). They also compare URL citations (using Google) with traditional bibliographic citation (using ISI).

In our Web co-citation method, we compute the co-citation similarity of two scientific papers by the frequency that they are "co-cited" on the Web. The notion of "co-citation" used here is also a "relaxation" in comparison with the traditional definition. If the Web document that mentions two scientific papers is another scientific paper then these two papers are normally co-cited. However, if this is a table of content of a conference proceeding, we could also say that these two papers are co-cited and have a relation because a conference normally has a common general theme. If these two papers appear in the same conference, they may have the same general theme. Similarly, if two papers are in the reading list for a course, they may focus on the same topic of this course. In summary, if two papers appear in the same Web document, we can assume that they have a (strong or weak) relation. The search engine used in our experiment is the Google search engine. To find the number of time that a paper is "cited" by Google we need only to send the title of this paper (as phrase search using quotation marks) to Google and note the number of hits returned. Similarly, to find the number of times that two papers are "co-cited", we send the titles of these two papers (as phrase search and in the same query) to Google and note the number of hits returned. This idea is illustrated in Fig. 2. In this example, the co-citation frequency of two examined papers is 11. In our experiments, we use a script to automatically query Google instead of manually using a Web browser.

Fig. 2. Illustration of the Web co-citation method

2.3   Similarity Measures

In this section we define some similarity measures that we use to compute the relatedness between scientific papers using citation approaches. We used some variants of the measure introduced in [23]. The co-citation similarity between two papers is defined as:

$$cocitation\_similarity(p_1, p_2) = ln(\frac{cocitation(p_1, p_2)^2}{citation(p_1) \cdot citation(p_2)}) \qquad (1)$$

or

$$cocitation\_similarity(p_1, p_2) = ln(\frac{cocitation(p_1, p_2)^2}{citation(p_1) + citation(p_2)}) \qquad (2)$$

In above formulae, $cocitation(p_1, p_2)$ is the number of times that these two papers are co-cited, $citation(p_1)$ and $citation(p_2)$ are respectively the citation frequency that papers p_1 and p_2 received.

The bibliographic coupling similarity between two papers is computed as:

$$bibcoupling\_similarity(p_1, p_2) = ln(\frac{coreference(p_1, p_2)^2}{reference(p_1) \cdot reference(p_2)}) \qquad (3)$$

In above formula, $coreference(p_1, p_2)$ is the number of co-references of p_1 and p_2, $reference(p_1)$ and $reference(p_2)$ are respectively the number of references of p_1 and p_2.

3   Experiments

As stated above, in this work we conduct experiments for evaluating performance of two methods: bibliographic coupling and co-citation with citation databases Web of Science and Google. The experiments described here are simulations of personalized search in a digital library using user profiles.

3.1   Test Collection

The test collection that we use in our experiments is the collection INEX 2005 (version 1.9[4].) This collection has 17000 XML documents extracted from journals and transactions of *IEEE Computer Society*. Thus this collection could be used as a medium-size digital library of Computer Science. This collection includes not only scientific papers but also other elements like *tables of content, editorial boards* ... Because we are interested only in scientific papers, so in the first step we have to remove these elements from the collection. We found that these elements either do not have a *title* (in <atl> tag[5]) or have simple titles like *News, About this Issue, Article summaries* ... which are not representative titles for scientific papers. Therefore we use a quick-and-dirty approach to eliminate these elements: we sent the titles of all documents in the collection to the Google search engine; if the title of a document receives more than 15000 matching documents then this title is considered not to be representative for a scientific paper and the correspondence document will be removed from the collection. Documents without title are also removed. After this process, the collection contains 14237 documents. Then we extract all necessary information for our experiments from these documents (title, journal, publication year, bibliography etc.)

There are also many topics with relevance assessments distributed with the collection. Each topic represents an information need. There are two types of topics used in INEX ad-hoc task: CAS topics (Content And Structure) which allows users to use structure of documents in their requests and CO topics (Content Only) which do not contain structure of documents (like in TREC). In our experiments we use only CO topics to form user queries. There are 29 original CO topics but only 20 topics that have more than 30 relevant documents will be used for experiments. The numbers of these topics are: 206 207 208 209 210 212 213 216 217 218 221 222 223 227 228 229 234 235 236 237.

As mentioned above, our experiments are simulations of personalized searching using user profiles. In this case, 20 topics represent different information needs of 20 different people. For each topic, we choose some relevant papers as "pseudo user profile" of this person (5 in average in our experiments). The selected papers are choosen among the highly relevant papers to the correspondence topic and those that receive many citations from other documents. We think that this approach is reasonable because in reality if a user of a digital library has to specify something as her/his profile, she/he may choose important papers in her/his research domain. The papers which are included in these profiles are removed from the collection to avoid effect on the experimental results.

3.2   Evaluation Procedure

After the preparation procedure, we use the **zettair** search engine[6] to index the INEX collection, then we send 20 queries (which are formed from above topics) to

---

[4] http://inex.is.informatik.uni-duisburg.de/2005/
[5] Article title.
[6] http://www.seg.rmit.edu.au/zettair

zettair; with each queries we take the first 300 documents for re-ranking using "user profiles" of correspondance topic. The similarity between a document d and a user profile p is computed as:

$$similarity(p, d) = \sum_{d' \in p} similarity(d', d) \qquad (4)$$

In Eq. 4, $similarity(d', d)$ is the similarity between a document d' in profile p and document d (see Sec. 2.3). Note that in the co-citation method, the similarity between a profile and a document is computed by using two different citation databases Web of Science and Google as we mentioned in Sec. 2. If we use Google, the only information we need is the titles of d and d'. If we use Web of Science, firstly we have to find the primary keys UT of d and d', then we use these primary keys to find their citing papers. In bibliographic coupling methods, the references (bibliography) of each paper are extracted from original textual content of documents. The $similarity(p, d)$ could be normalized to have value from 0 to 1. The final score of a document is obtained by combining its original score of **zettair** (the default model is *Dirichlet-smoothed* [24])and its similarity document-profile. We tried two combination function: a linear function and product function. However, in this experiment the product funtion seems to be better than the linear function. Therefore, it is used in final result which is presented below.

3.3   Results and Discussion

The experimental results are presented in Fig. 3 and Tab. 2. The metric used is precision/recall. From these experimental results, we can see that the co-citation method using Web of Science database do not bring any improvement, it even causes a slight performance decrease. The bibliographic coupling method performs better but not very clearly. The co-citation method using Google is the best, it brings 15.06% improvement for the precision at top 30 documents.

Now we will analyse the experimental data to explain these results. In the co-citation methods using Web of Science database, only 213 pairs of document (each pair consists of a document to be re-ranked and a document in the "pseudo user profile") are co-cited with the average co-citation frequency of each pair is 1.94. This small number of co-cited pairs is the reason why it could not bring any improvement and even becomes a noisy source which causes bad effect on the final result. There are many reasons that could have influence on the performance of the co-citation method. The most important reason is the capacity of the citation database that we use. The papers that are selected as "profiles" are also an important factor: the more famous they are, the more citation they receive, and the probability that they are co-cited with other papers will be higher. Although we tried to select important papers within the collection but there are no guarantee that they are also the most important papers in their domains. In the bibliographic coupling method, there are 1126 pairs of documents which have co-references with the average number of co-references of each pair is 1.69.

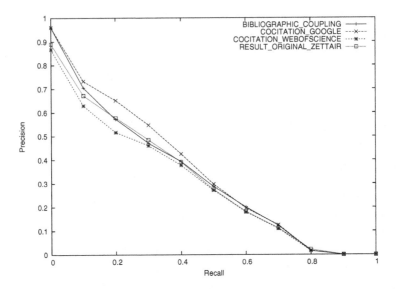

Fig. 3. Experimental results: re-ranking search results of zettair with different citation-based methods

This is a little better than the first case and it is able to make some improvement. Note that the bibliographic information (references) of papers in collection are extracted directly from the textual contents of these paper so they are not dependent on any citation database. In the co-citation method using Google as citation database, there are 4845 pairs of documents which are "co-cited" with the average co-citation frequency of each pair is 4.84. This is much better than the first two cases. That is why it gains the best performance.

Table 2. Precision at 5, 10, 20, 30 documents

	Original Result	Bibliographic coupling	Co-citation using WoS	Co-citation using Google
At 5 docs	0.6600	0.7300	0.6300	0.7100
At 10 docs	0.6150	0.6050	0.5900	0.6800
At 20 docs	0.5375	0.5600	0.5150	0.6025
At 30 docs	0.4867	0.4883	0.4567	0.5600

However, the number of pairs which are co-cited or have co-references and their average frequency are not unique factors. In the second experiment, we tried to increase the number of co-cited pairs with the co-citation method using Web of Science database: we adjust in each "pseudo user profile" all papers that cite at least one paper in this set and all references of these "pseudo user profile" (all these information come from Web of Science database). In this way, the size of each "pseudo user profile" becomes much larger. Then we conduct other experiment with this expanded profile. Result of this experiment is presented in Fig. 4:

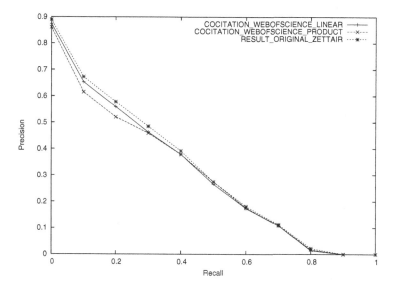

Fig. 4. Performance of co-citation method using Web of Science with the expanded profile

In this case, the number of co-cited pairs is 2715, which is much bigger than in the first experiment, and the average co-citation frequency of each pair is 1.66. However from this result, we can see that the co-citation method with expanded profile is still worse than the original result of zettair. The possible reason is that, in the expanded profile there are many papers that do not focus on the same topic with the original profile and hence produce noise which decrease performance of this method. This phenomenon is called "topic drift" [25]. Another thing that we want to note here is that the product combination is worse than the linear combination (although both of them are worse than original result). In the first experiment, the best results was obtained with the product combination, this make us conclude that in our experiments the product combination has more effect than linear combination, either this effect is improvement or deterioration.

4   Conclusions and Future Work

In this paper, we present different citation-based methods and their application for personalized search in digital libraries. From the experimental results, we can see that citation-based methods could be applied for this purpose. However, the performance of the co-citation method depends largely on the citation database that we used. We have some directions for the future work. For example, we could combine multiple citation databases to boost the performance of co-citation method; a k-fold cross-validation experiment is also being considered.

Acknowledgement

This work is done in the context of European Project CODESNET and supported by Web Intelligence Project of the "Informatique, Signal, Logiciel Embarqué" cluster of the Rhône-Alpes region.

References

1. Spink, A., Ozmutlu, S., Ozmutlu, H.C., Jansen, B.J.: U.S. Versus European Web Searching Trends. SIGIR Forum 36(2), 32–38 (2002)
2. Kelly, D., Teevan, J.: Implicit feedback for inferring user preference: a bibliography. SIGIR Forum 37(2), 18–28 (2003)
3. Shen, X., Tan, B., Zhai, C.: Implicit user modeling for personalized search. In: CIKM '05: Proceedings of the 14th ACM international conference on Information and knowledge management, pp. 824–831. ACM Press, New York (2005)
4. Chen, L., Sycara, K.: Webmate: a personal agent for browsing and searching. In: AGENTS 1998: Proceedings of the second international conference on Autonomous agents, pp. 132–139. ACM Press, New York (1998)
5. Speretta, M., Gauch, S.: Personalizing search based on user search histories. In: Thirteenth International Conference on Information and Knowledge Management (2004)
6. Seo, Y.W., Zhang, B.T.: A reinforcement learning agent for personalized information filtering. In: IUI 2000: Proceedings of the 5th international conference on Intelligent user interfaces, pp. 248–251. ACM Press, New York (2000)
7. Singh, A.: Hierarchical classification of web search results using personalized ontologies. In: Singh, A. (ed.) Proceedings of HCI International 2005, Las Vegas (2005)
8. Garfield, E.: Can citation indexing be automated? In: Statistical association methods for mechanized documentation: Symposium proceedings (1965)
9. Kessler, M.M.: Bibliographic coupling between scientific papers. In: Kessler, M.M. (ed.) American Documentation, pp. 10–25 (1963)
10. Marshakova, I.: System of document connections based on references. In: Marshakova, I. (ed.) Nauchno-Tekhnicheskaya Informatsiya Seriya 2 – Informatsionnye Protsessy i Sistemy, pp. 3–8 (1973)
11. Small, H.G.: Co-citation in the scientific literature: A new measure of the relationship between two documents. Journal of American Society for Information Science 24(4), 265–269 (1973)
12. Lai, K.K., Wu, S.J.: Using the patent co-citation approach to establish a new patent classification system. Information Processing and Management 41(2), 313–330 (2005)
13. Pitkow, J., Pirolli, P.: Life, death, and lawfulness on the electronic frontier. In: CHI 1997: Proceedings of the SIGCHI conference on Human factors in computing systems, pp. 383–390. ACM Press, New York (1997)

14. Dean, J., Henzinger, M.R.: Finding related pages in the world wide web. In: WWW 1999: Proceeding of the eighth international conference on World Wide Web, pp. 1467–1479. Elsevier, Amsterdam (1999)
15. Efron, M.: The liberal media and right-wing conspiracies: using cocitation information to estimate political orientation in web documents. In: CIKM 2004: Proceedings of the thirteenth ACM international conference on Information and knowledge management, pp. 390–398. ACM Press, New York (2004)
16. Couto, T., Cristo, M., Gonalves, M.A., Calado, P., Ziviani, N., de Moura, E.S., Ribeiro-Neto, B.A.: A comparative study of citations and links in document classification. JCDL 2006 (2006)
17. Jacso, P.: As we may search: Comparison of major features of the web of science, scopus, and google scholar citation-based and citation-enhanced databases. Current Science 89(9), 1537–1547 (2005)
18. Meho, L.I., Yang, K.: Multi-faceted approach to citation-based quality assessment for knowledge management. In: World Library and Information Congress: 72nd IFLA General Conference and Council (2006)
19. Isi web service (web site), http://scientific.thomson.com/support/faq/webservices/
20. Turney, P.D., Littman, M.L.: Unsupervised learning of semantic orientation from a hundred-billion-word corpus. CoRR cs.LG/0212012 (2002)
21. Vaughan, L., Shaw, D.: Bibliographic and web citations: what is the difference? J. Am. Soc. Inf. Sci. Technol. 54(14), 1313–1322 (2003)
22. Kousha, K., Thelwall, M.: Motivations for url citations to open access library and information science articles. Scientometrics 68(3), 501–517 (2006)
23. Prime-Claverie, C., Beigbeder, M., Lafouge, T.: Transposition of the cocitation method with a view to classifying web pages. J. Am. Soc. Inf. Sci. Technol. 55(14), 1282–1289 (2004)
24. Pehcevski, J., Thom, J.A., Tahaghoghi, S.M.M.: Rmit university at inex 2005: ad hoc track. In: Fuhr, N., Lalmas, M., Malik, S., Kazai, G. (eds.) INEX 2005. LNCS, vol. 3977, Springer, Heidelberg (2006)
25. Huang, S., Xue, G.R., Zhang, B.Y., Chen, Z., Yu, Y., Ma, W.Y.: Tssp: A reinforcement algorithm to find related papers. In: WI 2004: Proceedings of the Web Intelligence, IEEE/WIC/ACM International Conference, Washington, DC, USA, pp. 117–123. IEEE Computer Society, Los Alamitos (2004)

Personalized Information Access Through Flexible and Interoperable Profiles

Max Chevalier[1,2], Christine Julien[1], Chantal Soulé-Dupuy[1],
and Nathalie Vallès-Parlangeau[1]

[1] Université de Toulouse
Institut de Recherche en Informatique de Toulouse (IRIT - UMR5505)- SIG/D2S2
118 route de Narbonne
F-31062 Toulouse cedex 09
[2] Laboratoire Gestion et Cognition (LGC - EA2043)
129A, Avenue de Rangueil, BP 67701
F-31077 Toulouse cedex
{Max.Chevalier,Christine.Julien,
Chantal.Soule-Dupuy@irit.fr}@irit.fr,
nathalie.valles@univ-tlse1.fr

Abstract. When searching information, any user has to face huge cognitive efforts to obtain accurate and relevant results. The search task includes a set of complementary sub-tasks in which the user needs to be necessarily involved. But, the real place of the users is not obvious without an effective knowledge of their context, environment, and so on. So we assume that a better knowledge of the user and of available information should make it possible to implement techniques aimed at adapting the retrieved information contents, as well as the search process itself. This personalization mainly relies on the definition of profiles. Since applications principally manage specific user/information profiles (structure and content), we propose in this paper a generic and a flexible profile model. This latter facilitates the construction and the interoperability of various profiles coming from different applications and/or having different structure/content. This paper presents the way the different resources (user, information...) can be modeled within the information search process and its related tasks. Then, we discuss the usefulness of profiles in such processes/tasks. Finally we present the generic and the flexible profile model we propose.

Keywords: Generic Profile Model, Interoperability, Information Retrieval, Personalization.

1 Introduction

Searching for information in digital repositories is an activity which can be realized by anyone. Everybody nowadays, in our era of information, handles, exploits and retrieves information; a lot of information. To find information fitting his needs, the

M. Weske, M.-S. Hacid, C. Godart (Eds.): WISE 2007 Workshops, LNCS 4832, pp. 374–385, 2007.
© Springer-Verlag Berlin Heidelberg 2007

user makes use of different kinds of search tools in particular on the Web which is considered today as a privileged source of information. However which is the place of the user in such an activity? We propose to answer partially this question in this paper. In order to better understand the user's role in such an activity, we present initially the context of the information searching process. We underline then the user's implication during the different levels of the search process while taking into consideration previous presented elements. In order to allow the system to better know the user and to implement techniques of adaptation as well of the retrieved information contents as of the searching process itself, we present how this user is modeled through a user's profile[1]. We present finally our work linked to the user's profiles within the framework of the information searching process. Whereas each application proposes its own profile contents and structures, we propose to model any kind of profile through a generic profile. This model supports various contents coming from the literature but especially makes it possible to make interoperating different profiles defined in different applications, even if they do not have the same structure. This model is based on the assumption that the better the information is described, the better the system will exploit it. Thus, a semantic level, being used as support with techniques of inference, has been added to the profile. We also present the various tracks which we explore concerning the user within the framework of the searching for information process: for example the collective aspects or the Visualization Information Retrieval Interfaces. To end this paper, we propose a small discussion around the concept of "taking into account the user in the information searching process".

2   Searching for Information

Searching for Information is an activity in which the user has a prevalent place. Indeed, he is the only person to really know his needs and can reasonably appreciate the relevance and the usefulness of retrieved information. However, searching for information can be achieved through two specific modalities but used in a transparent way: querying and browsing (Figure 1).

Browsing consists in retrieving relevant information in "visiting" a document repository(ies) without knowing, a priori, the contents, the organization and the type

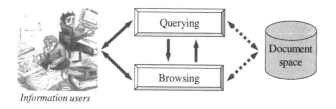

Fig. 1. Searching for Information

---

[1] In this paper, we are exclusively interested in the structure and the contents of the users' profiles. We do not treat the construction and the evolution of such profiles.

of the documents contained in the repository(ies). The most representative example surely is the Web.

Querying consists in seeking information thanks to a specific tool called a search engine. This latter tool gives information in response to a specific information need.

2.1 Browsing

Browsing makes it possible to go through the space of the documents of the collection without having to formulate needs. The main interest of this task is that the user acquires information without necessarily having to know, a priori, the contents and/or the structure (organization) of encountered information.

Three models [2] (Chap 2.10) have been defined to characterize the browsing concept (figure 2):

- The flat model. The documents are presented in a plan or a simple list;
- The structured model. By analogy with a file system, the documents are organized in tree-like structure. This model is interesting when we want to propose the documents according to their related topics to the user;
- The hypertext model. This model is based on the hypertext concept. This concept aims at extending the concept of linear (or sequential) text file by allowing a graph-like structure. This concept has been developed to propose a nonlinear consultation of the documents. The nodes are containers (granule or collections of granules) which are not limited to text but can also contain images (fixed or animated) and sound. A hypertext link is a referential link that establishes various non hierarchical semantic relations between the nodes.

In the Web context, browsing is primarily based on the hypertext model.

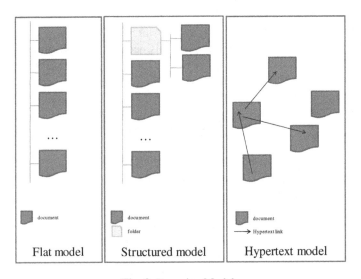

Fig. 2. Browsing Models

2.2 Querying

Querying makes it possible to obtain documents according to ad-hoc user's needs. He does not have to navigate blindly on through the space of documents as it is the case with browsing. We can classify querying tools according to two categories:

- Those which are based on a pull approach which consists in supplying information answering an explicit request of an individual. It is the case of systems such as search engines (Google for example);
- Those which are based on a push approach which consists in returning automatically to individual information which could interest him, taking into account his recurring needs. It is the case of information filtering tools and information recommender systems [11].

The duality between these two great categories of approaches was highlighted in [3]. Table 1 synthesizes the principal points of comparison.

Table 1. Pull Approaches versus Push Approaches

	Pull approach	*Push approach*
Information need	Short time (the time of the request)	Long time (persistent)
Document collection	Static	Dynamic
Interaction	high	low

3 The User Facing the Information Search Process

The user's implication within the information search process is very important. The more this implication is important, the more the results will be relevant from the user's point of view. So, the user gets involved in the following Information Retrieval (IR) steps:

- His human capacities (physical, mental, psychical....);
- His knowledge and particularly, as [7] underlines, his practical knowledge (Web experience) and his domain-specific background knowledge.

3.1 The User Facing Browsing

The main difficulty of the browsing paradigm from the user's point of view is the cognitive effort it requires. Indeed, even if browsing seems to be easy, [1], [2] indicate that the cognitive overload is inevitable. Indeed, an important cognitive effort is required to build a mental map of the visited hypertext graph. This latter map is completed whenever the user visits a new document since it memorizes the document organization. It also allows the user to browse the hypertext in an optimal way in reminding the location of non visited documents of interest. A cognitive overload occurs when the user can no more memorize the hypertext structure. Moreover, a second difficulty concerns the way the user's knowledge is implied in browsing. Indeed, every document the user visits must be evaluated to identify if it is relevant to the user's information need. The quality of the browsing is so correlated to his domain-specific background knowledge and his mental capacity. Moreover, browsing can be improved by his practical knowledge (URL content and interpretation, the use of browser …).

3.2 User Facing Querying

In the querying paradigm, two trends can be identified: *pull* and *push* methods. These two methods have the same goal but are different from the user's point of view.

3.2.1 Pull Approaches

Pull approaches are based on an important user interaction. First of all, the user has to transform his information need into a query (commonly a keyword list) that is understandable by the system. To do this, the user's domain-specific background knowledge is heavily implied in order to choose the keywords used to build the query. Indeed, the search results quality depends on the choice of the most accurate keywords. Too specific keywords make the number of retrieved documents decreasing. Too general keywords make the number of retrieved documents too high to be humanly managed. In this latter case, the search results are also general since selected keywords may be related to a too large concept. Even if the choice of keywords is crucial, the length of the query is also an issue because when the information search systems may have some difficulty to select the accurate context of the query. In practice, some studies show that the number of keywords of a query is relatively low (<3 keywords in average) [15], [8], [16]. We can imagine (and regret) that the length of queries should not evolve in the next years due to the high cognitive effort it requires to build a "good" query.

Secondly, at the opposite side of the information search process, the user is also involved when the system shows him the retrieved documents. His domain-specific background knowledge is another time heavily required associated to his practical knowledge to identify the relevant documents within the numerous retrieved documents (several million in most of case on the Web). In this case, the more the number of documents are retrieved, the more the possibility of cognitive overload is. Indeed, when having to manage the high number of retrieved documents within a ranked list requires patience and perseverance. To analyze the search results in its integrality, the user has to open and evaluate the relevance of every retrieved document, solution that cannot be imagined for a human. As a result, we can underline the fact that users are only visiting the 20/30 first retrieved documents ignoring other potentially relevant documents (from his own point of view).

3.2.2 Push Approaches

From the user's point of view, push approaches are more comfortable because the interaction is quite limited. Indeed, after giving his information needs (through a query or example of relevant documents) the user receives automatically relevant documents without having to interact with the system [17]. Moreover, the number of recommended documents is lower than in pull approaches because recommended documents are more filtered. The implication of the user is not limited to the interaction in push approaches; he must use his domain-specific background knowledge to evaluate the real matching between recommended documents and his information needs. Thanks to these relevance judgments, the internal representation of the information needs can be corrected and adapted to limit the number of "noisy" irrelevant documents like in collaborative filtering systems [13], [5].

The user is so the angular stone of any information search process. The quality of retrieved documents depends essentially on his experience and capabilities... To improve this quality, search systems have evolved and try to help the user in a more important manner. To achieve this goal, systems, qualified as "adaptive", must have the most complete information characterizing the user. Adaptive methods can be seen at different levels of the search process: query formulation, browsing, search results visualization... Commonly, information describing a user is called a "user profile/model".

4   User Modeling and Information Search Process

In the previous section, we underlined that the implication of the user is important at the different steps of the information search process. Now, we present the way the user can be modeled, via a user profile, in information search systems. A user profile can be defined as a set of characteristics allowing the system to identify or to represent the user. Being general this definition hides the reality. Indeed, even if this definition can be applied to all the adaptive information search systems, the structure and the content of a user profile is very different from one system to another. This paper deals with this issue but do not concern the profile evolution (in term of content or structure).

In most of adaptive information search system we identify in the literature, characteristics describing the user can be split into two classes:

-   Personal characteristics (last name, first name, age...);
-   Characteristics related to information needs. It can represent for instance the user's preferences (language, format, content, freshness, document visit history, last queries, relevance judgments, behavior information...). These characteristics are not exclusive and can be affined taking into account:
    o   Short and long term profile [10]. This distinction allows the system to react to an emerging information need. The construction of the long term user profile is commonly done in processing short term profile;
    o   Positive and negative user profile [6]. Most of the time, positive user profile are use to represent the user. Unfortunately, in this case, systems are missing important plus-value information that is to say information that the user is not interested in (negative profile). The integration of these two kinds of processes makes the relevance of retrieved documents higher.

Concerning the real user profile content, we can unfortunately underline that every system uses a unique user profile. For instance information needs can be described by a common keyword list [9] or by a more complex structure like a tree [5]. Some research initiatives tend to build a general information search user profile like APMD[2] (Accès Personnalisé à des Masses de Données) project [12]. Moreover, we have to notice that information search systems have only a partial view of the user since each system manages only a part of the search process (querying or browsing...). But, during a search process, the user manually switches from one activity (browsing/querying) to another. Thus, few works integrate browsing and querying.

Furthermore, the evolution of mentalities and technologies makes the raise of Web 2.0 which is based on participative/adaptive/"mash up"... In this context, to limit these drawbacks, we propose a new generic profile model that allows systems to integrate profiles coming from different applications and/or having a different structure. This model relies on a flexible description of resources (information/users/...).

5 Towards an Extended User's Model

Each application linked to the searching for information task has only a partial user model and specific information (the overlap of search engines database is low). For this reason, it is important to propose a model which allows the applications to interoperate and share information and representations. This approach is motivated by the raise of "mashup" technologies on which Web 2.0 relies. Each application could offer a specific API (Application Programming Interface) to give access to his services/information. In the same time, it is surely utopian to think that all the applications in an information search context will use exactly the same characteristics (name, type, semantics...) to characterize information and users. Thus, we have proposed a profile model which is generic, and nondeterministic. It supports the interoperability of the information search tools (matching between profiles coming from various applications) and it allows the system to model the user like any other element implied in the application (information for example). The interest of such a model consists in the integration of a semantic level making it possible the matching between profiles not being built in an identical way (coming from various applications for example). The model is not deterministic: the system can identify thanks to semantic information the links (matching) between the various profiles which it uses. This is why anybody could even imagine that any user could insert in his profile the characteristics which he wishes. Indeed, why the user could not described himself? He is however that who knows himself best! The suggested model thus does not describe what it is necessary to put in a profile but is used as a basis for management and matching between any types of profile. So the generic profile we propose is depicted in Figure 3 (in UML format). In this model four specific parts can be identified:

- The *Logical Structure* that corresponds to the real tree-like organization of attributes in the profile. In the model aggregation of existing models is possible via the Reusable attribute inherit;

---

[2] http://apmd.prism.uvsq.fr/

- The *Logical Structure Semantic* that associates semantic to the logical structure (described resource and relations between resources, concepts associated to each attribute...);
- The *Content* that corresponds to the real value of each attribute;
- The *Content Semantic* that associates semantic to the content of each attributes (type, logical expression).

The techniques of matching and reasoning are carried out using RDF associated with SparQL [14]. More details concerning this model and its exploitation are available in [4].

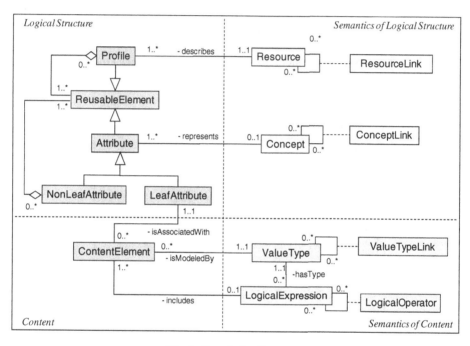

Fig. 3. Generic Profile Model

The semantic level of the model allows the system to find relationship between elements having different names or structures. This mapping is done owing to the 6 following steps (Figure 4).

The two first steps correspond to the conceptual matching between attributes from different profiles. It is done via a specific SparQL[3] query (Figure 5). This query verifies if two concepts C1 and C2 can be matched. It uses the list of available concepts from the file (Concepts.rdf). It verifies the matching between C1 and C2 through the sameAs and equivalentClass relations defined in OWL (many other relations could be used). <cf1> and <cf2> are the concepts for which the system has to verify the matching. Note that the "." in the *Where* clause corresponds to the logical AND.

---

[3] http://www.w3.org/TR/rdf-sparql-query/

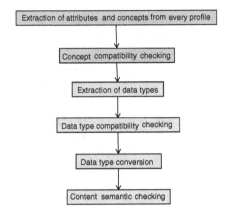

Fig. 4. Different Steps of the Profiles Automatic Matching Process

```
PREFIX owl:  <http://www.w3.org/2002/07/owl#>
PREFIX rdf:  <http://www.w3.org/1999/02/22-rdf-syntax-ns#>
SELECT ?c1  ?c2
FROM <Concepts.rdf>
WHERE
{
  ?c1 rdf:type sp:Concept .
  ?c2 rdf:type sp:Concept .
  { ?c1 owl:equivalentClass ?c2 }
   UNION
   { ?c1 owl:sameAs ?c2 } .
FILTER (
   (?c1=<cf1> || ?c1= <cf2>)
   &&
   (?c2=<cf1> || ?c2=<cf2>)  )
}
```

Fig. 5. SparQL Query to Check the Matching Between Concepts

The four last steps correspond to the data matching between attributes from different profiles. Three main processes can be identified:

-   **Extension of data types.** Lists are completed with missing values. It is important that all profiles have the same type. Note that the second value corresponds to the weight [0; 1] of this value in the attribute. For example, (French, 1) indicates that *French* for profile #1 is more important than *English* (English, 0.5);

Languages from Profil#1 {(French, 1), (English, 0.5)}

Languages from Profil#2 {(French, 0.5)}

Result of the transformation:

Languages from Profil#1 {(French, 1), (English, 0.5)}

Languages from Profil#2 {(French, 0.5), **(English, 0)**}

- **Modification of data reference.** The format of any attribute can be a list of values or can be a symbolic name that is expressed through a logical expression. In the following example, the freshness in profile #2 is partially characterized by the symbolic name "low" that has been expressed as the logical expression "greater or equal to 4 days";

> Freshness (in days) from profile #1{(4, 1)}
>
> Freshness (in days) from profile #2{(low, 1), (average, 0.3)}
>
> *(corresponding logical expression)* {(>=, 4)} {(>=, 1) AND (<=, 3)}
>
> *Result of the transformation:*
>
> Freshness (in days) from profile #1 **{(low, 1), (average, 0)}**
>
> Freshness (in days) from profile #2 {(low, 1), (average, 0.3)}

- **Data type implicit conversion.** Every type can be automatically converted thanks to a type translation library. For instance days can be converted into years;

> Freshness (in **days**) from profile #1{(4, 1)}
>
> Freshness (in **years**) from profile #2{(0.01, 1)}
>
> Both expressions represent the **same values** that is to say 4 days thanks to a naïve transformation (4/365=0.01).

Thanks to these different steps, the system can then compute similarity between two profiles. To do this, a weight is associated to each attributes having to be compared. Then, the weighted similarity is computed in a classical way (cosine measure for instance). A visual interface has been proposed to verify which attributes have been matched between two profiles.

As a conclusion, thanks to this model and particularly to the semantic level, each information search tools can exchange his information/user profiles with other tools via a specific API. Since search tools have a specific view of documents/users, they provide specific profiles having "proprietary" profile structure. To allow any search tool to manage these heterogeneous profiles coming from different other tool, the semantic level must be shared between all the search tools as a knowledge database. Indeed, if the different profiles are linked to the semantic level, the matching between heterogeneous profiles is easier.

In parallel, we carried out a thought concerning the types of profile features in searching for information. In addition to the features personal and related to the requirements in information, relational features emerge. Indeed, the user cannot be regarded any more as an isolated entity. He has relations with other individuals. These individuals are likely to have the same needs in information. These relations (formal or not) can be exploited at ends of recommendation, P2P searching, indexing… Within this framework, we study tracks allowing the system to take advantage of the information belonging to a group. They are there the premises of the exploitation of the collective intelligence to the service of the information search process.

6 Conclusion and Future Works

Taking into account the user within the framework of the information systems and the information search process is fundamental. The era of the applications not taking into account the users is completed. A current system must adapt itself to the user: it is necessary first of all that the system identifies the user needs and his context in the information search process. From this moment research works concerning the contents of the profile are in their early stage. At the present time, we can say that even if many projects try to standardize the information contained in a profile, each application has a partial profile i.e. a partial user's characterization.

In term of prospects, the taking into account of the user must be improved. In particular the capacity of the users (other that the domain-specific knowledge and the practice-specific knowledge) must be considered. Some features such as the age or a physical handicap influence the adaptation which is necessary to return useable results. Various works can be quoted in this framework such as *Sissi* (http://sissiprojet.free.fr/). Thanks to the study of the behavior of various users having different capacities, the Sissi project wants to identify the impact of some Web Services in any information search process. These Web Services would allow for example the adaptation of the contents compared to the various users' capacities.

The taking into account of social aspects (relationship between users) and of collective aspects should be still developed. Even if a collaborative information search emerges, the exploitation of the collective intelligence from our point of view is isolated. However the study of the social networks [18], the study of the uses of the documents by a group are sources of semantic information useful for an information search because it could thus integrate external information in addition to the simple contents of the documents; thus this information has a strong added value.

7 Discussion

It is important to underline an important limit when taking account of the users within the framework of the information search process. The user's profile must be sufficiently reactive. Indeed, it is necessary to take into account the new centers of interests. In addition, it is important that the user can control the profile which characterizes himself. It is of primary importance that he has the possibility of consulting or modifying information that relates to him. Lastly, an adaptive system within the framework of the information search process must take care not to be intrusive. Indeed, the current systems tend to be completely automated under pretext that they know the user through his profile. However, in the case of ambiguity for example, the tool will really refer "adapted" documents not meeting the user's needs. It however seems judicious that, in spite of the adaptation provided and necessary, the user must keep under control his search process. The system must act more as a companion than as a substitute.

References

1. Agosti, M., Smeaton, A.: Information retrieval and hypertext. Kluwer Academic Publisher, Dordrecht (1996)
2. Baeza-Yates, R., Ribeiro-Neto, B.: Modern information retrieval, ACM Press, Addison Wesley. ISBN 0-201-39829-X (1999)
3. Belkin, N.J., Croft, W.B.: Information Filtering and Information Retrieval: Two sides of the Same Coin? Communication of the ACM 35(12), 29–38 (1992)
4. Chevalier, M., Soulé-Dupuy, C., Tchienehom, P.: Profiles Semantics and Matchings Flexibility for Ressources Access. In: IEEE international Conference on Signal Image Technology and Internet based Systems, Yaoundé, pp. 224–231. IEEE Computer Society Press, Los Alamitos (2005)
5. Chevalier, M., Chrisment, C., Julien, C.: Helping people searching the web: towards an adaptive and a social system. In: Asaias, P., Karmakar, N. (eds.) IADIS/WWW Internet 2004 (2004)
6. Hoashi, K., Kazunori, M., Naomi, I., Hashimoto, K.: Document filtering method using non relevant information profile. In: Proceedings of the twenty third Annual International ACM SIGIR Conference on Research and Development in Information Retrieval: Distributed Retrieval, pp. 176–183 (2000)
7. Hölscher, C., Strube, G.: Web Search Behavior of Internet Experts and Newbies. In: Proceedings of the 9th international World Wide Web conference on Computer networks: the international journal of computer and telecommunications networking, pp. 337–346. North-Holland Publishing Co., Amsterdam (2000)
8. Jansen, B.J., Spink, A., Saracevic, T.: Real life, real users, and real needs: a study and analysis of user queries on the web. Information Processing and Management 36, 207–227 (2000)
9. Korfhage, R.R.: Information storage and retrieval. Wiley computer publishing, Chichester (1997)
10. Mizzaro, S., Tasso, C.: Personalization techniques in the tips project: The cognitive filtering module and the information retrieval assistant. In: Proceedings of the Workshop on Personalization Techniques in Electronic Publishing on the Web: Trends and Perspectives, Malaga (Spain) (May 2002)
11. Montaner, M., Lopez, B., Rosa, J.L.D.L.: A taxonomy of recommender agents on the internet. Artificial Intelligence Review 19, 285–330 (2003)
12. Naderi, H., Rumpler, B., Pinon, J.M.: An Efficient Collaborative Information Retrieval System by Incorporating the User Profile. In: 4th International Workshop on Adaptive Multimedia Retrieval, University of Geneva, Switzerland (2006)
13. Pazzani, M.: A framework for collaborative, content-based and demographic filtering. Artificial Intelligence Review 13(5-6), 393–408 (1999)
14. Prud'hommeaux, E., Seaborne, A.: Sparql query language for rdf. W3C Working Draft (July 2005), http://www.w3.org/TR/2004/rdf-sparql-query/
15. Silverstein, C., Henzinger, M., Marais, H., Moricz, M.: Analysis of a very large web search engine query log, SRC technical note #1998-014 (October 26, 1998), http://gatekeeper.dec.com/pub/DEC/SRC/technical-notes/abstracts/src-tn-1998-014.html
16. Spink, A., Jansen, B.J., Wolfram, D., Saracevic, T.: From e-sex to e-commerce: web search changes. Revue IEEE Computer 35(3), 107–109 (2002)
17. Teevan, J., Dumais, S.T., Horvitz, E.: Personalizing Search via Automated Analysis of Interests and Activities. In: proceedings of the 28th annual international ACM SIGIR conference on research and development in information retrieval, pp. 449–456 (2005)
18. Wasserman, S., Faust, K.: Social Network Analysis - Methods and Application. Cambridge University Press, Cambridge (1994)

A Tool for Statistical Analysis of Navigational Modelling for Web Site Personalization and Reengineering

F.J. Monaco, C.X. Sheng, and M.L.M. Peixoto

Instituto de Ciencias Matematicas e de Computcao,
Universidade de Sao Paulo Av. do Trabalhador SaoCarlense, 400, 13560-979,
Sao Carlos - SP - Brazil
monaco@icmc.usp.br

Abstract. A suitable navigational model, conveniently tailored for the needs and access characteristics of the intended public, is one of the basic ingredients for ensuring the usability of successful Web applications. Properly identifying the navigational requisites of large Web applications, though, is not always an easy task particularly face to structural changes introduced by Web site evolution, which is prone do deviate the effective navigational patterns from those conceived in the original model. This paper introduces an under development research tool for structural modeling and statistical navigational analysis of Web applications which traces link paths exercised by visitors and allow the identification of characteristic navigation patterns. Application in Web restructuring, pre-fetching cache techniques and adaptive content systems are discussed.

1 Introduction

Conceiving a suitable navigational model, conveniently tailored for the needs and access characteristics of the intended public, is one of the basic steps for ensuring the usability of successfull Web applications. Systematic Web design methodologies therefore relay on an adequate identification of navigational requisites for the careful organization of the application structure and well-planned emplacement of hyperlinks. Distinct navigation tools (indexes, guided tours etc.) apply to different purposes and the parsimonious positioning of contextually relevant links is important in order to avoid that either its excess subjects the user to disorientation or its insufficiency hinders the efficient retrieval of the information content [1] [2] [3].

Properly identifying the navigational requisites of large Web applications, though, is not always an easy task. Inherent intricacy of complex hyperconnection structures may be considerably difficult to map — and become even harder when we consider dynamic sites [4]. Alternate or overlaying navigation models may be provided for the diversity of interests of an heterogenous target public. Moreover, with the natural evolution process, continuous modifications in the application structure may deviate the navigation scheme from the one originally

M. Weske, M.-S. Hacid, C. Godart (Eds.): WISE 2007 Workshops, LNCS 4832, pp. 386–394, 2007.

proposed. Finally, users' intuition is something tricky to infer, thence it is not guaranteed that people will adhere to the navigation scheme as supposed at design time — access patterns may even change with time as users gain familiarity with the Web site or layout is altered.

If effective navigational patterns deviate too much from the ones for which the hyperlinks structure was designed for, usability may be compromised in unforeseen ways. It may happen that some resources are accessed through unnecessarily long paths, possibly leading the user to confusing meanders unrelated to the signaled context and prone to diverting visitors from the intended destiny. Overall performance of the application may be reduced due to inefficient resource utilization, user satisfaction may be affected and important information content may be left aside. Specially for complex and dynamic Web application it is thus interesting to monitor the accesses to the site and keep track of the navigational patters effectively exercised by the users, and compare then with the original design. If discrepancies are find, restructuring of the application may come in handy either to restore usage scheme to the original plan or to correct conceptual mistakes or changing navigational requisites.

Restructuring of Web application is an important topic of research under Web Engineering and is based on statistical analysis of user accesses. Most automated tools for such purpose relies on the information provided by the log facility of the Web server. Unfortunately, though, standard logging mechanisms of most popular servers are meant to register statistics of the number of accesses to single resources, for that this piece of information is used by application maintainers to estimate Web site popularity and to bill sponsors that "pay for hits"; no specific provision for tracing the navigation paths are available.

Typically, these access logs are provided as line-oriented text-files where each entry corresponds to a request-reply HTTP transaction with a few data fields that, for each event, informs the date of the occurrence, the address of the client, the requested resource, the server's response status code and possibly some extra detail concerning software and protocol version. With only this piece of information, existent tools for Web statistical analysis usually try to infer navigational data by judicious guesswork based, for example, on the interval between successive requests from the same origin. For instance, if it is known from the application structure model that a given page A has a link to another page B, then, if a request from a given IP is soon followed by a consecutive request of the same IP for the page B, then it is assumed that both requests correspond to the navigation of an unique user from A to B. While fairly reasonable for a very simplified scenario, such assumption may be too optimistic concerning real word utilization. It neglects, for example, that many distinct users behind a firewall — an increasingly common reality nowadays — may reach the Internet through the same IP, thereby leaving indistinguishable entries in the log file.

Likewise, it is not possible to unambiguously separate explicit requests (the user entered a new URL) from implicit requests (the user activated a link), nor the ad hoc usage of backtracking (browser back button). There is also the vague premise that a few successive close-in-time requests always denotes a single user navigation and the question of the empiric value of such time interval. Finally, even probabilistically, this guess can only tell something about page-to-page navigations; no clue regarding followed links are given. For instance, if page A has two navigation bars, one at the top and another at the bottom, there is no hint to know which bar was effectively used — and therefore the statistical navigation analysis that can be drawn is limited. An improved log facility which produces useful data for tracing user navigation through the Web application is the subject of this paper.

2   Navigational Log

While standard *resource access log* facilities collect information on the number of requests for each resource in a Web application, our *navigational log* is intended to register the paths effectively followed by users, identifying not only which pages were visited, but also which links within that pages were selected. The software NavLog, in development at the Laboratory of Distributed Systems and Concurrent Programming, ICMC-USP, is a tool for statistical navigation analysis of Web application intended to this goal. It comprises three main modules, illustrated in Figure 1, which interact as follows.

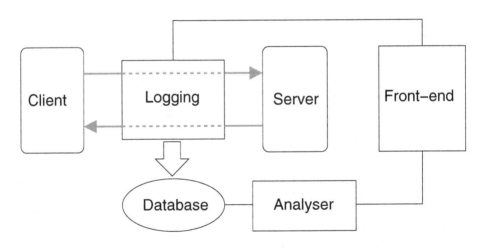

Fig. 1. Block diagram of the NavLog modules

2.1   Logging

Unless extra data is provided, the HTTP request sent by the Web client to the server carries no identification of either the mechanism (explicit or implicit) that generated the interaction or the hyperlink, if any, that was selected; if such information is needed, it must be artificially provided by the system — that is the duty of the logging module.

In the current prototype this module is implemented by a stand-alone TCP proxy daemon that intercepts the communication between the Web client and the Web server. It can run either in the client's machine to analyze local access to remote Web servers, or in the server machine to analyze external access to the web site. In the former case, the user can use the logging module by activating the proxy options of the Web browser, while in the latter the site manager can set the proxy to listen to the standard HTTP port and forward the incoming requests to the Web server running in an alternative port (future work may include the implementation of the logging system as a loadable Apache Web server module). The logging process can be summarized as follows:

- On the first access from the client to the server, the logging module intercepts the request and starts a new *user session* by assigning it an unique identification number (session id) and then forwards the message to the server. It also create a new entry for the session in its database associating it with a set of session data which include client address, requested resource etc.
- When the server responds, the proxy once again intercepts the communication and pass it through a parser algorithm that seeks for references to other documents. For each one it founds, it marks it with the corresponding session id, so that every link (URL) in the document includes an additional parameter identifying an specific navigation session. Before sending back the document to the client, the proxy also consults its database looking for an entry of the currently examined page. If none is found, a new entry is created; otherwise the time stamp of the most recent entry of the page is compared with the age of the file sent by the Web server. If the file is newer than the last log information, than a new entry with the current time stamp is created. In both cases, if the page is registered in the database, it is assigned an unique page identification (page id). Every link in this page is also examined by the parser, assigned an unique link identification (link id) and registered in the database along with its respective URL (Figure 2). Page id and link id are also added to every hyperlink along with the session id.
- When the client's browser renders the document, every link will have embedded in it the identification of the current session, page and link. Therefore, when the user issues a new implicit request by selecting one of the links, the composed URL that will be sent to the server carries those pieces of information.
- The logging module parses the new request message and, detecting the identification tokens, establishes that it is associated with an active session. It then adds a new entry for the session transaction database and forward the request to the server, but not before removing the identification tokens so that the communication proceeds transparently as though the data flow had been untouched.

Since every link is precisely marked with unique identifiers, when a new resource is requested the logging module knows exactly from which page and which link it was referred, and thus is possible to trace the user navigational paths along

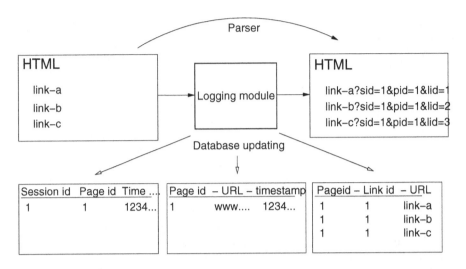

Fig. 2. Logging module

the site. HTML forms are treated analogously by artificially adding and removing hidden fields. In order to guarantee consistence for further analysis process, every time a page changes, a new snapshot of its hyperlink structure is added to the database. A new session is created when an unmarked request is received, and it can be ended after a specified time out period. In addition to the session, page and link id, an additional identifier adds a serial transaction number from which the proxy can detect duplicated requests such as those caused if the client uses the back button facility of the web browser or selects a stored URL from a bookmark collection. Current implementation uses an external SQL database (Postgresql[1]) server to manage logging data.

2.2   Analyzer

The analyzer is another stand-alone module intended for off-line processing of the log database. Given a time interval, it goes through the log entries examining user sessions and extracting information concerning hyperlink structure and navigational patterns, as well as their evolution along the time. Following the page-to-page navigation logging through the visited links the analyzer is capable of reconstructing the Web application structure. Therefore if a Web crawler is let to recursively scan the entire site, then the complete model can be build from the scratch based on the generated navigational log (all that is need is to tell the analyzer to process only the crawler transaction).

As to analyze the adequacy of the navigation model, instead of an automatic spider, the designer can start a session himself to exercise the planned navigation movements within the application, thence generating a reference log. The

---

[1] Postgresql: *www.postgres.org.*

remaining entries in the log database can then be processed to compile statistical data regarding the effective utilization of the reference paths. For example, if there is a planned path from A to D through the sequence $ABCD$ and the log shows that most users in reality go through $ABXYZD$, this may be an evidence that the original navigation model does not match the users' intuition or that the link from B to C is not clear enough to guide the user to the desired page — some structural change or layout modification may be needed.

The analyzer can also detect specific navigation patterns such as circular paths — if these are not foreseen in the model, their presence in the statistical log may be an indication of disoriented users wandering around in circles. Tree-like paths as in a depth-first search probe may suggest that users need to undergo an exhaustive exploration to locate target piece of information, whereupon the designer may try adding more explicit references to critical resources. Rarely followed paths and excessive traffic can be detected as well. Also, the designer may learn that the bottom navigation bar is seldom used, or that a given incidental link to some non-relevant information is misleading the users to wrong ways.

If a manual intervention is performed, such as the addition of a new link or a layout change to highlight a link, the analyzer can produce statistics of the utilization patterns before and after the modification, so that the designer can evaluate the impact of such a change on the site navigation.

2.3   Front-End

A graphical user interface implements a convenient front-end for accessing both the logging and analyzer modules and for the visualization of the application structure. An illustration of the interface under development is shown in Figure 3.

The diagram shows the retrieved structure of a simple Web site comprising three pages A, B and C; a forth page labeled "The Web" represent the rest of the Internet. Each page has a few links which are numbered in the order that they found by the logging module's parser. The solid arrows denote links from one page to another, while the dashed arrows indicate an artificial "jump" to other page by, for example, explicitly entering an URL. Dash-and-dot arrows from the Web to the other pages mean implicit or explicit movements from the outer world into the web site, while arrows from the pages to the Web show points where the user has left the application.

By specifying different time intervals for the analyzer, the visualizer can produce corresponding snapshots of the web site, evidencing structural changes. The functionality under implementation will also permit the designer to label the arrows with access frequency or to paint them in different colors to represent the same information. By clicking in consecutive links from one source page to one target page, the user can select an entire path instead of a single link — this can be useful to compare the traffic from A to C through B with the one that goes directly from A to C.

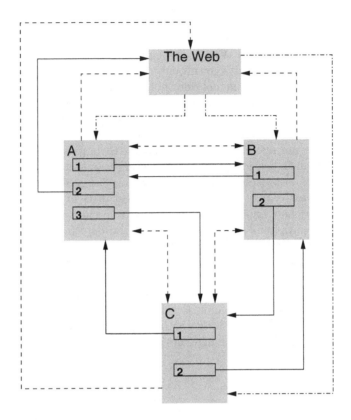

Fig. 3. Visualizer module

3   Implementation

NavLog is a prototype under development, several features are partially implemented and there is considerable work ahead before one fully operational tool
can be released. When ready, it will be distributed as free software under the
GNU GPL[5].

The logging module is the where the development has advanced most. It was
implemented from the scratch in C for GNU/Linux platform and makes use of
standard socket and thread APIs. As main limitations it has been noticed that
the module performance is below the expected, sensibly interfering in the normal server response time. It is believed that the rudimentary parser accounts
for this delay, and efforts to replace it with an optimized version are currently
been spent. As an alternative, it is been considered the replacement of the preliminary prototype with the free software *tinyproxy*[2], which has shown to be a
well-tested and good-performance solution into which the logging functionality
can be implemented.

---

[2] Tinyproxy: *http://tinyproxy.sourceforge.net.*

The analyzer is the second module in development stage. It is also been implemented in C and compiled for GNU/Linux. Basically, it connects to the Postgresql database server and extracts structural and statistical information. As for the first task it produces a text file describing the application model in a form which is easy for the visualizer to interpret. The statistic analysis, on the other hand, is less developed and uses some techniques that are subject of ongoing research.

Finally the visualizer is in the earliest stage of implementation. It is based on the GTK library and implement only basic features.

4   Conclusions and Future Work

NavLog is a tool for Structural Modeling and Statistical Navigational Analysis of Web applications. It extends the standard access logging facility of popular Web servers by including data on the paths accessed by users, thereby producing a navigational log. Based on this information, it allows the designer to compare the planned navigational model with effective navigation exercised by users, to detect abnormal or unwanted navigation patterns and to identify navigational changes with time. It also help the designer to analyze the impact of modifications in the structure of layout of the application regarding navigation scheme.

In addition to the analysis and restructuring of Web application, the augmented information provided by the navigational log can be useful also for other advanced purposes, some subject of active research in our research group. One of them is on policies of cache substitution for the Web. A more accurate knowledge of navigation patterns allows for better decisions on what documents should be cached and what should be discarded to release space. A step further in this direction concerns pre-fetching techniques [6]: if reliable statistics about the usual navigational paths are available, then from the list of last visited pages reasonable prediction for the next document to be requested can be made — and this data can be used either by the server or by the client to anticipate the user by pre-loading the probable document into the cache.

The implementation of differentiated services can also benefit from the navigational analysis. It is possible for example to detect distinct patterns of navigation and, based on then, identify classes of users with different characteristics and needs. Adaptive content can then be offered and special QoS levels can be selected. One advanced sophistication to be explored under this research project is the implementation of proactive dynamic Web applications, whose context-aware navigational structure is automatically tailored for the user characteristics. For instance, if its known from previous statistical analysis that most users that reach the page D through the path $ABCD$ do not detain themselves for longer at the intermediate pages before moving to E, then some automatic restructuring rule based on a suitable IA mechanism can decide to place a link in C directly to E for the convenience of the user.

Among known issues, is the fact that according to the current trends, advanced web design technology is moving toward the extensive use of embedded scripting language. This naturally poses new challenges for our content parser module, which is no more ambitious than a proof-of-concept implementation. To cope with novel technologies such as client-side dynamic content generation (including not only javascript but also all the new generation of scripting languages such as python and ruby), operational modules of NavLog have a long way ahead.

Statistical navigational analysis is a relevant subject of research in Web Engineering. This work report is intended to contribute to this area. The NavLog project is part of this effort.

Acknowledgements

This research project has been supported by FAPESP (*São Paulo State Foundation for Research Support*) - by CNPq (*National Council for Research and Development*) and by CAPES (*Foundation for the Coordination of Academic Staff Advancement*).

References

1. Ricco, F., Tonella, P.: Dynamic model extraction and statistical analysis of web applications. In: WSE 2002. International Workshop on Web Site Evolution, Montreal, Canada, pp. 43–52 (October 2002)
2. Ricco, F., Tonella, P.: Construction of the system dependence graph for web application slicing. In: SCAM 2002. International Workshop on Source Code Analysis and Manipulation, Montreal, Canada, pp. 123–132 (October 2002)
3. Ricco, F., Tonella, P.: Information Modeling for Internet Applications. In: van Bommel, P. (ed.) XI - Web Application Quality: Supporting Maintenance and Testing, pp. 231–258. Idea Group Publishing, USA (2002)
4. Monaco, F.J., Gonzaga, A.: Web Engineering: Managing Diversity and Complexity in Web Application Developmen. In: Murugesan, S., Desphande, Y. (eds.) Restraining Content Explosion vs. Constraining Content Growth. LNCS, vol. 2016, pp. 3–540. Springer, Heidelberg (2001)
5. GPL. Gnu general public licence. Free Software Foundation, `http://www.gnu.org/copyleft/gpl.html` (1991)
6. Oliveira, D.X.T.: Análise e avaliação de técnicas de pré-busca em cacnes para Web (Analysis and evaluation of techiques of pre-fetching in caches for Web). PhD thesis, Instituto de Ciências Matemáticas e de Computação de São Carlos, Universidade de São Paulo (2004)

"Watch the Document on the Wall!" An Analytical Model for Health Care Documents on Large Displays

Niels Windfeld Lund[1], Bernt Ivar Olsen[2], Otto Anshus[2], Tore Larsen[2],
John Markus Bjørndalen[2], and Gunnar Hartvigsen[2]

[1] Documentation Studies, Faculty of Humanities, University of Tromsø,
9037 Tromsø, Norway
[2] Department of Computer Science, University of Tromsø, 9037 Tromsø, Norway
`niels.windfeld.lund@hum.uit.no`,
`{bernt,otto,tore,jmb,gunnar}@cs.uit.no`

Abstract. Very large, high-resolutions displays, such as tiled display walls with resolutions ranging from tens of mega pixels up to 100+ mega pixels enables visualization of data in ways never before possible. Even if the technology is still emerging and faces challenges such as interaction problems and development of rendering technology, both hardware and software, the potentials for visualizing large data sets are still very exciting. Within this thematic framework, we provide an introduction to the current state of research on use of very large, high-resolution displays and a general discussion of the role of scale/size of documents as basis for discussion and a conceptual framework for analysis of "large documents", specifically for medical-imaging related documents.

Keywords: High-resolutions displays, health care documents, document model medical imaging, telemedicine.

1 Introduction

Does the size of documents matter? Does it make a difference whether a document is viewed on a cell phone, a laptop or on a large display wall? When the number of pixels and the size of the display grow, what happens with the user's way of working with the documents and with how efficiently the information contained in the documents are recognized by the users? Are there usage scenarios where wall-size, high resolution, normal pixel density displays offers possibilities not already attainable by conventional display technologies? Obviously, a wall-sized, high-resolution display with an order of magnitude larger size and number of pixels, allow for several users collaborating or working on disjunct problems simultaneously. Users can move close to the display wall to view details, and move away to see larger structures. The complexity and the number of documents displayed simultaneously also increase. Figure 1, Display wall shows a 22 mega pixel display wall at the University of Tromsø in use displaying meteorological data.

Software is generally categorized into one out of two categories: either software meant for single-users or multi-user (groupware). Trying to categorize the use of large

M. Weske, M.-S. Hacid, C. Godart (Eds.): WISE 2007 Workshops, LNCS 4832, pp. 395–406, 2007.
© Springer-Verlag Berlin Heidelberg 2007

Fig. 1. Display wall

displays can yield the same result. The large display surface can either provide one viewer with a very large work-surface for himself or provide viewing space in front of the screen for several people to view the same information at the same time. In addition, the use of the large display is either characterized by being a presentation in that someone wants to distribute a message or information to a larger audience, or it might be interactional where several users are interfacing the information on the display and authoring it. A large wall-sized high-resolution display also makes it interesting for several teams to work simultaneously on disjunct problems, or with a limited interaction between the teams.

A commodity projector typically has a native resolution of 1024*768 or 1200*1024 pixels. When the distance between the projector and the screen increases, the size of the projected image also increases. However, the number of pixels does not increase as the image size increases, and the information content carried by the pixels does not increase as the physical size of the displayed image increases. For a single-projector image the pixels per inch (PPI) decreases as the size of the projected image increases. A projected image of width 200 inches has a horizontal PPI around 5.

If you move closer to the projected image from a projector you will eventually reach a distance where the individual pixels become visible and the projected image lose coherence or becomes coarse. The image will apparently lose detail. Even when the image itself comprises more pixels than the projector is capable of displaying, no further details will become visible when moving up close to the image.

A wall-sized high-resolution display of the same size as a large single-projector image, has typically an order of magnitude more pixels within the same area. The PPI is at least 35. As a result a display wall is capable of both displaying a large image, and retain the detail inherent in the displayed image (or information). Current high-end cameras provide images with 10 or 12 megapixels resolution, while screens typically only provide up to 2 megapixels displayed resolution. This would imply that the image document has at least five times more pixels than what it is possible to display with a single display. A display wall with a resolution between 20-100Mpixels is today the only way we have to display images from high-end cameras if we want to see all the pixels simultaneously.

Doing a brief comparison between the different pixel-sizes provided by the different technologies brings forth the differences in granularity they offer. A single-projector projecting an image on about 200 inches across would yield a PPI of 5. On the display wall the PPI is above 35. High-end 30 inch display provide PPI as high as 120 or more. The human eye has large difficulties in discerning individual pixels on a screen with over 200 PPI. So even a display wall has a long way to go before being at 120 PPI. However, a display wall is scalable because it is tiled and it can get better PPI by using more projectors.

Before going any further in details, it might be relevant to talk about the concept of a "document" in relation to size. Is the displayed document on your smart-phone the same as the one when displayed on the wall sized display? Is it the same document anymore? Having the entire document available in your eyesight or in the periphery of your eyesight – how does this affect the perceived document? Is it still the same document? Another question is how to select the parts for the actual screen and frame the screen, in other words, create the visible document. It is a matter of parts and wholes/totalities. How should you be able to zoom and select specific images and still have an option to access an overview of the patient? It is a matter of composition of documents/interface. Where, when and what is a document? Is the document what you can see on the screen or is it the repository as a whole? From a technical viewpoint the answer is that it depends. -There are several transformations we can do to the data and to the visualization of the data:

(I). Transformations of the data itself
  a) The document data is scaled with loss to adapt them to the characteristics of the display it is displayed on. The actual displaying can happen in one of two ways:
    i. The document data is transmitted to the cell phone where it is unpacked and displayed by using software running on the cell phone.
    ii. The document data is read by software on some computer and the pixels are sent to the cell phone where the pixels are displayed. The actual document data is not sent to the cell phone.
  b) The data is not changed or loss-lessly changed. Otherwise as (a).
(II). Transformations of the visualization of the data: The data is not changed or transmitted to the cell phone. Instead, just the visualization of the data is sent pixel by pixel to the cell phone from one or several computers.
  a) The pixel representation of the visualization is scaled to adapt it to the cell phones display characteristics.
  b) The pixels are sent as is to the cell phone. The cell phone then

    i. display the pixels 1:1. If the image has more pixels than the cell phone display we must scroll around to see various parts of the image.

    ii. Or – scale them by running software on the phone. The image can be compressed to fit it into the phone display, but the image will be very small and hard to read.

As can be seen the document data can change or not, and the visualization can change or not.

A large display also allows for simultaneously viewing of several documents, and of more complex documents. A large, interactive, display surface also supports all types, or modes of documents, including text, images, video and audio. Comparing to the small display on a pda, the wall-sized display clearly can display both more documents and more complex documents simultaneously.

However, a document is not only a physical object containing a number of data. It is also a social phenomenon playing a certain role in a social environment [1], [2], [3], [4]. When data is displayed via an interface on a cell phone it is an individual document and when it is shown on the large display wall it becomes a collective document shared by a number of people located in the same room. However it is closely related to the documents on the cell-phone based on the same data. One may talk about a complex network of similar documents around for instance the same patient, person or place. Some of these documents may be large and only last a second; other may be small and last a year, if saved. But they are all documents. It has been suggested to reserve the notion of document and documentation for the field of perennial based documents capable of being retrieved and reused and talking about semiotic production when it comes to the very communicational event [5]. If one embraces the whole process of communication as a complementary process of documentation, it allows you to discuss how to design and create an instant document only supposed to last for a second and not necessarily planned to be reused. If you go to the roots of the Latin word: documentum, you get the verb doceo and the suffix mentum [3]. Doceo/docere refers to the action of demonstrating, teaching, to show something and not necessarily to preserve something. The suffix mentum refers primarily to the grammatical means of turning a verb into a substantive, but one may claim that it indirectly tells you that without some kind of means, you cannot demonstrate, teach or show anything. It leads to an option of choosing means, either small or large display and more or less lasting media, when you create a document. The important and defining question around a document becomes how you form and frame a document. It makes it reasonable to distinguish between small and large documents. -We will return to this in discussion and in the rest of this essay try to present an ontology for 'large documents', documents to be viewed on a large, high resolution display based on previous findings within research primarily in the field of human-computer interaction related to large displays and how the issues found within this research relates to some distinct use-scenarios within medicine: radiotherapy planning and radiology.

NOTE. – The display wall is only one of several ways of implementing large, high/resolution screens. It is in this context used as an illustrating example. There are other interesting technologies in this regard, as the Milan touch-board from Microsoft[1], but the issue we wish to discuss here is the one of size and resolution, which impacts the amount of- and format of the displayed information.

2 Research and Findings on Large Displays

As stated by Jonathan Grudin in [6], a typical large display today only occupies about 10% of your visual field available by moving your eyes. Consequently, quite a large part of your visual perception remains unused in front of a traditional screen. If we include the area visible when moving your head, the screen covers only about 1% of the visible area around you. As Grudin argues – in the natural world, having only 1 to 10% of your physical surroundings visible, you would probably feel quite restricted by your eyesight.

Taking two extremes into consideration, having either – say 10,2 inches on your ultra-portable laptop available – or having 230 inches on the wall – what are the affordances of the two displays in terms of perception, interaction and information access? A networked computer connected to the Internet, with a web-browser, is potentially an (increasingly-) infinite information source. Having a larger display available makes it possible to have more information available for viewing simultaneously. But anyone who has tried to search the web for information, or even worse, trying to gain overview of some topic has experienced the information-overload issue. What role does the size of the display have in this regard? And how large does the screen have to be in order to challenge the boundaries of our perception? In a recent experiment, Yost and North [7] set out to investigate this and compared a 2 megapixel display to a 32 megapixel display with regards to scalability of visualizations. What they found was that with a 32 mega pixel display the perceptual limits of humans was not challenged. This means that, for at least the kinds of visualizations investigated, increasing display size and resolution helped perception and the ability to solve problems based on information displayed.

The small physical size of typical commodity displays, ranging from 17" to about 30" makes it crowded in front of the screen if several people are to view the same information. Projectors, which allow for multiple viewers, have the drawback of losing detail when moving closer, although they allow for multiple users to view the same. Wall sized displays, and similar high resolution, large displays allow for both high detail (amount of pixels) and large physical size. Hence, the technology allows for both large amounts of information to be viewed concurrently and for multiple people seeing the same, almost from the same angle/viewpoint while being able to discuss and interact with the information. Single users will benefit from having more information viewable and be perhaps less restricted by the limitations of their working memory. Initial studies show that the use of large displays improve performance of certain tasks along with improved recognition memory and peripheral awareness [8]. Larger screens also provide a more immersive experience and as a result motivate different and more efficient mental strategies for the task of spatial orientation [9]. Spatial task orientation is an important sub-task in any work involving orientation

---

[1] http://www.microsoft.com/surface/ (accessed 31.8.2007)

related to 3-dimensional models or objects. How often such a task is involved in clinical decisions remain at this point speculations, but for now it suffices to mention that many tasks relate to visual models of the human body.

In their 1997 article, Swaminathan and Sato [10] recognize and classify three possible configurations for a large display, Distant-contiguous, Desktop-contiguous and Non-contiguous. As this classification is done based on a prototype system that was made out of a configuration of only six 29 inch CRT displays, not all of the points made are equally relevant for our discussion. However, both the distant-contiguous and desktop-contiguous are highly relevant. The distant-contiguous configuration means that the display is large and contiguous, but placed at a greater distance from the user, so that he can view all of the content on the screen without rotating the neck (visual angle of 20-40 degrees). The desktop-contiguous configuration is when the display is still large, but placed at reading distance from the user (without scaling the text). Advantages and characteristics of the two relevant configurations that they found in their setup were (we also add our own comments and subjective experiences):

(I). Distant-contiguous
      a. Advantage: a large contiguous drawing surface, all visible at once. This is substantiated in our experience from using a wall-sized 22Mpixel display wall over several years.
      b. Disadvantage: detailed work, such as word processing, drawing, etc. is done out of the so-called resting length of accommodation (in short: comfort zone for our eyes for such 'detailed' work). This is not in accordance with our experience. We have a display wall without any borders between the tiles, and we have software enabling us to scale any window, text, and drawings to compensate for a greater viewing distance.
      c. Useful for: large amounts of interrelated information (maps, design schematics, large models, graphs, etc.)
      d. Useful for: shared view for multiple users
(II). Desktop-contiguous
      e. Advantage: best of both worlds in large screen size and standard reading- or working distance
      f. Useful for: large amount of interrelated information where any part of the information can become useful for the user at any time
      g. Disadvantage: essentially a single-user configuration. This is also our experience. However, there is a rapid progress towards multi-user systems. We have developed an extensive multi-user support for our display wall including support for multiple mice, keyboards, sound and hand gestures from multiple users simultaneously.

Another interface issue mentioned by Swaminathan and Sato is pointer movement and control. This issue clearly needs different solutions once the display reaches a certain size. The issue of lost cursor is imminent even with a two-screen display setup, which is normal today, especially in radiology departments. Sketched solutions to this issue are: direct manipulation using laser pointers or touch, nonlinear mapping with sticky controls letting the cursor move quickly over 'empty' spaces and slowly over objects, and the dollhouse metaphor. The last strategy is to make a small model of the large display in one part of the display and treat manipulations on this "dollhouse" as manipulations on the larger display.

For our own display wall we have developed and implemented several techniques. These include using a cell phone, PDA, laptop or tablet computer as a magnifying glass into the display wall. The display on the smaller computer will always show a part of the display wall with centered around the cursor. This helps both in locating the cursor and in accurately positioning the cursor. Another technique we are using is to snap the fingers to make a sound wherever the cursor will be highlighted. The same technique is used to move a window automatically closer to the snapping user. A third technique we use is to have a large software magnifying glass on the display wall itself. When we need to increase accuracy or view details, we can rapidly bring it up and move it around.

Large displays providing a larger field of view also has proven to have a distinct and very interesting property regarding the gender biased task of (virtual) spatial navigation. Using computer screens to navigate in spatial models, or information spaces, men have the same advantage as in the real world. This means that when navigating in a virtual world, or manipulating and relating to 3-dimensional models on the screen, men will typically outperform females. However, when using large displays this advantage seems to diminish and even disappear [11]. This is quite an interesting fact, especially for situations where spatial orientation is a part of the daily work. Is this relevant for work done in hospitals and in healthcare? We will return to this question later.

Within the field of CSCW, the study of large displays has been a focus for quite a while, as the recent book on public and situated displays by O'Hara, et.al [12] states. This volume presents an overview of the current state and focus of research on the topic. As mentioned herein, the displays that have been studied over the years (primarily since the early 90s) have differed in their:

- form factor (small monitors to wall displays)
- form of content (text, images, multimedia)
- whether content is
  - authored (e.g. advertisement)
  - evolving (meeting notes, sketches, etc.)
  - ad hoc (e.g. video)
  - what kinds of content interaction is supported
- whether intended for
  - single use
  - multiple users
- whether displays are intended for
  - focused collaboration
  - more arbitrarily use/other

As is apparent from this list, display-walls in particular has not been the research issue in focus, but large displays in general – anything from a whiteboard to advertising displays in airports or along highways.

What has been studied within these subjects are social, physical and cognitive consequences of design (ibid., p. xix) and placement – what makes the displays attractive, how they can provide awareness of activities of others, how to provide shared access to the displays and more. Two particularly interesting aspects of large displays discussed are (1) publicity and (2) situatedness.

The issue of publicity is also of importance even in a room full of medical expertise. It is most likely the patient in question's data that are available to the room, and in general, all medical (hospital-) staff would be permitted to view the displayed data. However, there are important exceptions. Firstly – even when sharing a visualization, all of the actual data behind it is not shared. An example is a PowerPoint presentation where only a few transparencies are displayed on the display wall, while the rest of the presentation file is kept on a private laptop and not made available. We call this presentation style sharing. Secondly, when the visualization of data come directly from the desktop of personal laptops and PDAs, only select parts or windows of each person's desktop should be made visible on the display wall.

Some relevant, but in this discussion perhaps more peripheral issues regarding "public displays", are ownership and access. As stated in O'Hara et.al, these topics are complex and multi-layered concepts that are resolved in an on-going processes of negotiation between the users – whether these are individuals, groups or organizations. In the case of a hospital setting we expect the technology to have commodity status within a few years and consequently should not be a regarded a scarce resource. However, these displays – either used in a presentation mode – or in an interaction mode, will be subject of ownership-of-control issues, as for instance whiteboards are. The notion of "collaborating artifacts" is when the piece of chalk to write on the blackboard is used as the control-mechanism for whoever is currently authoring the content of the blackboard. The users are, in this case, involved in a quite complex – but socially "fluent" process of taking turns to author the content. At the same time, there are clearly hierarchies or power-structures inherent in the interaction, like in brainstorming meetings, where the most active parties may get the credit for the ideas presented. Whether or not this is an issue when designing technology will depend on many factors – like the culture in the working place or in the organization.

The issue of "situatedness" covers concepts such as Situated Action, Distributed Cognition and Activity Theory, in addition to ergonomic design of the technology. These issues concern both physical design and especially placement of the technology. The latter will for instance decide potential audience for the display and decide use and use-context of the display. Interaction opportunities with large, public displays differ based on the location for use: corridors and transitional places will be less inviting for interaction, as people tent to hurry through these places from one place to another, and more relevant for displaying small notes and announcements. On the other hand, displays situated in a communal kitchen area, or close to the coffee machine will be more attractive for interaction because of the way people are used to behave at these places.

While this issue of situatedness is relevant for the concept of large, public and interactional displays – within the context of a hospital ward they will perhaps be a little more peripheral in regards to the issue of usefulness of the technology. At first glance – the question of where to put the displays are obvious: it should be placed where the imaging or discussion or decision-making is done. Or – is this so obvious?

If the hypothesis is that large screens offer improvements cognitively for the process of shared decision-making (teamwork and –brainstorming arrangements) and that this hypothesis holds truth, then the technology should obviously be placed where the decisions are made. However, decision-making – and especially distributed decision-making (involving more than one individual) have proved to be quite a complex matter. The important question will be: where are the decisions really made? Perhaps are discussions performed in a particular room (where the patient data are displayed), but afterwards people meet in the hallway, or in the cantina – or any other convenient and perhaps arbitrary place – and a short talk could be basis of the final decision regarding treatment. These issues have been proven within CSCW-research to be non-trivial and not apparent before actually studied.

In the following section we will take a look a scenario with a potential of benefiting from the use of large-display technology and how these cases relate to the research findings already mentioned.

3 The Radiology Department and Radiation Therapy Planning

The radiology department in a hospital is responsible for all imaging of the patient's body that needs to be done. Requests for X-ray images, CTs, ultrasound or MRI-scan is done at this department, evaluated and a response is sent back to the unit that ordered it.

Typically, a Radiology department today is equipped with three monitors, two high resolution (3MP) greyscale LCD monitors that display images and one low-resolution (i.e. 1MP) colour monitor that displays patient data [13]. Before the digitization of images, the images were hanged in a rack in a particular way called "hanging protocol" on a rack. Although most radiology departments are digital today, much of the software presents the images in the same way, displaying series of related images in a similar grid on the screen.

The actual work performed within the radiology department makes this case perhaps not the ideal situation to test for introducing technology such as the display wall. This work is already facilitating (and requiring) multi-display technology with extremely high-end displays, which will be nearly impossible to match the quality of on a display wall. However, the radiology department also produces reports and interprets the images. They also communicate this information to the rest of the hospital. In this "interface" between the radiology department and the medical specialties that need the images, information is transferred in so-called "boards", meetings between the radiology departments and the specialty departments/clinics. It is these *boards* that are our primary interest. The next section gives an example where such meetings are required. Note that this is only one of the many problems where the mentioned "boards" are required between the radiology dept. and the other departments.

One process that is especially dependant on the work done in radiology departments is radiation therapy planning (RTP). Radiation therapy is the use of a certain type of radiation, ionizing energy beams, to kill cancer cells or shrink tumours on/within the body. The radiation kills or damages the cancer cells in the area treated or exposed to the radiation and damage the reproductive or genetic material, making the division of cancer cells and growth of tumours impossible.

The complexity of the decision process is what makes this case very interesting for testing large display technology. Many critical factors need to be accommodated in this process when choosing between several possible radiation-treatments. The planning-process, the simulation included, is a cognitively challenging task, which also includes several people in a team collaborating to make decisions about treatment. There is a number of professional specialties are included in the RTP process, as stated in [14]. These include a radiation oncologist, a radiation oncologist physicist, medical radiation dosimetrist and a radiation therapist. The sequence of events in a RTP process can be as follows:

1) *Physicians locate the region of interest (ROI, such as tumour) by using traditional X-ray fluoroscopy (diagnostic imaging).*
2) Move patient to the CT scanner. Place aluminium markers, visible on the CT slices, over the skin laser markers.
3) *Physicians analyze ROI and define the target volume(s) and the critical organ (s) on every CT slice. They then select treatment parameters.*
4) *Move patient to the Simulator again. Physicians simulate the treatment plan to verify its effectiveness by using X-rays instead of treatment rays. The area of treatment is documented on X-ray films. If the treatment plan is successfully verified, the patient proceeds to the treatment unit; if not, step 3 is repeated.*
5) Move patient to the treatment unit. Physicians carry out the actual RT treatment according to the RTP derived from the steps described above.

In italics are the parts of the process that involves reasoning on the provided images and information from the process. These are the parts where we see very large screens as a potential helpful tool. Each of the steps 1, 3 and 4 and their respective placements within the hospital provide for a "situatedness" for large displays. If the use of the displays is to be immediately initiated – and replace existing displays in these situations, this will most likely have to be stated within documents such as the mentioned quality assurance (QA) report and similar routine-descriptions at the local hospitals. The first apparent issue with a really large display as the display-wall is that it will be a very evident and probably 'invasive' artefact to bring into any process within quite streamlined processes. The use of the word "invasive" is here meant to imply that in that the display is so qualitatively and especially interactively a different experience than the available displays today, there would have to be either quite strong incentives for bringing the technology into the situations or that adoption will have to be done gradually, somehow.

At step 3 you might notice that the physicians will mark the critical organ(s) on every CT-slice. This is perhaps the most obvious place where one might wonder what it would mean for the quality of this procedure that all image-slices were available to the physician at once. Would it bring forth a better overview of the area and enable better markings of the critical organs – leading to fewer side effects of the treatment? Would the larger working-area ease the work of the physician making him/her work faster? These are the kinds of productivity benefits already proven when comparing work done on large screens compared to smaller ones [8].

In sum, the RTP planning process is a very interesting case for use-scenarios of very large, high-resolution displays. But – it is not the only one. The radiology department has similar responsibilities to a lot of the departments in a hospital. Investigating a case such as the RTP process in more detail, and figuring out the requirements for large-display technology and the potential benefit would give grounds for developing prototypes. We feel confident that the use of technology such as displaywalls (in some format and configuration) could make a difference in healthcare, especially for complex problems, like chronic diseases. Requirements for the use of such technology in healthcare would yield important feedback to the community that researches the development of large, high-resolution displays.

4 Discussion

There are two general ways of exploiting the wall sized display: (1) zoom on detail (e.g. show many CT-slices, zoom in on specific parts of model/image/...), (2) or show the big picture – "go broad", and collect data, information and knowledge from wherever relevant. The second option might also be aggregated data, visualized in some way (graphs, charts, etc.) The first task is more or less straightforward – the data is present within the local system(s) and it is only a matter of adjusting the displaying SW/HW to the large format (possibly facing some interface challenges). The second (2), is generally much more difficult, as information, knowledge and data might be scattered, unstructured, unavailable, etc. Pixel-count and scalability issues are relevant issues – how well one can enlarge/shrink sizes of the information/data. This issue goes into the core of the pre- and post perspective in relation to the document. In a pre-document perspective it is a matter of framing a set of data and thus – create a document. This is what we traditionally call interface-design. The definition of a document becomes what you have inside the frame! At any given time, you have the challenge to define what kind of document you want. Whether this is possible to define before the situation arises (automatically) or not – and whether this becomes a more difficult issue on a large, high-resolution display remains to be seen.

So the problem becomes the following: what kind of documents does the health care staff want to watch on the wall? It also becomes a matter of collectively "drawing the document on the wall". The situation changes from not only being a situation of watching but also to a situation of creating a document.

This acknowledgement makes it crucial to consider how you interact with the wall, making it a decisive interaction for what kind of document you get on the wall. Should you have a moderator in the group, a leader of the group, managing the cursor or something like a cursor? The large display seems, as all displays usable for two apparent tasks: displaying something (in a very large format), or allowing for manipulation (interfacing) and changing the contents viewed. The two tasks seem both intertwined and detached, in a way. Display and visualization seems to invite reflection and discussion about the topic at hand (the information content), but in the 'interaction mode', the large display might have different properties, especially as people are used to the mouse-and-keyboard interface to computing devices.

Through asking the doctors etc. you can design a number of relevant documents, defined by being "something" to watch within the frames of the wall. When you have defined a number of "relevant" document types, you can consider the specific problems in each case. This leads automatically to the issue of classifying the document. Is it relevant to save all the different temporary documents created out of the database? Some may only be useful for very short while other may be important. Never the less, it may be important to have a log of all kinds of documents you are creating.

References

1. Briet, S.: Qu'est-ce que la documentation? ÉDIT Paris (1951)
2. Brown, J.S., Duguid, P.: The Social Life of Documents. First Monday, Peer Reviewed Journal on the Internet (1996)
3. Lund, N.W.: Documentation in a Complementary Perspective. I: Aware and Responsible: Papers of the Nordic-International Colloquium on Social and Cultural Awareness and Responsibility in Library, Information and Documentation Studies(Scarlid), 93–102 (2004)
4. Pedauque, R.-T., Salaün, J.-M., Melot, M.: Le document à la lumière du numérique. C&F Editions, Paris (2006)
5. Zacklad, M.: Documentarisation Processes in Documents for Action (DofA): The Status of Annotations and Associated Cooperation Technologies. Computer Supported Cooperative Work (CSCW) 15, 205–228 (2006)
6. Grudin, J.: Partitioning digital worlds: focal and peripheral awareness in multiple monitor use. In: Proceedings of the SIGCHI conference on Human factors in computing systems, ACM Press, New York (2001)
7. Yost, B., North, C.: The Perceptual Scalability of Visualization. IEEE Transactions on Visualization and Computer Graphics 12, 837–844 (2006)
8. Czerwinski, M.: Toward Characterizing the Productivity Benefits of Very Large Displays. In: INTERACT 2003, IOS Press, Amsterdam (2003)
9. Tan, D.S., Gergle, D., Scupelli, P., Pausch, R.: With similar visual angles, larger displays improve spatial performance. In: Proceedings of the SIGCHI conference on Human factors in computing systems, ACM Press, New York (2003)
10. Swaminathan, K., Sato, S.: Interaction design for large displays. Interactions 4, 15–24 (1997)
11. Czerwinski, M., Tan, D.S., Robertson, G.G.: Women take a wider view. In: Proceedings of the SIGCHI conference on Human factors in computing systems: Changing our world, changing ourselves, ACM Press, New York (2002)
12. O'Hara, K., Perry, M., Churchill, E.: Public and Situated Displays: Social and Interactional Aspects of Shared Display Technologies (Cooperative Work, 2). Kluwer Academic Publishers, Dordrecht (2004)
13. Litchfield, J.: Color Your World: Diagnostic color displays take off. HealthImaging.com (2006)
14. Fraass, B.B., Doppke, K.K., Hunt, M.M., Kutcher, G.G., Starkschall, G.G., Stern, R.R., Van Dyke, J.J.: American Association of Physicists in Medicine Radiation Therapy Committee Task Group 53: quality assurance for clinical radiotherapy treatment planning. Medical physics 25, 1773–1829 (1998)

International Workshop on Web Usability and Accessibility (IWWUA)

Workshop PC Chairs' Message

Silvia Abrahão[1], Cristina Cachero[2], and Maristella Matera[3]

[1] Department of Information Systems and Computation,
Valencia University of Technology
Camino de Vera, s/n, 46022, Valencia, Spain
sabrahao@dsic.upv.es
[2] Department of Information Systems and Languages
University of Alicante
Campus de San Vicente del Raspeig. Apartado 99. 03080 Alicante, Spain
ccachero@dlsi.ua.es
[3] Dipartimento di Elettronica e Informazione
Politecnico di Milano
Via Ponzio, 34/5, 20133, Milano, Italy
matera@elet.polimi.it

According to recent studies, an estimated 90% of Web sites and applications suffer from usability and/or accessibility problems. As user satisfaction has increased in importance, the need for usable and accessible Web applications has become more critical. To achieve usability for a Web product (e.g., a service, a model, a running application, a portal), the attributes of Web artefacts must be clearly defined. Otherwise, assessment of usability is left to the intuition or to the responsibility of people who are in charge of the process. In this sense, usability models (describing all the usability sub-characteristics, attributes and their relationships) should be built, and Usability Evaluation Methods (UEMs) should be used during the requirements, design and implementation stages based on these models. Similarly, identifying the set of characteristics that make the Web more accessible for everybody, including those with disabilities is necessary to systematize the way practitioners face accessibility issues.

The previous motivations led us to organize the first edition of the International Workshop on Web Usability and Accessibility (IWWUA 2007), held in conjunction with the 8th International Conference on Web Information Systems Engineering (WISE), in Nancy, France, on December 2007. Our aim is to promote the discussion on Web usability and accessibility, bringing together professionals and researchers who are interested in discussing recent trends and perspectives in these two topics. The main purpose of this workshop is to access the effectiveness of existing approaches for Web usability and accessibility evaluation, as well as to provide an international forum for information exchange on methodological, technical and theoretical aspects of the usability and accessibility of Web applications. These proceedings collect the papers presented at IWWUA; they cover a wide range of topics that mainly refer to early usability evaluation based on design methods, evaluation techniques and empirical studies.

The majority of the papers aim at extending or enhancing the design process to pursue the usability and/or the accessibility of the final Web application. Moreno et al. discuss the benefits of using usability techniques during the analysis phase of a Web application development. The paper by Olsina et al. discusses how Web model refactorings can improve the external quality, and in particular the usability, of a Web application. Molina

M. Weske, M.-S. Hacid, C. Godart (Eds.): WISE 2007 Workshops, LNCS 4832, pp. 409–410, 2007.
© Springer-Verlag Berlin Heidelberg 2007

and Toval propose some extensions to the models used for the development of Web information Systems, and define a set of quality metrics to measure usability and accessibility during the design phase. Panach et al. apply some STATUS patterns to solve usability issues at the conceptual model level of the OOWS methodology. Sorokin et al. propose a framework for the model-driven development of accessible Rich Internet applications. Martin et al. illustrate how usability and accessibility requirements can be assured and validated in the whole life cycle of e-learning Web applications. Finally, in order to increase the accessibility for users belonging to deaf communities, Felice et al. propose an ontology for the Italian sign language, to support the construction of a visual interface on top of a sign language dictionary.

Some other papers discuss the evaluation of usability and accessibility. Centeno et al. propose the definition of XSLT templates addressing accessibility rules evaluated on the markup of Web pages, and introduce the WAEX Web accessibility evaluator. Xiong et al. investigate the support given by currently available tools for evaluating accessibility at different phases of the development process. Finally, the paper by Bolchini and Garzotto discusses the quality of usability evaluation methods, and proposes a method to evaluate their effectiveness and efficiency, also presenting some results achieved through an experimental study involving usability evaluators.

Finally, some papers report on empirical studies. Chen et al. introduce some criteria to evaluate the currency of free science information, and discuss the result of the analysis, based on the proposed criteria, of a sample of Web pages. The paper by de Castro et al. reports on an experiment that compares two Web information systems with the aim of validating a hypertext modelling method (HM3).

We would like to thank all authors for submitting their work to the workshop and contributing to shape up such a rich program, the members of the Workshop Program Committee for their efforts in the reviewing process, and the WISE organizers for their support and assistance in the production of the proceedings. We are also grateful to Nigel Bevan from the Professional Usability Services in England, who agreed to give a keynote speech. Finally, we would like to thank the European COST Action n°294 MAUSE (Towards the Maturation of IT Usability Evaluation – www.cost294.org) for sponsoring the workshop.

Silvia Abrahão
Cristina Cachero
Maristella Matera

IWWUA 2007 Chairs

Incremental Quality Improvement in Web Applications Using Web Model Refactoring

Luis Olsina[1], Gustavo Rossi[2], Alejandra Garrido[2],
Damiano Distante[3], and Gerardo Canfora[3]

[1] GIDIS_Web, Universidad Nacional de La Pampa, Argentina
[2] LIFIA, Universidad Nacional de La Plata and CONICET, Argentina
[3] RCOST, University of Sannio, Italy
olsinal@ing.unlpam.edu.ar,
{gustavo,garrido}@lifia.info.unlp.edu.ar,
{distante,canfora}@unisannio.it

Abstract. Web applications must be usable and accessible; besides, they evolve at a fast pace and it is difficult to sustain a high degree of external quality. Agile methods and continuous refactoring are well-suited for the rapid development of Web applications since they particularly support continuous evolution. However, the purpose of traditional refactorings is to improve internal quality, like maintainability of design and code, rather than usability of the application. We have defined Web model refactorings as transformations on the navigation and presentation models of a Web application. In this paper, we demonstrate how Web model refactorings can improve the usability of a Web application by using a mature quality evaluation approach (WebQEM) to assess the impact of refactoring on some defined attributes of a Web product entity. We present a case study showing how a shopping cart in an e-commerce site can improve its usability by applying Web model refactorings.

Keywords: refactoring, Web applications, usability, quality evaluation.

1 Introduction

The evolution of Web applications (WAs) is driven by a myriad of different factors: new requirements (stable and volatile), users' feedback, new technologies giving the chance to change the look and feel or the interaction style of the application, etc. In all cases, this evolution usually follows unpredictable patterns that imply a constant pressure on development teams. Agile methods have emerged to help developers cope with, and even welcome, continuous change in requirements [1]; as such, these methods are particularly suitable for developing WAs. Refactoring is one of the fundamental practices of agile development used to add flexibility and extensibility before introducing new functionality [3].

Refactoring was defined in the context of object-oriented systems to "factor out" new abstractions and perform other small transformations to the source code of an application without changing its behavior [11]. These transformations aim at improving the design of the code, making it more reusable and flexible to subsequent

M. Weske, M.-S. Hacid, C. Godart (Eds.): WISE 2007 Workshops, LNCS 4832, pp. 411–422, 2007.
© Springer-Verlag Berlin Heidelberg 2007

semantic changes. Though refactorings are performed in small steps, they are usually composable, yielding larger transformations that improve readability, reusability and maintainability of a system [15]. Refactoring to patterns has also been proposed to help keep the balance between under-engineering (continuously adding new functionality without a previous clean-up) and over-engineering (applying design patterns to create overly complex designs) [6].

In the context of WAs, refactoring may not only be applied to improve internal quality, but also to enhance navigability and presentation, which are external qualities influencing usability. In [4] we have defined the concept of Web model refactoring (WMR), i.e., refactoring applied to the navigation and presentation models of a Web application. WMRs aim at improving the application's usability by small transformations in the application's navigational topology, and/or interface look and feel. Additionally, WMRs guide the introduction of Web patterns [18] into the application's structure. In order to assess how WMRs improves usability, we use a well-known Web quality evaluation approach, namely WebQEM [8], and test the application's quality features before and after refactoring.

As an example that we will elaborate in this paper, Fig. 1.a shows a reduced version of the Amazon shopping cart and Fig. 1.b shows the same cart with some added information and operations. In our research we want to identify this kind of transformations and be able to measure the associated quality improvement, if any.

Fig. 1.a. Basic shopping cart   **Fig. 1.b.** Enhanced shopping cart

The main contributions of this paper are the following: (a) we propose WMR as a way to incrementally improve the external quality of a WA from the final user viewpoint; (b) we show how to incorporate quality assessment in the process of refactoring; and, (c) we demonstrate how WMR improves usability on a particular case study.

The structure of the paper is as follows. Section 2 summarizes the concept of WMR and the WebQEM quality evaluation approach. Section 3 shows how to apply WebQEM in the context of WMR by using a simple but meaningful case study. Section 4 presents related work and Section 5 concludes the paper and describes some further work.

2 Background

2.1 Web Model Refactoring

Refactoring is a technique that assists developers in the process of continuous improvement of source code or design models of an application [20]. We are interested in defining refactorings for the design models of a WA. Well-known design methods agree on the definition of a three-stage design process for WAs, resulting in the definition of three models: *application model, navigation model* and *presentation model* [2, 7, 16, 17]. While refactorings applied to the application model are similar to those already in the literature [19, 20], we have defined WMRs as those refactorings that can be applied to the navigation and presentation models of a WA [4].

2.1.1 Intent, Scope and Granularity

Refactoring was originally conceived as a technique that applies syntactic transformations to the source code of an application without changing its behavior but improving its maintainability [11]. Web code refactorings, e.g., those applied to the source code or HTML structures, are outside the scope of this paper but discussed elsewhere [13, 14]. Similarly, WMRs apply model transformations that affect the way in which the application presents contents, enables navigation through contents and provides interaction capabilities [4], but do not change the semantics as defined for these models.

Since WMRs transform entities that are perceived by the user, their changes are directly reflected in the way the final user may interact with the WA. Therefore, the intent of WMRs is to enhance usability [9]. These refactorings focus on those (small) changes that may improve comprehension, facilitate navigation, smooth the progress of operations and business transactions, etc. Furthermore, in the same way as traditional refactorings may be used to introduce design patterns [6], WMRs might be also driven by Web patterns [18] and therefore produce the same well-known benefits of the patterns they introduce.

Navigation refactorings aim at improving the application's navigability by small transformations of its navigation model. This model is usually described with a navigational diagram composed of nodes, links, indexes and other access structures. Behavior defined by the navigation model of a WA is represented by: (i) the set of available nodes and navigation links between nodes; (ii) the set of available user operations and the semantics of each operation. Navigation model refactorings may thus change, among others: the contents available in a node, the set of outgoing links of a node and the user operations accessible from a node. Alongside, navigation model refactorings have to preserve the set of possible operations and their semantics, and the navigability of the set of nodes [4]. Preserving the navigability of the set of nodes means that existing nodes may not become unreachable though the set may be augmented. Moreover, this type of refactorings should not introduce data, relationships or operations that are not in the application model.

The presentation model describes the look and feel of pages, the interface widgets they contain, the interface controls that trigger the application functionality and the interface transformations occurring as the result of user interaction. Presentation model refactorings may thus transform the look and feel of a page by changing the position of widgets or the interface effects, by adding new widgets or replacing them to enhance understanding, etc. At same time, the behavior defined by the presentation model, which must be preserved by the refactorings on this model, concerns with the actions that the user may trigger in a page, including both operations and links activation of the underlying nodes.

Most navigation model refactorings may imply other refactorings at the presentation model (e.g., new information added to a node requires the corresponding user interface to change in order to present the additional information). In contrast, presentation model refactorings must be neutral to the underlying navigation structure.

2.1.2 Examples

To illustrate the ideas presented above, we next describe some representative WMRs using a simplified template comprising: type of refactoring (navigation or presentation), motivation, mechanics and example. Other refactorings, including composite refactorings, can be found in [4].

ADD INFORMATION (NAVIGATION)

Motivation: we may eventually find the need to display more information than what is currently on a page. The information may come from different sources and have different purposes: it may be data extracted from the application model or obtained from the navigation model itself; it may be information added with the purpose of attracting customers or to help during navigation. This refactoring may be used to introduce patterns like *Clean Product Details* [18] to add details about products in an e-commerce site. With the aim of attracting customers, we may introduce *Personalized Recommendations* [18] or rating information.

Mechanics: the mechanics varies according to the different sub-intents above. In the most general case: add an attribute to a node class in the navigation model where the information is to be added. If the information is extracted from the application model, attach to the attribute the statement describing the mapping to the application model.

Example: this kind of refactoring can be applied to the node behind the page appearing in Fig. 1.a, to show information about price and savings of each product in the list; as a consequence we obtain the enriched version of the shopping cart shown in Fig. 1.b. It is worth noting that the information added to the shopping cart node is already available from the application model.

TURN INFORMATION INTO LINK (NAVIGATION)

Motivation: during the process of completing a business transaction, some Web pages may show intermediate results or a succinct review of the information gathered until a certain point. A common example occurs when checking the status of the shopping cart during the process of buying some products in an e-commerce website. Such Web pages should provide the user with the chance to review the information associated to the intermediate results (e.g., items in the shopping cart) by means of direct links to the pages showing details on them. When this does not happen, the page can undergo the kind of refactoring we propose here to improve it.

Mechanics: select the portion of the information about the target item that better distinguishes it. In the navigational diagram, add a link from the node representing the intermediate results to the nodes showing detailed information on the items to review; the anchor of the link could be the selected portion of information.

Example: this refactoring may be used to add links from names of products in a shopping cart to the pages showing detailed information about the products.

REPLACE WIDGET (PRESENTATION)
Motivation: the presentation model describes, for each node in the navigation model, the kind of widgets that display each data attribute and the widgets that permit to activate operations or links. Inspection of usage of the site may show that some information item or operation should be displayed with a different widget, to improve operability, usability or accessibility.

Mechanics: in the page of the presentation model that contains the widget found unsuitable, replace the current widget by a more appropriate one.

Example: check-boxes are best suited to enable users select one or more items from a list to perform an operation on them. A typical example is that of an email reader that allows selecting individual emails by means of check-boxes in order to apply, afterwards, operations like "delete". Thus, users do not expect check-boxes to dispatch an operation when they are clicked but just to show a check-mark in the box. In the case of Cuspide's shopping cart, which appears in Fig. 3, checking any box under the title "Borrar" automatically deletes the item from the cart, which is confusing and does not allow changing one's mind. In this case, a more suitable widget would be a button with label "Borrar" or the usual trash can icon (see Fig. 4).

2.2 The Web Quality Evaluation Method (WebQEM) Approach

WebQEM [8] is an evaluation method for WAs, i.e., a method for the inspection of characteristics, sub-characteristics (named calculable concepts and sub-concepts in Fig. 2), and attributes stemming from a quality model for WAs. WebQEM relies on a set of well-defined metrics and indicators for measurement and evaluation, in order to give recommendations for improvement. The main parts of the measurement and evaluation framework (named INCAMI [10], which stands for *Information Need, Concept model, Attribute, Metric* and *Indicator*), and the WebQEM method that instantiates it are, namely:

- The *non-functional requirements specification component,* which deals with the definition of the *Information Need* and the specification of requirements by means of one or more *Concept Models* -see Fig. 2. (Note that a concept model can be instantiated in external quality, quality in use models, among many others).
- The *measurement design and execution component,* which deals with the specification of concrete *Entities* to be measured, the metrics selection to quantify the attributes of the quality model, and the recording of the gathered measures; this component is centered in the *Metric* concept [10].
- The *evaluation design and execution component,* which deals with the definition of indicators, both elementary and global ones, decision criteria and aggregation models that will help to enact and interpret the selected concept model; this component is centered in the *Indicator* concept (see [10] for more details).

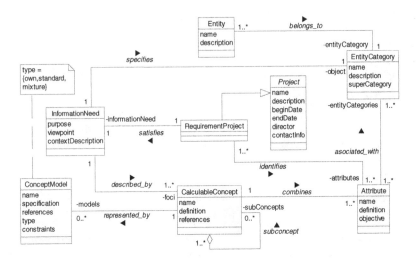

Fig. 2. Key terms and relationships that intervene in the non-functional requirements definition and specification component, which is instantiated by WebQEM

3   A Strategy for Incremental External Quality Improvement

Our proposal to continuously improve the external quality of a WA, such as its usability, is to enrich the lifecycle of a WA with a new activity: WMR. After each requirements/design/implementation cycle and before the next cycle begins, the design models of a WA should be inspected in order to find opportunities for refactoring. Thus, similarly to the well-known advantages of traditional refactoring [3, 20], the quality of the design will be incrementally improved, as well as the external quality. We also propose an additional activity to assess the benefits of applying WMR by evaluating the application's quality before and after refactoring. This is a first step to systematize the process of WMR as a quality-driven activity.

3.1   The Shopping Cart Case Study

The external quality (e.g., the usability and content) of the Cuspide shopping cart can be improved by systematically applying WMR. Instead of using a set of "good practices" for shopping carts, we used the Amazon.com shopping cart as a reference for desirable requirements and quality attributes of a shopping cart. In fact, Amazon is certainly a well-known instantiation of good practices for this type of application component. Referring back to Fig. 2, we can state that given an *entity* (the Cuspide shopping cart), the *information need* can be specified by its *purpose* (evaluate and compare), from the *user viewpoint* (developer), in a given *context* (using WMR in the context of an agile process). Moreover, the information need has the *focus* on a *calculable concept* (external quality) and *sub-concepts* (usability and content), which can be *represented by* a *concept model* (external quality model) and associated *attributes* (as shown in the left side of Table 1).

3.1.1 The External Quality Specification

To evaluate the impact of WMR, we have specified the external quality of the Cuspide shopping cart with regard to some usability and content attributes, contrasting it with some features of the Amazon shopping cart. In this sense, we carried out evaluation activities before and after applying WMRs. The external quality requirements that were assessed appear in the left column of Table 1. In addition, elementary, partial and global indicator values are shown for the Cuspide shopping cart before refactoring (as we analyse in the next sub-section). Many of these quality requirements were illustrated in [9], as well as the justification for the inclusion of the *Content* characteristic for assessing the quality of information in the Web.

Table 1. External quality requirements (with regard to usability and content) for a shopping cart. EI = Elementary Indicator value; P/GI = Partial/Global Indicator value.

External Quality Requirements	EI	P/GI
Global Quality Indicator		**61.97%**
1 Usability		**60.88%**
1.1 Understandability		83%
1.1.1 Shopping cart icon/label ease to be recognized	*100%*	
1.1.2 Information grouping cohesiveness	*66%*	
1.2 Learnability		51.97%
1.2.1 Shopping cart help	*50%*	
1.2.2 Predictive information for link/icon	*66%*	
1.2.3 Informative Feedback		41.5%
1.2.3.1 Continue-buying feedback	*66%*	
1.2.3.2 Recently viewed items feedback	*0%*	
1.2.3.3 Proceed-to-check-out feedback	*100%*	
1.2.3.4 User current status feedback	*0%*	
1.3 Operability		49.50%
1.3.1 Shopping cart control permanence	*100%*	
1.3.2 Expected behavior of shopping cart controls	*50%*	
1.3.3 Controls Accessibility		
1.3.3.1 Support for text-only version of controls	*0%*	
2 Content		**63.05%**
2.1 Information Suitability		63.05%
2.1.1 Shopping Cart Basic Information		50%
2.1.1.1 Line item information completeness	*50%*	
2.1.1.2 Product description appropriateness	*50%*	
2.1.2 Other Contextual Information		76.89%
2.1.2.1 Shipping costs information completeness	*100%*	
2.1.2.2 Applicable taxes information completeness	*100%*	
2.1.2.3 Return policy information completeness	*33%*	

3.1.2 Design and Execution of the Measurement and Evaluation

Following the WebQEM's steps (outlined in Section 2.2), evaluators should design, for each attribute of the instantiated quality model, the basis for the measurement and evaluation process. This step is accomplished by defining each specific metric and

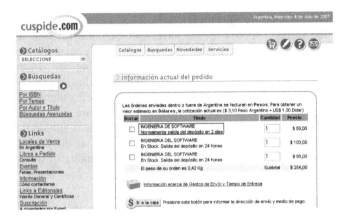

Fig. 3. Shopping cart in Cuspide (www.cuspide.com.ar) before refactoring

indicator for the given information need. Lastly, in the execution phase, we record the final values for each metric and indicator. The right columns of Table 1 show the indicators values, and Fig. 3 the entity to be measured.

For example, for the *"Line item information completeness"* attribute (coded 2.1.1.1 in Table 1), having the Amazon shopping cart as reference, we designed a direct metric named *"Degree of completeness of the Line item information"*. It specifies three categories considering an ordinal scale type, namely: 0. *Incomplete; 1. Partially complete*, (i.e., it only has title, price, quantity, and sometimes availability fields); and 2. *Totally complete* (it has 1 plus author, added on date, and availability).

Moreover, an elementary indicator can be defined for each attribute of the requirement tree. For instance, for the previous attribute, the elementary indicator *"Performance Level of the Line item information completenes"* interprets the metrics value of the attribute. Note that an elementary indicator interprets the level of satisfaction of this elementary requirement. After this, a new scale transformation and decision criteria (in terms of acceptability ranges) are defined. In our study, we use three agreed acceptability ranges in a percentage scale: a value within 40-70 (a marginal range) indicates a need for improvement actions; a value within 0-40 (an unsatisfactory range) means changes must take place with high priority; a score within 70-100 indicates a satisfactory level for the analyzed attribute. Table 1 shows a value of 50% for the 2.1.1.1 attribute of the Cuspide shopping cart, taking into account that a value of 1 mapped to 50% and a value of 2 mapped to 100% of satisfaction.

Furthermore, to design and execute the global evaluation, we should select and apply an aggregation and scoring model [8]. In this case study, we used an additive scoring model, so applying weights and the sum operator we related the hierarchically grouped attributes, sub-concepts, and concepts accordingly, yielding in the end the partial and global indicators (the rightmost column of Table 1). Thus, decision-makers can analyze the results and give recommendations. We can see that many indicators are below the threshold of the satisfactory acceptance range; thus, many attributes of the external quality of the Cuspide shopping cart may benefit from improvement.

3.1.3 Applying WMR to the Example

As discussed above, Cuspide should plan changes in the *Shopping Cart Basic Information* sub-characteristic (ranked 50%) mainly in attributes 2.1.1.1 and 2.1.1.2. For example, the 2.1.1.1 attribute should have at least the author's name besides the title of the item, since, as shown in Fig. 3, it is not possible to distinguish between two or more items with the same starting title (e.g., "INGENIERIA DE SOFTWARE"). We may apply the refactoring ADD INFORMATION (see Section 2.1.2) to the shopping cart node to incorporate the author's name to each item in the list. Moreover, note that it is not possible to navigate to the pages showing the detailed information on the items in the shopping cart for further information. We can apply the refactoring TURN INFORMATION INTO LINK to solve this problem. The outcome of applying these two refactorings is shown in Fig. 4. As a result we can predict the total satisfaction (100%) of both attributes mentioned above. Fig. 4 also shows the result of applying the refactoring REPLACE WIDGET (see Section 2.1.2) to correct the unexpected behavior of the delete item control.

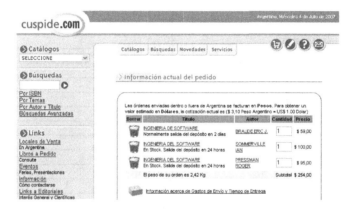

Fig. 4. The Cuspide shopping cart after refactoring

3.2 Discussion

As stated before, our research aims at: (a) presenting WMRs that may improve the external quality of a WA; and (b) integrating quality assessment in the refactoring process. There are two ways in which we can face the refactoring activity during the development cycle: as an informal improvement process or in the context of a structured evaluation framework. In the first case, we analyze our application (either by collecting users' feedback or by carefully inspecting its functionality) and find opportunities for refactoring. In fact, this is the way in which refactoring has been applied so far in the software community. When the designer is aware of a good catalogue of possible refactorings, the process is simplified. The second case (using a structured evaluation framework) arises when we are able to formally perform an evaluation before and after the refactoring. Moreover, by performing the evaluation of the entity to be refactored before and after the process, we can quantify and justify the quality gain, independently of the chosen lifecycle. Therefore the incremental quality improvement can be evaluated and/or predicted.

The most important aspect of this strategy is that, ultimately, we can map atomic or composite refactorings to attributes. In other words, associated with each meaningful attribute, there are one or more refactorings that may be applied to meet this requirement. Moreover, at the organization level, we can have a catalogue of WMRs and eventually a mapping to a catalogue of attributes. By knowing beforehand the impact of the transformations, we could estimate the improvement gain. In our case study, we were able to make a correspondence between refactorings and attributes that influence the external quality. Table 2 shows the result of this correspondence, with refactorings to the left of the table and the attributes they improve to the right.

Table 2. Mapping between refactorings and attributes (shown in Table 1) that can be applied to improve the external quality. NM = Navigation Model; PM = Presentation Model.

Refactorings that may apply	Attributes that may improve
Add Information (NM)	1.1.1 / 1.2.3.3 / 1.2.3.4 / 2.1.1.1 / 2.1.1.2 / 2.1.2.1 / 2.1.2.2 / 2.1.2.3
Add Category (NM)	1.1.2 / 1.2.1
Add Operation (NM)	1.2.3.3 / 1.3.1
Add Index (NM)	1.2.1
Add Guided Tour (NM)	1.2.1
Anticipate Target (NM)	1.2.2
Enrich Index (NM)	1.2.3.1 / 2.1.1.2
Introduce History (NM)	1.2.3.2
Multiply Category (NM)	1.1.2 / 1.2.1
Recategorize Item (NM)	1.1.2 / 1.2.1
Turn Info into Link (NM)	2.1.2.2
Add Widget (PM)	1.1.1
Replace by Text (PM)	1.3.3.1
Replace Widget (PM)	1.3.2

Of course we might not have to apply all the refactorings to all the Cuspide shopping cart attributes listed in Table 2. Some of them may not need real improvement, e.g., those in which the actual elementary indicator value is 100%. However, for those attributes which are weak or absent we can predict, after applying a focused cost-effective refactoring, a total level of satisfaction of all these requirements. Ultimately, our strategy for incremental improvement can be used both to predict the quality and to actually make the real assessment after refactoring.

4 Related Work

Our research differs from existing work in the refactoring field in three aspects: the subject, the intent, and the underlying strategy. Regarding the subject, we deal with the navigation and interface models of a WA, while existing literature works either at the code level or at the implementation design level (e.g. by refactoring on UML diagrams). Even for those Web design methods whose notations are based on UML-like diagrams [2, 7], our refactorings are different from conventional model refactorings [19]. Regarding the intent, WMRs aim at improving the user's experience with the WA and not internal attributes such as maintainability. Finally, regarding the underlying strategy, our approach differs from others in that it integrates an evaluation methodology (WebQEM) in order to improve non-functional information needs.

Ricca and Tonella [13] have worked on code restructuring for WAs. They define different categories of restructuring, like syntax update, internal page improvement, and dynamic page construction. Refactoring differs from restructuring in that the latter implies larger transformations that are usually run in batch mode by applying certain rules. Instead, refactorings are smaller and applied interactively. However, one main difference with our work is that their transformations apply on the source code, in this case, html, PHP and/or Javascript. Another difference is that WMRs are defined to improve operability, attractiveness, information suitability, among other non-functional characteristics, at the levels of characteristic and measurable attributes.

On the other hand, Ping and Kontogiannis apply refactoring at the level of hypermedia links, i.e., to the navigational structure of the application [12]. They propose an algorithm to cluster links into several types and group Web pages according to these link types. Applying this technique should provide a roadmap for the identification of controller components of a controller-centric architecture. Although their target is the navigational structure of a WA, they do not provide the mechanics to apply the transformation, but only a first step to recognizing where to apply them. In addition, in a recent work, Ivkovic et al. [5] include in their refactoring strategy a soft-goal hierarchy identification step. Even though this work represents a valuable contribution, a sound framework to justify measurement and evaluation results for analysis and recommendation of quality improvements is missing.

5 Concluding Remarks and Further Work

In this paper we have presented our proposal to continuously improve the external quality of WAs during their entire life-cycle. The approach is based on the use of WMR combined with WebQEM, a mature method for assessing the quality characteristics of a WA. We defined WMRs as those refactorings that can be applied to the navigation and presentation models of a WA, with the purpose of improving its external characteristics, while preserving its behavior. We showed how to incorporate a quality evaluation method in the process in order to assess the improvement gained by refactoring. We also presented a case study showing how a typical shopping cart in an e-commerce site can improve its usability by applying some WMRs from our catalog.

Our current line of research is being devoted to extend our catalogue of WMRs and their possible composition, and to map each of the refactorings to quality attributes of a WA. A further research issue is to develop tool support both for applying WMRs and for enabling assessment, based on the OOHDM design method.

References

1. Beck, K.: Test-Driven Development. Addison-Wesley, Reading (2002)
2. Ceri, S., Fraternali, P., Bongio, A.: Web Modeling Language (WebML): a Modeling Language for Designing Web Sites. In: Proc. of WWW9, Amsterdam, NL (2000)

3. Fowler, M.: Refactoring: Improving the Design of Existing Code. Addison-Wesley, Reading (2000)
4. Garrido, A., Rossi, G., Distante, D.: Model Refactoring in Web Applications. In: WSE 2007. 9th IEEE Int'l Symposium on Web Site Evolution, Paris, France, IEEE CS Press, Los Alamitos (2007)
5. Ivkovic, I., Kontogiannis, K.: A Framework for Software Architecture Refactoring using Model Transformations and Semantic Annotations. In: Proc. IEEE, CSMR 2006, pp. 135–144 (2006)
6. Kerievsky, J.: Refactoring to Patterns. Addison-Wesley, Reading (2005)
7. Koch, N., Kraus, A.: The Expressive Power of UML-based Web Engineering. In: Magnusson, B. (ed.) ECOOP 2002. LNCS, vol. 2374, pp. 105–119. Springer, Heidelberg (2002)
8. Olsina, L., Rossi, G.: Measuring Web Application Quality with WebQEM. IEEE Multimedia 9(4), 20–29 (2002)
9. Olsina, L., Covella, G., Rossi, G.: Web Quality (Chapter fourth). In: Mendes, E., Mosley, N. (eds.) Springer Book titled "Web Engineering" (2006) ISBN 3-540-28196-7
10. Olsina, L., Papa, F., Molina, H.: How to Measure and Evaluate Web Applications in a Consistent Way (Chapter Thirteenth). In: Rossi, Pastor, Schwabe, Olsina (eds.) Springer Book titled Web Engineering: Modelling and Implementing Web Applications. Human-Computer Interaction Series, vol. 12 (2007) ISBN: 978-1-84628-922-4
11. Opdyke, W.: Refactoring Object-Oriented Frameworks. Ph.D.Thesis, Illinois University at Urbana-Champaign (1992)
12. Ping, Y., Kontogiannis, K.: Refactoring Web sites to the Controller-Centric Architecture. In: CSMR 2004, Tampere, Finland (2004)
13. Ricca, F., Tonella, P.: Program Transformations for Web Application Restructuring. In: Suh, W. (ed.) Web Engineering: Principles and Techniques. ch. 11, pp. 242–260 (2005)
14. Ricca, F., Tonella, P., Baxter, I.D.: Restructuring Web Applications via Transformation Rules. In: SCAM, pp. 152–162 (2001)
15. Roberts, D.: Eliminating Analysis in Refactoring. Ph.D. Thesis, Illinois University (1999)
16. Schwabe, D., Rossi, G.: An Object Oriented Approach to Web-Based Application Design. Theory and Practice of Object Systems 4(4) (1998)
17. UWA Consortium: Ubiquitous Web Applications. Proceedings of the eBusiness and eWork Conference e2002: Prague, Czech Republic (2002)
18. Van Duyne, D., Landay, J., Hong, J.: The Design of Sites. Addison-Wesley, Reading (2003)
19. Van Gorp, P., Stenten, H., Mens, T., Demeyer, S.: Towards automating source-consistent UML Refactorings. In: Proc. of the 6th Int'l Conference on UML (2003)
20. Zhang, J., Lin, Y., Gray, J.: Generic and Domain-Specific Model Refactoring using a Model Transformation Engine. In: Beydeda, S., Book, M., Gruhn, V. (eds.) Model-driven Software Development. ch. 9, pp. 199–218. Springer, Heidelberg (2005)

Inclusive Usability Techniques in Requirements Analysis of Accessible Web Applications

Lourdes Moreno, Paloma Martínez, and Belén Ruiz

Computer Science Department, Universidad Carlos III de Madrid
Madrid, Spain
{lmoreno,pmf,bruiz}@inf.uc3m.es

Abstract. To follow accessibility standards does not guarantee complete accessible web applications. There are difficulties in web application development due to not consider accessibility in software life cycle together to forget important aspects in user interaction. A proposal to evaluate the benefits of using usability techniques with inclusion in the analysis phase of a web application development is presented.

Keywords: Accessibility, Software Engineering, Inclusive Design.

1 Introduction

The vertiginous growth in the use of Internet with more and more people using the Web makes necessary an advance of technology devoted to avoiding the exclusion of user groups. Due to the fact that not everyone has access to the Web in the same way, it is necessary to take into account this diversity to provide full accessibility to the Web contents.

Designers usually find difficulties in the design of accessible web pages because of scarce and inadequate methodologies and tools. The majority of the these tools aren't integrated in the development software but are designed to be used a posteriori, in evaluation phase, when detecting barriers increments costs, need of resources and sometimes it is even impossible to achieve a solution. It is necessary to include this in the cycle of process development, from its conception to the launch phase.

Also if a technical approach based on meeting accessibility standards regarding web code is followed, it has the inconvenience of going away from the users experience and therefore questions of the interaction of the Web with important users and the access to it could go unnoticed. All of this leads to the necessity on integrating usability and accessibility in software processes following a User Centered Design (UCD) using usability techniques.

The UCD [1] focuses on the idea of fulfilling user needs in every phase of the development process. It considers the following phases: analysis, design- prototype-iterative evaluation, implementation and maintenance. This iterative process allows evaluating the design during the development cycle and not only to evaluate the web page in its final stage.

M. Weske, M.-S. Hacid, C. Godart (Eds.): WISE 2007 Workshops, LNCS 4832, pp. 423–428, 2007.

The development process of accessible web applications does not require a specific methodology in itself; it only makes sense when the methodological approach includes accessibility criteria. Traditional standards that could be adapted are ISO/IEC 12207:1995, IEEE 1074-1997 for development processes for information system on the Unified Process for Object-oriented development. Proposals focused on web engineering with processes, models and adequate techniques to work with this type of information could be WSDM, SOHDM, OOHDM, UWE, OO-H, W2000, WebML amongst others.

2 Case Study: The Spanish Centre of Captioning and Audio Description (CESyA)

A web site is currently being developed for the recently created organism, CESyA [2]. It works towards the accessibility in audiovisual media using captioning and audio description services. One of the requirements that should be fulfilled is that the web has to be usable and accessible according to the standard WAI accessibility, coinciding in this particular case that amongst potential users disabled people are found.

Some experiments have been carried out on this web site to enable the evaluation of the consequences that would imply applying or not inclusion in the analysis phase. In this experimentation two development processes have been carried under the same domain, but with different ways to act in the capture of requirements. In the so-called "Inclusive Case" the work has been developed in the framework of Inclusive Design and not in the so-called "Non Inclusive Case".

The initial hypothesis was: the inclusive case needs fewer changes in the redesign of the interface, and consequently, has produced a reduction of costs in the development, avoiding new requirements to have in mind in the development phase. So, following the iterative model, the case study which needs fewer changes in this evaluation will have a smaller or inexistent increment in the next process iteration.

The analysts' team and the participants in the following phases in both experiments have similar characteristics. They do not have experience in design and development of accessible web applications but they have been given some training in accessibility items based on the WAI [4] documentation. The web applications that are being developed are based on similar technological profile with a basic core (WCAG1.0+XHTML+CSS with procedures for the dynamic contents).

2.1 User Modeling

In the user modeling tasks, several user groups have been taken into account in order to analyze their necessities in the CESyA area. These groups have been defined under the consideration of common attributes amongst users according to their access characteristics such as age, profession, frequent use of internet, information needs and software used, such as different browsers, players, etc. These common attributes which enable to model groups have been obtained through to investigation, interviews with clients and users, surveys, etc.

Once these attributes and values have been established, we have an approximation to all the users we want to reach to, and some User profiles considering common attributes.

In the Inclusive Case, more attributes are considered both in user's characteristics, for instance, whether user has a disability or not, and in access characteristics, for instance, special browsers as well as only text browsers, adapted assistive hardware technology amongst others.

Taking into account these new variables causes an increment of the number of users and contexts of use which need to be studied. Consequently, a new approach in modeling users is needed so that the study could be feasible. A first approximation of optimizing this task could be: a) the interaction of people without disabilities with the computer in an unfavorable context, b) the interaction of people with disabilities in normal environments or where the accessibility of the product does not vary. In this way, it will result more viable to give coverage to a greater number of users. As a second approximation, the application of scenarios, described in the next section, is considered.

2.2 Scenarios with Characters

With the approach of Scenarios with characters, the size of sample is minimized by studying various User profiles in only one scenario. The Inclusive case does not mean an exaggerated additional cost and additionally web designer becomes familiar with the user and designs taking into account his/her characteristics. The majority of the audience has been covered due to the creation of seven imaginative Scenarios in the case of the Non Inclusive Case and ten Scenarios in the Inclusive Case (see figure 1).

But in the Inclusive Case, as we have to keep in mind factors such as the use of magnifiers, screen readers, etc, we must investigate matters that will affect the future design of the user interface such as: how users with screen readers will access information, how to access it with a magnifier, how is someone supposed to design and write texts to make them more comprehensible, etc. These aspects of the accessibility in the Inclusive case have proactively been considered for the future design phases.

2.3 Card Sorting

Starting from the users´ behavior, the objective is to understand how users imagine the organization of the information and how they collect concepts and by means of this understanding the mental model of the user.

In the Non Inclusive Case we assume that it is enough to facilitate the access to the information by following the guidelines, and that mental organization of how to organize the web information is independent from the access characteristics of web users.

In the Inclusive Case it has been found, with the help of real users with disabilities, a diversity in some groups. Therefore, interviews allow us obtaining valuable information about (a) detecting possible barriers modifying the architecture of

IDENTIFICATION SCENARIO
Scenario code: 08
Name character: SOFÍA
Scenario description: Sofia is an University student; she is deaf and has a very bad vision but wears adequate glasses to improve her visual deficiency. Magnifying software also provides a style sheet with special contrasts which permits her to overcome slightly her visual problems. A subject of free election related to the audiovisual accessibility has been chosen. For this subject, they have asked her to do a work about CESyA. This work will be published on the university's web page. Sofia is at home working on it.

CHARACTERISTICS:
Demography : Young, 19 years old
Work responsibility/tasks: Student
Use frequency: weekly (frequency)
Hardware, Assistive technologies : a screen magnifier and sometimes combination of screen reader and Braille line
Environment: home
Software: Linux Debian, Opera
Experience: Low, little handling with of the environment and of the page
Disability: deafness blindness
Reason of use: Pedagogic.

BENEFICIARY USERS TYPE
- **First character:** Sofia, user of the CESyA web interested in general information about objectives, acts, and functions which are carried out in the CESyA.
- **Supplementary characters:** Users of the CESyA web such as own users or of organisms, companies, etc. as well as CESyA which look to contrast opinions and work fundaments in the audiovisual accessibility world.
- **Client characters:** CESyA and other entities such as the Ministry of Social Issues
- **Served characters:** Some fellow student of Sofia from the University, internet users who access contents of information about CESyA, who take as a reference the work that has been done..
Following the Inclusive Design framework:
- Disabled User: Deaf users, low hearing, blurry vision, blindness...
- Non-disabled User in an unfavourable context (users with certain limitations produced by unfavourable surroundings): forced situation of silence, noisy environments, audio impediment, blocked ears, existence of language audio elements not recognised (foreign language), visual difficulties caused by small screens, smoky environments, necessity to work at far distance from the visual device, forgetting your glasses, etc.

BARRIERS ACCESSIBILITY ON THE WEB AND ITS RESPECTIVE SOLUTIONS
The WCAG 1.0 Guidelines must be follow, and besides:
- When dealing with deafness, apart from providing textual contents and visual alternatives to sound contents as WCAG indicates, it would also be convenient if we were to value the difficulties some users have with the comprehension and writing of the language which makes the idea of applying easy reading rules to the texts as well as providing videos of sign language.
- For those with poor vision the barriers include pages where the font size cannot be changed, pages with complicated browser due to context loss when making bigger the font size, problems in the colour contrast when it is not permitted to personalise the sheet style. If someone with poor vision requires a screen reader to be able to access the information is it very important that the presentation is separated from the contents, and the page is to be framed properly, shortcuts and short keys to access the contents as well as downloadable documents are to be accessible to a screen reader.

Fig. 1. A scenario for the CESyA web according to the Inclusive case

information and (b) the design of post phases knowing that the elements of the architecture could have problems which would require attention and reinforcing its design.

3   Analysis of Testing Results

There are many factors that have been detected in the Inclusive Case in relation to the Non Inclusive one that are necessary to have in mind in the design phase. These included: (1) different types of hardware and software access can produce accessibility barriers, especially in the indirect access to the web by users who need assistive technologies, (2) different levels of browsing in the in different groups of users, etc. (3) Concerning the information architecture, different mental models where information is clustered.

Table 1. Some problems and improvements in user testing

barrier/ improvement	NON INCLUSIVE CASE	INCLUSIVE CASE
description link	Non-adequate attribute title . The links have to be more descriptive	Few observations. Contextual information is given, the attribute title with value
Icon and decorative image	The decorative images are labelled in XHTML with the alt = " text equivalent " attribute	No observations, the decorative image with the alt =" " and icon is included in CSS with a correct marking in the XHTML
Invisible shortcuts	Inexistent. The users have suggested them as an improvement.	No observations, they do exist to skip the initial browsing and go to the main content
Visibility edges interface	Presence of navigation menus in the corners but the users with magnifiers do not see them.	No observation. There are no elements situated in the corners or edges of the interface
Types of text Font	The font type used is Verdana. The users would prefer bolder fonts.	No observations. Arial is the font type used from the Sans Serif family, very legible according to literature.
Language accessibility	Many problems have been found in the comprehension of the texts.	Few problems have been detected but it is an area for improvement.
Structural marking	Disabled users have asked for a better structural marking, better definition of the header elements	Few observations. Good structural marking and good contents in relation to defined the Arquitecture of the information
re-dimensions design	Although relative units according to WCAG 1.0 have been implemented users have observed that precision is lost in the design of the presentation when making bigger or smaller the font size	No observation. Precision is lost in the design of the presentation when changing the font size in 3 levels, considered an exception on the web.
Multimedia elements	Users have asked for control options for the user, of information, format types, time, weight, etc. Some users could not access this resource	Few observations: The following procedures have been tried: accessible audiovisual content (with caption and audio description) and a facilitated access (with download, progressive download or streaming) + intuitive and usable access to the user (control, format, connection, size, time information).
Panoramic screen	There is a loss in the precision in the design of the presentation due to the flexible design	No observation. Precision is not lost in the design of the presentation or in the use of the hybrid model
Different browsers	Loss of precision of the presentation in some browsers	No observations. The same presentation is assured in the different browsers

As a hypothesis, the two prototypes implemented according to the two analysis cases, Inclusive and Non Inclusive Case are to comply the WCAG 1.0 (level AAA).

Regarding usability, on a earlier stage of the prototype, there were two heuristic evaluations. According to obtained results, barriers and improvements were found and, in both cases, they were corrected.

About these prototypes, a more extensive accessibility evaluation was taken following the WAI methodology and the usability, such as:

1) Expert and Manual revision following Validation Methodology WAI.

2) A test was made with users in both cases, including people with or without disabilities and diverse conditions of usage to evaluate usage aspects and access in both prototypes. This test has been based on the questionnaires and forms that users have filled in:

- Forms where the accessibility characteristics and context of use are reflected, as well as software and hardware characteristics in tests development.
- Questionnaires to evaluate usability according to heuristic evaluation, and accessibility of the different areas of the Web.

The most important enhancements in accessibility, as was suggested by users according to the test carried out, are shown in table 1. These items have been translated into new needs included in the Non Inclusive Case in this stage of the development process. This provokes a new re-design phase different to the inclusive case, which confirms the initial hypothesis.

4 Some Conclusions

The benefits that the WCAG 1.0 guidelines are well known, and their help is undoubtedly fundamental for designers. The UCD framework with the use of usability techniques allows us to come closer to the necessary user, but the benefits of the inclusion [3] in the analysis phase after this experimentation are distinguishable.

From WCAG 1.0 we have seen that accessibility guides provide better knowledge about detecting barriers and accessibility evaluation applicable a user interface prototype in an advanced phase of Design-Evaluation, than knowledge about how to design in analysis phase. For professionals without previous experience in the development of accessible web applications, the guides make knowledge obtained in the inclusive case an added value. Validation carried out with user's participation has shown this, due to the fact that guidelines do not make explicit which are the success factors in the compliance of accessibility and usability requirements.

The Analysis Phase of the Inclusive Case evidently requires more effort and is more costly than the Non Inclusive Case. On the other hand, this cost is viable as it has been previously explained, due to the approximations to the modeled user. But in the experimentation, it has been proved that this knowledge, not obtained in the Non Inclusive Case, requires in the iterative process of development an increment of accessibility much more costly with more adjustments in the design.

We are working in a methodology that includes the accessibility into Software engineering process. Precise methodologies to help those professionals in designing and developing accessible Web applications are necessary.

References

[1] Bevan, N.: Usability Net Methods for User Centred Design. Human-Computer Interaction: theory and Practice (vol 1). http://www.usabilitynet.org/tools/13407stds.htm (2003)
[2] Centro Español de subtitulado y Audiodescripción (CESyA), http://www.cesya.es, http://www.rpd.es/cesya.html (2006).
[3] Newell, A.F., Gregor, P.: User Sensitive Inclusive Design: in search of a new paradigm. In: En: CUU 2000. First ACM Conference on Universal Usability (2000)
[4] W3C, Web Accessibility Initiative (WAI) http://www.w3.org/WAI/ (2006)

A Visual Ontology-Driven Interface
for a Web Sign Language Dictionary

Mauro Felice[1], Tania Di Mascio[2], and Rosella Gennari[3]

[1] EURAC, Viale Druso 1, 39100 Bolzano, Italy
Mauro.Felice@eurac.edu
[2] L'Aquila U., I–67040 Monteluco di Roio, L'Aquila, Italy
tania@ing.univaq.it
[3] KRDB, CS Faculty, FUB, Piazza Domenicani 3, 39100 Bolzano, Italy
gennari@inf.unibz.it

Abstract. Sign languages are visual-gestural languages developed mainly in deaf communities; their tempo-spatial nature makes it difficult to
write them, yet several transcription systems are available for them. Most
sign languages dictionaries interact with their users via a transcription-
based inteface; thus their users need to be expert of their specific transcription system. The e-LIS dictionary is the first web bidirectional dictionary
for Italian sign language-Italian; using the current interface, the dictionary users can define a sign interacting with intuitive iconic images, without knowing the underlying transcription system. Nevertheless the users
of the current e-LIS dictionary are assumed to be expert of Italian sign language. The e-LIS ontology, which specifies how to form a sign, was created
to allow even the non-experts of Italian sign language to use the dictionary.
Here we present a prototype of a visual interface based on the e-LIS ontology for the e-LIS dictionary; the prototype is a query-oriented navigation
interface; it was designed following the User Centred Design Methodology,
which focuses on the user during the design, development and testing of
the system.

Keywords: Ontology Visualisation, Querying Task, User Centred Design Methodology, Sign Languages.

1   Introduction

A sign language (SL) is a visual language based on body gestures instead of sound
to convey meaning. SLs are commonly developed in deaf communities and vary
from nation to nation; for instance, in Italy we have Italian sign language (*Lingua Italiana dei Segni*, LIS). As highlighted in [12], SLs can be assimilated to
verbal languages "with an oral-only tradition"; their tempo-spatial nature, essentially 4-dimensional, has made it difficult to develop a written form for them.
"However" — as stated in [12] — "Stokoe-based notations can be successfully
employed primarily for notating single, decontextualised signs".

M. Weske, M.-S. Hacid, C. Godart (Eds.): WISE 2007 Workshops, LNCS 4832, pp. 429–440, 2007.
© Springer-Verlag Berlin Heidelberg 2007

The sign components singled out by a Stokoe transcription system can be classified in the following *Stokoe classes*:

1. the *handshape* class collects the shapes the hand/hands takes/take while signing; this class alone counts more than 50 terms in LIS;
2. the *palm orientation* class gives the palm orientations, e.g., palm up;
3. the *movement* of the hand/hands class lists the movements of the hands in LIS;
4. the *location* of the hand/hands class provides the articulation places, i.e., the positions of the hands (e.g., on your forehead, in the air).

These classes are used to decompose and group signs in the Electronic Dictionary for Italian-LIS (e-LIS). The e-LIS dictionary is part of the homonymous research project lead by the European Academy of Bozen-Bolzano (EURAC). The e-LIS project commenced at the end of 2004 with the involvement of the ALBA cooperative from Turin, active in deaf studies.

Initially, the e-LIS dictionary from LIS to verbal Italian was intended for expert signers searching for the translation of an Italian sign. At the start of 2006, when the development of e-LIS was already in progress, it was realised that potential users of a *web* dictionary would also be non-experts of LIS. Then the idea of an ontology and the associated technology for the dictionary from LIS to Italian took shape. The e-LIS ontology [9] introduces novel classes and relations among classes of sign components, thereby making explicit relevant pieces of information which were implicit and somehow hidden in the e-LIS dictionary. For instance: it makes explicit that each one-hand sign is composed of at least one handshape by introducing an appropriate relation among the corresponding classes, one-hand sign and handshape.

The e-LIS ontology can serve as the input of a DIG-enabled query tool like [1]; this allows the dictionary users to browse parts of the ontology and query the e-LIS database. The visualisation of the browsing and querying should meet the needs of the different users of the dictionary — in particular, deaf users, who are essentially visual learners [14]. However neither the current e-LIS dictionary nor the query tool support this: the former implements a visual interface but is not integrated with the ontology; the latter can be integrated with the ontology but does not implement a visual interface convenient for all kinds of users of e-LIS.

In this paper we present an innovative visual interface for the e-LIS dictionary: it integrates the e-LIS ontology and implements a novel visual metaphor for browsing and querying parts of the ontology. First, we give the necessary background on the e-LIS project and the current dictionary interface. Our novel interface and its development are then presented in details. We conclude with an assessment of our work.

2   The e-LIS Project and Dictionary

The e-LIS project commenced in 2004, focussing on the creation of a web bidirectional dictionary for Italian-LIS; information is shown in both verbal Italian and LIS, thus giving LIS the rank of any other language which can talk about itself. The e-LIS dictionary is available at [6]. In this paper, we concentrate on the dictionary from LIS to Italian, for which our interface is developed.

2.1   The e-LIS Database

Currently, the e-LIS dictionary stores data in XML files; this format forces a specific structure and organisation of information without providing knowledge concerning the semantics of sign decomposition. Thus the dictionary users can easily make mistakes during their search for a sign, e.g., they can easily specify a combination of components that corresponds to no LIS sign. A rather simple search engine allows the user to retrieve signs from the e-LIS database. The engine performs a translation of the user's selection in ASCII strings, and then queries the database looking up for that specific string. If no match occurs, the engine searches for similar strings, namely, for signs with one of the parameters equal to those defined by the user. In this manner, the engine always shows one or more results, avoiding an empty result set.

2.2   Interface and Interaction

The e-LIS dictionary [6] contains two modules: one for translating words into signs and the other for translating signs into words; in the dictionary, the modules are respectively labelled ITA>LIS and LIS>ITA. The LIS>ITA module allows the user to search for the translation of a specific sign. The user has to specify at least one of the Stokoe-based classes of sign components: handshape, palm orientation, location and movement, which in the dictionary are respectively labelled *configurazione, orientamento, luogo, movimento* as shown in Fig. 1.

When the dictionary user selects a class (e.g., the handshape), the system shows all the elements of the chosen class (e.g., a specific handshape); once the user chooses an element of the class, this element represents a search parameter. After that, the user can choose to either trigger the search engine or to set another sign component.

The interaction between the system and the user is a wizard-like process: the four Stokoe-based classes of sign components are shown, each with their own elements. A visual feedback is shown, representing the element chosen by the user. Figure 1 shows the core part of the LIS>ITA module after the user has chosen a specific handshape (*configurazione*), with the palm orientation class (*orientamento*) as the current choice.

The current interface of the e-LIS dictionary has some positive characteristics: it provides an iconic representation of sign components and it renders the sign translation with videos. The choice of using icons instead of sign transcriptions (e.g., with a Stokoe transcription system) improves the user's search process; in fact, the average *web* dictionary user is likely not to know any sign transcription system; thus the current interface allows even non experts of sign transcription systems to use the dictionary.

Moreover, the use of LIS videos is likely to increase the user's satisfaction: firstly, LIS videos allow to clearly render LIS signs; secondly, they are also employed to provide information concerning the translation of a sign, such as examples, thus stressing that LIS is an autoreferential language as much as verbal Italian is.

However, the design of the current interface of the e-LIS dictionary needs to be improved. Firstly, the current e-LIS interface cannot be integrated with the e-LIS ontology [9], thus it does not provide its users with the benefits of the ontology and the related technology. For instance the e-LIS ontology, which specifies how sign components are related, would allow for the creation of a dynamic interface for the e-LIS dictionary, that is, the interface would change according to the user's choices; by traversing parts of the ontology, the dictionary users could watch how their choices are related to other sign components. Without the ontology, the dictionary users cannot acquire any new knowledge on sign components and their relations; thus the navigation of the current e-LIS dictionary does not train users to become expert users of the dictionary. Moreover, without the ontology, the dictionary users can arbitrarily combine sign components and specify gestures that do not exist in LIS or the e-LIS database.

Secondly, the current e-LIS interface lacks powerful undo tools; this implies that, if a user commits to an erroneous choice, the user does not have efficient tools to backtrack to the previous state, that is, to undo the last choice. Therefore the current interface does not effectively support the decision-making process.

Finally, the definition of a sign with the current interface is long and tedious, e.g., the user may need to perform several mouse clicks. In this manner, web users are likely to soon get tired and abandon their search rather soon.

In the remainder, we focus on the design of a brand new interface which can exploit the e-LIS ontology; our proposal aims at overcoming the aforementioned drawbacks of the current e-LIS interface.

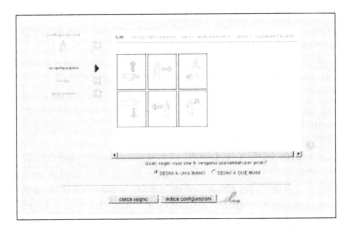

Fig. 1. The core part of the LIS>ITA module, after the user has chosen a specific handshape (*configurazione*), with the palm orientation (*orientamento)* as the current choice; all the available orientations are shown

3   The Interface Proposal

The interface proposed in this paper was designed following a standard methodology in *Human-Computer Interaction*, which focuses on the system users as the focal point of the whole development process. This is the *User Centred Design Methodology* (UCDM) [5], which consists of the following activities:

1. understanding and specifying the context of use,
2. specifying user and organisational requirements,
3. producing design solutions,
4. and evaluating design against requirements.

Such activities are always performed referring to the system users so as to achieve effective, efficient and satisfying results. By achieving effectiveness, efficiency and user satisfaction we improve the *usability* of the dictionary. Several definitions of usability exist [5]; in particular, [4] defines usability as "the extent to which a product can be used with efficiency, effectiveness and satisfaction by *specific users* to achieve *specific goals* in a *specific environment*". From this perspective, usability is the quality of interaction between the system and its users.

In the remainder of this section, we present our interface prototype, explaining how its design followed the UCDM activities.

3.1   The Context of Use

The context of use is the environment in which the project is developed, necessary to define the set and the type of intended users, tasks and environments in sufficient details so as to support the system design [5]. Therefore we analysed the background of the e-LIS project and the current online dictionary, studying the existing application domain, in particular, the e-LIS ontology. We decided not to entirely adhere to the e-LIS ontology [9], but to base our prototype on a smaller and more intuitive taxonomy resulting from the e-LIS ontology. Our simplification makes the new interface of the e-LIS dictionary more usable and closer to that of the current e-LIS dictionary. More precisely, we extracted and rendered the taxonomy of the following concepts of the e-LIS ontology:

1. handshape, used to define the configuration of the hand(s);
2. palm orientation component, which formalises the initial position of the hand(s);
3. location, used to define the part of the body with which fingers or hands contact;
4. one-hand movement component and relational movement component, used to define the movement of the hand(s) for one hand signs and two hand signs, respectively.

These concepts are grouped into intermediate ones; hence we have a multilevel hierarchical structure, in which every element has one parent and possibly one or more children, like in classical trees. The concepts associated with the Stokoe classes (e.g., handshape) are always 0-level elements; their direct children are 1-level elements; and so on.

Our interface visually renders only the taxonomy of these concepts, without any concern about relations different from *is-a* ones.

As for the handshape concept, the e-LIS ontology follows and extends the classification of [13]; this is hardly intelligible by the average user of the dictionary. We thus decided to group the handshapes following the more intuitive criterion of the current e-LIS dictionary: the number of extended fingers. In our simplified ontology, we have 0-finger handshapes, 1-finger handshapes, and so on.

Such a simplified version of the e-LIS ontology is the basis of our interface.

3.2   User and Organisational Requirements

Starting from the analysis of the context of use, we determined functional, user and organisational requirements with two main goals in mind:

1. the *profiling of the dictionary users*, in order to highlight their social, cultural, physical and psychological characteristics;
2. the analysis of the *tasks* that the users of the dictionary perform.

Users of the e-LIS dictionary are mainly deaf or hard-of-hearing people; according to some research findings, their ability of reading does not often go beyond that of a eight-year old child [11]; in particular, abstract concepts [3, 7] seem to be problematic for some deaf users. These are critical observations for the design of our e-LIS interface; e.g., our e-LIS interface must be highly visual. Moreover, not all deaf people have the same knowledge of LIS; in general, the average web user of the dictionary is likely not to know LIS at all; thus we did not assume that the dictionary web users have any prior knowledge of LIS.

Given such profiles, we turned to the design of our prototype interface and the tasks it supports: its users should be able to

1. specify the components of the sign they are searching for in an intuitive and clear manner,
2. transparently query the e-LIS database in order to retrieve the signs that match the sign components they selected,
3. interact with the results, that is, they can browse the results.

All the aforementioned tasks are carried on by exploiting the ontology; to this end, they are mapped into ontology interactions. At the current stage of the project, we decided to focus on the first task: the aim of our current interface is to simplify the composition of signs and to reach an interface usable by all the e-LIS users, whether hearing or deaf. The composition of sign can be performed by navigating the ontology; with our interface, the dictionary users see only the concepts related to their current choice.

Starting from the dictionary user profiles and the tasks of our interface, we set up the *usability goals* of our interface:

effectiveness: the dictionary users should be expertly guided during their search for a sign, thus minimising errors and obtaining satisfactory results out of their search;

efficiency: the interaction with the interface should be fast (e.g., mouse clicks are minimised) and the decision-making process should be effectively supported by the interface reducing the need of undo tools;

users' satisfaction: the interface should be well organised and plain, thus minimising the cognitive effort of users — a satisfactory interface keeps the users' attention alive.

Next, in explaining the design solution activity, we show *how* our interface aims at meeting such usability goals by using the e-LIS ontology and implementing a treemap visual technique.

3.3   Design Solutions

During the design solution activity, several mock-ups and prototypes are realised; designers and users analyse each version in order to highlight its pros and cons with respect to the usability goals of the interface.

Usually, designers start producing several mock-ups in order to evaluate the system from the functional perspective. When the final mock-up is usable, designers deploy the first prototype, representing the first version of the overall system. Designers, along with users, evaluate this prototype against several criteria. The idea of the design solution activity is to incrementally deploy the final system, since editing a prototype is better than editing the whole system.

In this paper we focus on the current prototype of our ontology-based interface, which is the result of successive refinements of the initial mock-up. Two screen-shoots of the current interface prototype are shown in Figs. 2 and 3.

The prototype presented in this paper is a visual interface designed to support the composition of a sign by navigating the e-LIS ontology. It is an *information visualisation* system, thus it is characterised by three main components [2]:

1. the visual metaphor, i.e., the graphic elements used to render information;
2. the number of dimensions, either 2D or 3D;
3. the *space-saving* strategy, that is, a tradeoff between the amount of information to be represented and the available space.

Visual metaphor. We adopt the tree metaphor as the visual metaphor, and in particular the *treemap*. The treemap visual technique [15] is shown in Fig. 2. Such a technique allows us to show a tree in a space-constrained layout, that is, the tree is turned into a planar space-filling map. Treemap uses the Shneiderman algorithm to recursively fill the available space with several areas. In our interface, the treemap technique visually renders the ontology classes and their subsumption relations.

Number of dimensions. We decided to use 3D as the number of dimensions in order to show the four Stokoe-based classes simultaneously: each treemap represents one Stokoe-based class or a subclass of its. Each treemap is embedded in a 3D plane, that is, a 3D treemap; the third dimension saves space and suggests the idea of a link between the classes.

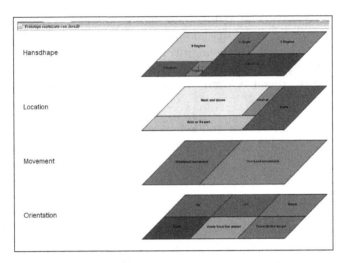

Fig. 2. The interface of the current prototype

Space-saving strategy. The space-saving strategy we adopt is the *focus +* *context* strategy [2]. In particular, in order to maintain the context, we introduce the miniatures of planes: by referring to Fig. 3, the four big planes in the right part of the interface represent the focus, whereas the four miniatures in the left part of the interface represent the context.

The visualisation of the ontology is done progressively, as explained in the following.

1. In the first step, in the left part, the interface visualises the four labels of the Stokoe-based classes, namely, handshape, location, movement, and orientation. In the right part, the interface visualises four planes; each plane is a treemap representing the direct children of the main concepts. In Fig. 2, which represents the first step, the treemap in the high part is associated to the handshape concept of the e-LIS ontology; each area of this treemap represents the direct descendant concepts of the handshape concept in the e-LIS ontology: 0-finger, 1-finger, 2-finger, 3-finger, 4-finger, 5-finger.

2. From the second to the fourth and last step, the layout of the interface is the same as that of the first step except for the presence of four smaller planes on the left. These are the *miniatures*. For instance, see Fig. 3. In the left part, the interface visualises the four labels of the Stokoe-based classes; in the right part, the interface visualises the treemaps representing some of the direct descendants as in the e-LIS ontology; in the centre, the interface visualises four smaller planes which are the *miniatures* representing the choice performed in the previous step.

In general, each treemap is composed of several areas, whose extents are computed with the Shneiderman algorithm [15]; the area's dimension is proportional to the number of child concepts the area represents. Each area in each treemap

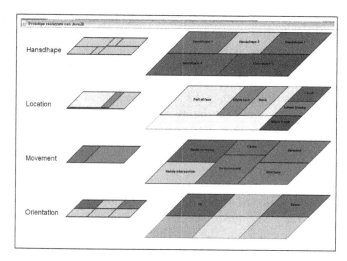

Fig. 3. The interface after the user has chosen the 1-finger handshape class

has a unique colour, because this visual feature is not used to convey a specific meaning — similar colours do not highlight related areas.

The interaction with the interface aims at being simple and intuitive, so as to allow the dictionary users to minimise their cognitive effort. As explained above, the navigation of the e-LIS ontology is organised in steps. In order to move to the next step, the user has to commit to a choice, that is, to select a specific area in a specific treemap. The choice is divided into two stages: a *preliminary* selection and a *definitive* one. Thanks to this two-stage choice, the system can show how the choice of an area propagates on the other areas of the four treemaps; thus the interface supports the decision-making process. This is the starting point for realising a dynamic visual interface.

The preliminary selection allows the dictionary users to watch how their choice of a sign component affects their search path. To this end, a transparency effect is applied to all the areas which are inconsistent with the current selection. Consistency is evaluated against the ontology. To improve efficiency and users' satisfaction, the transparency effect is applied when the user moves the mouse over a specific area — the same effect could be applied to the click over a specific area. With the mouse-over effect, we gain the following benefits:

1. from the user's perspective, moving the mouse is less obtrusive than clicking with it, since the latter is the consequence of a conscious will to make a click;
2. the effects produced by moving the mouse over a specific area can be discovered accidentally.

If the dictionary users consider the current selection as the right one, they can make it definitive by clicking on it, thus moving to the next step. Otherwise, they can easily move the mouse over a different area. Let us assume that in the first step, illustrated in Fig. 2, the user chooses the 1-finger concept. The second step is shown in Fig. 3; here miniatures are smaller representations of the 3D planes of the previous step. In the current prototype the context is given by the immediately previous step. Since miniatures store the information contained in the previous step, they can be used as an undo tool: a click on them will bring the user back to the previous step. In miniatures, inconsistent concepts are coloured in grey. The grey colour is thus associated to the notion of inconsistency. Such a feature helps users to remember the selection made in the previous step.

When the users terminate the interaction with the treemaps and hence the navigation of the ontology, they have created a query with sign components.

4   Related Work

Electronic dictionaries for SLs offer numerous advantages over conventional paper dictionaries; they can make use of the multimedia technology, e.g., video can be employed for rendering the hand movements. In the remainder, we review available electronic dictionaries from an SL to the verbal language of the country of origin, which are of interest to our work.

The Multimedia Dictionary of American Sign Language (MM-DASL) [16] was conceived in 1980 by Sherman Wilcox and William Stokoe. The innovative idea was the enrichment of the textual information with digital videos showing signing people. MM-DASL developed a special user interface, with film-strips or pull-down menus. This allows users to look up for a sign only reasoning in terms of its visual formational components, that is, the Stokoe ones (handshape, location and movement); search for signs is constrained via linguistic information on the formational components. Users are not required to specify all the sign's formational components, nevertheless there is a specific order in which they should construct the query. Since the e-LIS ontology embodies semantic information on the classes and relations of sign components for the e-LIS dictionary, the ontology can be used as the basis for an ontology-driven dictionary which forbids constraint violations. The MM-DASL project was never merchandised for several reasons, explained in [16]. For instance, platform independence of the system was a problem for MM-DASL; this is an issue the e-LIS team is taking into account, thus the choice of having the e-LIS dictionary as a web application. The profile of the expected user was never analysed, whereas e-LIS aims at a dictionary non-experts of LIS can use.

Woordenboek [8] is a web bilingual dictionary for Flemish Sign Languages (VGT). Users search for a sign by selecting its sign components, as in the current e-LIS dictionary. However, in the current version of Woordenboek:

1. users are not guided through the definition of the sign, thus users can specify a gesture which corresponds to no VGT sign or a sign that does not occur in the dictionary database;
2. the sign components are not represented via iconic images as in e-LIS; they are represented with symbols of the adopted transcription system; thereby the dictionary from VGT to Flemish is hardly usable by those who are not expert of VGT or the adopted transcription system.

Our ontology-driven interface to the e-LIS dictionary allows us to tackle such issues. To the best of our knowledge, ours is the first ontology-driven dictionary for an SL.

5   Conclusions

The current e-LIS dictionary has innovative features. As explained in Sect. 2, e-LIS uses LIS as a self-explaining language, e.g., LIS videos are also employed to propose examples of the use of a sign, or its variants. The dictionary can also be used by non experts of the Stokoe-based transcription system of e-LIS, since the elements of the so-called Stokoe classes introduced in Sect. 1 are represented via expressive icons, hence neither transcribed nor explained in verbal Italian. The mix of these features makes the e-LIS dictionary usable also by people with literacy difficulties, as it may be the case of deaf people [11].

In Sect. 3, we presented an innovative interface for the e-LIS dictionary that retains the advantages of the current e-LIS interface and integrates the e-LIS ontology. By using this as the "hidden shepherd" in the search for a sign, our interface prototype allows for the minimisation of the user's selection errors: the user can only specify a sign which is consistent with the composition rules encoded in the ontology. In this manner, expert knowledge of neither the Stokoe transcription system nor LIS are required: even non experts can masterly look for signs in the dictionary.

Thanks to our visual interface, all the dictionary users can *transparently* navigate the e-LIS ontology. Our interface can also be integrated with a query tool such as [1]. Notice that the visual metaphor we adopted improves considerably on the text-based interface of this query tool; that is, our treemap interface is essentially visual hence closer to the needs of deaf people, who are visual learners [14]. In this manner, the interface can equally support deaf as well as hearing users during the interaction process; in this sense, we designed an expert visual system, following the UCDM criteria.

Our ontology-based interface allows our users to watch the propagation of a selection of sign components, that is, our users can watch how their selections affect the search path. In this manner efficiency gets improved; users "learn by navigating" hence the interaction process gets faster. Visually showing the effects of the users' choices can also minimise the need of undo tools: the dictionary users will start a search path only if the prospected next choices are suitable to them. Such a dynamic navigation interface effectively supports the decision-making process: the dictionary users can watch the propagation of their choices before committing to them.

The visual interface presented in this paper is still a prototype. As such, it has to be fully evaluated and studied with end-users, as stated in the UCDM [10]. In particular, we must perform usability studies with a sample of end-users, thus testing the robustness of the prototype, its drawbacks and its advantages.

At the start of this paper, we stressed why we cannot assume literacy in verbal Italian of all the dictionary users, thus the need of effective intuitive icons in our treemap-based interface. Future work includes a deeper analysis of this topic, with an evaluation of the choice of such icons with our end-users.

References

[1] Catarci, T., Dongilli, P., Di Mascio, T., Franconi, E., Santucci, G., Tessaris, S.: An Ontology Based Visual Tool for Query Formulation Support. In: Proceedings of the 16th Biennial European Conference on Artificial Intelligence, ECAI 2004, Valencia, Spain (2004)

[2] Chen, C.: Information Visualization: Beyond the Horizon, 2nd edn. Springer, Heidelberg (2006)

[3] Chesi, C.: Inferenze Strutturali. Analisi sull'Uso degli degli Elementi Funzionali nel Linguaggio Verbale dei Bambini Sordi. Master's thesis, Siena U, 1999/2000, 2000 (An updated version is available as Il Linguaggio verbale non-standard dei bambini sordi, 2006. Ed. Univ. Romane) (1999)

[4] ISO Consortium. Software Product Evaluation: Quality Characteristics and Guidelines for Their Use. (Last visit: December 2005),
http://www.tbs-sct.gc.ca/its-nit/standards/tbits26/crit26_e.asp

[5] Di Mascio, T.: Multimedia Information Retrieval Systems: from a Content-Based Approach to a Semantics-based Approach. PhD thesis, Università di L'Aquila (2007)

[6] Dizionario Elettronico di Base Bilingue Lingua Italiana dei Segni-Italiano. Last visit: (May 2007), http://elis.eurac.edu/diz/

[7] Fabbretti, D.: L'Italiano Scritto dai Sordi: un'Indagine sulle Abilità di Scrittura dei Sordi Adulti Segnanti Nativi. Rassegna di Psicologia xviii(1), 73–93 (2000)

[8] Flemish Dictionary. URL: http://gebaren.ugent.be/visueelzoeken.php. Last visit: November 2006.

[9] Gennari, R., di Mascio, T.: An Ontology for a Web Dictionary of Italian Sign Language. In: WEBIST 2007. Proc. of the 3rd International Conference of Web Information Systems (2007)

[10] Norman, D., Draper, S.: User Centered System Design. LEA Hillsdale, N.J (1986)

[11] Paul, P.V.: Literacy and Deafness: the Development of Reading, Writing, and Literate Thought. Allyn & Bacon (1998)

[12] Pizzuto, E., Rossini, P., Russo, T.: Representing Signed Languages in Written Form: Questions that Need to Be Posed. In: LREC 2006. Proceedings of the Workshop on the Representation and Processing of Sign Languages (2006)

[13] Radutzky, E.: Dizionario Bilingue Elementare della Lingua italiana dei Segni. Kappa (2001)

[14] Sacks, O.: Seeing Voices: a Journey into the World of the Deaf. Vintage Books (Italian version: "Vedere Voci", Adelphi, 1999) (1989)

[15] B. Shneiderman.: Treemaps for Space-Constrained Visualization of Hierarchies. (Last visit: November 2006), http://www.cs.umd.edu/hcil/treemap-history/

[16] Wilcox, S.: The Multimedia Dictionary of American Sign Language: Learning Lessons about Language, Technology and Business. Sign Languages Studies iii(4), 379–392 (2003)

Improvement of a Web Engineering Method Through Usability Patterns*

José Ignacio Panach, Francisco Valverde, and Óscar Pastor

Department of Information Systems and Computation
Technical University of Valencia
Camino de Vera s/n, 46022 Valencia, Spain
Ph.: +34 96 387 7000; Fax: +34 96 3877359
{jpanach,fvalverde,opastor}@dsic.upv.es

Abstract. Usability is a feature of software quality that has traditional significance in the Human Computer Interaction (HCI) community. Recent works that
have been proposed by the Software Engineering (SE) community are intended
to improve the usability of software applications. This paper combines aspects
that are defined in both these communities to produce usable web applications.
To achieve this goal, a well-known strategy to improve usability is used: usability patterns. However, many usability patterns and guidelines could only be applied when the final system is implemented. In this work, STATUS patterns
have been chosen because they solve usability issues at conceptual level. The
main purpose of this paper is to improve the usability of Web Applications
automatically generated by OOWS (a model-based web engineering method)
applying the STATUS patterns.

Keywords: Web engineering, Web usability, MDA, model-driven engineering,
automatic code generation, usability patterns, Presentation Model.

1 Introduction

Usability has become increasingly important in web engineering methods, even more
important than in conventional desktop applications [6]. Some works have proposed
methods for measuring usability, like Olsina's work [11]. Moreover, recent works incorporate usability as part as an MDA [9] development process [1].

Following this last emergent research line, this paper is focused on how to deal
with the required usability aspects of web applications that are generated automatically in a model-driven web development process. Specifically, the objective is to improve usability in OOWS [5] (Object Oriented Web Solutions) web engineering
method. OOWS has an automatic code generation process based on the MDA paradigm that produces a web application from its corresponding web conceptual schema.

OOWS is complemented by OlivaNova [3], the industrial tool that implements the
methodology called OO-Method [12]. OOWS generates the code corresponding to the
specific, web oriented user interface, preserving the business logic layer and the persistence layer generated by OlivaNova.

---

* This work has been developed with the support of MEC under the project DESTINO
TIN2004-03534 and cofinanced by FEDER.

M. Weske, M.-S. Hacid, C. Godart (Eds.): WISE 2007 Workshops, LNCS 4832, pp. 441–446, 2007.

Currently, there are several web engineering methods that model the web interaction in an abstract way and distinguish between navigation and interface like OOWS does. Some of these methods are WebML [4], OOHDM [14] or OOH [2]. The interaction aspects related to usability are considered in all these web engineering methods by means of a specific model or a quality framework. However their proposed approaches to usability are coupled with the particular method. Therefore, applying the same concepts to another web engineering method is a difficult task.

To solve this problem the use of patterns is proposed in this work. A usability pattern suggests an abstract solution to a usability problem without taking into account platform constraints.

Several authors have written about usability patterns, like Welie [17], who makes an explicit distinction between the user's perspective and the designer's perspective. Tidwell [16] defined other patterns that are very similar to the patterns proposed by Welie. She proposes using patterns to help the design of the Conceptual Model behind the interface. Moreover, some authors, such as Kimberly Perzel [13], have been working to define usability patterns for applications oriented to the World Wide Web. However, none of the mentioned above approaches implements a true Model Compiler, meaning that the specification of the usability aspects is done at the modeling step, and is properly converted into the required software components through the corresponding transformation process.

Therefore, the main contribution of this paper is to include the usability of web applications generated with OOWS as an essential aspect to be considered. This fact is motivated by the experiences provided by users of Web applications generated by OOWS. With the purpose of solving OOWS usability problems, we have selected a set of usability patterns defined in a European project called STATUS (SofTware Arquitectures That support USability) [15]. For this purpose, the current OOWS Presentation Model (the part of the Conceptual Model that models the interaction between the user and the system) is extended. This usability improvement can be divided into two steps: 1) to select the STATUS patterns that solve the OOWS usability problems, and 2) to enrich the OOWS Presentation Model with the required expressiveness to model the functionality of the patterns that are not currently supported.

To accomplish these goals, the paper is structured as follows. Section 2 presents the usability problems in web applications generated by OOWS and which STATUS pattern provides a solution. Section 3 shows an extension of the OOWS Conceptual Model to incorporate the functionality of the usability patterns that OOWS does not currently support. Finally, section 4 presents the conclusions.

2   Analysis of OOWS Usability

STATUS patterns are defined in Juristo et al [8] as a set of generic solutions to solve common usability issues. The solution for each problem is described by means of several UML Diagrams (Class Diagram and Sequence Diagram) that can be applied in a specific methodology. The use of these patterns provides two main advantages:

- The usability mechanism is described in an abstract way from an object-oriented perspective. As a consequence, applying a STATUS pattern to OOWS is a task that can be easily performed.

- The solution proposed in the pattern is neither designed for a particular method nor a specific software platform (Web, Desktop etc.). The same principles can be applied in another Web Engineering methods

As OOWS and OO-Method are based on UML, the solution described by the STATUS pattern can be easily introduced inside the software development process. Since the set of STATUS patterns is very extensive, this paper is only focused on two STATUS patterns that, due to their functionality, proposes a solution to the usability problems that have been detected. The usability problems are related to data entry mistakes when users try to perform an operation in an OOWS Web Application. In addition, in order to make the most appropriate choice, the usability recommendations on data entry stated in [10] have been followed.

2.1 User Input Errors Prevention: Structured Text Entry

The main objective of this pattern is to anticipate possible mistakes caused by invalid user actions. To do this, the pattern proposes the use of different input mechanisms and default values. Several widgets can perform the same task, even though their visual representations are different. The goal of the usability analyst is to choose the correct widget for a concrete input data. Currently, OOWS does not allow the user or the analyst to choose the widget type, for example "list boxes" for a concrete set of values or "edit masks" to insert data in a specific format.

Another way to avoid user errors is to provide default values to the user that can be changed when they are not appropriate. OOWS supports this functionality by means of the OO-Method Structural Model. However, frequently, the list of possible values in a widget may depend on the values inserted in other widgets, thereby creating dependency relationships between widgets. OOWS does not have any primitives to model this behaviour.

2.2 Wizard: Step by Step

This pattern helps users to execute a complex action that requires several steps. Using this pattern, the analyst can define a wizard that will help users with complex actions that require them to introduce information in several steps. The fact of splitting the operation into different steps improves the user guidance. This functionality cannot be modelled in OOWS yet.

3 Improving the OOWS Presentation Model: A Case of Study

This section explains how the OOWS Presentation Model can be improved with usability patterns whose functionality is not currently supported. The solution proposed is to use UML stereotyped elements in order to abstract STATUS patterns functionality. These new UML elements are introduced inside the OOWS Presentation Model extending the current conceptual primitives. Since many web engineering methods are based on UML, this approach can be used to define new usability concepts into their models. A prototype of model compiler that includes the functionality of *Step by Step* and *Structured Text Entry* patterns is used to generate the code of the case study.

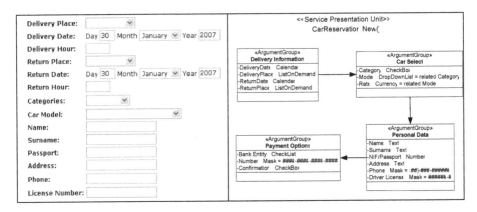

Fig. 1. a)Service Reservation Interface before applying usability patterns; b) SPU for car reservation service

To test the advantages of the proposed usability patterns, we have developed a case study based on an on-line car rental service. This paper focuses on the **reservation** service to simplify. A part of the current automatically generated web interface is shown in Figure 1a.

To support both patterns, the *Service Presentation Unit* (SPU) primitive has been introduced into the OOWS Presentation Model. The purpose of the SPU is to model how the interface that executes a service (usually a web form) will be shown to the user. Each SPU is related to a service defined in the Structural Model and is composed of the set of correspondent arguments. Figure 1b shows the SPU for the case study presented here. In the following subsections, we detail the conceptual primitives that the SPU has to model our service interface.

3.1 Supporting the "Step by Step" Pattern

This paper defines an *Argument Group* represented as a stereotyped UML Class. Thanks to this primitive, it is possible to model how the service arguments are grouped in a SPU. The sequence, in which each group of arguments should be introduced, is represented by means of arrows. Moreover, each *Argument Group* can include a text description to inform the user. Briefly, an *Argument Group* represents a Wizard step (Figure 1b).

At the implementation level, the result is a set of web pages (a Wizard) to collect the argument values. In the proposed case study four *Argument Groups* are defined: Delivery information; Car Selection; Personal Data; Payment Options.

3.2 Supporting the "Structured Text Entry" Pattern

With the current OOWS Presentation Model, it is not possible to delimit the correct set of values for an argument or its input mechanism. The compiler takes into account the data type of an argument to render an appropriate widget. To solve this problem, three mechanisms have been added to the *Service Presentation Unit*:

- **Argument Widget:** It specifies the widget type that will receive the value. This primitive is defined in the attribute type for a particular argument. If no widget type has been defined, the default widget (Text) is used to input any kind of alpha-numerical string. In our case study (Figure 1b), the delivery and return dates are typed as Calendar, the car categories are rendered as checkboxes and phone and credit cards are masks.

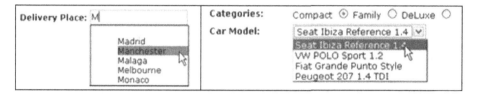

Fig. 2. a) List On Demand input Widget; b) Argument Car Model related to Category

- **List on Demand:** A *List On Demand* is a type of dropdown list with an input text whose values are retrieved dynamically. When the user writes a string, a list of instances which matches the text written will be shown. This primitive is used in the arguments "Delivery Place" and "Return Place" show in Figure 1b. An example of the interface generated from this primitive is shown in Figure 2a.
- **Related Argument:** It is useful to restrict the values that the user can insert in a widget depending on the values that the user has previously inserted in other widgets. For example, if the user has selected the desired car category, only the car models that are related to that category must be shown. Figure 1b shows this primitive in the arguments "model" and "rate". Figure 2b shows the final interface.

4 Conclusions

This research work presents a usability improvement of web applications built with an automatic code generation process. The proposed solution has the following steps:

1. **Choice of STATUS usability patterns.** STATUS usability patterns have been selected because these patterns can be incorporated in the architecture of the system throughout the entire software development process. As a consequence, this is a suitable approach to improve usability in another web engineering methods.
2. **Selection of a subset of STATUS patterns.** Of all the STATUS patterns, this paper is centered only on the patterns whose functionality has been considered to be more appropriate for web usability according to [10] and users' experiences.
3. **OOWS Conceptual Model Extension.** Two STATUS Patterns, *Step by Step* and *Structured Text Entry*, are introduced into OOWS as UML elements.

A prototype version of the new Model Compiler has been developed in order to introduce the changes in the OOWS Presentation Model. As an example of an application of this approach, this paper presents a little case study that includes the usability patterns presented. Users feedback verifies that the new web interface is more usable than previous one because it has been generated using the new conceptual primitives.

As future research, the rest of the STATUS patterns that are not currently supported by OOWS must also be considered. In addition, the Conceptual Model should include a set of metrics to measure the usability before generating the system. Finally, an empirical evaluation of usability, with industrial web applications generated by OOWS, will be carried out to validate the usability improvement of the generated code.

References

[1] Abrahao, S., Insfrán, E.: Early Usability Evaluation in Model-Driven Architecture Environments. In: 6th IEEE International Conference on Quality Software (QSIC 2006), Beijing, China (2006)

[2] Cachero, C., Genero, M., Calero, C., Meliá, S.: Quality-driven Automatic Transformation of Object-Oriented Navigational Models. In: Embley, D.W., Olivé, A., Ram, S. (eds.) ER 2006. LNCS, vol. 4215, Springer, Heidelberg (2006)

[3] Care Technologies: Last visited: (Febuary 2007), http://www.care-t.com

[4] Ceri, S., Fraternali, P., Bongio, A.: Web Modeling Language (WebML): a modeling language for designing Web sites. In: WWW9 Conference, Amsterdam, pp. 137–157 (2000)

[5] Fons, J., Albert, P.V.: Development of Web Applications from Web Enhanced Conceptual Schemas. In: Song, I.-Y., Liddle, S.W., Ling, T.-W., Scheuermann, P. (eds.) ER 2003. LNCS, vol. 2813, pp. 232–245. Springer, Heidelberg (2003)

[6] Hitz, M., Leitner, G., Melcher, R.: Usability of Web Applications. In: Web Engineering, Wiley, Chichester (2006)

[7] ISO/IEC 9126-1, Software engineering - Product quality - 1: Quality model (2001)

[8] Juristo, N., López, M., Moreno, A., Sánchez, I.: Improving software usability through architectural patterns. In: International Conference on Software Engineering. Workshop Bridging the Gaps Between Software Engineering and Human-Computer Interaction, Portland, USA (2003)

[9] MDA: Last visited: (Febuary 2007), http://www.omg.org/mda

[10] Nielsen, J.: Prioritizing Web Usability, 1st edn. New Riders Press, Indianapolis (2006)

[11] Olsina, L., Rossi, G.: A Quantitative Method for Quality Evaluation of Web Sites and Applications. In: IEEE Multimedia Magazine, pp. 20–29 (2002)

[12] Pastor, O., Gómez, J., Insfrán, E., Pelechano, V.: The OO-Method Approach for Information Systems Modelling: From Object-Oriented Conceptual Modeling to Automated Programming. Information Systems 26(7), 507–534 (2001)

[13] Perzel, K., Kane, D.: Usability Patterns for Applications on the World Wide Web. In: PloP 1999 Conference (1999)

[14] Schwabe, D., R.G., Barbosa, S.: Systematic Hypermedia Design with OOHDM. In: ACM Conference on Hypertext, Washington, USA (1996)

[15] STATUS Project: Las visit: (Febuary 2007), http://is.ls.fi.upm.es/status

[16] Tidwell, J.: Designing Interfaces, O'Reilly Media (2005)

[17] Wellie, M., Traetteberg, H.: Interaction Patterns in User Interfaces. In: PLoP 2000, Allerton Park Monticello, Illinois, USA (2000)

Flex RIA Development and Usability Evaluation

Lenja Sorokin[1,2], Francisco Montero[2], and Christian Märtin[1]

[1] Augsburg University of Applied Sciences
Faculty of Computer Science
D-86161 Augsburg (Germany)
l.sorokin@web.de, maertin@informatik.fh-augsburg.de
[2] Universidad de Castilla-La Mancha
Department of Computer Science
02071 Albacete (Spain)
fmontero@dsi.uclm.es

Abstract. A key factor for successful Rich Internet Applications (RIA) is to provide an interface which is effective and efficient to use. This paper proposes a framework for model driven development of Adobe® Flex™ RIAs which uses task models and international standard task-based metrics to integrate usability characterization and evaluation from the very beginning. It supports the generation of alternative user interface versions which can be employed as prototypes in usability testing to get early user feedback.

Keywords: automated usability evaluation, remote usability testing, web usability, model-based evaluation, Flash, Flex, model-driven development.

1 Introduction

Adobe® Flex™ is one of the major tools for creating cross-platform Rich Internet Applications (RIA). Like Adobe® Flash™, it uses SWF files as output format and thereby takes advantage of the widely used Flash Player [3]. The difference to Flash is its application area: Flash supports the creation of animated interactive content, while Flex focuses on application development.

Flash sites have always struggled with bad reputation regarding usability [11]. One reason is that the high potential for graphical design and animations as well as missing interaction conventions encouraged Flash developers to emphasize more on design than on usability [8]. There have been major improvements when Adobe formed a cooperation with the Nielsen Norman Group and released Flash MX which introduced standard user interface elements. Yet usability is a key factor in the development of successful Flash websites and its evaluation is an essential need [14].

Realizing the importance of this issue, Adobe has designed Flex to get closer to the usability of desktop applications. It provides a large set of predefined user interface components and design principles, thus supporting usability goals as e.g. consistency, efficiency of use, etc. [2]. The potential to create effective and easy to use interfaces is a fundamental quality of Flex, particularly considering its application area. As it is

M. Weske, M.-S. Hacid, C. Godart (Eds.): WISE 2007 Workshops, LNCS 4832, pp. 447–452, 2007.

designed to support the development of web-based business processes and applications (examples: http://www.flex.org/showcase/), there is a strong focus on creating usable sites and there will be a need for methods and tools to support developers of Flex applications in achieving this goal.

In this paper our main goal is to support the development and evaluation of Flex applications within an iterative design process by enabling the creation of different user interfaces from a single model and gathering information for usability evaluation using standardized metrics. This paper is organized as follows: In section 2 we discuss existing approaches to automated usability evaluation. Section 3 provides an overview of the framework and section 4 discusses conclusions and future prospects.

2   Related Work

A variety of different tools and methods for evaluating the usability of websites and applications has been developed. Ivory [7] provides a good overview and classification of existing methods. Our approach focuses especially on automated and semi-automated usability testing.

The WebTango prototype [6] uses predictive models to analyze the structure of individual pages and the overall site by computing 157 quantitative measures. WebQuilt [5] is a framework for remotely recording and analyzing user interaction, which captures clickstream data and visualizes usage traces. A similar approach is used by WebVip [15], which is also based on an interpretation of user navigation. AWUSA [16] is a task based approach which combines log file analysis, automated usability evaluation and web mining. By analyzing the website and users' behavior it provides information about usage patterns, problem resources, task usage and user groups. RemUsine [13] and later WebRemUsine [12] define task and page-related metrics which are useful to detect some critical aspects of evaluated websites, for example particularly long tasks. Other proposals [1] and [9] try to combine model-driven development and usability by using conceptual and interaction patterns respectively. In these product-oriented proposals, usability is defined by using the ISO 9126 and additional usability criteria. However, these proposals don't work with metrics, but with desirable facilities of development and interaction. Also the RAD-T toolset [4] works with a development-oriented approach. Its proposition is to extend rapid application design of websites to include usability testing.

In short, there is plenty of literature and tools available on automated usability evaluation. However there are no proposals on how to evaluate the usability of Flex applications. Furthermore most of the tools use clickstream data, which does not allow conclusions about task-related issues. Yet such information would provide a deeper insight on user actions and the possibility to use internationally accepted metrics. Those who do work with a task-based approach as e.g. RemUsine don't use standardized metrics. Though it is widely acknowledged that it is important to integrate usability into the development process as early as possible, only few approaches support it. RAD-T is the only tool described, which works with development-oriented evaluation. But it does not provide support for comparing alternative designs, which is an essential task when creating web applications.

3 Proposed Framework

Our framework features a combination of automated usability evaluation based on standardized metrics and a development-oriented approach, which allows to support the comparison of different concrete user interface versions and thus alternative Flex RIAs. It integrates the prerequisites to measure task-based metrics into a model-driven development (MDD) environment using a set of four layers and appropriate transformations: Task model (TM), Abstract User Interface (AUI) Model, Concrete User Interface (CUI) Model and finally a specific RIA.

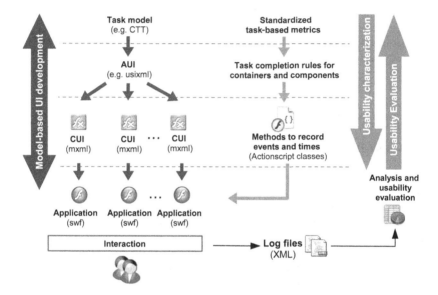

Fig. 1. Proposed framework

Figure 1 presents an overview of the framework. The left side shows the development from a task model (TM) to different versions of concrete user interfaces. The TM contains a hierarchy of tasks, currently CTT [13] is used for its specification. From the TM the abstract user interface (AUI) is created automatically e.g. by using IdealXML [10], which creates a UsiXML file. In a next step, the AUI is converted semi-automatically into several different concrete user interfaces (CUIs), which can then be compiled into an swf file, i.e. a Flash application.

The right side of the framework shows how usability evaluation is supported at the respective layer, relating events of user interactions to a set of metrics. Corresponding to the development based on a TM, the evaluation relies on task-based metrics. To be able to apply these metrics on the elements of a user interface, that is containers and components, we identified some rules to define task completion and times. On the level of the CUIs we developed ActionScript classes, which implement these rules and allow to record events and times of user actions. When a user is interacting with one of the Flash applications, the ActionScript classes will produce log files. These are analyzed for evaluating the usability of the alternative user interface versions.

The set of metrics (Table 1), which is used by our framework, is based on international standards (ISO 9241-11 and ISO 9126-4). We are concerned with quantitative usability measures such as task completion times, error frequency, number of mouse clicks or number of errors, etc.

Table 1. Metrics from international standards and usability factors

Metric	ISO 9241-11	ISO 9126-4
Task completion		X
Error frequency		X
Number of persistent errors	X	
Percentage of relevant functions used	X	
Percentage of tasks completed successfully on first attempt	X	
Number of accesses to help	X	
Task time		X
Time taken on first attempt	X	
Time spent on correcting errors	X	
Frequency of reuse	X	

3.1 A Model-Based User Interface Development Environment

The translation of the TM to an AUI is already provided by idealXML [10]. Therefore our focus is on creating different CUIs from the AUI. As Flex provides an XML based language (MXML) for defining user interface elements, the translation from the UsiXML file of the AUI can be done using XSLT. To define the necessary rules, we assigned a list of possible Flex components (Table 2) to each type of AUI component, described by a combination of facilities (*input, output, control, navigation*).

When creating a CUI the developer is prompted for each AUI component to select one of the corresponding Flex components.

Table 2. Association of AUI components, described by a set of UsiXML facilities, with Flex components

Combination of UsiXML facilities	Flex component
Input	TextInput, TextArea
Output	Text, Label, Tooltip, Image, VideoDisplay, ProgessBar
Control	Button, PopupButton
Navigation	LinkButton
Input + Output	Editable TextInput, Editable TextArea
...	...
Input + Output + Control	DateField, DateChooser, ComboBox, List, DataGrid

To integrate the ability to log user interactions into the CUI, we take advantage of the Flex programming model. It is based on MXML for defining the user interface components and ActionScripts for the logic. Flex takes the MXML files and compiles them with all associated ActionScript files to build an swf file. For each Flex component we implemented a Proxy Class. Using these custom components, the

ActionScripts, which capture events, measure times and store these data in a log file, are automatically included into the resulting Flex application.

3.2 Enabling the Evaluation of Task-Based Metrics

The base unit used to measure the metrics is a basic task, i.e. a task that cannot be decomposed further. Currently we focus on interaction tasks, as user tasks could not be logged in a fully automated way by our framework. Interaction tasks are directly related to specific user interface elements. Therefore we developed a rule for each interactive element, to define the completion of the related task. Applied to Flex applications these rules are expressed by a combination of events for a specific Flex component, e.g. for a TextInput "FocusEvent.FOCUS_IN" defines the beginning of the task and the last "Event.CHANGE" its completion. Additionally we integrated the *validator* concept of Flex in order to provide a more powerful checking of user entries and task completion. We established these rules for all classic components provided by Flex. When applying our framework on more complex RIA's with custom components, developers can include these by defining the respective rules.

Using the resulting log entries, which are related to the basic tasks, and the task model, we can also evaluate task completion, task times, errors etc. of high-level tasks (tasks composed of subtasks). For this purpose the compliance of user actions with the temporal relationships defined in the task model has to be checked. This can be done in three ways: One would be to use the same approach as WebRemUsine [12] and traverse the task tree. Another one would be to implement a finite state machine. But if we think of interacting with the interface by using its specific language, we could also apply a far more interesting approach: We could translate the TM as an abstract specification of the user interface into a grammar. The events are understood as symbols and their combination as strings of the language of the specific interface. Thus we could check their validity using the TM based grammar.

4 Conclusions and Future Work

In this paper we propose a development environment for Flex applications, which allows to create alternative user interface versions from a single initial model. The resulting Flex applications automatically include facilities for gathering usage data related with usability metrics from international standards (ISO 9126-4 and ISO 9241-11). Thereby the framework supports involving users into the design process to identify usability problems and to compare different interface versions.

Currently we are working on the analysis of the data obtained from recording user interactions. A possible approach could be to translate the TM into a grammar and use the result to evaluate the interaction of the users. Future work for improving our framework includes the support for automatic transition from an AUI to CUIs. Also we will try to validate our initial set of metrics (Table 1) and we want to associate them with patterns of usability problems to improve our analysis of data.

Acknowledgements. We would like to gratefully acknowledge the support of EU-COST Action no. 294: MAUSE (http://www.cost294.org).

References

1. Abrahao, S., Pastor, O., Olsina, L.: A Quality Model for Early Usability Evaluation. In: International COST294 workshop on User Interface Quality Models (UIQM 2005) in Conjunction with INTERACT 2005, Rome, Italy, (2005)
2. Adobe Systems Incorporated: Flex 2 technical overview – technical whitepaper (2006)
3. Adobe: Flash Player Version Penetration, http://www.adobe.com/products/player_census/flashplayer/version_penetration.html
4. Becker, S.A., Berkemeyer, A.: RAD-T: Rapid Application Design and Testing of Web Usability. In: Multimedia, IEEE, Los Alamitos (2002)
5. Hong, J.I., Landay, J.A.: WebQuilt: A Framework for Capturing and Visualizing the Web Experience. In: Proceedings of The Tenth International World Wide Web Conference (WWW10), Hong Kong, China, pp. 717–724 (2001)
6. Ivory, M.Y., Hearst, M.A.: Improving web site design. IEEE Internet Computing, Special Issue on Usability and the World Wide Web 6(2), 56–63 (2002)
7. Ivory, M.Y., Hearst, M.A.: The State of the Art in Automating Usability Evaluation of User Interfaces. ACM Comput. Surv. 33(4), 470–516 (2001)
8. MacGrebor, C., Waters, C., Doull, D.: The Flash Usability Guide, New York, pp. 7–10. Springer, Heidelberg (2003)
9. Montero, F., González, P., Lozano, M., Vanderdonckt, J.: Quality Models for Automated Evaluation of Web Site Usability and Accessibility. In: International COST294 workshop on User Interface Quality Models (UIQM 2005) in Conjunction with INTERACT 2005, Rome, Italy, 12.-13. (September 2005)
10. Montero, F., Lopéz-Jaquero, V.: IdealXML: an interaction design tool. In: Proceedings of the Sixth International Conference on Comuter-Aided Design of User Interfaces (2006)
11. Nielsen, J.: Flash: 99% Bad (2000), http://www.useit.com/alertbox/20001029.html
12. Paganelli, L., Paternó, F.: Intelligent Analysis of User Interactions with Web Applications. In: Proceedings of the 7th international conference on Intelligent user interfaces (2002)
13. Paternó, F.: Model-Based Design and Evaluation of Interactive Applications, pp. 152–172. Springer, Heidelberg (2000)
14. Piyasirivej, P.: Towards usability evaluation of Flash Web sites. In: World Forum Proceedings of the International Research Foundation for Development (2006)
15. Scholtz, J., Laskowski, S.: Developing Usability tools and techniques for designing and testing web sites. In: Proceedings of the Fourth Conference on Human Factors & the Web (1998)
16. Tiedtke, T., Märtin, C., Gerth, N.: AWUSA – A Tool for Automated Website Usability Analysis. In: PreProceedings of the 9th Int. Workshop DSV-IS 2002, Rostock, Germany, pp. 251–266 (2002)

Usability and Accessibility Evaluations Along the eLearning Cycle

Ludivine Martin[1], Emmanuelle Gutiérrez y Restrepo[2], Carmen Barrera[1],
Alejandro Rodríguez Ascaso[1], Olga C. Santos[1], and Jesús G. Boticario[1]

[1] aDeNu Research Group, Artificial Intelligence Department, Computer Science School,
UNED, C/Juan del Rosal, 16. 28040 Madrid, Spain
{ludivine.martin,arascaso,ocsantos,jgb}@dia.uned.es,
cbarrera@invi.uned.es
http://adenu.ia.uned.es/
[2] SIDAR Foundation-Universal Access.
C/ Sancho Dávila, 35, 28028 Madrid, Spain
emmanuelle@sidar.org
http://www.sidar.org

Abstract. In the paper we present how usability and accessibility requirements can be assured and validated in the whole life cycle of eLearning. On a 4-phase eLearning cycle, which includes adaptation features to cover the particular needs of each learner, we have included the evaluation of usability and accessibility requirements. We have performed this evaluation on an accessible standard-based open-source learning management system called dotLRN. The validation methodologies proposed are applied in three research projects, EU4ALL, ALPE and ADAPTAPlan.

Keywords: Usability guidelines, Accessibility guidelines, eLearning, Evaluation user's satisfaction, Automatic evaluation of Web accessibility.

1 Introduction

eLearning removes time and space barriers. This should benefit to people with disabilities but, they can sometime face more problems in eLearning environments that in face-to-face education. Current research in eLearning strongly focuses on providing adaptation support to users. aLFanet project (IST-2001-33288) developed an adaptive Learning Management System (LMS) which combined design and runtime adaptation along the full life cycle of the learning process with a pervasive use of educational standards [1]. EU4ALL project (IST-2006-034778) develops the delivery of learning services along the life cycle of the learning process to cover the needs of people with disabilities. Thus, usability and accessibility (U&A) have also to be considered in the eLearning lifecycle (eLLC), on top of existing adaptation support.

This paper proposes some guidelines to apply U&A criteria along the eLLC, following the adaptation cycle as defined by aLFanet project. We identify the particularities of LMS when studying their U&A and detail how evaluations and

M. Weske, M.-S. Hacid, C. Godart (Eds.): WISE 2007 Workshops, LNCS 4832, pp. 453–458, 2007.
© Springer-Verlag Berlin Heidelberg 2007

requirements are integrated into the eLLC, the learning experience and the theme of adaptation. Using the framework of our current European projects ALPE and EU4ALL, we describe the U&A methodologies used, some key results and findings to date as well as the challenging areas.

2 Learning Management Systems, Usability and Accessibility

A variety of commercial or open-source LMS exist, offering a wide range of integrated features. LMS can be classified in different categories: general LMS with tools for creating and managing courses, collaborative learning support systems, questionnaire and test authoring systems, people and institutions resources management systems and virtual classrooms. One LMS can belong to several categories at the same time.

Before adopting a LMS, it is essential that the institutions identify their needs and criteria in a first place and later go through a comparative analysis of the available LMS. Several sites[1] and articles provide reviews of LMS, mostly concentrated in comparative study of their technical features. At the time of choosing a LMS in year 2000, UNED (Spanish National University for Distance Education) chose ArsDigita Community System (ACS) developed at the Massachusetts Institute of Technology (MIT) for the technical support and collaborative features. ACS evolved into dotLRN[2] open source eLearning platform, currently a reference in terms of support for adaptation, accessibility and educational standards [2].

LMS have particularities when evaluating U&A:

- **Variety of needs.** LMS have to be flexible enough to address a variety of teaching and learning styles, as well as interaction preferences and devices.
- **High level of customization.** LMS usually offer a wide range of configuration options at the admin level, the tutor level and the student level. This flexibility affects the overall usability of the system.
- **Captive audience.** The end-users (students and tutors) don't choose the LMS. Instead, they choose a learning institution and then have to work with the institution's LMS. If users get frustrated by the LMS, they cannot leave it; they have to bear with it. One advantage is that at the time of evaluation, users might be more frank on their opinion of LMS and the results might be more valid.
- **Educational standards.** In order to facilitate the reusability of the author work and the adaptability of the course to the learner, the LMS has to comply with educational standards (IMS, SCORM). It is common that the complexity of educational standards impacts the interface, therefore the user experience.
- **Container/contained relationship.** When evaluating the overall U&A of a LMS, we need to look at three different elements: the *platform*, where the course materials are stored and delivered; the *formal content*, or packaged course materials, compliant to educational standards and the *content generated by users*.

---

[1] EduTools: http://www.edutools.info
[2] Officially named .LRN, pronounced 'dotlearn' and usually referenced as dotLRN to avoid confusion. More information available on the website: http://dotlrn.openacs.org

3 Adaptation, Usability and Accessibility in the eLLC

To support an adaptive eLearning approach based on educational standards, design guidelines cover the instructional design for learning resources and activities as well as the bridging elements (hooks) to dynamically and individually support learners at runtime. The main characteristic of aLFanet is to deliver adapted courses based on pervasive use of educational standards and several user-modelling techniques in a multi-agent architecture, which applies machine-learning techniques to learn the model attributes and provide recommendations to learners [1].

From the point of view of adaptation, the four steps in the eLLC (see Fig. 1) focus on providing a learning experience adapted to the particularities of each learner. aLFanet focuses on the learners and their needs, and not the LMS capabilities [3]. To provide support in these four phases, and especially for the design phase, we need authoring tools that produce standard-based educational material that can be understood by LMS, managed by the existing assistive technologies and accessible to end users regardless of their abilities. In respect to this, U&A requirements must be thought out beforehand and planned ahead for the each phase of the cycle.

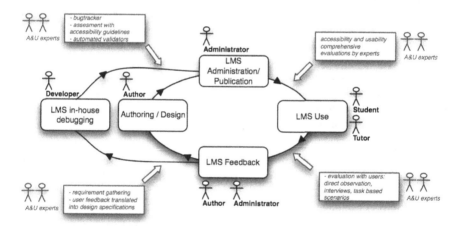

Fig. 1. aLFanet eLLC including interactions with U&A

(a1) LMS in-house debugging phase

Currently U&A features are not guaranteed in any LMS and there is a need to evaluate them. Before releasing a platform version, developers should have a habit of revising the latest developments conformity to the relevant U&A guidelines, including automatic validators along their development cycle. However, when installing the corresponding platform, in-house evaluations have to be made to detect issues not covered. In-house evaluations can be documented in a Bug Tracker tool, with priority level and dates tracking. This way, improvements on the platform can be made (internally or externally) to cope with the U&A requirements.

(a2) Authoring/Design phase
This phase includes the design and creation of content, questionnaires and activities, as well as the definition of the services used to reach defined learning objectives for defined learners' profiles. The design phase integrates the U&A requirements defined via requirements gathering, elicitation methods and evaluation results from the feedback phase. The author needs to be aware of U&A guidelines. Before delivering the course, a strict quality evaluation on accessibility, usability and standards compliance is performed.

(b) LMS Administration/Publication phase
This phase includes the management of all data, e.g. enrolling particular users with their corresponding users' roles and access rights. The environment is prepared to provide an adapted experience to the learners, such as pedagogical, interface, content and services customization. U&A expert reviews are performed to detect the changes and errors appearing when the course is imported in the LMS. Methods include heuristic evaluation and cognitive walk-through. External audit is preferable since it usually detects failures that escape to the developers (developers may not perceive them as failures).

(c) LMS Use phase
The use phase supports the actual interactions of users (tutors and learners) with the contents and services available. Dynamic adaptations and support based on the users' interactions (both individual and collaborative) are offered. User's behaviour and interactions are tracked by the system to learn more about the users. Empirical testing or user-based methods also takes place, including direct observation, using the thinking aloud and contextual inquiry techniques, with task-based scenarios. Communication channels to users can also be provided to allow them report failures or suggest improvements.

(d) LMS Feedback phase
Author and administrator get reports on the actual use of the course, mainly statistical information showing how learners proceeded. During this phase, results from experts, users' evaluations and automatic processing of interactions are analyzed and translated into i) technical requirements for the next LMS release and ii) guidelines for course authoring.

4 Accessibility and Usability Evaluations

The LMS accessibility evaluation considers not only the Web Content Accessibility Guidelines (WCAG) [4] but also the other accessibility guidelines in order to cover all LMS processes and components, including interface and any integrated interaction elements or applications, such as Authoring Tool Accessibility Guidelines (ATAG) [5], User Agent Accessibility Guidelines (UAAG) [6] and Guidelines for Developing Accessible Learning Applications (GDALA) [7]. Both technical and end-user assessment can be performed.

The technical assessment requires, as advised by the W3C Web Accessibility Initiative (WAI) "knowledge and familiarity with Web mark-up languages, more than one scripting and interface languages, a variety of evaluation tools and approaches, and strategies and assistive technologies used by disabled people". The team must evaluate all templates, content, and templates and content combined. At least two web accessibility evaluation tools should be applied to the selected sample of pages and at least one tool should run across the entire website. Automatic evaluation tools, such as HERA [8] facilitate the generation of reports and include pointers on methods and techniques for manual revision.

An accessibility expert methodologically tested dotLRN LMS through different interaction strategies. It detected 22 problems: 27 % of Priority 1, 64% of Priority 2 and 9% of Priority 3. 75% of those accessibility problems have been fixed so far. The remaining issues are under study for the next release. Experts from the Tenuta project[3] ran an analysis based on an automatic revision plus a manual revision of 15 out of the 65 WCAG 1.0 checkpoints. The result was very positive.

Regarding end-users U&A, the WAI advise the inclusion of users with disabilities in the review process [9]. We follow this methodology to verify the application of the ATAG, UAAG and GDALA. Since a LMS is constantly evolving, accessibility evaluation must be continuous. In particular, within the ALPE project we are experiencing ALPE platform with 300 learners from 3 countries (UK, Spain and Greece) from our target group (adult learners with visual or hearing impairments) to evaluate the effectiveness, efficiency and satisfaction of each user group.

For the usability assessment, we started with 2 experts performing a heuristics evaluation, using an extensive checklist based on the Ten Usability Heuristics [10] declined in 250 checkpoints [11]. The platform complied particularly well on 'Aesthetic and Minimalist Design' and 'Pleasurable and Respectful Interaction with the User'. Improvements must be made in 'Match Between System' and 'Real World and Help and Documentation'. It is now essential to focus on improving those categories even more considering our target users (adult learners with disabilities) and the overall specificity of the platform (its accessibility). However heuristics data must be used with caution because they do not address several usability issues. Also the Tenuta project carried out a revision of ALPE platform using task scenarios, expert evaluation and HHS guidelines [12].

Finally, as part of the ALPE project, we conduct an end-user market validation focusing on the user satisfaction of the product. In a first trial, online users fill a background questionnaire before going through series of the tasks and a feedback questionnaire after the session. In a second trial, qualitative evaluations with a representative sample of users involve direct observation of users interacting with the application. Results will be analyzed and translated into design recommendations and technical specifications to be implemented as part of the iterative design process.

5 Conclusions, Challenges and Future Works

From the point of view of U&A, LMS have specificities that must be taken into account. U&A evaluations must be planned ahead for the entire eLLC, taking place at different steps of the cycle but considered as a continuous process.

One of the main challenges regards the accessibility of the content generated by users. Learners become content authors, but they are not aware of WCAG when

---

[3] Tenuta provides U&A support to eTEN funded projects: http://www.etenuta.org

producing their contributions. The LMS provides support, such as WYSIWYG editors but they have limitations, do not fully guarantee accessibility and must be improved. One usability challenge is to ensure the usability of a highly flexible system which offers lots of customization options at every level and integrates components (platform, course, content generated by users) generated from different sources (developers, administrators, authors, students, tutors) for different purposes.

Evaluations with end-users are conducted as part of ALPE (from September 2007) and EU4ALL (year 2008) projects. ADAPTAPlan project focuses on the automatic generation of the learning design taking into account learners' needs and preferences. Moreover, Web 2.0 accessible authoring and the validation of processes to assess automatically web contents created by users are part of future works to be undertaken.

Acknowledgments. The authors would like to thank the EC for funding of EU4ALL (IST-2006-034778) and ALPE projects (eTEN 029328), and the Spanish Ministry of Science and Technology for ADAPTAPlan (TIN2005-08945-C06-00).

References

1. Boticario, J.G., Santos, O.C.: An open IMS-based user modelling approach for developing adaptive learning management systems. Journal of Interactive Media in Education (in press)
2. Santos, O.C., Boticario, J.G., Raffenne, E., Pastor, R.: Why using dotLRN? UNED use cases. In: Proceedings of FLOSS (Free/Libre/Open Source Systems) International Conference (2007)
3. Van Rosmalen, P., Boticario, J.G., Santos, O.C.: The Full Life Cycle of Adaptation in aLFanet eLearning Environment. Learning Technology newsletter 4, 59–61 (2004)
4. W3C-WAI Web Content Accessibility Guidelines for creating accessible web pages (WCAG) v1.0, available at: http://www.w3.org/TR/WAI-WEBCONTENT/ WCAG v2.0 (draft at time of writing), available at: http://www.w3.org/TR/WCAG20/
5. W3C-WAI Authoring Tool Accessibility Guidelines 1.0 (ATAG), available at: http://www.w3.org/TR/WAI-AUTOOLS/
6. W3C User Agent Accessibility Guidelines 1.0 (UAAG), available at: http://www.w3.org/TR/WAI-USERAGENT/
7. IMS (2002). IMS Guidelines for Developing Accessible Learning Applications (GDALA), available at: http://www.imsglobal. org/accessibility/ accessiblevers/ index. html
8. Benavidez, C., Fuertes, J.L, Gutiérrez, E., Martínez, L.: Semi-Automatic Evaluation of Web Accessibility with HERA 2.0. In: Miesenberger, K., Klaus, J., Zagler, W., Karshmer, A.I. (eds.) ICCHP 2006. LNCS, vol. 4061, pp. 12–14. Springer, Heidelberg (2006)
9. W3C-WAI. Involving Users in Web Accessibility Evaluation. W3C-WAI (2005), available at: http://www.w3.org/WAI/eval/users.html
10. Nielsen, J.: (2005). Heuristic Evaluation. Retrieved (April 05, 2007), http://www.useit.com/papers/heuristic/
11. Pierotti, D., Heuristic Evaluation - A System Checklist. Retrieved (April 05, 2007), http://www.stcsig.org/usability/topics/articles/he-checklist.html
12. U.S. Health and Human Services, Research-Based Web Design & Usability Guidelines Book. U.S. Health and Human Services (2006)

Web Accessibility Evaluation Via XSLT

Vicente Luque Centeno, Carlos Delgado Kloos, José Mª Blázquez del Toro[1],
and Martin Gaedke[2]

[1] Carlos III University of Madrid
{vlc,cdk,jmb}@it.uc3m.es
[2] University of Karlsruhe
gaedke@tm.uni-karlsruhe.de

Abstract. Web accessibility rules, i.e., the conditions to be met by Web
sites in order to be considered accessible for all, can be (partially) checked
automatically in many different ways. Many Web accessibility evaluators
have been developed during the last years. For applying the W3C guidelines, their programmers have to apply subjective criteria, thus leading
to different interpretations of these guidelines. As a result, it is easy to
obtain different evaluation results when different evaluation tools are applied to a common sample page. However, accessibility rules can be better
expressed **formally** and declaratively in rules that assert conditions over
the markup. We have found that XSLT can be used to represent templates addressing many accessibility rules involving the markup of Web
pages. Even more, we have found that some specific conditions relaying
in the prose of the XHTML specification not previously formalized in
the XHTML grammar (the official DTD or XML Schemas) could also be
formalized in XSLT rules as well. Thus, we have developed WAEX as a
Web Accessibility Evaluator in a single XSLT file. Such XSLT file contains 70+ singular accessibility and XHTML-specific rules not previously
addressed by the official DTDs or Schemas from W3C.

1 Introduction

Most Web pages nowadays are plenty of accessibility barriers, i.e., characteristics that avoid some people having a correct access to Web contents or Web
functionality. Most of these Web barriers cause serious problems for people with
some sort of disabilities. Examples of disabilities and usual problems found by
those users are:

1. **Sensorial disabilities:** Quite often, images or other multimedia elements
   in Web pages don't provide descriptive texts for blind people. Written transcriptions of audio tracks are rarely found by people who can't hear. Other
   people may not properly read nor resize small fonts. Others don't have the
   ability to recognize some color combinations.
2. **Physical disabilities:** People who can't properly manage a keyboard or
   a mouse use to have problems for interacting with links and forms in Web
   pages. Mouse-only or keyboard-only navigation, or the form's small active

M. Weske, M.-S. Hacid, C. Godart (Eds.): WISE 2007 Workshops, LNCS 4832, pp. 459–469, 2007.
© Springer-Verlag Berlin Heidelberg 2007

zones lacking a conveniently associated descriptive text are usually a challenge for people with reduced mobility.

3. **Neurological disabilities:** Navigation through a Web site can be difficult if it has terms being difficult to understand, it lacks a clear map of the Web site, form fields lack a proper description that explains how each form field expects to be filled in, people find difficulties on trying to repeat or explain a Web navigation path, or, simply if pages inside the same Web site present inconsistencies.

4. **Technological disabilities:** Users with old operating systems, alternative browsers, a limited keyboard or mouse, a low bandwidth Internet connection, or hardware limitations like a small display or memory, or lacking a specific plugin, usually find problems when navigating. Also users with their brand new hand-held wireless devices find problems as well. Many Web sites have been conceived only for a concrete screen size and resolution, thus raising layout problems on devices with other capabilities.

WAI (Web Accessibility Initiative)'s WCAG (Web Content Accessibility Guidelines) 1.0 [1] from W3C has become an important reference for Web accessibility in the Web community. It has been accepted as a set of guidelines that improve accessibility and eliminate barriers on Web sites. The lack of accessibility affects a large amount of people (between 10% and 20%, according to [10]) that frequently find important barriers when trying to navigate through today's Web sites. Barriers reduce accessibility not only for people with some sort of personal disability. Accessibility is also a major step towards device independent Web design, allowing Web interoperability to be independent from devices, browsers or operating systems. Web accessibility allows having cheaper Web site maintenance and a wider target public. Some governments are also requiring that some Web sites become accessible.

The set of the 65 WCAG's checkpoints that accessible documents have to pass is a very heterogeneous set of constraints, which are difficult to evaluate and repair. Both WCAG 1.0 [1] and the HTML Techniques for the new WCAG 2.0 draft [2] specifications are written at a high abstraction level, which is frequently quite open to subjective interpretation (as *too long texts* or *too many* elements), including implicit multi-evaluation conditions or, simply, containing constraints whose detection cannot be easily automated.

There are several tools nowadays which can help us to detect accessibility barriers on Web sites. Watchfire [11], Tawdis [12] or HERA [13] are only a few of them. It is well known that all these tools require a person to **supervise** and **complete** the results of the evaluation tools because a lot of rules are relayed on manual checks by the user. However, even regarding automatable evaluations only, we still have the problem of different *particular interpretations* on each tool. This is due to the fact that they were built on top of different subjective interpretations of the W3C's open-to-interpretation guidelines, without using a formalized version as depicted at [16]. As a result, we can easily find several tools reporting different evaluation results for a same sample page. Even further, it is difficult indeed to find two evaluation tools that evaluate the same conditions.

One of the main reasons for this heterogeneity is due to the fact that these tools have been implemented as **black boxes** whose internal evaluation mechanisms are hidden to the public. Since they have been mostly implemented using **procedural** programming languages, the conditions being checked are usually unknown for their users. Our approach to solve that problem consists on providing non-hidden **declarative** rules, readable and reusable by anyone, even for those with very little programming background. The procedural approach for checking document consistency is powerful, but difficult to bring consensus. On the other hand, the declarative and transparent rule-based approach provides consensus rules that many people might easily understand and agree with. This is one of the advantages of using declarative rules in a DTD or XML Schema. Instead of using self-developed software for XML validation, it's better to use rules that declare elements as optional or required and how they must be nested, then leaving validation as a rule-based checking process. Declarative rules might be used to validate many Web accessibility conditions easily.

Our approach lead us to build WAEX [17]: a Web Accessibility Evaluator in a single XSLT file. Because of its XSLT nature, it can only be applied to well formed HTML pages. Fortunately, HTML reparation software like Tidy [15] and the HTML parser of libxml [9] allows us to apply WAEX to almost any Web page.

The rest of this paper depicts the most important WAEX's rules, and is organized as follows: Section 2 deals with accessibility conditions already expressed formally in the XHTML grammar (the DTD or XML Schema) and compares them with the corresponding XSLT templates in WAEX. Section 3 deals with those accessibility conditions expressible in a XHTML grammar, but not expressed indeed in a DTD or Schema. Section 4 covers accessibility conditions not expressible in a XHTML grammar. Section 5 coverts mobileOK Basic tests [5] from W3C. Section 6 contains some conclusions and future work.

2   Accessibility Conditions Already Expressed in the XHTML Grammar

Some of the most important accessibility issues already involve validation against a public grammar. In fact, the XHTML grammar already represents some well known accessibility checkpoints. Specific rules in the XHTML's DTD and Schema already require that specific mandatory markup properly appears in XHTML documents. For instance, table 1 contains rules that state that every image must always have an `alt` attribute (a textual description of the image). Both the DTD and the XML Schema declare the `alt` attribute as **required** for every image. However, non XHTML documents still can use XSLT to spot as a barrier those images having no such `alt` attribute.[1,2]

---

[1] A similar set of expressions can be used to state that the alt attribute is also required for every `area` element.

[2] Ornamental images containing no information, should have an empty string as the textual alternative, but this attribute must be present anyhow according to WCAG. The fact that the textual alternative is an adequate alternative for the image is another checkpoint which is outside of our scope.

Table 1. Images without alternative text. The `img`'s `alt` attribute is mandatory.

Sample barrier	<img src="mypic.jpg" /> <!-- No alt! -->
Sample corrected	<img src="mypic.jpg" **alt="My picture"**/>
DTD	<!ATTLIST img alt CDATA **#REQUIRED**>
XML Schema	<xs:element name="img"> <xs:complexType> <xs:attribute name="alt" use="**required**" type="xs:string"/> </xs:complexType> </xs:element>
XSLT	<xsl:template match="//img[not(@alt)]"> <xsl:message>Images with no alt attribute</xsl:message> </xsl:template>

Table 2. Document without a unique title. Document's title is mandatory.

Sample barrier	<html><body> ... </body> </html> <!-- No head & no title!-->
Sample corrected	<html> <head><title> ... </title></head> <body> ... </body></html>
DTD	<!ELEMENT head (%head.misc;, ((**title**, %head.misc;, (base, %head.misc;)?) | (base, %head.misc;, (**title**, %head.misc;))))>
XML Schema	<xs:element name="head"><xs:complexType> <xs:sequence> <xs:group ref="head.misc"/> <xs:choice> <xs:sequence> <xs:element ref="**title**"/> <xs:group ref="head.misc"/> <xs:sequence minOccurs="0"> <xs:element ref="base"/> <xs:group ref="head.misc"/> </xs:sequence> </xs:sequence> <xs:sequence> <xs:element ref="base"/> <xs:group ref="head.misc"/> <xs:element ref="**title**"/> <xs:group ref="head.misc"/> </xs:sequence> </xs:choice> </xs:sequence> </xs:complexType></xs:element>
XSLT	<xsl:template match="/html[count(./head/title)!=1]"> <xsl:message>Document without a unique title</xsl:message> </xsl:template>

Not only mandatory attributes, but also mandatory elements can be required by a grammar. For instance, table 2 contains rules to indicate that every document must have a unique `title` inside the `head` element.

Validation is also useful for detecting forbidden or deprecated markup within a document. It is important to avoid deprecated markup in order to properly separate structure from presentation, thus allowing users to customize presentational features (colors, font sizes, ...) according to their personal needs using only CSS rules without modifying the HTML markup. Deprecated elements like

Table 3. Deprecated elements no longer in use

Sample barrier	<font color="red"> <! − − Deprecated, use CSS instead − − >
Sample corrected	<... style="**color: red**"> <! − − As external CSS − − >
DTD	*No such* <!ELEMENT font ...> *nor*
	<!ELEMENT blink ...> ... *(in new XHTML versions)*
XML Schema	*No such* <xs:element name="font"/> *nor*
	<xs:element name="blink"/> ... *(in new XHTML versions)*
XSLT	<xsl:template match="//font|//blink|//b|//i|//tt|//center">
	<xsl:message>Deprecated elements no longer in use</xsl:message>
	</xsl:template>

font, b (bold), i (italic), tt (tele-type) or center, as well as forbidden elements not in the HTML specification like blink or marquee, can be spotted (as presentational markup that should be left out in favour of CSS) by either a grammar which does not recognize them as valid elements or with an XSLT rule like the one at table 3. Fortunately, these rules have been considered in all examined tools in a similar way.

3 Accessibility Conditions Expressible (But Not Expressed) in a XHTML Grammar

XHTML is not a single language. Since its birth, it has had several versions, starting from Transitional and Strict [3], launching XHTML Basic [4] and ending by XHTML 1.1. Those different languages have different restrictions, some of them had been previously declared in the 1999's WCAG [1]. For example, the alt attribute is mandatory since XHTML Transitional 1.0. Deprecated elements like font or center were removed in XHTML Strict 1.0 (but still remains in Transitional). XHTML 1.1 introduced some new rules and removed some deprecated features from XHTML Strict 1.0. XHTML Basic (specifically oriented for small devices) does not allow frames, plugins, both server or client side image maps or nested tables for layout.

The refinement process from XHTML 1.0 through XHTML 1.1 has already included some accessibility rules into their successive DTDs and/or XML Schemas. However, some refinements still remain uncompleted. For example, WCAG 1.0 states that all frames should have a mandatory title attribute describing the frame's purpose. Even though this rule is very similar to the one exposed at table 1, the XHTML grammar still defines this attribute as optional, instead of required. For this reason, rules from table 4 are required.[3,4]

---

[3] Unless we locally redefine this rule in our own DTD or XML Schema.

[4] A similar approach (DTD or Schema redefinition) could be taken for the usage of the noframes element, because the public DTD does not properly require such element even though it is required to be present as an alternative for frames for browsers that can't support frames.

Table 4. Frames without description. The **frame**'s **title** attribute is mandatory.

Sample barrier	\<frame src="toc.html"/> \<! − − Title attribute is required! − − >
Sample corrected	\<frame src="toc.html" **title="Table of Contents"**/>
DTD	\<!ATTLIST frame title CDATA **#REQUIRED**> *However, it is defined as* **#IMPLIED** *(not required)*
XML Schema	\<xs:element name="frame"> \<xs:complexType> \<xs:attribute name="title" use="**required**" type="xs:string"/> \</xs:complexType> \</xs:element> *However, it is defined as* **optional** *(not required)*
XSLT	\<xsl:template match="//frame[not(@title)]"> \<xsl:message>Frames without description.\</xsl:message> \</xsl:template>

Table 5. Restrictions in XHTML specification (prose) not expressed in the DTD

Container element	Forbidden contents	Condition
pre	img, object, big, small, sub, sup	Always forbidden
a, label	a, area, label, input, select, textarea, button	Always forbidden
form	form	Always forbidden
table	table	In XHTML Basic [4]
ins	Any *block-type* element	If the **ins** element is inside an *inline-type* element

Table 6. Sample barriers from XHTML specification (prose) not expressed in the DTD

Forbidden elements in pre	\<pre>\<img src="a.jpg" alt="An image" />
	 \</pre> \<! − − Images forbidden in pre! − − >

Nested focusable elements	\<a href="p1.html">\<label>... \</label> \</a> \<! − − Label forbidden inside a link! − − >
Nested forms	\<form>\<form>... \</form> \</form> \<! − − Nested forms forbidden! − − >
Forbidden insertions	\\<ins> \<p>New paragraph.\</p> \</ins> \ \<! − − Bad insertion! − − >

Table 5 contains a summary of XHTML prohibitions stated in the prose of XHTML specification [3] but **not formally expressed** in any XHTML grammar. Such restrictions could be easily expressed formally in the XHTML DTD. However, up to the present moment, they have not been formalized by any rule yet. Table 6 contains sample barriers breaking rules from table 5. Even though those snippets validate the official DTD and Schemas, according to the XHTML specification prose, they are not valid XHTML and some of them involve serious accessibility problems.

The best of having checkpoints guaranteed by a grammar is that they are **very cheap and easy to detect**: just choosing a modern grammar to validate documents against to, and using an existing XML-ized validator, may easily spot where barriers belonging to this category appear within a document.

4   Accessibility Conditions Not Expressible in a XHTML Grammar

There are some accessibility-related conditions that are not expressible indeed in a DTD or XML Schema. And this happens even though they are conditions concerning the syntax of the documents. Many tools have their own internal implementation of these conditions hard-coded within lines of a program coded in an **procedural** programming language like Java, C or PHP. This easily leads to complex algorithms which (usually) implement something different from the desirable checkpoint (thus providing different evaluation results). Since they are built as black boxes, people find difficulties on understanding these differences. Approaches like [14] have tried to represent such rules in a set of **declarative expressions** declared in self-developed XML files that represent rules. However those rules can only be evaluated by a self-developed ad-hoc program. Using already existing software was the aim of the WCAG formalization approach [16]. Translating the formalized approach to XSLT implies not only more precise, but also **cheaper** evaluation because no specific software needs to be developed indeed. Just an XSLT engine needs to be applied. As a result, we have an XSLT-based declarative ruleset which is **smaller, more reusable** and easier to understand and homogeneize than the equivalent procedural routines.

As a starting example, let's begin with attributes that enhance accessibility, but whose presence is required only under certain circumstances. For instance, table 7 states that every image input, (i.e., input form fields whose **type** attribute is **image**) requires the **alt** attribute (as any other image). However, it would not be nice to declare the **alt** attribute as mandatory for every **input** element, because input form fields having a **type** attribute different of **image** don't really expect such **alt** attribute. Conditions that indicate whether some markup is mandatory or not, are not expressible in DTDs or XML Schemas.

In fact, conditions that involve some comparison or calculation over element's or attribute's content, are not expressible on DTDs or XML Schemas. Other accessibility-related attributes are also required under certain conditions as well,

Table 7. Image inputs without alternative text. The **alt** attribute is mandatory.

Sample barrier	\<input type="image" src="submit.gif"/> \<! − − No alt! − − >
Sample corrected	\<input type="image" src="submit.gif" **alt="Submit"**/>
DTD	*NOT FEASIBLE!!*
XML Schema	*NOT FEASIBLE!!*
XSLT	\<xsl:template match="//input[@type='image'][not(@alt)]"> \<xsl:message>Image input with no alt\</xsl:message> \</xsl:template>

Table 8. Attributes conditionally required

Container element	Required attribute	Condition to be mandatory
input	alt	If input's type is "image"
html	lang	If document's xml:lang is missing
th	abbr	If table header's text is too long
table	summary	If table contains tabular data, i.e, it is not a *layout-only* table

as depicted at table 8. They are summarized, but they can be used just like the expressions in table 7. For instance, the `abbr` attribute must be used on any `th` element having a too long text in order to facilitate efficient text-to-speech transformation. Not all tools check these conditions properly.

Conditionally required elements are represented in table 9. For instance, any multimedia resource (`object` element) should have some alternative contents. In order to have manageable data items, `fieldset` elements should group form fields sharing common characteristics. When the number of options in a combo box (`select` element) is high, it is a good idea to group them with the `optgroup` element. Non-layout tables should have a `caption` element describing the table ... The expressions from table 9 are summarized.

Table 9. Elements conditionally required

Container element	Required element	Condition to be mandatory
object	*some alternative inside*	If object does not contain alternative text
form	fieldset	If too many form fields
select	optgroup	If too many options
table	caption	If table contains tabular data, i.e, it is not a *layout-only* table

Some other important accessibility rules don't really involve existence or absence, but **misusage** of HTML markup. These rules state that some HTML attributes should not be improperly used. For example, it is a very bad practice to trigger JavaScript routines from the `href` attribute of active elements like links or client-side map areas. It is much better to use event-focused attributes like `onclick` for that purpose, and leave the `href` attribute usable for users that have browsers with no JavaScript support. The XSLT template from table 10 detects this bad practice. (Several Web accessibility evaluation tools do not).

Other bad practices involving typically misused attributes are collected in table 11. As we can see, it is bad practice to use client side auto-refresh or auto-redirect, unadvised emerging windows, improper keyboard short-cuts or unclear tabulation orders.

However, the expressive power of XSLT becomes clearer when we leave out these previous simple conditions involving the relationship between elements and their contents and we start looking at **relationships** among elements placed **anywhere** in the document. For example, let's write a rule for *"each client-side*

Table 10. Not only JavaScript-based navigation. JavaScript code should appear in event attributes.

Sample barrier	<a href="javascript:load('a.html')"><! − − JavaScript-based navigation − − >
Sample corrected	<a **href="a.html" onclick**="load('a.html'); return **false**">
DTD	*NOT FEASIBLE!!*
XML Schema	*NOT FEASIBLE!!*
XSLT	<xsl:template match="(//a|//area)[starts-with(@href,'javascript:')][not(@onclick)]"> <xsl:message>Improper place for JavaScript.</xsl:message> </xsl:template>

Table 11. Attributes typically misused

Container element	Misused attribute	Misusage condition
a, area	href	JavaScript code at `href` attribute, rather than at `onclick` attribute
meta	http-equiv	Badly used for auto-refresh or auto-redirect
input	alt	If input's type is not "image"
* (*any tag*)	target	Badly used for emerging windows
* (*any tag*)	tabindex	Not a set of unique consecutive numbers
* (*any tag*)	accesskey	Not a unique character

Table 12. A redundant link is required for each client-side map area

Sample barrier	<area href="a.html"/> <! − − Missing redundant link!− − >
Sample corrected	<area href="a.html"/> <**a href="a.html"**>...</a>
DTD	*NOT FEASIBLE!!*
XML Schema	*NOT FEASIBLE!!*
XSLT	<xsl:template match="//area"> <**xsl:variable** name="area" select="."/> <xsl:if test="not(//a[@href = $area/@href])"> <xsl:message>Missing redundant area's link</xsl:message> </xsl:if> </xsl:template>

map area should have a redundant link (somewhere in the document) for those users that cannot use client-side maps". This implies that, for every **area** element, at least one link in the same document should share the same destination (i.e. point to the same **href**). In other words, some link in the document should replicate the **href** attribute for each area.The XSLT expression from table 12 states that any **area** whose **href** attribute is not being replicated by a link somewhere else in the document will be spotted as an accessibility barrier (not suited for people who can't use maps). Similar templates have been implemented to detect form fields without a label element referring them.

5   Suitability for Handheld Devices (W3C's MobileOK)

Specific conditions for handheld devices have recently been defined by the W3C at [5]. Many previous conditions apply for handheld devices, but there are also some specific rules only for this domain, as depicted in table 13. For instance, not only maps, frames, and nested tables are forbidden. The number of external resources is limited for each Web page, and the size of images is mandatory to be specified inside the HTML markup.

Table 13. W3C's mobileOK specific conditions

Avoid maps	<xsl:template match="//*[@usemap or @ismap]"> <xsl:message>Maps are forbidden</xsl:message> </xsl:template>
Avoid missing dimensions for images	<xsl:template match="//img[not(@height) or not(@width)]"> <xsl:message>Image dimensions are mandatory</xsl:message> </xsl:template>
Avoid frames	<xsl:template match="//frame|//frameset|//iframe"> <xsl:message>Frames are forbidden</xsl:message> </xsl:template>
Avoid nested tables	<xsl:template match="//table[.//table]"> <xsl:message>Nested tables are forbidden</xsl:message> </xsl:template>
Avoid too many external resources	<xsl:if test="count(//style|//img|//link|//object) > 20"> <xsl:message>Too many external files</xsl:message> </xsl:if> <! − − 20 is the W3C's defined limit − − >

6   Conclusions and Future Work

The recently developed XHTML's DTDs and XML Schemas from W3C could collect some more accessibility-related conditions than currently do. They would be a great place to formalize syntax-related Web accessibility issues, because many Web authors already know how to validate their documents against those formal grammars. However, DTDs and XML Schemas are not powerful enough to formalize Web accessibility syntax-related rules. Anyway, we found that a declarative language like XSLT could be used to fill in the gap where DTDs and XML Schemas didn't fit.

We started reverse-engineering most Web accessibility evaluation tools [6]. We rewrote in the XSLT rules of WAEX all the checkpoints which were according to the W3C specification, rejecting all checkpoints which were not. Those very few ones out of the scope of XSLT, like CSS parsing, image analysis, color combinations, or JavaScript parsing, have been partially implemented by calling already available server-side tools hosted at W3C [7,8].

We soon became aware of the expressing power of this XSLT approach. We became aware that we could **easily express syntax-related conditions from already built Web Accessibility evaluators** in a fine and clean way, no matter if they were simple or complex. We also realized that we could express conditions not previously being checked. Thus, we included conditions not previously addressed by previous tools:

1. XHTML conditions from the XHTML specification not previously formalized in the corresponding DTD or Schema [3,4]
2. Conditions from the recently published W3C's MobileOK Basic test suite [5] (suitable for mobile devices).

Acknowledgements

We gratefully acknowledge support from the ITACA *TSI-2007-65393-C02-01*, ITACA 2 *TSI-2007-65393-C02-01* and InCare *TSI-2006-13390-C02-01* projects of the "Ministerio de Educación y Ciencia" (Spanish ministry).

References

1. W3C Web Content Accessibility Guidelines 1.0 (Recommendation, May 1999) `www.w3.org/TR/WCAG10`
2. W3C Techniques and Failures for Web Content Accessibility Guidelines 2.0 -W3C Working Draft (May 17 2007),
   `www.w3.org/TR/2007/WD-WCAG20-TECHS-20070517/`
3. W3C XHTML 1.0 ^{TM} 1.0 - The Extensible HyperText Markup Language (Second Edition), A Reformulation of HTML 4 in XML 1.0, W3C Recommendation 26 January 2000, revised (August 1 2002), `www.w3.org/TR/xhtml1`
4. W3C XHTML ^{TM} Basic 1.1 - W3C Candidate Recommendation (July 13 2007), `www.w3.org/TR/xhtml-basic`
5. W3C W3C mobileOK Basic Tests 1.0 W3C W3C Working Draft (May 25 2007), `www.w3.org/TR/mobileOK-basic10-tests`
6. W3C Web Accessibility Evaluation Tools: Overview, `www.w3.org/WAI/ER/tools/Overview.html`
7. W3C CSS Validation Service, `jigsaw.w3.org/css-validator`
8. W3C HTTP HEAD service, `cgi.w3.org/cgi-bin/headers`
9. Veillard, D.: The XML C parser and toolkit of Gnome, `www.xmlsoft.org`
10. WebAim Introduction to Web Accessibility, `www.webaim.org/intro`
11. Watchfire WebXACT Accessibility tool, `webxact.watchfire.com`
12. CEAPAT, Fundación CTIC, Spanish Ministry of Employment and Social Affairs (IMSERSO) Online Web accessibility test, `www.tawdis.net`
13. Fundación SIDAR Accessibility testing with Style, `www.sidar.org/hera`
14. Vanderdonckt, J., Beirekdar, A.: Monique Noirhomme-Fraiture Automated Evaluation of Web Usability and Accessibility by Guideline Review. In: Koch, N., Fraternali, P., Wirsing, M. (eds.) ICWE 2004. LNCS, vol. 3140, pp. 28–30. Springer, Heidelberg (2004)
15. Sourceforge HTML parser and pretty printer in Java, `jtidy.sourceforge.net`
16. Centeno, V.L., Kloos, C.D., Gaedke, M., Nussbaumer, M.: WCAG Formalization with W3C Standards. In: The 14th International World Wide Web Conference (WWW2005), Chiba, Japan. ISBN 1-59593-051-5, pp. 1146–1147 (May 11-14 2005)
17. Centeno, V.L.: WAEX: Web Accessibility Evaluator in a single XSLT file, `www.it.uc3m.es/vlc/waex.html`

Analyzing Tool Support for Inspecting Accessibility Guidelines During the Development Process of Web Sites

Joseph Xiong, Christelle Farenc, and Marco Winckler

LIIHS-IRIT, Université Paul Sabatier,
118 route de Narbonne, 31062 Toulouse Cedex 4, France
{xiong,farenc,winckler}@irit.fr

Abstract. Whilst accessibility is widely agreed as an essential requirement for promoting universal access of information, many Web sites still fail to provide accessible content. This paper investigates the support given by currently available tools for taking care of accessibility at different phases of the development process. At first, we provide a detailed classification of accessibility guidelines according to several levels of automation. Then we analyze which kind of automated inspection is supported by currently available tools for building Web sites. Lately, by the means of a case study we try to assess the possibility of fixing accessibility problems at early phases of the development process. Our results provide insights for improving current available tools and for taking accessibility at all phases of development process of Web sites.

Keywords: accessibility, guidelines inspection, evaluation tools, Web sites, development process, Web design.

1 Introduction

Accessibility is an essential requirement for promoting universal access of Web sites. Accessibility gained in importance not only because of ethical issues (e.g. e-inclusion) but also because of the opportunities for creating new markets (e.g. elderly customers) and for increasing audience (e.g. accessible Web not only benefits impaired users but users in general).

In more recent year Accessibility became a legal requirement as many countries have enacted for Accessibility responsibility of content published on the Web. In the United States, the regulation Section 508 of the Rehabilitation Act 1973 [17] was amended in 1998 to address the new Information Technologies and the Web. In Europe, the European Council encourages state members to enact laws for accessibility of public Web sites at all levels of government. Many member states such as France [15], Germany [4], Portugal [12], and UK [6], among many others, have created laws for the accessibility of digital content. The European Council has also supported initiatives for promoting the accessibility such as the creation of a European e-Accessibility Certification (EuraCert) [8] and the development of a Unified Web Evaluation Methodology (UWEM 1.0) [21, 22] which aim is to provide a set of guidelines and a standard procedure for manual and/or automated accessibility

M. Weske, M.-S. Hacid, C. Godart (Eds.): WISE 2007 Workshops, LNCS 4832, pp. 470–480, 2007.

inspections. Whilst UWEM 1.0 and EuraCert represent important steps forward measurement and certification of Web sites, from a technological point of view they are similar to W3C's Web Accessibility Initiatives [26] which already provide guidelines and tools for accessibility checking and digital certificates (e.g. WAI-A, WAI-AA, WAI-AAA) for accessible-compliant Web sites.

Currently, there are many tools for automating guidelines inspection of Web sites such as WebXACT [23], TAW [18], WAVE [24], Ocawa [14], and A-Prompt [3], most of them include the verification of usability and accessibility guidelines. Even though not all guidelines can be automatically assessed by these tools, they provide a valuable help by reducing costs and time of manual inspections. The main inconvenient with these tools is that the evaluation is mainly done on the source code (i.e. HTML, XHTML, CSS) of Web sites which are only available at latter phases of the development process when the Web site is ready to be published. At that step the cost of fixing related to guidelines violation are high due to the number of possible occurrences of the problem in all pages.

The main goal of this paper is to determine in which extension accessibility guidelines can be assessed in earlier steps of the development process and how such early assessments might contribute to reduce costs of inspections and improve accessibility of Web sites in general. For that purpose we present hereafter a case study which combines several methods including analysis of tool execution over real Web sites, analysis of artifacts produced by tools, and analysis of support provided by tools during edition of Web pages.

The rest of the paper is organized as follows: Section 2 presents several issues related to the automation of accessibility guidelines, including interpretation of high-level guidelines and levels of automated inspection. Section 3 presents an overview of development tools and evaluation tools currently used to build Web sites. Tools presented in section 3 will be later employed in the case study which is fully described in section 4. Section 5 discusses the findings and point out some conclusions.

2 Accessibility Guidelines

Different guidelines exist [19], most of them mixing usability and accessibility recommendations. As far accessibility is the concern, the W3C/WAI WCAG 1.0 set of accessibility guidelines [26] is still the most widely agreed. WCAG 1.0 contains 14 guidelines which express general accessibility principles. They are divided into 65 checkpoints. A checkpoint is assigned a priority level that ranges from 1 (high priority) to 3 (low priority). Based on these priorities, a Web application is said to be level A conformant if all priority 1 checkpoints are satisfied, level AA if all priority 1 and 2 checkpoints are satisfied, and level AAA conformant if all checkpoints are satisfied. In the rest of the paper we use the term 'guideline' to both denote the terms 'checkpoint' and 'guideline'.

2.1 Guidelines Interpretation

Guidelines are expressed in natural language and interpretation is often necessary to translate them into a computing language. A guideline's rule can be more or less abstract [16] and is not appropriate to an automatic translation in all the cases. If so, rules should be decoded in one or more concrete rules. For example, the rule *"Provide accessible content for visual impaired users"* is too imprecise to be automatically implemented. So it should be interpreted in a more concrete way having a single meaning. For example: *"Provide alternative text for images"*.

Quite often, many concrete rules must be created in order to cover all possible facets a guideline might have with respect to the UI contents or widgets. For example, what does the rule "Provide accessible content for visual impaired users" mean for images, menus, buttons, animation, and so on?

The translation process might also create a semantic distance between the original guidelines and the concrete rules created to support automated inspection. Considering the example above, the concrete guideline *"Provide alternative text for images"* only helps to check if a text description is associated to images; this can be considered as a kind of syntactic verification. Currently, there is no available tool supporting the verification of the semantics of alternative text and image's content, thus creating a gap between the rule and what is possible to verify concerning the rule.

2.2 Guideline Automation Level

Due to their nature guidelines may be more or less suitable for automation. Characterizing the levels of automation for guidelines will help us to estimate the impact of correcting errors when guidelines are violated. Hereafter we present an extended version of Ivory's taxonomy [11] for describing the levels of automation:

- **None:** the guideline has no level of automation, i.e. the guideline is supported by the software tool but no automatic analysis is performed (this is typically the case of warnings that are always thrown, e.g. guideline 14.1: "Use the simplest and clearest language");
- **Capture:** the software tool records data that are related with the guideline;
- **Analysis:** the software tool automatically detects guideline violations;
- **Critique:** the software tool automatically identifies guideline violations and suggests improvements;
- **Suggestion[1]:** the software tool suggests accessibility and/or usability options when designing the application so as to automatically conform to the guideline. For example, an accessibility option can be checked to automatically prompt the user to give a textual alternative every time he inserts an image in a page. Another example is the automatic suggestion and consequently creation of accessible tables.

3 Tools for Developing and Evaluating Accessible Web Sites

Tools for Web development are numerous [9] and can be used at different steps of the development process. The **development** tools are used during the design and

---

[1] The category *Suggestion* has been added to the former categories proposed by Ivory [11].

implementation of applications while the latter denotes tools used to evaluate the final application. The following categories of tools are considered in this work.

(i) **Visual editors and site managers** concerns authoring tools that make developing Web sites easier by designing without HTML programming, and also provide features dedicated to site management. Tools in this category typically propose a WYSIWYG editor. Adobe® DreamWeaver®[2] and Microsoft® FrontPage®[3] are such products.

(ii) **Database-centric tools** do not offer more accessibility features than visual editors and site managers. Although Fraternali [9] divided this category into 4 classes of tools we grouped them in a single category as they are all database-driven. Tools in this category (a) were originally dedicated to offline hypermedia applications and later adapted for the Web (Web-enabled hypermedia authoring tools); (b) were conceived to simplify the integration of database queries within a Web page (Web-DBPL integrators); (c) support the deployment of data-intensive Web applications (Web form editors, report writers, and database publishing wizards); and (d) some of them gather (a), (b), and (c) categories of tools in a single framework in order to support the sum of their functionalities (multi-paradigm tools).

(iii) **Model-driven application generators** employ several conceptual models to generate the Web applications. Thus, such tools generally cover most of the lifecycle activities and provide a high level of automation. Some examples of tools in this category are: e-Citiz [7], WebRatio [1,25] and VisualWade [10]. Except e-Citiz, accessibility is not currently supported by these tools.

(iv) **Evaluation tools**, typically, are used at the end of the development process to assess the final Web application in order to detect usability and/or accessibility problems. There are several categories of evaluation tools [11]: log analysis tools, automated inspection tools, HTML syntax validators, etc. If all these categories may support accessibility evaluation or some aspects of accessibility evaluation, guidelines inspection tools are certainly the most powerful and the most widespread for assessing Web accessibility. Among the many products in this category are WebXACT [23], TAW [18], WAVE [24], Ocawa [14], and A-Prompt [3].

4 Case Study

Only the WCAG 1.0 guidelines were considered in the study. Moreover, the analysis was limited to the level AA guidelines (i.e. priority 1 and 2 checkpoints) making a total of 49 individual checkpoints.

The analysis of guidelines, inspection of tools and the artifacts produced by tools allowed the estimation of the number of guidelines covered by tools, number of guidelines that could be evaluated at each phase of development process, estimation of the potential effort for correcting accessibility problems in early phases of the

---

[2] Available at: http://www.adobe.com/products/dreamweaver/

[3] Available at: http://www.microsoft.com/frontpage/. Frontage has been recently replaced by two new tools (Office SharePoint® Designer 2007 and Expression™ Web Designer) but this updates has no impact on the discussion carried out on this paper.

development process. So that, three different methods of assessment were used according to the category of tools, as follows:

1. Tool execution over real Web sites known by accessibility defects. This method was used with evaluation tools (i.e. TAW, Ocawa and WebXACT) and it reflects the results that can be obtained by the current practice.
2. Estimation of inspections that could be done using information available on artifacts (e.g. conceptual models, metrics, HTML code, etc). This method was used with tools following a model-driven approach (i.e. SketchiXML, VisualWade, WebRatio and e-Citiz). All artifacts produced/manipulated by these tools were analyzed in order to determine if the information they contain can be used (or not) for supporting inspection of accessibility guidelines.
3. Analysis of support provided during edition of Web pages. This method takes into account the support (e.g. active help, automatic correction, etc) to treat accessibility checkpoints during the edition of Web pages (i.e. HTML, XHTML and CSS). It was used with FrontPage 2003 and Dreamweaver.

In the analysis presented hereafter, we provide a ranking (left to right) of tools performing better according to the criteria measured.

4.1 Tools Evaluated

The tools selected for this study are representative of the categories presented in section 3, except by the category *data-centric tools* which concerns generic tools (such as database systems, compilers, report writers, etc.) that quite often do not have much impact on the development of the user interface. One might considerer the use of the underlying relational model of the database applications to drive the user interface definition including description of functionalities and navigation. However, since the use of metadata as input for the user interface definition requires a specific work on both transformation methods which is out the scope of this paper.

Figure 1 presents a distribution of tools according to phases of development process. Yet simplified, a development process is required to understand when tools are used in the development process. Although there is no consensus on phases of development process, one might agree on generic phases such as *specification* (abstract modeling of application), *design* (models refinement including design constraints, integration of content, visual design of elements, etc.), *implementation* (production of application code, database creation, etc.), *post-implementation* (testing and maintenance).

As depicted in Figure 1, some tools can be used in a specific phase of the development process (e.g. WebXACT, TAW and Ocawa, at post-implementation phase; SketchiXML, at Design phase), whilst other span several phases (e.g. Frontpage, Dreamweaver, VisualWade) or provide some support during the entire development process (e.g. WebRatio and e-Citiz). The notion of activities performed and tool support at each phase is important to understand where guidelines should be taken into account in the development process. Tools, like VisualWade, which generates HTML code from specifications should implement accessibility policies very early in the development process to ensure the generation of accessibility problems on generated HTML code.

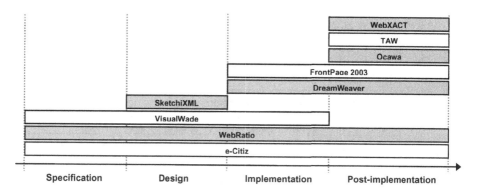

Fig. 1. Tools support according to steps in development process

As e-Citiz [7], WebRatio [1,25], and VisualWade [10] follow a model-driven approach which make them suitable in early phases of development process. WebRatio and VisualWade are generic tools for specification of Web sites whilst e-Citiz is a framework dedicated to the specification and deployment of electronic administrative procedure (or e-procedure). One of the raisons e-Citiz was included in this survey is the fact it implements accessibility policies to generate e-procedures compliant with WCAG 1.0 guidelines.

SketchiXML [5] is a representative tool for the activities carried out in the design phase. This tool allows a designer to rapidly draw an interface such as with a classic graphics painting program. An automatic identification of the different components is then performed to reconstruct the interface: this is made possible with a text recognition engine. The interface is specified using the UsiXML notation [20]. Thus, a mock-up reflecting the final interface of a Web page is easily obtained with SketchiXML. Even if SketchiXML has no features dedicated to accessibility conformity, evaluations can be performed on the UsiXML source code.

Macromedia DreamWeaver [2] and Microsoft FrontPage 2003 [13] are among the most popular tools that facilitate the development of Web sites. The main advantage of such tools is a high level of productivity even for a non expert in Web technologies. This is due to many facilities such as a WYSIWYG programming style, CSS styles editing, semi-automatic generation of code for complex structure (tables, forms, etc.), pre-designed block of scripts code (Javascript, Flash, etc.). Dreamweaver and FrontPage both include accessibility features that will help conforming to accessibility guidelines while developing the application.

The purpose of tool such as WebXACT, TAW and Ocawa is to evaluate the accessibility of any Web application. All these tools are online Web services that parse the HTML source code of pages and produce an evaluation report where accessibility errors and warnings are mentioned.

4.2 Number of WCAG 1.0 Checkpoint Supported

The number of checkpoints actually supported by tools devoted to evaluation is higher than those supported by development tools, as expected. As depicted by Figure 2, TAW presents the best performance, followed by WebXACT, covering 44 of the 49 WCAG 1.0 checkpoints (i.e. priority 1 and 2 checkpoints, namely AA guidelines). We recall

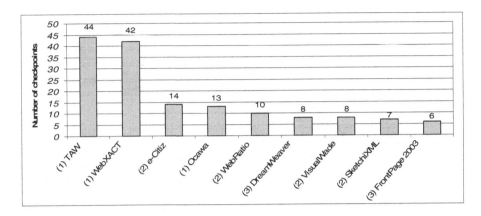

Fig. 2. Number of WCAG 1.0 checkpoint supported

that results were obtained by different methods. In Figure 2, the numbers in parenthesis indicate the tool category and the method of analysis employed (i.e. (1) execution of evaluation tools, (2) estimation based on artifacts produced by model-driven tools, (3) analysis provided during edition of pages).

As one can notice in Figure 2, Ocawa is only able to check 13 of the 49 checkpoints, which means the worst performance among evaluation tools (i.e. category 1). This is explained by the fact Ocawa only consider checkpoints which can be automatically detected (i.e. automation level *analysis*). However, in Figure 3 when comparing Ocawa performance to TAW and WebXACT, there is no slight difference in number of checkpoints inspected for the automation level *analysis*.

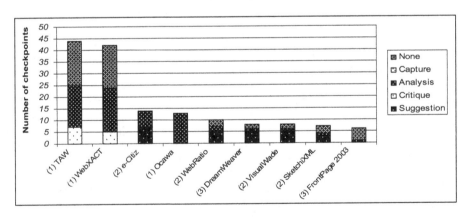

Fig. 3. Number of WCAG 1.0 checkpoint according the automation Level

4.3 Analyzing Quality of Checkpoint Support

It is noteworthy that the support provided by tools can be different according to the information available in artifacts. The checkpoint *"Provide alternative text for images"* can be fully supported over HTML but it becomes non-verifiable over navigation models, for instance. In addition, some tools automatically detect errors (i.e. guideline violation) whilst others just raise warnings informing designers to proceed manual inspection.

In order to measure the quality of inspections we establish some criteria based on empirical evidence using the tools. Thus, Figure 4 provides a distribution of checkpoints according to such a qualitative evaluation. Hereafter we provide a detailed description of criteria:

- *Fully supported*: all facets of the guideline are taken into account. Not only inspection is done but extra features such as examples and/or best practices are provided.
- *Correctly supported*: the guideline is supported by the tool but does not tackle all the evaluations that are possible. Example: only highlighting errors/warning when automatic correction could be done;
- *Poorly supported*: only errors are thrown, not warnings; no explanations are given; no additional documentation for corrections is given; etc.
- *Non-supported*: the guideline can be verified but is not implemented by the tool.
- *Non-verifiable*: the guideline cannot be verified even by a human expert, because it requires information that is missing. An example of such guidelines is 14.1: "Use the clearest and simplest language appropriate for a site's content". At the specification phase this rule is not verifiable because texts contained in Web pages are not yet available.
- *Non-relevant for the domain*: guidelines are not taken into account, deliberately, because elements never appear in the application domain addressed by the tool.

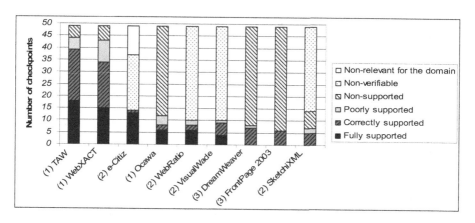

Fig. 4. Distribution of checkpoint according to qualitative support

4.4 Assessing Guidelines Support Throughout the Development Process

In order to provide a rough estimation of the potential for accessibility evaluation of tools in this survey we keep in Figure 5 the best score of tools at each phase. So that, we have 14 checkpoints supported by e-Citiz at specification phase, 7 checkpoints supported by SketchiXML at the design phase, 10 checkpoints supported by both

VisualWade and WebRatio at the implementation phase and 44 checkpoints supported by TAW at post-implementation phase. Such estimation should not be taken too seriously because it does not reflect any real development process combining the synergistic use of these tools. However, it provides some insights about the potential of currently available artifacts to support automated inspection.

It is noteworthy that abstract models can provide enough information to support the inspection (at some level of automation) of 28,5% checkpoints. At design phase it is actually possible to inspect 15,9% of the guidelines. This lower number of guidelines inspected is due to the kind of artifacts used in the analysis. More interesting is the performance of tools at the implementation phase, which indicates a poor support in terms of guidance and active help for accessibility during the edition of pages. This poor performance is very surprising because tools at implementation phase manipulate the same artifacts (i.e. HTML, XHTML, CSS code) used in inspection for evaluation tools which perform the best.

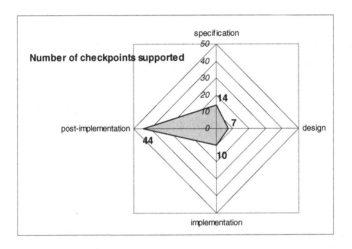

Fig. 5. Number of checkpoints supported during the development process

5   Discussions

Using accessibility guidelines is not straightforward for developers and evaluators because some expertise is required to correctly understand and apply guidelines. In addition, manual inspections are tedious and costly. Currently, some tools for automated inspection exist but most of them only provide support at post-development phases of development process. However ensuring accessibility conformity should be a concern at each phase of the development process and consequently it should be integrated as soon as possible in the lifecycle.

Currently only a few tools allows to support accessibility inspection in earlier phases than post-implementation. So we have combined analysis of existing evaluation tools and estimate the possibility of inspection over artifacts produced by tools. Concerning the analysis over artifacts, it is important to say that results will certainly change according to the kind of information artifacts contain. A feasibility study could determine which information could be added into artifacts (i.e. abstract models) to increase the number of accessibility checkpoints inspected.

The performed analysis demonstrated that it is possible to evaluate some guidelines in very early phases of the development process. Even if the fully automation of all guidelines inspection is not possible in early phases, some automation is possible and it could help to reduce the costs of correction over final implementation. We are aware that we need better method to estimate costs, but the analysis performed seems enough to conclude that "do it right from the beginning is cheaper than do it again because it was wrong". We also suggest that this proactive strategy applied in early phases of development would help designers to think about accessibility which would have a positive effect on the general quality of the user interface.

We found that most checkpoints could be inspected in early phase and could also be integrated into development tools, thus preventing designers to introduce by mistake accessibility problems into design. A typical example of this is the checkpoint "creation of associations between labels in fields" which can be either automatically inspected but also used as start point to generate accessible final code of Web sites.

Not all checkpoints can be fully automated and some manual inspections are still required. But this case study shows that an important number of guidelines could be taken into account in early phases of the development process.

The results presented in this paper are part of a preliminary study on cost/benefits of accessibility guidelines inspections at early phases of the development process. Currently, it is very difficult to estimate correction effort of accessibility problems because there is a lack of metrics related to activities required to fix bugs on the user interface. Some problems could be easily fixed, for example by changing an attribute of a tag (i.e. add alternative text to images), while others would require an entire reengineering of Web page (ex. remove tables misused to do page layout). The correction effort should also take into account the multiplicative effect of errors (that could be detected in early phase, for example on templates) that are repeated in all pages of the Web site. One of the future works is to develop an estimation model for assessing cost/benefits of correction of accessibility problems in early phases.

Acknowledgements

This research is partially supported by the action COST294-MAUSE (*Maturation of Usability Evaluation Methods,* http://cost294.org), WebAudit project, ANRT-CIFRE and Genigraph (http://www.genigraph.fr), Genitech Group.

References

1. Acerbis, R., Bongio, A., Butti, S., Ceri, S., Ciapessoni, F., Conserva, C., Fraternali, P., Carughi, G.T.: WebRatio, an Innovative Technology for Web Application Development. In: Koch, N., Fraternali, P., Wirsing, M. (eds.) ICWE 2004. LNCS, vol. 3140, Springer, Heidelberg (2004)
2. Adobe Dreamweaver CS3, http://www.adobe.com/fr/products/dreamweaver/
3. A-Prompt, Web Accessibility Verifier, http://aprompt.snow.utoronto.ca/

4. Bühler, C.: Barrier Free Access to the Internet for all in Germany. In: Accessibility for all Conference, Nice, France, pp. 27–28 (March 2003)
5. Coyette, A., Vanderdonckt, J., Limbourg, Q.: SketchiXML: A Design Tool for Informal User Interface Rapid Prototyping. In: Guelfi, N., Buchs, D. (eds.) RISE 2006. LNCS, vol. 4401, Springer, Heidelberg (2007)
6. Disability, Available at: http://www.disability.gov.uk/
7. e-Citiz, Le monde des téléservices à portée de clic. http://www.e-citiz.com/
8. European eAccessibility Certification, Available at: http://www.euracert.org/
9. Fraternali, P.: Tools and approaches for developing data-intensive Web applications: A survey. ACM Computing Surveys 31, 227–263 (1999)
10. Gomez, J.: Model-Driven Web Development with VisualWADE. In: proceedings of ICWE'2004, Munich, Germany, pp. 611–612, July 28-30 (2004)
11. Ivory, M.Y.: Automated Web Site Evaluation: Researchers' and Practitioners' Perspectives. Kluwer Academic Publishers, Dordrecht (2003)
12. Martins, A., Monteiro, C. (reporters) Petition for the Accessibility of the Portuguese Internet. Report approved by the Portuguese Parliament, on 30th of June, 1999. Also available at: http://www.acessibilidade.net/petition/parliament_report.html
13. Microsoft FrontPage (2003), www.microsoft.com/frontpage/
14. Ocawa, Web Accessibility Expert, http://www.ocawa.com/
15. Official Journal of French Republic from February 11th, Regulation number At: (2005), http://www.legifrance.gouv.fr/imagesJOE/2005/0212/joe_20050212_0036_0001.pdf
16. Scapin, D., Leulier, C., Vanderdonckt, J., Bastien, C., Farenc, C., Palanque, P., Bastide, R.: Towards automated testing of web usability guidelines. In: Kortum, Ph.E. (ed.) Proc. of 6th Conf. on Human Factors and the Web HFWeb'2000, University of Texas, Austin (2000) (Aus-tin, June 19 2000)
17. Section 508: The Road to Accessibility, Available at: http://www.section508.gov/
18. TAW, Web Accessibility Test, http://www.tawdis.net
19. Vanderdonckt, J.: Development Milestones towards a Tool for Working with Guidelines. Interacting with Computers 12(2), 81–118 (1999)
20. Vanderdonckt, J., Limbourg, Q., Michotte, B., Bouillon, L., Trevisan, D., Florins, M.: UsiXML: a User Interface Description Language for Specifying Multimodal User Interfaces. In: Proc. of W3C Workshop on Multimodal Interaction WMI'2004, Sophia Antipolis (July 19-20 2004)
21. Vellemann, E., Strobbe, C., Koch, J., Velasco, C.A., Snaprud, M.: A Unified Web Evaluation Methodology using WCAG. In: Universal Access in HCI (to appear, Proceedings of the 4th International Conference on Universal Access in Human–Computer Interaction), Beijing, China, Lawrence Erlbaum Associates, New Jersey (2007)
22. WabCluster. The EU Web Accessibility Benchmarking Cluster, Evaluation and benchmarking of Accessibility, Available at: http://www.wabcluster.org/
23. Watchfire WebXACT, http://webxact.watchfire.com/
24. Wave 3.0, Web Accessibility Tool, http://wave.webaim.org
25. WebRatio, You think You get, http://www.webratio.com
26. World Wide Web Consortium (W3C), Web Accessibility Initiative (WAI), http://www.w3.org/WAI/

Quality of Web Usability Evaluation Methods:
An Empirical Study on MiLE+

Davide Bolchini[1] and Franca Garzotto[2]

[1] TEC-Lab, Facoltà di Scienze della Comunicazione,
Università della Svizzera Italiana
Via G. Buffi 13 TI - 6900 Lugano (Switzerland)
[2] HOC – Hypermedia Open Center
Department of Electronics and Information, Politecnico di Milano
Via Ponzio 34/5, 20133 Milano (Italy)
davide.bolchini@lu.unisi.ch, franca.garzotto@polimi.it

Abstract. What are the quality factors that define a "good" usability evaluation method and contribute to its acceptability and adoption in a real business context? How can we measure such factors? This paper investigates these issues and proposes to decompose the broad, general concept of "methodological quality" into more measurable, lower level attributes such as *performance*, *efficiency*, *cost effectiveness*, and *learnability*. We exemplify how to measure such attributes, reporting an empirical evaluation study of a usability inspection method for web applications called MiLE+.

Keywords: web usability, quality, empirical study, inspection, heuristics.

1 Introduction

In spite of the large variety of existing usability evaluation methods, both for interactive systems in general [4, 6, 12, 13], and for web applications in particular [2, 10, 11, 14], the factors that define their *quality* are seldom discussed in the literature, and relatively few empirical studies exist that attempt to measure them [5, 9, 15]. Consider for example heuristic evaluation, one of the most popular methods to inspect the usability of web sites [11, 12]. It is claimed to be "simple" and "cheap", implicitly assuming that these attributes are quality factors. Still, little empirical data supports these claims, which are mainly founded on informal arguments (e.g., "few simple heuristics", "no user involvement", "no need of special equipment").

Understanding the quality factors for usability evaluation methods, defining proper measurement procedures, and developing sound comparative studies, not only represent a challenging research arena, but may also pave the ground towards the *industrial acceptability* of these methodological "products": the empirical evidence of quality is a key force to promote a method and to have it accepted and adopted in a real business context.

This paper investigates the concepts of quality and quality measurement for web usability evaluation methods, aim at raising a critical reflection on these issues. We propose to decompose the general concept of methodological quality into lower level,

M. Weske, M.-S. Hacid, C. Godart (Eds.): WISE 2007 Workshops, LNCS 4832, pp. 481–492, 2007.
© Springer-Verlag Berlin Heidelberg 2007

more measurable attributes such as *performance, efficiency, cost effectiveness,* and *learnability.* We also discuss an empirical study in which we measured the above factors for a specific web usability inspection method called MiLE+.

2   MiLE+ at a Glance

MiLE+ (Milano Lugano Evaluation Method – version 2) is the evolution of two previous inspection techniques for the usability of hypermedia and web applications - SUE [10] and MiLE [1, 14] - developed by the authors' research teams. It also borrows some concepts from various "general" usability evaluation methods (heuristic evaluation, scenario driven evaluation, cognitive walkthrough, task based testing).

The main purpose of MiLE+ is to be *more systematic* and *structured* than its "inspirators", and to be particularly suited for *novice* evaluators. A key concept of MiLE+ is that an interactive application can be evaluated along *two main perspectives* (see figure 1): from a "technical", "neutral", "application independent" perspective, and from a "user experience", "application dependent" perspective.

An application independent evaluation is called *Technical Inspection* in MiLE+; it considers the design aspects that are typical of the web and can be evaluated independently from the application's domain, its stakeholders, user requirements, and

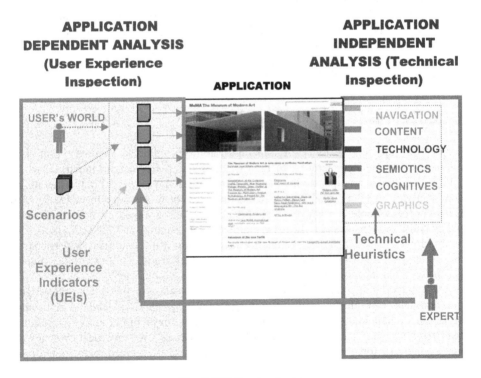

Fig. 1. MILE+ at a glance

contexts of use. A Technical Inspection exploits a built-in *library of* (82) *Technical Heuristics,* coupled by a set of operational *guidelines* that suggest the inspection tasks to undertake in order to measure the various heuristics. These are organized according to various *design dimensions* (see examples in Table 1)*:*

- *Navigation*: (36) heuristics addressing the website's navigational structure
- *Content*: (8) heuristics addressing the information provided by the application,
- *Technology/Performance*: (7) heuristics addressing technology-driven features of the application
- *Interface Design*: (31) heuristics that address the *semiotics* of the interface, the *graphical layout*, and the *"cognitive"* aspects (i.e., what the user understands about the application and its content or functionality).

Table 1. Classification of MiLE+ Technical Heuristics

Dimension		Examples of Heuristics
Navigation		Consistency of navigation patterns
		Index Backward Navigation
Content		Text accuracy
		Multimedia consistency
Technology/Performance		System reaction to user errors
		Operations management
Interface design		
	Cognitive	Information overload
		Scannability
	Graphics	Background contrast
		Text layout
	Semiotics	Ambiguity of link labels
		Conventionality of interaction images

For example, the Interface Design/Graphics heuristic "Background contrast", states a general principle of web visual design "The contrast between the page background and the text should promote the legibility of the textual content". The Navigation heuristic "Index Backward Navigation") claims that "When a user reaches a topic page from a list of topics ("index page"), (s)he should be able to move back to the index page without resorting on the back button of the browser".

An application dependent evaluation is called *User Experience Evaluation in MiLE+*. It focuses on the aspects of the user experience that can be assessed only considering the actual domain of the application, the profiles of the intended users, the goals of the various stakeholders, or the context of use. The usability attributes that are evaluated during this activity are called *User Experience Indicators (UEIs)*. MiLE+ provides a library of 20 UEIs, organized in three categories (see Table 2):

- *Content Experience Indicators*: 7 UEIs focusing on the quality of the *content*
- *Navigation & Cognitive Experience Indicators*: 7 UEIs focusing on the naturalness of the navigation flow and how it meets the user cognitive model
- *Operational Flow Experience Indicators:* 6 UEIS considering the naturalness of single user operations (e.g., data insert or update) and their flow.

Table 2. Examples of MiLE+ User Experience Indicators

Categories	Examples of User Experience Indicators
Content Experience UEIs	Completeness
	Multilinguism
Navigation & Cognitive Experience UEIs	Predictability
	Memorability
Operational Flow Experience UEIs	Naturalness
	Recall

Consider for example the Content Experience UEI *Multilinguism,* which states that "the main contents of the web site should be given in the various languages of the main application targets". Obviously, there is no way to assess if *Multilinguism* is violated or not, without knowing the characteristics of the application targets. A similar argument holds for *Predictability,* which refers to the capability of interactive elements (symbols, icons, textual links, images…) to help user anticipate the related content or the effects of an interaction [6]. Being predictable or not depends at large degree on the user familiarity with the application domain, with the specific subject of the application, and with the application general behaviour.

MiLE+ adopts a *scenario-based approach* [3,4] to guide User Experience Evaluation. In general terms, scenarios are "stories of use" [3]. In MiLE+, they are structured in terms of a "general description", a user profile, a goal (i.e., a general objective to be achieved) and a set of tasks that are performed to achieve the goal (see Table 3). During User Experience Inspection, the evaluator behaves as the users of the scenarios that are relevant for the application under evaluation; he performs the tasks envisioned in these "stories", tries to image the user thoughts and reactions, and progressively scores the various UEIS on the basis of the degree of user satisfaction and fulfillment of scenarios goals and tasks.

Table 3. A MiLE+ scenario for a museum website

Scenario description	A well-educated American tourist knows he will be in town, he wants visit the real museum on December 6th 2004 and therefore he/she would like to know what special exhibitions or activities of any kind (lectures, guided tours, concerts) will take place in that day.
User profile	Tourist
Goal	Visit the Museum in a specific day
Task(s)	• Find the exhibitions occurring on December 6th 2004 in the real museum • Find information about the museum's location

In principle, scenarios should be extracted from the documentation built during user requirements management or design (the application development phases in which scenarios are frequently used). In practice, in most cases such documentation is missing and scenarios are defined by the evaluators in cooperation with the different stakeholders (the client, domain experts, end-users, …).

3   Quality Attributes for a Usability Evaluation Method

Quality is a very broad and subjective concept, oftentimes defined in terms of "fitness to requirements" [7], and should to be decomposed into lower level factors in order to be measured.

For usability evaluation methods, a possible criterion to identify such factors is to consider the *requirements of usability practitioners* and to focus on the attributes that may contribute to *acceptance* and *adoption* of a method in the practitioners' world [8]. Our experience in academic teaching and industrial training and consulting heuristically indicates that "practitioners" want to become able to use a method after an "acceptable" time (1-3 person-days) of "study"; they want to detect the largest amount of usability "problems" with the minimum effort, producing a first set of results in few hours, and a complete analysis in few days.

We operationalize such requirements in terms of the following factors: performance, efficiency, cost-effectiveness, and learnability, defined as follows.

Definition 1: Performance
Performance indicates the degree at which a method supports the detection of all existing usability problems for an application. It is operationalized as the average rate of the number of different problems found by an inspector (P_i) in given inspection conditions (e.g. time at disposal) against the total number of existing problems (P_{tot})

$$Performance = avrg\ (P_i)/P_{tot}$$

Definition 2: Efficiency
Efficiency indicates the degree at which a method supports a "fast" detection of usability problems. This attributes is operationalized as the rate of the number of different problems identified by an inspector in relation to the time spent [5], and then calculating the mean among a set of inspectors:

$$Efficiency = avrg\left(\frac{P_i}{t_i}\right)$$

where P_i is the number of problems detected by the *i-th* inspector in a time period t_i.

Definition 3: Cost-effectiveness
Cost-effectiveness denotes the *effort* - measured in terms of *person-hours* - needed by an *evaluator* to carry on a complete evaluation of a significantly complex web application and to produce an evaluation documentation that meets professional standards, i.e., a report that can be proficiently used by a (re)design team to fix the usability problems.

Definition 4: Learnability
Learnability denotes the ease of learning a method. We operazionalize it by means of the following factors:

-     the *effort,* in terms of *person-hours,* needed by a *novice,* i.e., a person having no experience in usability evaluation, to become "reasonably expert" and to be able to carry on an inspection activity with a reasonable level of performance

- the novice's *perceived difficulty of learning,* i.e., of moving from "knowing nothing" to "feeling reasonably comfortable" with the method and "ready to undertake an evaluation"
- the novice's *perceived difficulty* of *applying application*, i.e., of using the method in a real case.

All the above definitions use, explicitly or implicitly, the notion of *usability problem,* which deserves a precise definition for web applications. Clearly, *a* usability problem has to do with a violation of a usability principle (heuristic, user experience indicator...) in some pages of the application. We must consider that most pages might be "typed", i.e., they share content structure, lay-out properties, and navigational or operational capabilities as defined by their "type" or "class". If a usability violation occurs in one page of a given type, it may occur in other, if not all, pages of the same type, which share the same design. Thus we will count the violations of the *same* principle in *a set* of *pages of the same type* as *one* usability problem.. In contrast, when we consider untyped, or "singleton", pages that represent a "unique" topic or functionality and cannot be reduced to a class, we should count *each* violation in *each* singleton page as one problem. This approach is expressed by the following definition:

Definition 5: Usability Problem
A Usability Problem is a violation of a usability principle in a singleton page, or the equivalence class of the violations of the same usability principle in any set of pages of the same type.

4 An Empirical Study on MiLE+

The purpose of our empirical study was to measure the "quality" of MiLE+ evaluation process in terms of the factors defined in the previous section: performance, efficiency, cost-effectiveness, and learnability. The study involved *two* sub-studies – hereinafter referred as *Study 1* and *Study 2* - that focused on different quality aspects and used different procedures.

4.1 Participants

The overall study involved *42* participants, selected among the students attending two Human Computer Interaction classes of the Master Program in Computer Science Engineering at Politecnico di Milano, hold respectively in the Como Campus and in the Milano Campus. The participant profile was homogeneous in term of age and technical or methodological background. All students had some experience in web development but no prior exposure to usability. They received a classroom training on usability and MiLE+ during the course, for approximately *5 hours* consisting   of an introduction to MiLE+, discussed of evaluation case studies, and Q&A sessions. All students were provided with the same learning material, composed of: a MiLE+ overview article [1], the "MiLE+ Library of Technical Heuristics and User Experience Indicators" (including guidelines and examples), the complete professional evaluation reports in two industrial cases, course slides, an Online Usability Course developed by the University of Lugano (http://athena.virtualcampus.ch/webct/logonDisplay.dowebct).

4.2   Procedure of Study 1: MiLE+ "Quick Evaluation"

The purpose of Study 1 was to measure the *efficiency* and *performance* of our method. We also wanted to test a *hypothesis on learnability:* the *effort* needed by a novice to study the method (besides the 5 hours classroom training) and to become able to carry on an inspection activity with a reasonable level of performance is *less than 15 persons/hours.*

Study 1 involved the *Como group* (*16 students*), who were asked to use MiLE+ to evaluate a portion of an assigned web site (Cleveland Museum of Art website - www.clevelandart.org/index.html) and to report the discovered usability problems, working individually in the university computer lab for *three hours*. The scope of the evaluation comprised the pages from "home" to the section "Collection", which describes the museum artworks, and the whole "Collection" section, for a total of approximately 300 pages (singletons or of different types). Students did not know the assigned website in advance. Before starting the evaluation session, they received a brief explanation of the application's goals and of the general information structure of the web site, and a written specification of two relevant scenarios. Students were asked to report one "problem" (as defined in the previous section) for the same heuristic or UEI, to force them to experiment different heuristics and UEIs. They used a reporting template composed of: *Name* and *Dimension* (of the violated heuristic or UEI), *Problem Description* (maximum three lines), *url* (of a sample page where the violation occurred). The students' evaluation sessions took place one week after MiLE+ classroom training, so that, considering the intense weekly schedule of our courses, we could assume that the students had at disposal a maximum of 15 hours to study MiLE+.

4.3   Procedure of Study 2: Mile+ Evaluation "Project"

The purpose of Study 2 was to investigate the *perceived difficulty* of *learning* and *using MiLE+*, and the *effort* needed to perform a *professional* evaluation. We also wanted to explore the effort needed for the different MiLE+ activities, i.e., technical inspection, user experience inspection, scenario definition, "negotiation" of problems within a team, and production of the final documentation.

Study 2 involved the *Milano group* (26 students) for a two months time period, from the mid to the end of semester 2. Since we wanted to investigate an as much as possible *realistic* evaluation process using MiLE+, i.e., similar to the one carried on by a team of usability experts in a professional environment; participants had to evaluate an entire, significantly complex web site, to work in team (of 3-4 persons), and to deliver an evaluation report of professional quality. The subject of evaluation was freely selected by the teams within a set of assigned web sites that had comparable complexity and suffered of a comparable amount of usability problems (detected by means of a preliminary professional evaluation). To ensure an acceptable and homogeneous level of knowledge on MiLE+ in all participants, study 2 involved only students who had successfully passed an intermediate written exam about the method. The evaluation documentation delivered by the study participants was acknowledged as a course "project" and considered for exam purposes. All teams were scored quite high (A or B), meaning that they produced a complete evaluation report of good or excellent quality.

The data collection technique for measuring the different attributes was an online *questionnaire*. It comprised closed questions about the degree of *difficulty* of studying and using MiLE+ and about the *effort* needed to learn the method and to carry on the various evaluation activities. The questionnaire was explained to the students before they started their project and was delivered at the course exam together with the final project documentation.

4.4 Results

For lack of space, we discuss here only the main results of the two empirical studies. The analysis of the 16 problem reports produced by Como students in study 1 shows that the average number of problems was *14,8,* with an *hourly efficiency* of *4,9* (average number of problems found in one hour). Since the total number of existing problems (discovered by a team of usability experts) is *41*, the *performance* is *36%*. If we consider the profile of the testers and the testing conditions, these results can be read positively. They confirm our hypothesis on learnability and indicated that after 6 hours of training and a maximum of 15 hours of study, a novice can become able to detect more than one third of the existing usability problems!

Some key results of the analysis of the *questionnaire* data collected during study 2 are presented in the following figures.

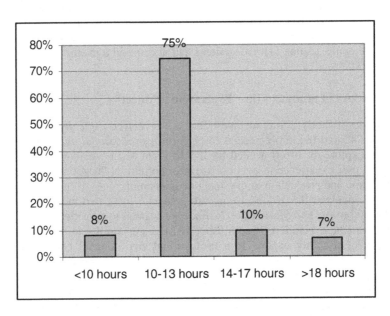

Fig. 2. MiLE+ Learning Effort

Concerning the *learning effort*, participants invested in the preliminary study of MiLE+ an average amount of time of *10-13 hours* (see Fig. 2), which is comparable with the estimated effort of Como students. Concerning *learning difficulty*, a large majority of participants (*73%*) found MiLE+ study activity *rather simple*- see Fig. 3.

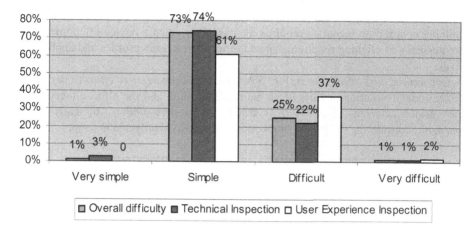

Fig. 3. Perceived Difficulty of Learning MiLE+

Fig. 4 highlights that students perceived the *use* of MiLe+ in a real project as more complex than studying it. Only *47%* of the students scored *"simple"* the use of MiLE+, while *53%* judged it *difficult* or *very difficult*.

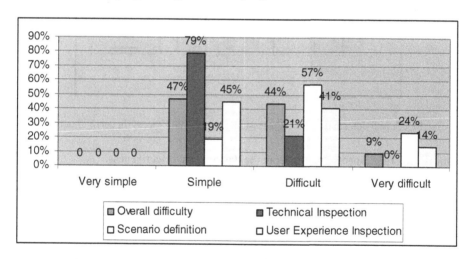

Fig. 4. Perceived Difficulty of Using MiLE+

Fig. 4 also shows that the User Experience Evaluation was perceived slightly more difficult than it was expected from the study (compare fig. 4 with fig. 3). These data may indicate a weakness of the MiLE+ method: although the number of User Experience Indicators (32) is smaller than the number of Technical Heuristics (82), the definition of the former is more vague and confused, and their measurement may result more difficult for a novice. Another reason for the difficulty of performing User Experience Inspection might be related to the difficulty of defining "good" *scenarios*. Fig. 4 pinpoints that a significant amount of participants (*81%*) estimated this activity *difficult* or *very difficult.* Indeed, if the concept of scenario is simple and intuitive, defining appropriate scenarios requires the capability - that a novice oftentimes does not possess - of eliciting requirements and reflecting on users profiles and application goals.

Concerning *cost-effectiveness,* Fig. 5 & 6 highlight the *average effort* to perform a professional evaluation process of an entire application, and the effort allocation on the various activities. The effort is calculated in person/hours, by each single evaluator, considering the time spent working both individually and in team.

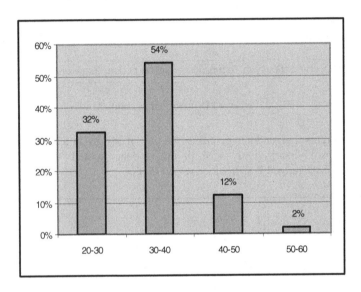

Fig. 5. Individual Effort for a Professional Evaluation Process (in person/hours)

Some interesting aspects emerge from the data on cost effectiveness:

- *54%* of the participants invested from *30 to 40 hours* in the *overall* evaluation process; this means that a team of 2-3 evaluators can deliver a professional report of a significantly complex web application in one week at a total cost of 0.5-0.75 person/month, which is a reasonable timing and economic scale in a business context
- consistently with the results in Fig. 4, the activity of *scenarios* definition is an effort demanding task: *69%* of the participants invested *5-10 hours* in this work
- *5-10 hours* is also the effort invested by 41% of the students in *reporting*; if we consider that all team declared that the reporting work was shared among team members, we can estimate as approximately 1,5-1 person-week the global team effort for the reporting task
- the "negotiation activity" (i.e., getting a team agreement about the final results to be reported) resulted quite fast (3-5 hours for 94% of the persons), which suggest that MiLE+ supports the standardization of the inspection process and the homogenization of results.

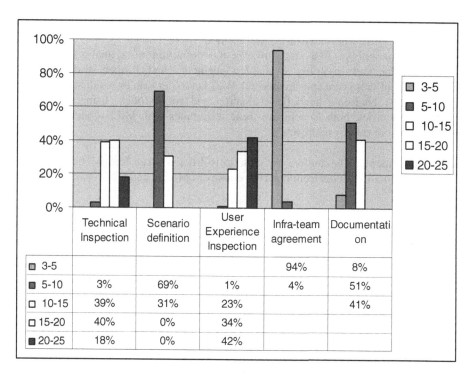

	Technical Inspection	Scenario definition	User Experience Inspection	Infra-team agreement	Documentation
▣ 3-5				94%	8%
▮ 5-10	3%	69%	1%	4%	51%
☐ 10-15	39%	31%	23%		41%
☐ 15-20	40%	0%	34%		
▮ 20-25	18%	0%	42%		

Fig. 6. Individual Effort distribution per Task in a Professional Evaluation Process

In summary, the analysis of the experimental results has proved that MiLE+ meets the need of "practitioners" stated in section 3. Our empirical has proved that the learnability of the method is good, since after a short training (5 hours), understanding MiLE+ basics requires an acceptable workload of study (10-15 hours). The method has also proved to support inexperienced inspectors in performing an efficient and effective inspection both in the context of a short term, quick evaluation (3 hours) and in the context of a real project. Still, our study has also shown that shifting the inspection scope from a (relatively) small-size web site to a full-scale complex application, requires higher levels skills and competence (e.g., for scenario definition) that go beyond usability know how in a strict sense, and can only be gained through experience.

5 Conclusions

Quality is a very broad and generic term, especially if applied to methodological products, and can be defined along many different perspectives. In this paper, we suggest that *learnability, performance, efficiency,* and *cost effectiveness* are possible measurable attributes for methodological quality of web usability evaluation techniques, since they are critical factors for the *acceptance* and *adoption* of methodological products in the practitioners' world. We have discussed how the above factors can be measured, presenting an empirical study that evaluated the quality of the MiLE+ usability inspection method.

Our work is only a first step towards the definition of a quality assessment framework for web usability evaluation methods, and further discussion and investigation of these concepts are needed. We plan to perform the evaluation of other methods (e.g., Nielsen's heuristic evaluation and walkthrough) using our quality criteria and metrics, both to compare these techniques with MiLE+, and to test the soundness of our quality approach.

Acknowledgments. The authors are grateful to Luca Triacca for his contribution in study 1, and to all students from our HCI classes in Milano and Como participating in our research.

References

1. Bolchini, D., Triacca, L., Speroni, M.: MiLE: a Reuse-oriented Usability Evaluation Method for the Web. In: Proc. HCI International Conference 2003, Crete, Greece (2003)
2. Brinck, T., Gergle, D., Wood, S.D.: Usability for the web. Morgan Kaufmann, San Francisco (2002)
3. Carroll, J.: Making Use – Scenario-based design of Human-Computer Interactions. MIT Press, Cambridge (2002)
4. Cato, J.: User-Centred web Design. Addison Wesley, Reading (2001)
5. De Angeli, A., Costabile, M.F., Matera, M., Garzotto, F., Paolini, P.: On the advantages of a Systematic Inspection for Evaluating Hypermedia Usability. In International Journal of Human Computer Interaction, Erlbaoum Publ. 15(3), 315–336 (2003)
6. Dix, A., Finlay, J., Abowd, G., Beale, R. (eds.): Human Computer Interaction, 2nd edn. Prentice-Hall, Englewood Cliffs (1998)
7. Fenton, N.E. (1991) Software Metrics: A Rigorous Approach, 2nd edn. Chapman & Hall, Sydney, Australia (2002)
8. Garzotto, F., Perrone, V.: Industial Acceptability of Web Design Methods: an Empirical Study. Journal of Web Engineering 6(1), 73–96 (2007)
9. Lim, K.H., Benbasat, I., Todd, P.A.: An experimental investigation of the interactive effects of interface style, instructions, and task familiarity on user performance. ACM Trans. Comput. Hum. Interact., 3(1), 1–37 (1996)
10. Matera, M., Costable, M.F., Garzotto, F., Paolini, P.: SUE Inspection: An Effective Method for Systematic Usability Evaluation of Hypermedia. IEEE Transactions on Systems, Men, and Cybernetics 32(1) (2002)
11. Nielsen, J.: Designing Web Usability. New Riders, Indianapolis (1999)
12. Nielsen, J., Mack, R.: Usability Inspection Methods. Wiley, Chichester (1994)
13. Rosson, M.B., Carroll, J.: Usability Engineering. Morgan Kaufmann, San Francisco (2002)
14. Triacca, L., Bolchini, D., Botturi, L., Inversini, A.: MiLE: Systematic Usability Evaluation for E-learning Web Applications. In: ED Media 04, Lugano, Switzerland (2004)
15. Whiteside, J., Bennet, J., Holtzblatt, K.: Usability engineering: Our experience and evolution. In: Helander, M. (ed.) in Handbook of Human-Computer Interaction, pp. 791–817. North-Holland, Amsterdam (1988)

An Assessment of the Currency of Free Science Information on the Web

Chuanfu Chen, Qiong Tang, Yuan Yu, Zhiqiang Wu, Xuan Huang, Song Chen, Haiying Hua, Conjing Ran, and Mojun Li

School of Information Management, Wuhan University, 430072, Wuhan, China
cfchen@whu.edu.cn, tangqiong01@163.com, yuyuan1978@126.com, wuzhiqiang518@tom.com, yuccaer2000@163.com, songchen_cs@hotmail.com, hrice77@sina.com, rancongjing@163.com, mojunlee@gmail.com

Abstract. As the Internet has become a ubiquitous tool in modern science, it is increasingly important to evaluate the currency of free science information on the web. However, there are few empirical studies which have specifically focused on this issue. In this paper, we used the search engines Google, Yahoo and Altavista to generate a list of web pages about 32 terms. Sample pages were examined according to the criteria which were developed in this study. Results revealed that the mean of currency of free science information on the web was 2.6482 (n=2814), only 982 (34.90%) of samples got higher mean scores than the average. Sample pages with different domain names or subjects had difference with significance (P< 0.05). In conclusion, the currency of free science information on the web is unsatisfactory. The developed criteria set here could be a useful instrument for researchers and the public to assess information currency on the web by themselves.

Keywords: Internet, free science information on the web, currency, timeliness, assessment instruments.

1 Introduction

The networked world contains a vast amount of information. According to the estimate of International Data Corporation (IDC), the number of surface Web documents has grown from the 2 billion in 2000 to 13 billion in 2003 [1]. And XanEdu, a ProQuest subsidiary, reported that 5.5 billion pages of copyright cleared articles were currently available for use in course management materials in 2003. Even by conservative estimates, this figure was expected to double by 2007 [2]. In modern science, as a ubiquitous tool, the Internet plays an important role in scholarly communication and is significant for researchers to access academic information [3]. It was reported by Electronic Publishing Initiative at Columbia (EPIC) online use and cost evaluation program that faculty and students use electronic resources on a regular basis for research, teaching, coursework, communicating with colleagues, or just look up general information related to their academic work. Virtually the entire faculty who accepted the survey used electronic resources a few times a week or more (91.8%) [4]. However, because of the speed and the lack of centralized control with which the

M. Weske, M.-S. Hacid, C. Godart (Eds.): WISE 2007 Workshops, LNCS 4832, pp. 493–504, 2007.

information is accumulating, the quality of Web information is not optimistic; a lot of time is wasted to seek reliable and valid things on the Internet [5]. Evaluating the quality of information on the Web is particularly significant.

Several researchers have proposed different Information Quality (IQ) Frameworks to assess Internet information, and currency (also called timeliness [6]) is a prior consideration and one of the most important quality attributes [7a, 8, 9]. Fogg, B.J. (2002) revealed that "Sites that are updated 'since your last visit' are reported to be more credible"[10]; and according to the report of a national survey of Internet users conducted by Princeton Survey Research Associates (2002), users "want Web sites to be updated frequently"[11]. A lot of organizations and individuals have developed guidelines and tools to assess the currency of web information. Sacramento library of California State University (2003) [12], Athina, et al. (2003) [7b] and Cornell University Library (2004) [13] proposed several criteria to assess the currency of information, including: creation date, regularly revised or updated, valid for people's topics. Paul (2003) [14] pointed out currency can be measured by available links either. Most of the authors considered date of creation or update disclosed as the most important criteria to measure currency [15a, 16, 17]. Previous studies indicated that the currency of web information varied widely. Jürgen et al. (2001) found the date of the last update is low available on the web sites of surgical departments, as "62% of the valuations did not find an indication of the last update, 30% valued 'update in the last six months ' , 7% 'more than six months ago' , and 1% 'more than 1.5 years ago'" [18a]. Peter and David (2006) also found that only 12% of the 75 sites related to prostate cancer indicated a date of last update within 6 months, and concluded that there were numerous shortcomings especially related to currency [19a]. Nevertheless, in terms of the survey conducted by Young and Babara (2006), 48.7% (56 of 115) of web sites on food safety issues are updated daily. The next most frequent category was "latest update date is from January 2005 to July 2005." This finding showed that national health-related web sites, in general, presented current information [20]. Although researchers have studied the quality of several types of information on the web, and revealed the status quo of currency of them, none of these previous empirical studies have specifically focused on the currency of free online science information issues. Moreover, results may be various in different approaches, target groups, and environments. The purpose of this study is to determine appropriate terminology, criteria, implementation, and develop a theoretical framework of currency, which can be used by researchers, students and other people to better understand, select, and make use of this information on the web, improving the life quality of the public and promoting scientific progress.

2   Method

In this paper, we defined currency as the quality or state of being current; it refers to the extent to which the information is sufficiently up-to-date for the task at hand [21a, 22a]. Free science information on the web means that online academic information can be accessed by Internet users freely and with no strings attached, such as abstract of academic journals, academic resources published on the web pages established by government, public organizations, business organizations, personal web sites, blogs and so on.

2.1 Criteria of Currency and Weights

Based on empirical studies, we proposed a framework to assess the currency of free science information on the web. Meanwhile, in order to assign the weights of each criterion, we send questionnaire to researchers through E-mail, responders including: authors who had published papers in authoritative scholarly journals abroad and domestic, such as Science, Acta Physica Sinica, Chemical Journal of Chinese Universities, Acta Ecologica Sinica, Journal of Geographical Sciences, Chinese Journal of Computers, Chinese Journal of Mechanical Engineering, etc. The others were professors and doctoral students from School of Water Resource and Hydropower, School of Power & Mechanical Engineering, School of Chemistry, School of Life Science, School of Computer Science, School of Information Management of Wuhan University. Seventy-nine experts filled out the questionnaire, and six didn't pass the Consistency Test. So there are seventy-three effective questionnaires collected. We took Analytic Hierarchy Process (AHP) as our method to assign weights of criteria, which includes four steps: (1) development of judgment matrices by pair-wise comparisons; (2) Synthesis of priorities and the measurement of consistency; (3) ranking the total level; (4) group decision making. And we used a five-item Likert Scale to estimate the priority of each sub-criterion. The criteria set and weights are shown in the following table:

Table 1. Criteria of currency and the weights

Criteria category (weights)	Sub-criteria (weights)
A Date of creation / publication disclosed and very recent information provided (0.23)	
B Update (0.21)	
	B1 Date of last revision disclosed and very recent information provided (0.10)
	B2 Reviewed regularly according to requirements (0.11)
C Valid for topic at hand (0.30)	
D Viable links (0.26)	

Gunther et al. (2002) put forward that among 170 articles about assessing the quality of health information on the web, "almost all studies sought for provision of a date of creation or last update (rather than actual currency of the content), which is a technical criterion" [23a]. However, from Table 1 we can see that among all of the criteria of currency, responders focused on whether the information is still valid for people's topic at hand (0.30). "Viable links" is of second importance (0.26), while the date of creation or last update which adopted by several researchers have the lowest weight. This may due to responders in this investigation are experts or doctoral students, who prefer to find more background information rather than just recent information when carrying out scientific researches. In this case, even though information is not updated frequently, it is still valid for their study. Secondly, some topics in the Humanities or Social Science often require material written many years ago [24a]. Furthermore, in face of the volatility of information on the web ("44% of the sites available on the Internet in 1998 had vanished one year later") [25], viable links reflect on a large scale of the status quo of currency, and provide more additional information which is valid for researcher's topics.

2.2 Search Methodology

We chose 32 search terms from medicine & health (n=5), Chemistry & Biotechnology (n=5), Earth Science (n=3), Humanities (n=3), Material & engineering science (n=3), Computer & Information Science (n=3) and Social Science (n=7). Such as "Tamiflu and bird flu", "Open access and scholar communication", "Genetically modified food and safety ", "Neural selection and human being", " Teflon and health", "Copyrights and software" and so on. These topics were chosen abiding by following principles: (1) they are closely connected to people's life; (2) they are scientific knowledge, comprehensible to ordinary understanding or knowledge; (3) references can be found in peer-reviewed periodicals or reports, encyclopedia and so on.

A keyword search was conducted from January 4, 2007 to January 11, 2007. The first 50 results from three search engines: Google (www.google.com), Yahoo (www.yahoo.com), and Altavista (www.altavista.com) were selected; these search engines represent some of the most common options for Internet surfing by general consumers. A ceiling of 4800 web pages was chosen because of time limitation and considering that most searches performed by individuals on the Internet rarely examine

Table 2. Characteristics of free open science information evaluated (n=2814)

Characteristics	Web pages (%)
domain names:	
.com	1166 (41.44%)
.org & .int	686 (24.38%)
.edu & .ac	489 (17.38%)
.gov	212 (7.53%)
.net & .info	91 (3.23%)
else	170 (6.04%)
subject:	
Social Science	696 (24.73%)
Medicine & Health	511 (18.16%)
Chemistry & Biotechnology	408 (14.50%)
Computer & Information Science	269 (9.56%)
Mathematics & Phisics	268 (9.52%)
Material & Engineering	259 (9.20%)
Earth Science	215 (7.64%)
Humanities	188 (6.67%)

beyond 50 sites [7c, 15b, 26]. In this initial sample, 1986 (41.38%) web pages were discarded, because they were either duplicated web pages, dead links or irrelevant. This resulted in a sample of 2814 unique web pages for analysis in this study (see Table 2).

2.3 Evaluation of Web Pages

Site name, URL, type of domain names, and the country each web page belonged to were recorded. Nine doctoral students and six graduate students came from Wuhan University participated in the evaluation between January 20, 2007 and April 30, 2007. Their disciplinary areas were chemistry, sociology, biology, physics, computer science, and library and information science. In order to maintain the results to be more objective, web pages of each search term were assessed by two reviewers.

Each criterion was scored on a five-point Likert scale: 5 = completely satisfy, 4 = mostly satisfy, 3 = basically satisfy, 2 = partially satisfy, 1 = failure to satisfy. Furthermore, if a criterion was not applicable, it will be categorized as "N/A". The total score for the scale can range from 1 to 5. Reference standards were established for reviewer when assessing information, including: (1) Information is recent or valid for topic at hand or not is determined on the basis of whether data or facts are updated timely and accord with information presented in peer-reviewed periodicals or reports, Encyclopedia Britannica Online and so on. (2) In case the information was updated in six month, the currency would be excellent [18b] and scored as "5" when examined whether information is updated regularly. Besides, reviewers should also consider whether information is needed to be revised regularly. For some types of information, such as topics in history, they are still sufficient to satisfy the criterion even if created many years ago.

Analyses were conducted in SPSS12.0; P-values were two-tailed.

3 Results

3.1 Currency of Free Science Information on the Web

With a histogram portrayed to describe the value distribution of currency (taking N/A as missing value which was replaced by the mean of certain criterion), frequencies analysis was used to discuss the mean scores of currency of all samples.

As a result, the mean value of currency of 2814 samples was come up to 2.6482. Only 34.90% of the sample web pages were higher than the average, while 65.10% lower than it. From Fig. 1 we can found that most of the web pages (66.10%) got mean values distributed between the score "2" and "3", only 140 (4.98%) web pages had mean scores of "4"or above, which proved that the currency of free science information on the web is unsatisfactory.

For information get score higher than "3" means that the information mostly or completely satisfies criteria, we calculated samples with "4" or "5" point scores on each criterion (See Table 3). There are 1153 (40.97%) web pages provided creation date, but merely 13.54% performed well on it; 532 (18.91%) web pages had last updated date, while the overall number of Web pages performed well on it was less than one tenth; 22.78% of web pages revised regularly; about one quarter of online samples were mostly helpful to meet researcher's needs, which is higher than the others, whereas the

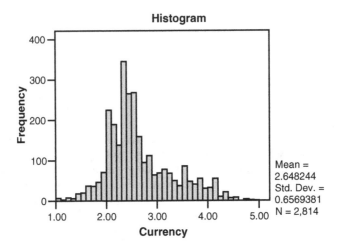

Fig. 1. Histogram of currency values

Table 3. Number of web pages meeting criteria of currency

Criteria category	number of samples scored as "4" or "5"	proportion of total
A Date of creation/publication disclosed and recent information provided	381	13.54%
B1 Date of last update disclosed and recent information provided	271	9.63%
B2 Revised regularly according to requirements	641	22.78%
C Valid for topic at hand	714	25.37%
D Viable links	519	18.44%

Annotation: 1589 sample web pages didn't provided web links related to the information.

proportion was comparatively low; and the amount of web pages provided enough effective links were less than 20%. When further discussing on those with last updated date, only 8.24% were updated in the last six months, 2.84% were reviewed in the last one year, while 7.85% were updated more than 1.5 years ago. It seemed that online free scientific information was not timely updated which was basically consistent with some of the findings mentioned above [18c].

3.2   Comparison of the Currency of Free Online Scientific Information with Different Domain Names

All the samples' domain names had been classified into six kinds (com, gov, edu&ac, org&int, net&info and else). Is there any difference between online free science information with different domain names? Analysis is as follows.

3.2.1 Currency of Free Online Scientific Information with Different Domain Names

Analysis of variance was conducted by using the Kruskal-Wallis Test for K Independent Samples. As the P value was less than 0.05, there was a statistically significant difference between the mean scores of various domain names.

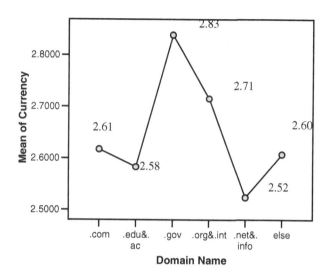

Fig. 2. Mean of currency of free online scientific information with different domain names

From Fig. 2 we can see that samples with .gov domain scored significantly higher overall than the others (P<0.05); while web pages with .net & .info domain scored considerably lower than the others. Compartmentalizing samples by the mean value into two parts, one's currency was lower than it with domain of .com, .edu & .ac , .net & .info and else, timeliness of which should be improved as quickly as they could; while the others including web pages with .gov and .org & int domain were better than the average (P<0.05).

3.2.2 Performances on Each Criterion of Currency with Different Domain Names

We had calculated further of each criterion with different domain names as follows (samples got the score "4" or "5" on each criterion were counted):

Less than 20% of all kinds of domain names performed well on providing date of creation and recent information, web pages with .org & .int, .net & .info and .com domain names were comparatively better. As to information updated, "Date of last revision disclosed and recent information provided" was the lowest met criterion of several kinds of domain names since no proportion of which had passed one-tenth; Whereas, it was founded the most in .gov domain (31.60%) which was also the best (38.68%) on being regularly revised. Besides, merely about 15% of web pages with .edu & .ac, .net & .info domains performed well on revising information regularly.

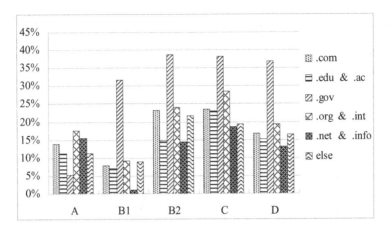

Fig. 3. Performances on each criterion of currency with different domain names

The number of web pages associated with good scores on "valid for topic at hand" in all kinds of domain names (except .net & .info and else) went beyond 20%. It seems that free scientific web pages can provide researchers a certain extent of valuable information. Among them, .gov and .org & .int domain names were better than others, while the domain .net & .info still the worst. More viable links were founded in .gov domain (36.79%), and the percent of other types of domain names mostly satisfied it were around 15%.

In conclusion, currency of .gov domain web pages was the best in total, as most of sample web pages in our survey were managed by U.S., UK and Canada which have better e-governmental web sites construction, generating more releasing rules on government web information, paying more attention on the quality of web information.

3.3 Comparison of the Currency of Online Free Scientific Information with Different Subjects

3.3.1 Currency of Free Online Scientific Information with Different Subjects
In analysis of variance, the P value was close to 0, less than 0.05, proved that the currency of free online science information with different subjects had a statistically significant difference.

The mean scores of currency of online free scientific information with different subjects were showed in Fig. 4. The currency of information related to Humanities scored significantly higher overall than the other subjects (mean score was 2.7607, $P<0.05$). And the web pages of Social Science, Material & Engineering and Chemistry & Biotechnology also scored relatively well; while Mathematics & Physics, Computer & Information Science, Earth Science and Medicine & Health got lower scores than the average level, and the Medicine & Health related web pages showed the worst quality on currency among them. Many scholars have assessed the currency of the different information related to Medicine & Health, it turned out that the information didn't perform well on currency, which matched basically with ours [18d, 19b].

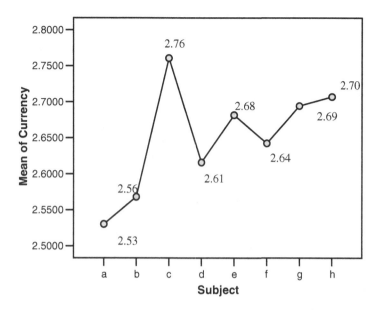

Fig. 4. Mean of currency of free online scientific information with different subjects Abbreviations: a=Medicine & Health, b=Earth Science, c=Humanities, d= Computer&Information Science, e=Chemistry & Biotechnology, f= Mathematics & Physics, g= Material & Engineering, h=Social Science

3.3.2 The Performances on Each Criterion of Currency with Different Subjects

Counting samples which got score "4" or "5" on each criterion as regards eight subjects; Fig. 5 shows the results as follows:

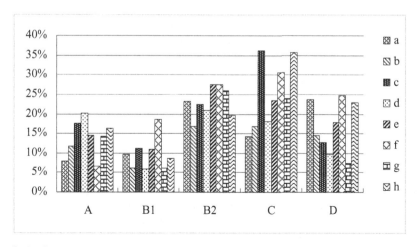

Fig. 5. Performances on each criterion of currency with different subjects Abbreviations: a=Medicine & Health, b=Earth Science, c= Humanities, d=Computer & Information Science, e=Chemistry & Biotechnology, f= Mathematics & Physics, g= Material & Engineering, h=Social Science

About twenty percent of the web pages of Computer & Information Science found the date of creation and provide more recent information. Nevertheless, there were only less than ten percent of web pages of Medicine & Health and Mathematics & Physics satisfied the criterion well.

There were few web pages mentioning the date of last update and reflecting recent information well. More web pages of Mathematics & Physics mostly satisfied the criterion than others, but the percentage was only 18.66%. Over 1/5 web pages of many subjects (except Earth Science and Social Science) revised regularly according to requirements. Matthew indicated that "information must be updated as frequently as the source value changes or else it will age. However, such real-time updating may not be necessary for the user's purpose, nor practical, feasible or cost-effective", in his view, "If information is updated frequently enough for the user's purposes then it is current" [22b]. Therefore, whether the information is current should be combining with the users' research purposes, information currency is a relative concept.

The criterion, "valid for topic at hand", is the important aspect that exactly combines the users' purposes with currency. Fig. 5 shows that among these subjects, web pages related to Humanities met users' real-research needs the best (36.17%), this may be due to the features of it. Information that has been written for ages is still valid for some topics of Humanities. For example, the two search terms, "the Pearl Harbor attack" and "discovery of Peking Man", were all related to history, which did not require timely updated.

Web pages of Material & Engineering, Medicine & Health and Social Science provided more available links. On the contrary, no more than 10% percent Computer & Information Science and Material & Engineering web pages updated timely according to the changes of the links.

4 Conclusion

Empirical studies have been performed to evaluate the quality of information on the web, and currency is one of the "core standards" [24b]. The Internet provided plenty of free science information which has played an important role in scholarly communication and researches. People may be misled by outdated Internet information, and certain scientific researches may be disrupted by it. Therefore, it is important to determine appropriate terminology, criteria, implementation, and develop a theoretical framework of currency, which can help researchers, students and other people to better understand, select, and make use of the information on the web. To our knowledge, ours is the first study to specifically focus on the currency of the free science information issues.

This study indicated that the mean score of currency of the free science information on the web was 2.6482 (n=2814), which shows that the information merely partially satisfied criteria we proposed. Among all of the samples, only 982 (34.90%) got higher mean values than average. Calculating samples scored as "4" or "5" on each criterion, we found 13.54% of total samples performed well on "date of creation disclosed and recent information provided", while web pages disclosed last updated date and provided more recent information were the least (9.63%); 22.78% of web pages revised regularly; about one quarter of online samples were mostly helpful to meet researcher's needs, and the amount of web pages offering enough effective links were less than

20%. In a word, the currency of free science information on the web is unsatisfactory, which still needs to be improved.

We also found that the currency of free science on the web with different domain names and subjects had difference with significance. Samples with .gov domain were the best in total. Web pages of Humanities and Social Science provided more current free science information; while the currency of information of Medicine & Health was worse compared to the other subjects.

Although most of empirical studies sought for provision of a date of creation or last update when assessing the currency of web information, our investigation revealed that responders focused on whether the information is still valid for people's topic at hand.

The research in this paper may have some limitations. First of all, we picked the top 50 ranked results from three search engines. As many search engines can rank the better sites first, the search tool could influence the results [23b]. Secondly, most studies of online health information, including ours, are limited by the constantly changing nature of the Internet. Therefore, if our study were repeated, the findings may be different [15c, 21b]. Furthermore, in spite of we had trained reviewers before evaluation, and they have good information literacy, the results may still be influenced by subjective factors.

Multiple previous studies found that Internet users do not assess currency when searching for information online [27, 28], which is not beneficial for them to get more valuable information. The criterion set developed here could be a useful instrument for researchers and the public to assess the currency of web information by themselves. Future research will focus on developing instruments and techniques, aiming at automatically predicting currency of free science information on the web for scientific researchers and the public.

Acknowledgement. This paper is supported by Program for New Century Excellent Talents in University, the Ministry of Education of China.

References

1. Michael, K.B.: The Deep Web: Surfacing Hidden Value. The Journal of Electronic Publishing, 7, 1 (2001), available at: http://www.press.umich.edu/jep/07-01/bergman.html
2. OCLC. Five-year Information Format Trends (2003), available at: http://www5.oclc.org/downloads/community/informationtrends.pdf
3. Franz, B.: The Role of the Internet in Informal Scholarly Communication. Journal of the American Society for Information Science and Technology 57(10), 1350–1367 (2006)
4. Kate, W., et al.: EPIC Online Use and Costs Evaluation Program (2004), available at: http://www.epic.columbia.edu/eval/eval04frame.html
5. Athina, T., et al.: Important Drug Safety Information on the Internet: Assessing its Accuracy and Reliability. Drug Safety 26(7), 519–527 (2003) (5a, 5b, 5c)
6. Diederik, G., et al.: Comparison of Search Strategies and Quality of Medical Information of the Internet: A Study Relating to Ankle Sprain. Injury, Int. J. Care Injured 32, 473–476 (2001)
7. Gunther, E., Thomas, L.D.: Towards Quality Management of Medical Information on the Internet: Evaluation. Labeling, and Filtering of Information. BMJ 317, 1496–1502 (1998)
8. Shirlee-ann, K., Janice, B.: Developing a Framework for Assessing Information Quality on the World Wide Web. Journal of Information Science 8, 159–172 (2005)

9. Carmine, S., Stephen, B.: Towards a Weighted Average Framework for Evaluating the Quality of Web-located Health Information. Journal of Information Science 31, 260–272 (2005)

10. Fogg, B.J., et al.: Stanford-Makovsky Web Credibility Study 2002: Investigating What Makes Web Sites Credible Today (2002), available at: http://captology.stanford.edu/pdf/Stanford-MakovskyWebCredStudy2002-prelim.pdf

11. Princeton Survey Research Associates. A Matter of Trust: What Users Want from Web Sites (2002), available at: http://www.consumerwebwatch.org/pdfs/a-matter-of-trust.pdf

12. California State University, Sacramento. Evaluating Internet Sources (2003), available at: http://library.csus.edu/services/inst/ICCS/infocomp/tutorials/module5/web/timeliness/index.htm

13. Cornell University Library. Critically Analyzing Information Sources (2004), available at: http://www.library.cornell.edu/olinuris/ref/research/skill26.htm

14. Stapleton, P.: Assessing the Quality and Bias of Web-based Sources: Implications for Academic Writing. Journal of English for Academic Purpose 2, 229–245 (2003)

15. Elmer, V.B., Dawn, M.S., Muhammad, W.: Instruments to Assess the Quality of Health Information on the World Wide Web: What can Our Patients Actually Use? International Journal of Medical Informatics 74, 13–19 (2005) (15a, 15b, 15c)

16. Carmine, S., Stephen, B.: Towards a Weighted Average Framework for Evaluating the Quality of Web-located Health Information. Journal of Information Science 31(4), 260–272 (2005)

17. Laura, O.G.: Future Directions for Depicting Credibility in Health Care Web Sites. International Journal of Medical Informatics 75, 58–65 (2006)

18. Jürgen, S., et al.: Health Care Providers on the World Wide Web: Quality of Presentations of Surgical Departments in Germany. Med. Inform., 26(1), 17–24 (2001) (18a, 18b, 18c, 18d)

19. Peter, C.B., David, F.P.: Prostate Cancer on the Internet—Information or Misinformation? The Journal of Urology 175(5), 1836–1842 (2006) (19a, 19b)

20. Young, N., Barbara, A.A.: Analysis of Governmental Web Sites on Food Safety Issues: A Global Perspective. Journal of Environmental Health 69(5), 10–15 (2006)

21. Wang, Y., Liu, Z.: Automatic Detecting Indicators for Quality of Health Information on the Web. International Journal of Medical Informatics 4, 1–8 (2006) (21a, 21b)

22. Matthew, W.B.: Information Quality: A Conceptual Framework and Empirical Validation [D]. The University of Kansas , 102–103 (2004) (22a, 22b)

23. Gunther, E., et al.: Empirical Studies Assessing the Quality of Health Information for Consumers on the World Wide Web: A Systematic Review. JAMA 287(20), 2691–2700 (2002)

24. Environmental Education and Training Partnership. Evaluating the Content of Web Sites: Guidelines for Educators (1999), (24a, 24b) available at: http://www.epa.gov/enviroed/pdf/evalwebsites.pdf

25. The library of Congress. National Digital Information Infrastructure and Preservation Program, available at: http://www.digitalpreservation.gov/importance/

26. Peter, S., Peter, Z., Mark, K.P.: The Internet and Patient Education — Resources and their Reliability: Focus on a Select Urologic Topic. Adult Urology 53(6), 1117–1120 (1999)

27. Meric, F., et al.: Breast Cancer on the World Wide Web: Cross Sectional Survey of Quality of Information and Popularity of Web Sites. BMJ 324, 577–581 (2002)

28. Eysenbach, G., Kohler, C.: How do Consumers Search for and Appraise Health Information on the World Wide Web? Qualitative Study Using Focus Groups. Usability Tests, and In-depth Interviews, BMJ 324, 573–577 (2002)

Including Routes in Web Information Systems as a Way to Improve the Navigability: An Empirical Study*

Valeria de Castro[1], Marcela Genero[2], Esperanza Marcos[1], and Mario Piattini[2]

[1] KYBELE Research Group, Rey Juan Carlos University
Tulipán S/N, 28933, Móstoles - Madrid, Spain
{valeria.decastro,esperanza.marcos}@urjc.es
[2] ALARCOS Research Group, University of Castilla La Mancha
Paseo de la Universidad 4, 13071, Ciudad Real, Spain
{Marcela.Genero,Mario.Piattini}@uclm.es

Abstract. Simplifying the achievement of the user tasks is a factor that determines usability in Web development. In order to better reflect the best paths that may drive the user through the Web Information Systems (WISs) to search the desired information/services, navigation models have been widely adopted by the Web Engineering community. However, the design of WISs often relies only on a domain model leaving many decisions, which may directly affect usability, to the designer skills. In order to limit this arbitrarily in the navigation design process, we have proposed a hypertext modeling method (HM^3) which explicitly bases the construction of the navigation model on the services required by the user. This way, the navigation model built with HM^3, includes *routes*, which help the user to properly carry out the required service. The main goal of this paper is to present an experiment we carried out to corroborate the hypothesis that "using *routes* it is possible to obtain more navigable WISs".

1 Introduction

As the complexity of the WISs grows, the difficulty in using them systems grows as well. Usually, the users of traditional information systems spend a lot time becoming familiar with the features and design of these systems. Conversely, on the Web, users usually do not want to read any manuals or help instructions for individual sites [8]. Several authors claim that design oriented to the user needs is one of the key aspects of usability: "users are never going to even get close to the correct pages unless the site is structured according to user needs and contains a navigations scheme that allows people to find what they want" [8].

It is widely recognized in the Web community that usability is significantly associated with navigation [8,9]. Good navigation aids let users acquire the information they are seeking quickly and efficiently; and therefore contribute to the

---

* This research is partially granted by the next projects: GOLD (TIN2005-00010), CALIPO (TIC2003-07804-C05-03), FOMDAS (URJC-CM-2006-CET-0387) and MESSENGER (PCC-03-003-1).

M. Weske, M.-S. Hacid, C. Godart (Eds.): WISE 2007 Workshops, LNCS 4832, pp. 505–510, 2007.

perceived success by site users [9]. Such navigability is usually expressed in Web Engineering methods through navigational models whose design often depends on the skills of the designer. We want to be more rigorous in that sense, so we proposed a method to model WIS navigation from a user needs oriented perspective, called Hypertext Modeling Method of MIDAS (HM^3) [6,7]. This method has the particularity that allows identifying *routes*. A ***route*** is defined as "the sequence of steps established for the WIS that the user must follow to execute a user service". After implementing the *routes* in a WIS, signposting the sequence of steps that the user must follow to properly carry out each service, they will guide users to navigate through the WIS. Moreover, the method proposes to identify a main menu including the services required by the user which represent the beginning of the *routes*. Drawing an analogy with the road system, the main menu is analogous to a signal indicating possible destinations in the origin of several ways, and the route is analogous to the signposting of the road. Thus, in the same way that these characteristics in a road could help the drivers to get their destination, these characteristics in a WIS could also help the users to perform the task they need.

But, although intuitively, it seems obvious to us that a WIS with signposted routes is more easily navigable, as Zelkowitz et al. [13] pointed out, a new proposal in software engineering lacks credibility if there is no empirical evidence of its usefulness. For that reason, we have carried out an experiment in order to corroborate our hypothesis: "Using routes it is possible to obtain more navigable WIS". In order to test our hypothesis, we have decided to compare the navigability of two WISs for conference organization: *ConfMaster* [3], a well known WIS in this field and widely used in prestigious conferences; and *WebConference* [11], a similar WIS built using *routes*. Both WISs have the same functionality and identical interface style, the only difference is in their navigation model; only the second one (*WebConference*) has *routes* and a main menu indicating the beginning of each route. So, our aim is to measure the impact of the *routes* in the final users. Note that, the aim of the experiment presented in this paper is not to test that HM^3 is better than other methodologies for WISs development. In fact, the methodology used for developing *ConfMaster* is unknown for us, and it is not important in this phase. In this experiment we just want to measure the impact of the *routes* in the final users. For this reason, we have evaluated two WISs in which the only difference is that one of them includes routes as well as a main menu indicating the beginning of the routes, whereas the other WIS does not include them.

The rest of the paper is structured as follows. In Section 2 we describe the controlled experiment. Section 3 presents the analysis and interpretation of the empirical data. Finally, the conclusion in Section 4 underlines the main contribution of the paper and future work.

2 Experiment Description

The main objective of the experiment expressed using GQM [1] template is: **Analyze** the *WebConference* and *ConfMaster* WISs, **For the purpose of** evaluating, **With respect to their** navigability, **From the point of view of** the researchers, **In the context of** undergraduate students enrolled in the fourth-year of the Computer Science at the Rey Juan Carlos University Hereafter, we briefly describe the experimental process, using the format (with minor changes) proposed by Wohlin et al. [12]

Subjects. 84 students enrolled in the fourth-year of the Computer Science at the Rey Juan Carlos University (Spain) carried out the experiment. The subjects were selected for convenience i.e. they are students who had on average 5 years of experience in using web applications and whom we considered competent enough to perform the level of experimental tasks required.

Variables. We considered the WIS used to be the independent variable. This variable has two levels: WebConference and ConfMaster. On the other hand, the dependent variable was navigability, measured through the following measures, taken from [5]:

- Perceived Ease of Navigation, defined as the degree to which a person believes that using a particular tool facilitates the navigation through the WIS.
- Effectiveness, defined as how well the usage of a particular WIS allows the required tasks to be achieved.
- Efficiency, is defined as the effort required to use a particular WIS correctly..

Hypotheses. The following hypotheses investigated were shaped by our experience with WIS modelling:

- $H_{0,1}$: There is no difference in Perceived Ease of Use of subjects using WebConference and ConfMaster. $H_{1,1}$: $\neg H_{0,1}$
- $H_{0,2}$: There is no difference in Efficiency of subjects using WebConference and ConfMaster. $H_{1,2}$: $\neg H_{0,2}$
- $H_{0,3}$: There is no difference in Effectiveness of subjects using WebConference and ConfMaster. $H_{1,3}$: $\neg H_{0,3}$
- $H_{0,4}$: The subjects prefer using ConfMaster over WebConference.
- $H_{1,4}$: The subjects prefer using WebConference over ConfMaster.

Experimental material. For each participant, we had prepared a folder containing for each WIS: a debriefing questionnaire regarding subjects' experience; four tasks to be carried out using the WIS; a survey of 8 questions related to the Perceived Ease of Use of the WIS (survey 1), and; a survey where the subjects had to compare both WISs (survey 2).

Execution. The experiment was carried with two groups of subjects (G1 with 44 students and G2 with 40 students) located in two different laboratories. In the first laboratory we first gave the material related to ConfMaster and secondly WebConference. In the second laboratory we changed the order to cancel out learning effects. Subjects were given an intensive training session before the experiment took place. In this session, we gave them a test similar to those we used in the experiment and we explained to them the tasks they had to carry out. The students worked under examination conditions, with no speaking among themselves or asking questions about doubts to professors supervising the experiment. The subjects had to perform the following experimental tasks for each WIS:

- To fill out a debriefing questionnaire (including personal details and experience).
- To perform, using the corresponding WIS, the four tasks required, writing down the time when they began doing the first task and the time when they finished the

fourth task. Moreover, they had to write down the name of the links they navigated while performing each required task. From these tasks, we obtained values for:

- Effectiveness = Nº of Correct Clicks Done / Nº of Clicks the task required
- Efficiency = Nº of Correct Clicks Done / Time

- To answer the eight questions in survey 1, which had to rate them using a 5 point Likert scale.
- On finishing the experimental tasks with both WISs, the subjects had to fill out survey 2, where they had to express their preference between both WISs.

Data validation. We collected the material filled out by the subjects, checking if they were complete. There was a bit of incomplete data detected and rejected with the statistical analysis.

Threats to validity. In our opinion the greatest threats are to the internal validity of our experiment; i.e. the degree to which conclusions can be drawn about the causal effect of the independent variable on the dependent variable [2]. One possible threat to internal validity is the accuracy of subject responses, given that they have to write down manually the time spent on doing the tasks and the name of the links they navigated. Even though we placed special emphasis on the relevance of the accuracy of these data in the training part of the experiment, we never could be sure about this and we have to trust them. The students were motivated to participate in the experiment by a "prize", 0.5 points of the final mark for participating, and another 0.5 points for performing the required tasks correctly. With respect to the external validity, i.e. the ability of generalize the obtained findings to the population under study and other research settings [2], we consider that the functionality of the WISs selected was probably simple. For that reason the results need to be confirmed by replication experiments.

3 Analysis and Interpretation

For testing the hypotheses we merged the empirical data of groups G1 and G2. All the data analysis was carried out by means of SPSS [10]. The debriefing questionnaire allows us to obtain the following data that reflects the profile of the participants: 24 year old on average, with 5 years of experience using WISs, 2 years of experience designing WISs and 4 years of experience designing traditional ISs.

The data used to test the *first hypothesis* are the subjective ratings given by the subjects in the first survey. First, we checked the inter-rate reliability (Cronbach's alpha [4]). to determine how consistent the results of the rates were with what order. The Cronbach's alpha obtained for the responses about WebConference was 0.81 and for ConfMaster 0.91. Both coefficient values were above 0.7, the suggested value to consider the results reliable. We tested the first hypothesis for each question (Q1..Q8) and also considered the median (M) of the eight responses using the Wilcoxcon test (a non-parametric test for for ordinal measures).As all the significance levels were lower than 0.05 we can reject $H_{0,1}$. This means that the Perceived Ease of Use is different in ConfMaster and WebConference. Moreover, comparing the median values of the responses obtained for each WIS, we can conclude that the Perceived Ease of Use is better for the WebConference. Moreover, we found that 67% of subjects have rated WebConference with a value greater than 4 (as a median). This fact demonstrates that the majority of subjects perceive WebConference to be easier to use.

To test the **second hypothesis** we calculated the mean of the efficiency between the values of efficiency for carrying out the four tasks required for each WIS. As both measures are ratio scale measures, we will carry out an ANOVA, considering the WIS as a within-subject factor and the order in which the subjects receive each application as a between-subject factor. As the significance levels is not less than 0.05 we can not reject the $H_{0,2}$. Thus, it seems that there is not a significant difference between the efficiency of subjects when using WebConference or ConfMaster. We also carried out an ANOVA for investigating the efficiency obtained in each task separately, and we only found differences between the efficiencies of both WIS for task 4. We also investigated the behavior of the mean of the Time spent on carrying out the tasks. ANOVA indicated there was difference between the times spent when using each WIS, and we discovered that the order when using each WIS influenced the results. Table 1 reveals that on average subjects spent more time using ConfMaster.

Table 1. Descriptive statistics for the mean of Time (seconds)

	Min	**Max**	**Mean**	**St. Dev.**
WebConference	18.8	305.44	89.5119	57.0183
ConfMaster	25.8888	261.78	108.4840	50.5317

As the order influences on time, we compared the means by order, and results indicated that, independently of the order, the subjects spent less time using WebConference (see Table 2).

Table 2. Comparison of the mean of Time considering the order (seconds)

WebConference(1)	105.9890	WebConfernce(2)	71.3005
ConfMaster(1)	111.4058	ConfMaster(2)	86.8628

To test the **third hypothesis** we calculated the mean of the Effectiveness between the values of effectiveness in carrying out the four tasks required for each WIS. As both measures are ratio scale measures, we carried out an ANOVA, considering the WIS as a within-subject factor and the order in which the subjects received each WIS as a between-subject factor. The ANOVA results allow us to reject $H_{0,3}$, which means that there exists a difference between both WISs with respect to Effectiveness.

To test the **fourth hypothesis** we used the data obtained in the second survey, assigned to the subjects after they performed the experimental tasks with both WISs..As a result we found that 34 subjects preferred WebConference, 15 ConfMaster and only 4 did not have a preference. Analyzing the probabilities of preferences we obtained a p-value < 0.001 that suggests to reject $H_{1,4}$, confirming thus there exist greater probability that the subjects prefer WebConference over ConfMaster.

4 Conclusions and Future Work

In this work we have presented an experiment to corroborate if effectively, using *routes*, it is possible to build more navigable WISs. The most important conclusions obtained through the empirical study are that the subjects perceive WebConference, (the WIS that was built using *routes*) is easier to use. They were more effective using it, i.e., using WebConference leads them to perform the required tasks in a more correct way.

Even though the results obtained are encouraging, we consider them to be preliminary. Further validation is needed to obtain conclusive results about whether HM^3 really leads to WISs which are easier to use, more effective and more efficient. For that reason, we are planning to carry out a replication of this experiment. Moreover, due to the WISs evaluated are very simples, we're also planning to make the experiment with more complex WISs, in which the results should be more conclusive.

References

1. Basili, V.R., Rombach, H.D.: The TAME project: towards improvement-oriented software environments. IEEE Transactions on Software Engineering 14(6), 758–773 (1998)
2. Briand, L.C., Bunse, C., Daly, J.W.: A Controlled Experiment for evaluating Quality Guidelines on the Maintainability of Object-Oriented Designs. IEEE Transactions on Software Engineering 27(6), 513–530 (2001)
3. ConfMaster. Available in (2005), http://confmaster.net/phpwebsite_en/index.php
4. Cronbach, L.J.: Coefficient alpha and the internal structure of tests. Psychometrika 16(3), 297–334 (1951)
5. Davis, F.D.: Perceived Usefulness. Perceived Ease of Use and User Acceptance of Information Technology. MIS Quarterly 3(3) (1989)
6. De Castro, V., Marcos, E., Cáceres, P.: A User Service Oriented Method to model Web Information Systems. In: Zhou, X., Su, S., Papazoglou, M.M.P., Orlowska, M.E., Jeffery, K.G. (eds.) WISE 2004. LNCS, vol. 3306, pp. 41–52. Springer, Heidelberg (2004)
7. Marcos, E., Cáceres, P., De Castro, V.: An approach for Navigation Model Construction from the Use Cases Model. In: Persson, A., Stirna, J. (eds.) CAiSE 2004. LNCS, vol. 3084, pp. 83–92. Springer, Heidelberg (2004)
8. Nielsen, J.: Design Web Usability. New Riders Publishing (2000)
9. Palmer, J.: Designing for Web Site Usability. IEEE Computer 35(7), 102–103 (2002)
10. SPSS, 2002 SPSS 11.5. Syntax Reference Guide. Chicago. SPSS Inc (2002)
11. WebConference (2005), http://kybele.escet.urjc.es/webconference/
12. Wohlin, C., Runeson, P., Host, M., Ohlsson, M.C., Regnell, B., Wesslen, A.: Experimentation in Software Engineering: An Introduction. Kluwer Publishers, Dordrecht (2000)
13. Zelkowitz, M., Wallace, D.: Experimental validation in software engineering. Information and Software Technology 39(11), 735–743 (1997)

A Generic Approach to Improve Navigational Model Usability Based Upon Requirements and Metrics*

Fernando Molina Molina and Ambrosio Toval Álvarez

Software Engineering Research Group
Department of Informatics and Systems
University of Murcia (Spain)
{fmolina,atoval}@um.es

Abstract. In recent years, the fast evolution of Internet and the Web
has caused an exponential increase in the number of Web Information
Systems (WIS) developed. This has led to the appearance of a new discipline, Web Engineering, which has served as a framework for the development of numerous methodologies and tools which seek to contribute to
the development of WIS with the quality parameters required by their
users. Among the quality attributes of a WIS are accessibility, usability and the easy of navigation offered by the system. These attributes
are usually analyzed when the WIS has been developed, using strategies like the analysis of the HTML code of the WIS or the evaluation
of the system by a group of users. This paper presents a proposal to
extend the models used in the methodologies for WIS development and
to define a set of quality metrics so that modelers can consider usability requirements during WIS modelling. The automatic support for this
proposal and the metamodel extension necessary for its integration into
the existing methodologies for WIS development are also presented.

1 Introduction

The evolution of Internet and the WWW in recent years has caused an exponential increase in the number of WIS developed and has motivated the appearance of a new discipline called Web Engineering [1]. Recent intensive research
in this area has led to the appearance of numerous methodologies, languages,
techniques, design patterns and tools specially focused on WIS development as
HDM, RMM, OOHDM, WebML, OO-H or MIDAS. However, when current WIS
development methods are analyzed, we observe that they do not offer support
to deal with usability requirements during the development process and they
delay this task until the system has been completely developed. Some guidelines

---

* This work was partially supported by the projects DEDALO (TIC2006-15175-C05-
03) financed by the Ministry of Science and Technology of Spain and FoMDAs
(URJC-CM-2006-CET-038) financed by URJC and Madrid Community. The first
author was partially supported by the Fundación Séneca (Spain).

M. Weske, M.-S. Hacid, C. Godart (Eds.): WISE 2007 Workshops, LNCS 4832, pp. 511–516, 2007.
© Springer-Verlag Berlin Heidelberg 2007

have appeared that collect properties and features that WIS must fulfil to improve their usability. In the same way, a number of tools called *usability and accessibility validators* has arisen with the aim of analyze a WIS and to validate these guidelines for it. Some examples of tools are EvalIris, TAW, WebTango or WebXACT. [2] presents a detailed study about methods for automatic WIS evaluation. These tools validate the HTML and CSS code of the WIS, that is, they analyze the code of the system when it has been developed but they do not consider the possibility of moving some of the validations towards the navigation or presentation models proposed by WIS development methodologies. The potential of model-based usability evaluation and its advantages are analyzed in [3,4]. In addition, there exists a gap between tools for automatic validation and the methodologies for WIS development, because neither the usability validation processes nor the tools that automatize the validation are integrated in the methodologies for WIS development or in the CASE tools that support them.

This paper proposes to consider as far as possible usability and accessibility requirements during WIS modeling focusing on navigational models. The improvement of the quality of the conceptual models used in WIS modeling will contribute to the quality of the WIS finally developed. We propose to extend the metamodels used for WIS development methodologies with the aim of providing modelers capacity to reflect their own usability requirements for the WIS models. In addition, we propose a set of metrics to help modelers to determine the quality of the WIS models. The possibility of expressing these features on models permits the validation of usability requirements during WIS modeling. A tool for supporting this approach will be presented too.

2   Usability in Navigational Models

There exist different definitions of usability in the literature as the proposed by the standards ISO 9126-1 [5] or ISO 9241-11 [6]. However, usability is an attribute that depends on numerous factors. Thus if we want to improve the usability of a WIS we must pay attention to multiple features, like the intuitive navigation in the system, simplicity to carry out tasks or a comfortable and attractive presentation. The approach presented in this work is centered on improving the usability of a WIS offering an intuitive navigation that makes it easy for the user to carry out the tasks for whom the WIS was conceived. Our proposal centeres on navigational models which model the interaction between users and the WIS. Navigational models are usually composed by nodes and links. A node is used to represent a set of information or functionality that will be presented to WIS users. Links are used to join nodes, indicating the possibility of navigating from one node to another. In addition, other navigation structures like menus or indexes can appear in these models. The quality of navigational models is an important feature because these models represent the possible paths that users can follow during the navigation through the WIS. Thus errors in these models or less useful navigation designs have influence on the usability of the WIS finally developed.

Studies exist focused on the evaluation and improvement of the quality of navigation models. [7,8] define metrics that inform modelers about the quality of WIS models. [9] proposes a strategy for verification of properties for navigational models. Other studies focused on early usability and quality evaluation of WIS are [10,11]. All these studies permit the evaluation and improvement of WIS models. However, this work tries to extend these models to offer modelers the capacity to express and to validate automatically their own usability requirements because, as Section 3 will show, at the moment there exist usability requirements that can not be expressed in the models proposed by the different methodologies. Thus we expect to contribute to WIS usability from modelling, by contributing to the overall quality of the system finally developed.

3   Usability Requirements for Navigational Models

During the requirements elicitation process for a WIS, a set of requirements that at the moment can not be expressed in the navigational models proposed by WIS development methodologies are collected. For example, it is possible that a modeler wants to express requirements related to the maximum number of clicks that a user must need to carry out a task. Other requirements could be related to the access to the information of the WIS. This kind of requirements are usually forgotten and, for this reason, the WIS finally developed do not permit an intuitive navigation and users usually feel lost and disoriented. The approach presented in this work will permits modelers to express these kind of requirements in navigational models and it will permit to verify that navigational models fulfil them. The following section presents a set of usability requirements that can be represented in navigational models and classify them in two groups: requirements related to the access to the information and related to connectivity between nodes.

3.1   Information Accessibility

Modelers can divide the information and functionality of a WIS into many levels of importance depending on the main aim of the system. Our proposal will permit modelers to establish different importance levels to classify the information and functionality of a WIS. In this way, each node in the model can be labeled with one of the defined importance levels and we obtain capability to support the following requirements:

R1. Maximum/minimum distance from the entry point to the WIS for each importance level. Modelers can establish that the nodes labeled with a level of importance do not need more than X clicks from the entry point to the system and, in the same way, at least Y clicks from this entry point. These maximum and minimum distances will be defined by modelers for each importance level. In this way, we offer the possibility to detect nodes labeled as important that are too far from the entry point and, in the same way, nodes labeled as less important excessively near this point.

R2. Maximum/minimum distance from the entry point to the WIS for each node. Although a node will be labeled with a importance level, it is possible that a modeler wishes to label it with a different maximum/minimum distance to the rest of nodes of its level due to the information or functionality that the node represents. Moreover, it may be that modelers do not want to use the *importance level* concept but want to establish distance requirements to individual nodes of their models.

R3. Distances between nodes. In a WIS, if two nodes represent informations or functionalities related to each other, modelers probably wish that they will be near in the system. The capacity to express the maximum distance between a node and another related can be useful for modelers.

R4. Average distances. Navigational models can be analysed to obtain the average distances between the nodes of each level and the node that represents the point entry. In this way we obtain a measure for each level that can serve to detect shortfalls in the navigation design, for example, if the average distance for the nodes of a level is too high.

3.2   Connectivity Between Nodes

This section analizes requirements related to the necessity of interconnectivity between the different elements in the WIS. Some of these requirements are:

R5. Direct connectivity between nodes. It is useful to express the requirement that two nodes must be directly connected. The system can check that a link between nodes exists for all the nodes desired by the modeler. This property can be generalised in the sense that if we consider a node we may wish that a set of nodes can be reachable from the first one. Modellers will define the set of reachable nodes and the tool that support the proposal will check this reachability.

R6. Connectivity between nodes. Other requirements will be related to ensuring that the users can reach other node from a node that represent an information or service.

R7. Obligation to previous crossing by a node. Some requirements of the WIS could establish constraints on the order in which tasks are carried out. Our proposal will permit modelers to define, for each node, a set of nodes that must be previously visited during the navigation through the system.

R8. Obligation to later crossing by a node. In the same way as in the previous property, the modeler can establish that if a user carries out a task (that is, visit a node) he must then carry out other set of tasks (this is, to visit a set of nodes) before the overall task will be considered as finished.

4   Metamodel Extension for Usability Requirements Support

This section shows an extension of the metamodel of navigational models which aim to offer support for the expression of usability requirements over these models. In our metamodel extension, shown in Figure 1, we only show nodes and links for two reasons. First so that readers can understand better our proposal and, secondly, to show the generality of our proposal, because nodes and links are presented in numerous methodologies and our approach could be integrated in any that includes these elements. Our approach extends navigational metamodels with the following elements:

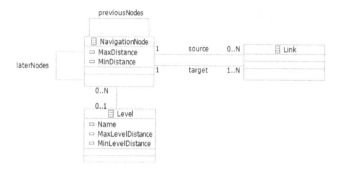

Fig. 1. Navigation metamodel extension

- **Adding new attributes and links to the element *NavigationNode*.** These attributes called *MaxDistance* and *MinDistance* are used to support requirements like *R2*. The links called *previousNodes* and *laterNodes* represent, for each node, a list of previous and later nodes that a user must visit if he visits the node. These lists can be empty and they are useful to support the kind of requirements shown in *R7* and *R8*.
- **Adding new elements.** The *Level* entity and its attributes are used to support the requirements related to the importance of the functionality and information shown in Section 3.1. This entity represents the concept of *importance of a node*. The link between *NavigationNode* and *Level* permits modelers to label each node with an importance level. Each *Level* has three attributes that define the minimum and maximum distance between the nodes labeled with this degree of importance and the node that represents the entry point to the WIS.

5   Automatic Support

The next aim of our work was the design of an automatic support for our approach that will allow modelers to evaluate and to improve the quality for their navigational models in a comfortable way. This tool supports the specification of the requirements previously shown and it permits the verification of these requirements on navigational models. It is important to emphasize that our aim is not to design an isolated tool for usability evaluation but a tool that will be integrated in the CASE tools used for WIS development methodologies. For reasons of space the prototype of the tool is not shown in depth. In general, the tool can load models represented in the standard XMI format and it offers a list of usability requirements and verification proofs that can be executed over the models to verify the fulfillment of these requirements and to determine the quality of the navigational models. When the tool executes these proofs, information about the fulfilment of the requirements previously defined is shown, which can be used by the modelers to improve their models.

6   Conclusions and Further Work

This approach shows the viability of considering usability requirements in the first stages of WIS development. A set of these requirements that can be considered in navigational models has been presented. In addition, an extension of navigation metamodels has been presented to illustrate our approach and the automatic support that has been developed so that modelers can obtain benefits easily. As further work, we will look for new requirements that should be expressed and validated in the models used in WIS development and the metamodels and the automatic support for our proposal will again be extended to support them. Moreover, the possibility of defining automatic transformations to improve the quality of models designed from the results obtained in the requirements validation process will be studied. With regard to the tool support for our approach, it will be integrated in CASE tools for WIS development.

References

1. Ginige, A., Murugesan, S.: Guest editors' introduction: Web engineering - an introduction. IEEE MultiMedia 8, 14–18 (2001)
2. Ivory, M.Y., Hearts, M.: An Empirical Foundation for Automated Web Interface Evaluation. PhD thesis, University of California at Berkeley (2001)
3. Atterer, R., Schmidt, A.: Extending web engineering models and tools for an automatica usability validation. Journal of Web Engineering 1, 43–64 (2006)
4. Briand, L.L., Basili, M.S.: Defining and Validating Measures for Object-based High-level Design. IEEE Transactions on Softw. Eng. 25(5), 722–743 (1999)
5. I.S.O.: Iso/iec 9126-1. softw. eng.-product quality - part 1: Quality model (2000)
6. I.S.O.: 9241-11. ergonomic requirements for office work with visual display terminals (vdt)s - part 11: Guidance on usability (1998)
7. Dhyani, D., NG, W., Bhowmick, S.: A survey of web metrics. ACM Computing Surveys 34(4), 469–503 (2002)
8. Abrahão, S., Condori, N., O.L., O., P.: Defining and validating metrics for navigational models. In: METRICS 2003, Sydney, Australia, pp. 200–210. IEEE Press, Los Alamitos (2003)
9. Lucas, F., Molina, F., Toval, A., De Castro, M., Cáceres, P., Marcos, E.: Precise WIS Development. In: International Conference on Web Engineering, ICWE 2006, Menlo Park, California (USA), ACM, New York (2006)
10. Cachero, C., Calero, C., Poels, G.: Metamodeling the quality of the web development process' intermediate artifacts. In: ICWE 2007. LNCS, vol. 4607, Springer, Heidelberg (2007)
11. Abrahao, S., E., I.: Early usability evaluation in model-driven architecture environments. In: 6th IEEE International Conference on Quality Software, Wiley, Chichester, IEEE Press, Los Alamitos (2006)

Author Index

Lecture Notes in Computer Science

Sublibrary 3: Information Systems and Application, incl. Internet/Web and HCI

For information about Vols. 1– 4439
please contact your bookseller or Springer

Vol. 4656: M.A. Wimmer, J. Scholl, Å. Grönlund (Eds.), Electronic Government. XIV, 450 pages. 2007.

Vol. 4655: G. Psaila, R. Wagner (Eds.), E-Commerce and Web Technologies. VII, 229 pages. 2007.

Vol. 4654: I.-Y. Song, J. Eder, T.M. Nguyen (Eds.), Data Warehousing and Knowledge Discovery. XVI, 482 pages. 2007.

Vol. 4653: R. Wagner, N. Revell, G. Pernul (Eds.), Database and Expert Systems Applications. XXII, 907 pages. 2007.

Vol. 4636: G. Antoniou, U. Aßmann, C. Baroglio, S. Decker, N. Henze, P.-L. Patranjan, R. Tolksdorf (Eds.), Reasoning Web. IX, 345 pages. 2007.

Vol. 4611: J. Indulska, J. Ma, L.T. Yang, T. Ungerer, J. Cao (Eds.), Ubiquitous Intelligence and Computing. XXIII, 1257 pages. 2007.

Vol. 4607: L. Baresi, P. Fraternali, G.-J. Houben (Eds.), Web Engineering. XVI, 576 pages. 2007.

Vol. 4606: A. Pras, M. van Sinderen (Eds.), Dependable and Adaptable Networks and Services. XIV, 149 pages. 2007.

Vol. 4605: D. Papadias, D. Zhang, G. Kollios (Eds.), Advances in Spatial and Temporal Databases. X, 479 pages. 2007.

Vol. 4602: S. Barker, G.-J. Ahn (Eds.), Data and Applications Security XXI. X, 291 pages. 2007.

Vol. 4601: S. Spaccapietra, P. Atzeni, F. Fages, M.-S. Hacid, M. Kifer, J. Mylopoulos, B. Pernici, P. Shvaiko, J. Trujillo, I. Zaihrayeu (Eds.), Journal on Data Semantics IX. XV, 197 pages. 2007.

Vol. 4592: Z. Kedad, N. Lammari, E. Métais, F. Meziane, Y. Rezgui (Eds.), Natural Language Processing and Information Systems. XIV, 442 pages. 2007.

Vol. 4587: R. Cooper, J. Kennedy (Eds.), Data Management. XIII, 259 pages. 2007.

Vol. 4577: N. Sebe, Y. Liu, Y.-t. Zhuang, T.S. Huang (Eds.), Multimedia Content Analysis and Mining. XIII, 513 pages. 2007.

Vol. 4568: T. Ishida, S. R. Fussell, P. T. J. M. Vossen (Eds.), Intercultural Collaboration. XIII, 395 pages. 2007.

Vol. 4566: M.J. Dainoff (Ed.), Ergonomics and Health Aspects of Work with Computers. XVIII, 390 pages. 2007.

Vol. 4564: D. Schuler (Ed.), Online Communities and Social Computing. XVII, 520 pages. 2007.

Vol. 4563: R. Shumaker (Ed.), Virtual Reality. XXII, 762 pages. 2007.

Vol. 4561: V.G. Duffy (Ed.), Digital Human Modeling. XXIII, 1068 pages. 2007.

Vol. 4560: N. Aykin (Ed.), Usability and Internationalization, Part II. XVIII, 576 pages. 2007.

Vol. 4559: N. Aykin (Ed.), Usability and Internationalization, Part I. XVIII, 661 pages. 2007.

Vol. 4558: M.J. Smith, G. Salvendy (Eds.), Human Interface and the Management of Information, Part II. XXIII, 1162 pages. 2007.

Vol. 4557: M.J. Smith, G. Salvendy (Eds.), Human Interface and the Management of Information, Part I. XXII, 1030 pages. 2007.

Vol. 4541: T. Okadome, T. Yamazaki, M. Makhtari (Eds.), Pervasive Computing for Quality of Life Enhancement. IX, 248 pages. 2007.

Vol. 4537: K.C.-C. Chang, W. Wang, L. Chen, C.A. Ellis, C.-H. Hsu, A.C. Tsoi, H. Wang (Eds.), Advances in Web and Network Technologies, and Information Management. XXIII, 707 pages. 2007.

Vol. 4531: J. Indulska, K. Raymond (Eds.), Distributed Applications and Interoperable Systems. XI, 337 pages. 2007.

Vol. 4526: M. Malek, M. Reitenspieß, A. van Moorsel (Eds.), Service Availability. X, 155 pages. 2007.

Vol. 4524: M. Marchiori, J.Z. Pan, C.d.S. Marie (Eds.), Web Reasoning and Rule Systems. XI, 382 pages. 2007.

Vol. 4519: E. Franconi, M. Kifer, W. May (Eds.), The Semantic Web: Research and Applications. XVIII, 830 pages. 2007.

Vol. 4518: N. Fuhr, M. Lalmas, A. Trotman (Eds.), Comparative Evaluation of XML Information Retrieval Systems. XII, 554 pages. 2007.

Vol. 4508: M.-Y. Kao, X.-Y. Li (Eds.), Algorithmic Aspects in Information and Management. VIII, 428 pages. 2007.

Vol. 4506: D. Zeng, I. Gotham, K. Komatsu, C. Lynch, M. Thurmond, D. Madigan, B. Lober, J. Kvach, H. Chen (Eds.), Intelligence and Security Informatics: Biosurveillance. XI, 234 pages. 2007.

Vol. 4505: G. Dong, X. Lin, W. Wang, Y. Yang, J.X. Yu (Eds.), Advances in Data and Web Management. XXII, 896 pages. 2007.

Vol. 4504: J. Huang, R. Kowalczyk, Z. Maamar, D. Martin, I. Müller, S. Stoutenburg, K.P. Sycara (Eds.), Service-Oriented Computing: Agents, Semantics, and Engineering. X, 175 pages. 2007.

Vol. 4500: N.A. Streitz, A.D. Kameas, I. Mavrommati (Eds.), The Disappearing Computer. XVIII, 304 pages. 2007.

Vol. 4495: J. Krogstie, A. Opdahl, G. Sindre (Eds.), Advanced Information Systems Engineering. XVI, 606 pages. 2007.

Vol. 4480: A. LaMarca, M. Langheinrich, K.N. Truong (Eds.), Pervasive Computing. XIII, 369 pages. 2007.

Vol. 4473: D. Draheim, G. Weber (Eds.), Trends in Enterprise Application Architecture. X, 355 pages. 2007.

Vol. 4471: P. Cesar, K. Chorianopoulos, J.F. Jensen (Eds.), Interactive TV: A Shared Experience. XIII, 236 pages. 2007.

Vol. 4469: K.-c. Hui, Z. Pan, R.C.-k. Chung, C.C.L. Wang, X. Jin, S. Göbel, E.C.-L. Li (Eds.), Technologies for E-Learning and Digital Entertainment. XVIII, 974 pages. 2007.

Vol. 4443: R. Kotagiri, P. Radha Krishna, M. Mohania, E. Nantajeewarawat (Eds.), Advances in Databases: Concepts, Systems and Applications. XXI, 1126 pages. 2007.